EFFECTS OF LAND USE ON FRESH WATERS
Agriculture, Forestry, Mineral Exploitation, Urbanisation

ELLIS HORWOOD BOOKS IN WATER AND WASTEWATER TREATMENT

This collection of authoritative works reflects the world awareness of the importance of water, wastewater treatment, and closely-related subjects, to the world. The titles are written and edited by experts from a wide range of countries closely concerned with research and development, monitoring and improving water quality and supplies, and treatment and disposal of wastewater.

Titles published in collaboration with the WATER RESEARCH CENTRE, UK

AQUALINE THESAURUS 2
Editor: G. BOTHAMLEY, Water Research Centre
SEWAGE SLUDGE STABILISATION AND DISINFECTION
Editor: A. M. BRUCE, Water Research Centre
STABILISATION, DISINFECTION AND ODOUR CONTROL IN SEWAGE SLUDGE TREATMENT
Editors: A. M. BRUCE and E. S. CONNOR, Water Research Centre
BULKING OF ACTIVATED SLUDGE
Editors: B. CHAMBERS and E. J. TOMLINSON, Water Research Centre
BIOLOGICAL FLUIDISED BED TREATMENT OF WATER AND WASTEWATER
Editors: P. F. COOPER, Water Research Centre, and B. ATKINSON, University of Manchester Institute of Science and Technology
WATER RESEARCH TOPICS
Editor: I. M. LAMONT, Water Research Centre
DESIGN AND OPERATION OF SMALL WATER POLLUTION CONTROL PLANTS
E. H. NICOLL, Scottish Development Department, Edinburgh
EFFECTS OF LAND USE ON FRESH WATERS
Editor: J. SOLBÉ, Water Research Centre
RIVER POLLUTION CONTROL
Editor: M. J. STIFF, Water Research Centre

Other books in Water and Wastewater

WATER-ABSTRACTION, STORAGE, TREATMENT, DISTRIBUTION
J. JEFFERY, General Manager, North Surrey Water Company
POLLUTION CONTROL AND CONSERVATION
Editor: M. KOVACS, Hungarian Academy of Sciences
NATURAL WATER SAMPLING
J. KRAJCA *et al.*
HANDBOOK OF WATER PURIFICATION Second Edition
W. LORCH, Cranfield Institute of Technology
CHEMICAL PROCESSES IN WASTE WATER TREATMENT
G. MATTOCK, Process Plant and Chemical Ltd, Slough, and W. J. EILBECK, University College of North Wales
BIOLOGY OF SEWAGE TREATMENT AND WATER POLLUTION CONTROL
K. MUDRACK, University of Hanover, FRG and S. KUNST, Institute for Sanitary Engineering, Hanover, FRG
GROUND WATER PROTECTION
V. PELIKAN, Research Institute of Geological Engineering, Brno, Czechoslovakia
IMPROVING WATER QUALITY
M. RAPINAT, Compagnie Generale des Eaux, France
DRINKING WATER AND MATERIALS: Field Observations and Methods of Investigation
D. SCHOENEN and H. F. SCHÖLER, Hygiene-Institut der Universität, Bonn, FRG
BIOTECHNOLOGY OF WASTE TREATMENT AND EXPLOITATION
J. M. SIDWICK and R. S. HOLDOM, Watson Hawksley, Buckinghamshire
INTRODUCTION TO LIMNOLOGY
J. SCHWOERBEL, University of Freiburg, West Germany

ELLIS HORWOOD BOOKS IN AQUACULTURE AND FISHERIES SUPPORT

Series Editor: Dr. L. M. LAIRD, University of Aberdeen

FISH DISEASES: Bacteria as Fish Pathogens
B. AUSTIN, Heriot-Watt University, and D. AUSTIN, Fresh Water Biological Association, East Stoke, Dorset
FISH FARMING: Salmon and Trout
L. M. LAIRD and T. NEEDHAM, Fish Technology and Information Services, Aberdeen
MARINE FISH FARMING: Warm Water Marine Culture
J. F. MUIR, University of Sterling

EFFECTS OF LAND USE ON FRESH WATERS
Agriculture, Forestry, Mineral Exploitation, Urbanisation

Editor:
J. F. de L. G. SOLBÉ, B.Sc.Hons.
Manager, Conservation and Fisheries Management
Water Research Centre, Medmenham

1986

WRc

Published for the
WATER RESEARCH CENTRE
by

ELLIS HORWOOD LIMITED
Publishers · Chichester

First published in 1986 by
ELLIS HORWOOD LIMITED
Market Cross House, Cooper Street,
Chichester, West Sussex, PO19 1EB, England
The publisher's colophon is reproduced from James Gillison's drawing of the ancient Market Cross, Chichester.

Distributors:
Australia and New Zealand:
JACARANDA WILEY LIMITED
GPO Box 859, Brisbane, Queensland 4001, Australia
Canada:
JOHN WILEY & SONS CANADA LIMITED
22 Worcester Road, Rexdale, Ontario, Canada
Europe and Africa:
JOHN WILEY & SONS LIMITED
Baffins Lane, Chichester, West Sussex, England
North and South America and the rest of the world:
Halsted Press: a division of
JOHN WILEY & SONS
605 Third Avenue, New York, NY 10158, USA

British Library Cataloguing in Publication Data
Effects of land use on fresh waters:
agriculture, forestry, mineral exploitation, urbanisation. —
(Ellis Horwood series in water and waste water technology)
1. Water — Pollution
I. Solbé, J. F. de L. G.
363.7'394 TD420

Library of Congress Card No. 86–10695
ISBN 0–7458–0054–8 (Ellis Horwood Limited)
ISBN 0–470–20363–3 (Halsted Press)

Phototypeset in Times by Ellis Horwood Limited
Printed in Great Britain

Table of Contents

Editor's Preface

This book is an account of a conference 'The Effects of Land Use on Fresh Waters' organised by the Water Research Centre and the World Health Organization and held at the University of Stirling in June 1985.

The conference focused on the effects of four major uses of land on fresh water above and below ground — agriculture, forestry, urbanisation and the exploitation of minerals. In each case, presentations dealt with the types of problem associated with each class of use and any remedial measures available to tackle the problems.

The conference programme aimed to be international: one third of the authors were from overseas and 14 countries were represented among the 149 delegates. It was devised to give expression to a wide range of interests and to encourage a debate (reported in this book) on the competitive interactions and impacts of the chosen uses of land on the freshwater environment and on water supply.

The Opening Address and the Introduction were given, respectively, by Michael Ancram, Under Secretary of State, Scottish Office, and Ian Waddington, Director of the WHO Environmental Health Service. Their texts appear in full on pages 17 to 24. Urbanisation was the first land use to be discussed. Chapters 1 to 6 draw on experience from the UK, Italy and the Netherlands. Descriptions are given of the sewerage and waste-water treatment of two new towns, problems with the quality of motorway drainage and long-term contamination of aquifers below cities. Remedies to this last problem, although difficult, are showing promise. In the treatment of landfill leachates and of the complete flow of rivers draining conurbations, advances are being made.

To illustrate Mineral Exploitation Chapters 7 to 10 ranged widely, as well as focusing on case histories, some with happy endings, from the coal, china clay and metalliferous mining industries.

Agriculture was seen as the subject requiring most attention, and 13 chapters were devoted to it. Papers from Denmark, the European Commission, France, Hungary, the UK and the USA covered general aspects of the use of water in food production, as well as the specific areas of pesticides and nitrates (both of which led to detailed discussions of legislation), acidification and gross organic pollution. Authors' opinions of the needs for and types of remedies were given, with real examples of successful solutions.

In Forestry, (Chapters 24 to 29) authors from Finland, West Germany, the UK and the USA gave accounts of the results of the wood production and processing industries on water yield and surface-water problems. Ecological and operational solutions to these difficulties were then described.

The final discussion session was chaired by Dr Steve Warren, Director of WRc Environment. The edited transcript (pp 453 to 459) shows that it was principally concerned with planning, control and costs.

Hugh Fish, Chairman of the Natural Environment Research Council gave the Summing-Up Address.

Appendix I to this book consists of the poster displays which contributed to the breadth of coverage of the conference.

The book should interest a wide readership — those involved with the politics, legislation, science and operational management of the broad categories of land use selected for debate, as well as those concerned with the protection of the environment and water supplies.

To all the authors and the chairmen of sessions and to the many other people who helped to make the conference a success the Water Research Centre and the World Health Organization extend their gratitude.

I, as Editor, could not have completed this work without the dedicated help of my colleagues and the meticulous care of Mary Kennedy-Ping.

John Solbé
Medmenham, January 1986

Directory of Contributors

		Chapter
Adamson, J. K.	Institute of Terrestrial Ecology, Merlewood Research Station, Grange-over-Sands, Cumbria. LA11 6JU	Appendix I
Alföldi, L.	General Director, Research Centre for Water Resources Development, Budapest, PO Box 27, H-1453 Budapest 52, Hungary.	16
Anderson, A. R.	Forestry Commission, Northern Research Station, Roslin, Midlothian. EH25 9SY	Appendix I
Barbour, Dr A. K.	RTZ Limited, York House, Bond Street, Bristol. BS1 3PE	7
Baumann, J. S.	Wisconsin Department of Natural Resources, PO Box 7921, Madison, Wisconsin 53707.	19
Bayes, C. D.	WRc Scotland, 1 Snowdon Place, Stirling. FK8 2NH	Appendix I
Beveridge, M. C. M.	Institute of Aquaculture, Stirling University, Scotland. FK9 4LA	Appendix I
Binns, Dr W. O.	Head of Site Studies, Forestry Commission, Research and Development Division, Alice Holt Lodge, Wrecclesham, Farnham, Surrey. GU10 4LH	24
Blackie, J. R.	Head of Exploratory Catchment Studies Section, NERC Institute of Hydrology, Maclean Building, Crowmarsh Gifford, Wallingford, Oxon. OX10 8BB	26
Bruenig, Professor E. F.	University of Hamburg, and Director. Institute for World Forestry, Institute for World Forestry, Federal Research Centre for Forestry and Forest Products, Leuschner Strasse 91, D2050 Hamburg 80 Germany.	25
Cavallaro, Professor A.	PMIP, Laboratory of Public Health, Chemical Section, Juvara 22, Milano, Italy.	4
Chiaudani, Dr G.	Department of Biology, University of Milan.	Appendix I
Colbourn, P.	Ministry of Agriculture, Fisheries and Food, Field Drainage Experimental Unit, Anstey Hall, Maris Lane, Trumpington, Cambridge. CB2 2LF	Appendix I
Cominetti, M. E.	Pollution Officer (West), South West Water, Victoria Square, Bodmin, Cornwall. PL31 1EF	9
Comyns, R. A.	Process Development Manager, English Clays Lovering Pochin & Co Ltd, John Keay House, St Austell, Cornwall. PL25 4DJ	10
Cooke, A.	Nature Conservancy Council, Godwin House, George Street, Huntingdon, Cambs. PE18 68U	20
Coombes, C.	Anglian Water, Ambury Road, Huntingdon, Cambs. PE18 6NZ	1

Name	Address	Chapter
Corradi, C.	PMIP, Laboratory of Public Health, Chemical Section, Juvara 22 Milano, Italy.	4
Croll, Dr B. T.	Regional Scientist (Treatment), Anglian Water, Ambury Road, Huntingdon. PE18 6NZ	13
De Felice, G.	PMIP, Laboratory of Public Health, Chemical Section, Juvara 22 Milano, Italy.	4
Edwards, Dr A. M. C.	Regional Environmental Scientist, Yorkshire Water, West Riding House, 67 Albion Street, Leeds. LS1 5AA	Appendix I
Ellis, J. B.	Middlesex Polytechnic, Centre for Urban Pollution Research Queensway, Enfield, Middlesex. EN3 4SF	Appendix I
Evision, L. M.	Lecturer, University of Newcastle, Department of Civil Engineering, Claremont Road, Newcastle. NE1 7RU	Appendix I
Fawell, J. K.	Section Head Toxicology, WRc Environment, Henley Road, PO Box 16, Marlow, Buckinghamshire. SL7 2HD	Appendix I
Forslund, Dr J.	Chemical Engineer, Ministry of the Environment, National Agency of Environment Protection, 29 Strandgade, DK-1401 Copenhagen K, Denmark.	17
Fry, G.	Nature Conservancy Council, Great Britain Headquarters, Northminster House, Peterborough. PE1 1UA	20
Grassi, P.	PMIP, Laboratory of Public Health, Chemical Section, Juvara 22, Milano, Italy.	4
Gray, A. V.	Water Quality Branch, 17 Antrim Road, Lisburn, Co. Antrim. BT28 3Al	Appendix I
Green, M. B.	Severn Trent Water, Biological Laboratory, Kingsbury Road, Minworth, Sutton Coldfield, West Midlands. B76 9DP	6
Hall, J. E.	WRc Environment, Henley Road, PO Box 16, Marlow Buckinghamshire. SL7 2HD	Appendix I
Hamaker, P. H.	Institute for Land and Water Management Research, The Netherlands.	12
Hamilton, R. S.	Middlesex Polytechnic, Centre for Urban Pollution Research, Queensway, Enfield, Middlesex. EN3 4SF	Appendix I
Hammerton, D.	Director, Clyde River Purification Board, City Chambers, Glasgow. G2 1DU	8
Hardwick, Dr D. C.	Chief Pollution Scientist, Ministry of Agriculture, Fisheries and Food, Great Westminster House, Horseferry Road, London. SWIP 2AE	11
Harris, G. L	Hydrologist, Ministry of Agriculture, Fisheries and Food, Field Drainage Experimental Unit, Anstey Hall, Maris Lane Trumpington, Cambridge. CB2 2LF	Appendix I
de Heer, H.	Agro-Hydrologist, Institute of Pesticide Research, Marijkeweg 22, 6709 Wageningen, The Netherlands	12
Heinonen, P.	Limnologist, National Board of Waters, Water Research Office, PB 250, 00101 Helsinki 10, Finland.	27
Hellawell, Dr J. M.	Severn Trent Water, Biological Laboratory, Kingsbury Road, Minworth, Sutton Coldfield, West Midlands. B76 9DP	6
Henin, Professor S.	88 Rue de Javel, 75015 Paris, France.	14
Homonnay, Dr A.	Research Centre for Water, Resource Development, VITUKI, PO Box 27, H-1453 Budapest 52, Hungary.	16
Howells, Dr W. R.	Director of Scientific Services, Welsh Water, Penyfai House, 19 Penyfai Lane, Llanelli, Dyfed. SA15 4EL	18
Konrad, Dr J. G.	Chief,Nonpoint Source Section, Wisconsin Department of Natural Resources, PO Box 7921, Madison, Wisconsin 53707, USA	19
Macdonald, T. D.	Principal Civil Engineer, Scottish Development Department, Room 330, Pentland House, 47 Robb's Load, Edinburgh. EH14 1TY	Appendix I
McIntyre, Dr A. E.	Now at: Consultants in Environmental Sciences Ltd, Yeoman House, 63 Croyden Road, London. SE20 7TW	3

Melanen, Dr M.	National Board of Waters, Water Research Office, PB 250, 00101 Helsinki 10, Finland.	Appendix I
Melkas, M.	National Board of Waters, Water Research Office, PB 250, 00101 Helsinki 10, Finland.	Appendix I
Merriman, R. P.	Water Quality Officer, Welsh Water, Hawthorn Rise, Haverfordwest, Dyfed, South Wales. SA61 2BH	18
Mills, Dr D. H.	Department of Forestry and Natural Resources, University of Edinburgh, King's Buildings, Mayfield Road, Edinburgh, EH9 3JO	28
Newbold, Dr C.	Chief Scientist's Team, Nature Conservancy Council, Great Britain Headquarters, Northminster House, Peterborough. PE1 1UA	20
Newson, Dr M. D.	Institute of Hydrology, Maclean Building, Wallingford, Oxon. OX10 8BB	26
Ott, J. A.	Wisconsin Department of Natural Resources, PO Box 7921, Madison, Wisconsin 53707, USA.	19
Parkinson, E. C.	Forth River Purification Board, Colinton Dell House, West Mill Road, Edinburgh. EH13 0PH	Appendix I
Parsons, P. J.	Solid Waste Management, Department of the Environment, Romney House, 43 Marsham Street, London. SW1P 3PY	5
Perry, Professor, R.	Professor of Public Health & Water Technology, Imperial College of Science and Technology, Imperial College Road, London. SW7 2BU	22
Phillips, Dr M. J.	Institute of Aquaculture, University of Stirling, Scotland. FK9 4LA	Appendix I
Phillips, Dr V. R.	Group Leader, Waste Engineering, National Institute of Agricultural Engineering, Wrest Park, Silsoe, Beds. MK45 4HS	21
Ponce, Dr S. L.	Hydrologist, National Park Service, Federal Building Rm 343, 301 S. Hawes Street, F. Collins, COt 8521. USA.	29
Premazzi, G.	Commission of the European Communities, Joint Research Centre, Ispra Establishment, 21020 Ispra (Va) Italy.	Appendix I
Pyatt, D. G.	Head of Site Studies (North) Branch, Forestry Commission, Research Division, Northern Research Station, Roslin, Midlothian. EH25 9SY	Appendix I
Revitt, D. M.	Middlesex Polytechnic, Centre for Urban Pollution Research, Queensway, Enfield, Middlesex. EN3 3SF	Appendix I
Shutes, R. B. E.	Middlesex Polytechnic, Centre for Urban Pollution Research, Queensway, Enfield, Middlesex. EN3 4SF	Appendix I
Smith, Dr R. V.	Department of Agriculture, Freshwater Biological Investigation Unit, Greenmount Road, Muckamore, Antrim. BT41 4PX	Appendix I
Stewart, J. A.	Institute of Aquaculture, Stirling University, Scotland. FK9 4LA	Appendix I
Tuinstra, L. G. M. Th.	State Institute for Quality Control of Agricultural Products, The Netherlands.	12
Uunk, E. J. B.	IJsselmeerpolders Development Authority, PO Box 600, AP Lelystad, The Netherlands.	2
van der Burg, A. M. M.	Glasshouse Crops Research and Experimental Station, The Netherlands.	12
van de Ven, F. H. M.	DBW/RIZA Maerlant 6, Lelystad, The Netherlands.	2
Vighi, M.	Faculty of Agriculture, University of Milan	Appendix I
Walker, S.	North West Water, Dawson House, Great Sankey, Warrington. WA5 3LW	Appendix I
Walsh, Dr P. D.	Pincipal, North West Water, Dawson House, Great Sankey, Warrington. WA5 3LW	Appendix I
Williamson, T.	Deputy Director, Forth River Purification Board, Colinton Dell House, West Mill Road, Colinton, Edinburgh. EH13 0PH	Appendix I

Worthington, P.	Commission of the European Communities, 200 Rue de la Loi, B 1049 Brussels, Belgium.	23
Young, C. P.	Manager, Waste Disposal to Land, WRc Environment, Medmenham Laboratory, Henley Road, PO Box 16, Marlow, Buckinghamshire. SL7 2HD	15
Youngman, R. E.	WRc Environment, Medmenham Laboratory, Henley Road, PO Box 16, Marlow, Buckinghamshire. SL7 2HD	Appendix I

Opening address

MICHAEL ANCRAM, Under Secretary of State, Scottish Office

Mr Chairman, Distinguished Guests, Ladies and Gentlemen,
It gives me great pleasure to give the opening address at this very important conference and I am particularly pleased to welcome delegates from beyond the United Kingdom. After all, international co-operation is now very much an essential feature of our endeavours to protect and improve all the different sectors of our environment and the conference organisers are to be congratulated, if I may say so, for having the good sense to select Scotland, and Stirling University in particular, as your venue. It has crossed my mind occasionally that gatherings of this kind might best be located in areas of environmental stress as an aid to generating a proper sense of urgency, but in this case we find ourselves in surroundings which may do the opposite and generate feelings of contentment and relaxation. But, on reflection, I think that this is right because it is evident from your programme that there will indeed be much to occupy your minds in the expert presentations and discussion that will demand a good deal of close and serious attention. The opportunity to relax in these rather fine surroundings and in the pleasant neighbouring countryside I think will provide you with a welcome and essential contrast.

Mr Chairman, the importance of clean fresh waters to our community cannot be exaggerated. We in Scotland are used to an abundance of good quality fresh water. Indeed, we could frequently do with rather less, especially when we are on holiday or pursuing a favourite outdoor activity. Nonetheless, we are not complacent. The government attaches considerable importance to environment issues and does keep the quality of our environment under constant review. Indeed, over the last decade we have increasingly been forced to question our established conceptions about water quantity and water quality as we have discovered the hitherto unsuspected effects of other activities: for example, the effects of afforestation on the yield of public water-supply catchments and of the intensification of agricultural practices on the quality of surface and ground waters. Also the development of more sophisticated analytical techniques and equipment has alerted us to the presence of substances both natural and man-made, previously unsuspected. The significance of many of these substances to the environment is now rightly a major cause of concern and enquiry.

In the United Kingdom, the bringing into force of the water pollution control provisions of the Control of Pollution Act and the introduction of a Code of Good Agricultural Practice is evidence of our determination to maintain and improve the quality of our surface and underground waters. There is, however, a growing international dimension to the formulation of environmental policies. The importance of contributing to this formulation is highly regarded by us in the United Kingdom. Already we have implemented a number of European Community directives setting standards of water quality for various uses. The directive on the quality of drinking water comes into force next month and there will be others.

Directives, as you will be aware, list specific requirements about selective substances, set maximum values for their concentration and specify methods of sampling and analysis, and this approach, in many ways, complements the United Kingdom traditional method of setting general Environmental Quality Objectives which, in the case of water, relate to its particular use. Thus the adoption of such detailed standards and/or guidelines as those specified in directives, requires a change in our traditional approach to the control of water quality. Nevertheless, it should be borne in mind that spectific numerical standards for specific substances may not always line up with health or environmental criteria. They can provide a useful discipline and lead to a better understanding of previously ill-defined situations but may be counter-productive if too rigidly applied and it is, therefore, important that international standards allow for attention to be paid to the overall suitability of a water for a particular use as well as being concerned with particular parameters.

Acid deposition, rising levels of nitrate, particularly in ground water, ferruginous discharges from active and abandoned mines and the apparently endless list of persistent toxic chemicals could generate an attitude of despair but I believe that this would be both foolish and unjustified.

The history of the last two or three decades caused a dramatic improvement in the quality of our inland and coastal waters, so much so that, as you will know, Atlantic salmon have returned to rivers such as the Thames and the Clyde from which they have been absent for all of this century and indeed for longer. Surveys of river-water quality show that a very high percentage of our rivers can now be classified as clean. For the others, plans and programmes are afoot to upgrade them as quickly as resources allow and we are therefore entitled to be somewhat confident that however intractable new problems may appear to be, solutions and the will to implement them will be forthcoming. The improvements of the last decade will continue, albeit the results may not always be dramatic or immediately apparent to the general public.

Just as yesterday's problems were solved by careful investigation and good practical measures, today's and tomorrow's problems will be solved or, better still, anticipated by ensuring that action plans are based on sound scientific foundations. Conferences like this make invaluable contributions to the building of such a foundation.

Realistic goals, usually based on use or potential use, need to be set for water quality and such goals are implicit in the approach to water-pollution control practised in the United Kingdom, but technical feasibility and cost-effectiveness must also be a part of the decision-making process and we have to be sure that in solving one problem we do not create a yet more intractable problem in another sector of the environment.

Important and fascinating as they are, scientific discoveries and developments tend to be taken for granted by communities who now accept rapid technological developments as commonplace. What is now apparent, however, is the groundswell of general interest by the public in environmental matters and while this public awareness is very much to be welcomed, it does create public sensitivity which must be recognised in the search for solutions to the problems. Scientists, whether pure or applied, must pay regard to this public dimension if they are to be successful in convincing the decision-makers of the relevance of their scientific advice.

International co-operation, involving exchange of information and co-operative investigations and research is also increasing. We now known that without this international pulling together, many of our problems and potential problems could not be solved. This conference is unequivocal and reasurring evidence of the healthy international dimension.

Your programme is ambitious in the extent of its coverage and the authors of the papers clearly widely experienced in their chosen subjects and there is no doubt that the conference will be successful in terms of exchange of information and focusing thinking and action on specific subject areas. I hope that contacts which will be made in the course of this conference will endure in the coming years to the mutual benefit of the community at large and the water environment in particular.

I note, Mr Chairman, from your programme, that on Friday you are to visit relevant sites demonstrating the subject of the conference. If I may say so, to reach these sites, some of which are located in areas of considerable beauty, you will be travelling through some of Scotland's finest countryside and I hope that you will take the opportunity to see as much as possible of this very striking part of Scotland and visit some of our historic places such as Stirling Castle.

Mr Chairman, the care with which the organisers have prepared for the conference has, I am sure, ensured its success and this success will be enhanced as delegates actually participate in the formal and informal discussions. This conference strikes me very much as being a participating affair and I know that the opportunities will not be missed. I am only sorry that my parliamentary duties require me to return to London immediately I have finished addressing you but I hope that I will get a full report on the deliberations of this conference and I will read it with interest.

Mr Chairman, it is our common duty, to husband, to conserve and to improve our environment and to pass on this heritage in good order to those who come after us. It is on occasions such as this that we can find and renew the inspiration to develop and carry on the essential work which will enable us to do so. I believe this conference will be no exception and I take this opportunity of not only opening it but wishing you every success in your deliberations.

Introduction

J. I. WADDINGTON, Director, Environmental Health Service, World Health Organization Copenhagen

Ladies and Gentlemen,
It is now a little over eleven years since I left the United Kingdom to take on international responsibilities and it is a particular pleasure to be involved in a meeting in the delightful Stirling University campus and I think I have to congratulate the Water Research Centre not only for their choice of venue but the timing when the rhododendrons and azaleas are at their best.

Since I left, in my relatively frequent but irregular visits to this country I have been more than a little interested in the changes that have taken place. As a nation perhaps we are not noted for taking too much account of what foreigners think of us but sometimes one perhaps gets a clearer impression of trends or changes as an outsider than if one is living continually in a particular ambience and experiencing them day by day and month by month. In visual terms I have seen very obvious changes to the British landscape, for example when driving through the eastern half of England and experiencing the visual effects of the rapidly expanding cornbelts. Another obvious factor is that compared to some other countries there is often a feeling of overused and worn out facilities and one is conscious of the large population density, not in a large part of Scotland, but certainly taking the country as a whole, each person requiring more room for his activities than a generation ago. The density of road traffic has more effects on our tight little island than on many countries on the Continent and these are only too apparent to the visitor driving particularly in the southern half of Britain.

The reasons for the recently reported damage to the German forests apparently contributed to by air pollution are not as yet fully understood but there is increasing evidence of the effect of vehicle emissions. These are related not only to maximum allowable road speeds and fuel policies but also to the whole question of land use in relation to motorway development compared to, for example, expanding rail networks.

Many of the environmental problems which this and other countries are experiencing are resulting from the increasing rapidity of land-use change and one of the critical facts about changes in land use is that the effects may be long-term and sometimes irreversible. We are thus making decisions which affect our children and our children's children. What is done is not easily undone. It is clear that in densely populated European countries we have to think in terms of multi-purpose development and land use. For this there is a need for overall planning strategies and then appropriate control over individual developments. There has been great progress in the last thirty years. Long gone are the days when the mere purchase of land could guarantee the owner's right to develop it as he wished.

These policies have been fairly successful in securing rational locations for major developments such as power stations, industry, new towns. They have also hopefully led to an overall improvement in the urban environment, although this is not clear from the appearance of some of our inner cities. However, in some European countries, including this one, agriculture has been largely excluded from planning controls and it is a relatively new thought that ploughing vast areas of grassland over a considerable timescale might have more profound long-term effects on the environment then say building a major industrial complex. We now need a better understanding of the long-term changes which may accompany major alterations in land use and to build these in to physical and economic planning. It is, of course, a form of environmental impact assessment and the capacity to predict. If we are going to produce a new mining complex we should think that mine is not going to last forever, it may last 10 years or 100 years. What is going to be the effect on the environment and water quality when it finally comes in to disuse? Perhaps we should have been a little cleverer when we installed 300m stacks on some of our power stations and given a little thought to the long-term and far-reaching effects of what we were doing instead of merely congratulating ourselves on the maintenance of a healthy local environment.

There is undoubtedly a new consciousness among the agricultural community of their responsibilities, as I hear, as a regular listener to the BBC early morning farm programme. I have the benefit that there is a one hour time difference between Denmark and here so my listening to that programme is not quite as impressive in terms of getting up early as you might imagine. The daily talk now on the early morning farm programme is on the effects of intensive production on soil structure, conservation and the heritage reponsibilities of farmers and I am sure other sectors of society have similarly increased consciousness during the last few years.

Land use is closely involved with national deeply-felt values and it cannot be looked at purely in economic terms, vitally important though these be. The culture that neglects its long-term land use responsibilities for the sake of a fast buck is perhaps on the decline. I am not, of course, referring in any respect to the effects of the Common Agricultural Policy.

There is no more practical example of the importance of land use considerations than their relationship towards water quality because no

European country can afford to despoil its resources of fresh water. The nitrogen cycle provides a striking example of some of the factors involved where substantial changes have been brought about by ploughing, increased use of fertilizers, afforestation, intensive animal husbandry and, of course, by the increased release of nitrogen oxide into the atmosphere from the burning of fossil fuels. Perhaps one-third of applied nitrogen fertilizers is lost by leaching on average. In many areas nitrate levels in drinking water are well in excess of those recommended by the World Health Organization and this is undoubtedly a disturbing trend even though the health effects to the general population having moderate levels of nitrate intake may have been exaggerated.

Perhaps one of the most pressing problems is related to the protection of groundwater. Any effects on aquifers can persist for many years. At a recent WHO working group the Danish participant indicated that even if nitrogen fertilizers were prohibited today, and this is clearly impossible, it would take 30 years for the nitrate present in the aquifers to be released and result in a decrease in their concentration in drinking water, and in a small country like Denmark there are no alternative sources.

The nitrate case is a good one from the economic standpoint, involving agricultural support systems of the real cost of food. It is interesting to speculate that if extra water-treatment costs are required as a result of agricultural activities the burden might fall on the agricultural sector by virtue of the 'polluter-must-pay' principle; but life is not as simple as that.

Ladies and Gentlemen, we look forward to an interesting and balanced debate on this topic tomorrow. One could, of course, discuss in this meeting the macro effects of land use on climate, for example the destruction of the rain forests or the extensive cutting down of trees for domestic fuel in Africa, both of which have profound effects on the hydrological cycle and hence on water quality. One could have reviewed the effects on water quality of using land for dumping hazardous wastes and the erosion of sensitive soil cover by excessive tourist densities. Some of the most important land-use problems in relation to water quality are undoubtedly of a trans-frontier nature and development on one side of a national boundary may seriously affect the usability of water resources in a neighbouring country. All these things could have been discussed but the agenda is already a full one for a four-day meeting.

I believe we can look forward to a very useful conference and I have high hopes that the outcome will be specific, realistic and practical. I hope its deliberations will be carefully noted by authorities within all relevant sectors of government.

Your closing session is being presided over by Hugh Fish, Chairman of the Natural Environment Research Council. He will expect from you something of a structured outcome. We would like practical, brief, succinct recommendations which can easily be disseminated, understood and acted on and this gives a responsibility to Session Chairmen to challenge authors to sum up in a clear, succinct, practical manner.

The solutions to these and other environmental problems can best be reached by a partnership. We must not have antagonism between farmers, water authorities, foresters, developers, industrialists. All must understand that in the overcrowded situation in European countries a partnership is required, and a partnership also is an international one because these solutions can best be found by co-operation among the international community.

The World Health Organization is very happy to be associated with this event and wishes to thank the Water Research Centre for their excellent and thorough preparations.

Part 1
Urbanisation

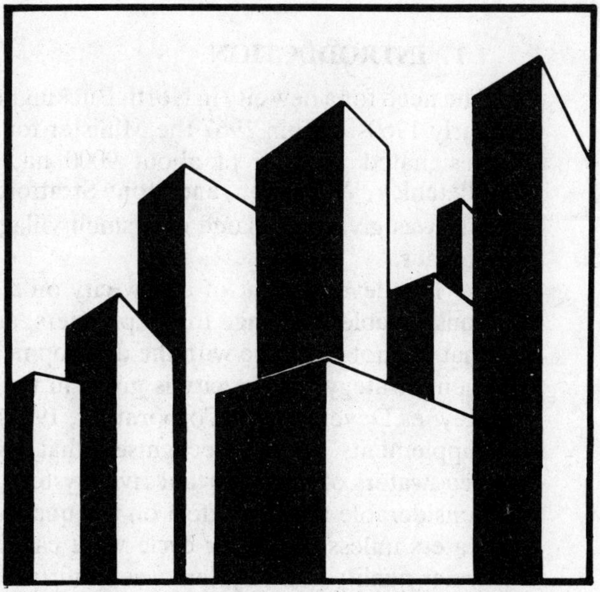

1

Current practices in controlling the effects on fresh waters of urbanisation in the Milton Keynes area

C. Coombes

1.1 INTRODUCTION

The need for a new city in North Buckinghamshire was first recognised in the early 1960s, and in 1967 the Minister for Housing and Local Government designated an area of about 9000 ha, including the existing towns of Bletchley, Wolverton, and Stony Stratford, as a suitable site. The proposed city was given the name of a small village in the designated area, Milton Keynes.

The development of a new city on a virtually green-field site posed a considerable challenge to the planners, and also gave rise to opportunities that are not available with the development of existing cities. The development strategy for the city is given in the plan for Milton Keynes (Milton Keynes Development Corporation, 1970) and the accompanying technical supplements. It was recognised that the new city being placed at the headwaters of an important river system, the Bedford Ouse, could have a considerable adverse effect on the quality and subsequent use of the fresh waters unless the water cycle were carefully managed. The concern over water quality in the system was confirmed by the establishment, in 1969, of the Associated Committee for the Great Ouse Area. This committee recommended that a three-year study should be carried out to provide a means by which the river water quality could be predicted until the end of the century. This resulted in the Bedford Ouse Study (Final Report, 1979) which commenced in October 1972.

This chapter summarises the progress in the development of Milton Keynes, examines the methods which have been used to protect water quality, and assesses their effectiveness. It has been necessary to draw heavily on the information contained in the two reports already mentioned, and these are recommended reading for anyone interested in detail rather than generalities.

1.2 MILTON KEYNES

The devoloping city is situated at the headwater of the Great Ouse immediately upstream of the confluence of the River Ouzel and River Ouse. Figure 1.1 shows the location of the city with the extent of urbanisation in 1969 and the extent of the proposed ultimate urbanisation. In addition to the new development, the existing towns of Bletchley, Stony Stratford, and Wolverton, and a number of small villages, have been incorporated into the new city. The plans include commerce and manufacturing industry as well as residential property. Around 50% of the gross area will be used for housing with only about 10% being used for employment. The city is expected, however, to be self-sufficient in terms of employment for its residents, in the long term. The population of the area was expected to rise from 45 000 in 1970 to 250 000 by 1994, the latter being the planned ultimate population for the city.

1.3 THE EFFECTS ON FRESH WATER

It was expected that Milton Keynes would affect both the quality and the flows of the Bedford Ouse river system. The growth in population and industry would correspondingly increase the volume of effluent discharged to the river. This would necessarily raise the input of conservative pollutants and could also, if allowed, result in an additional input of non-conservative pollutants. Similarly the increase in non-permeable area would give rise to greater runoff giving increased risk of flooding and a larger amount of oil and road grime from surface water. These potential problems were recognised particularly with respect to the water quality of the Bedford Ouse and the possible effects on the abstraction of water downstream of Milton Keynes at Clapham, serving the city of Bedford.

1.4 MEASURES TAKEN

Following a report by consulting engineers (J. D. & D. M. Watson) it was decided that sewage and industrial effluents should be treated at a single sewage-treatment works situated at the downstream end of the city. The

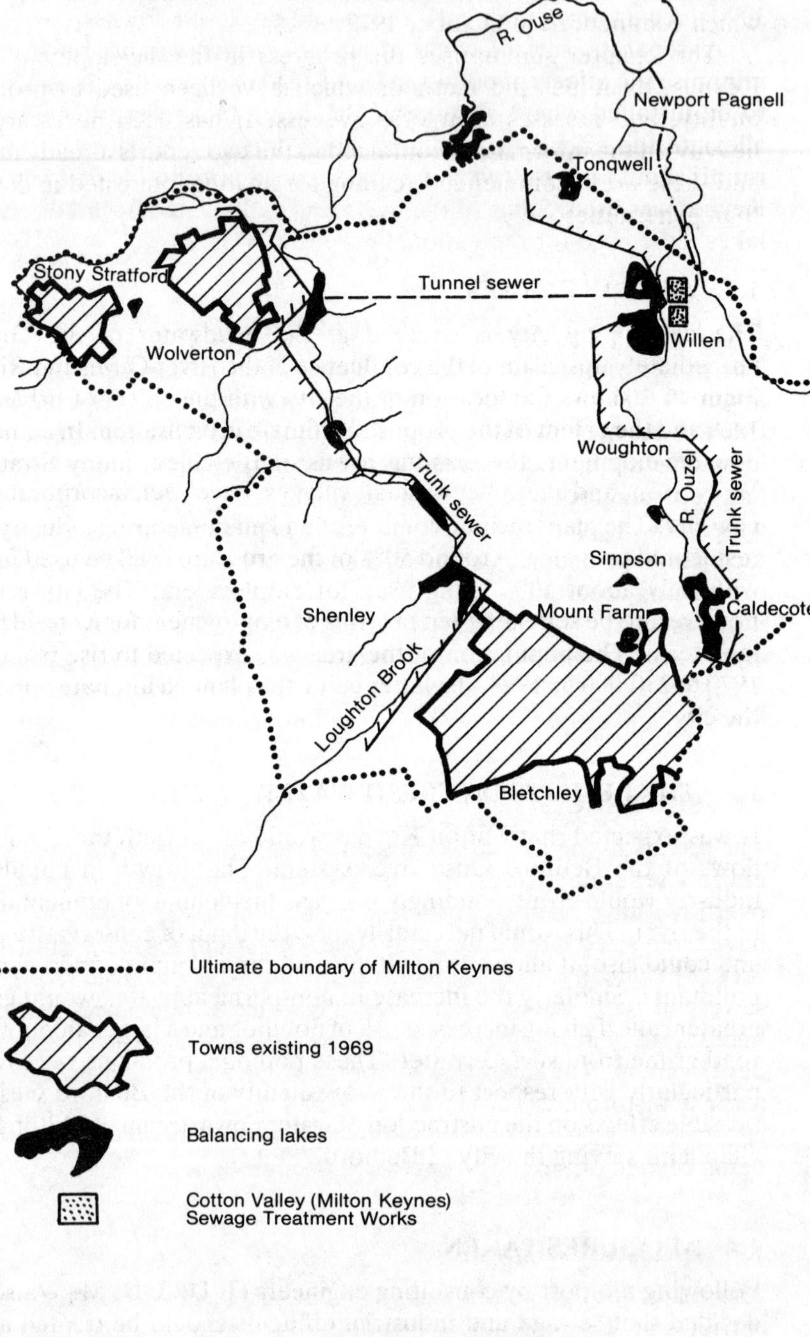

Fig. 1.1 — Milton Keynes

works was to be served by trunk gravity foul sewers, these being fed by pumped mains where necessary. The towns of Bletchley, Newport Pagnell (just outside the city), Stony Stratford, and Wolverton would be drained to the new sewage treatment works (STW). The existing, inadequate (even in 1970), sewage-treatment facilities would be closed down.

A separate surface-water drainage system was to be installed. To minimise the effects of pollution by oil and road grime, oil interceptors were to be installed where necessary. Balancing tanks were to be included to alleviate the effects of flushes of storm water. The criterion that rates of runoff should not exceed those experienced prior to the building of Milton Keynes was used. Some of the balancing tanks would be built as permanent lakes to be used for recreational purposes.

1.5 CURRENT STATE OF DEVELOPMENT

The present popultion of Milton Keynes is approximately 120 000. This is about 40 000 lower than, or three years behind, that predicted in the plan (Milton Keynes Development Corporation, 1970). Development of the infrastructure has been according to plan but of course housing development has kept pace with population demands. The required foul- and surface-water sewers, oil interceptors, balancing lakes, and sewage-treatment facilities have been constructed. The overall picture is of the development proceeding according to plan but at a slower pace. Housing development is now expected to be completed during the late 1990s with industrial development about two years later. The ultimate population is likely to be somewhat lower than originally predicted at about 220 000. This is mainly due to the housing density and occupancy being lower now than in the late 1960s.

1.6 SEWERAGE

The sewers in the new development areas for Milton Keynes are based on entirely separate systems for foul and surface water. Thus no storm-water overflows have been incorporated in the foul-sewage systems. The sewers have been sized large enough to cope with storm sewage from the existing areas of Bletchley, Newport Pagnell, Stony Stratford, and Wolverton. The sewers in these developments are a mixture of combined, partially combined, and separate depending on local policy at various stages of their development. The main trunk sewers for the foul system are complete and form a distorted H as shown in Fig. 1.1. The local sewers have been constructed as required by the building developments. A present work is being carried out in the Shenley and Woughton areas.

1.7 BALANCING LAKES

The major works on the surface-water sewerage system have also been carried out. Considerable expenditure has been incurred (approximately £20 m at current prices) by the construction of flood-water balancing lakes and reservoirs. These have been designed on the basis that runoff and flooding should be no worse than during the 1947 flood (the worst recorded in the area in recent times). Thus the balancing system has been designed to cope with and control the runoff from the increased impervious area created by the development of the city.

The balancing function has been divided into two systems (Fig. 1.1). The first is based in the Ouzel Valley and comprises mainly off-line wet lakes of which Willen and Caldecote are the largest. When river flows reach a given level some of the flow is diverted into the lake. The outflow from the lake is computer-controlled at a fixed percentage of inflow. Mount Farm, Tongwell, and Walton are also off-line wet lakes and it is proposed to extend the computer-control system to them during 1985/86. The remaining 'lake' in the Ouzel Valley is Simpson: a wet/dry reservoir which only fills at times of excessive runoff. It empties when the river system is able to cope with the flow. The second system of lakes is in the Loughton Brook Valley and is on-line, that is the brook runs through the middle of them. The control system is much simpler, the outflow being controlled by fixed discharge orifice, the level in the lake increasing as the inflow increases. Many of the lakes are used for recreational activities and have been constructed so that they enhance the environment in addition to protecting the river system. Between 30 and 40 per cent of the costs of the lakes can be attributed to providing amenity value. Activities such as angling, sailing, board-sailing and bird-watching are carried out. All the lakes have footpaths and cycle paths to allow general amenity and access.

1.8 OIL INTERCEPTORS

The plan for Milton Keynes provided for areas of industrial development. It was considered, therefore, that the river system and balancing lakes would be at risk from oil pollution. This was expected to arise from accidental (and possibly deliberate) discharges of oil into the surface-water sewers and the flushing of oil from vehicle hard-standing and car parks. This led to a policy of installation of oil interceptors on the surface-water drains and sewers in the industrial areas of the catchment. A three-stage policy of oil interception was developed.

The first stage involved the installation of oil interceptors at individual industrial sites. These are required at high-risk sites or where there is extensive car or vehicle parking (30 to 50 cars). Over one hundred such interceptors have been installed at the owner/occupier's expense.

The next stage of interception is at an area level, that is oil interceptors suitably placed on the surface-water sewers serving an industrial estate or part of an industrial estate, the idea being to trap oil that is flushed out of poorly-maintained private interceptors or oil that has been spilled on, or flushed from, non-intercepted areas. Three different types of interceptors have been installed on an area bases; two are of the American Petroleum Institute pattern, six are tilted plates and two are of the flow-foil type.

The third stage of interception is at the inlet to the flow-balancing lakes, and consists of oil booms. Two have been installed at Linford (which is just outside the Milton Keynes boundary) and Mount Farm Lake. The policy of oil interception is currently under review especially with regard to the second and third stages, the reasons for this being the enormous cost and ill-defined need for interceptors. Approximately 60% of the ultimate sewerage system for Milton Keynes has been built. To complete the installation of oil interceptors on this alone, in accordance with the original policy, would require approximately 20 additional area interceptors and six third-stage interceptors. The cost of doing so can be assessed from the capital expenditure of £1.5–2m that has been incurred to install the interceptors mentioned above. The revenue costs for maintaining the interceptors is approximately £5000 per annum. During 1984 about three cubic metres of oil were removed from the interceptors. Before further capital is expended on interception, the extent of the problem and need will be reassessed. Storm-water sewers that have been constructed without interceptors have been built in such a way as to allow them to be installed at a later date if necessary. A further consideration is that housing and shopping areas and public car parks have not been protected by interceptors.

1.9 SEWAGE-TREATMENT FACILITES

The sewage-treatment facilities were planned to be phased to keep pace with the development of the city. The first stage was completed at the end of 1974 and was designed as follows:

Population	117 000	
Dry-weather flow (DWF)	36 300 m^3/d (420 l/s)	
Effluent standard (mg/l)	suspended solids	:10
	biochemical oxygen demand (BOD)	:10
	ammoniacal nitrogen	:10

Figure 1.2 shows the plan of the sewage-treatment works.

Sewage enters the works via a gravity main and collects in the well of the inlet pumping station. Three screw pumps have been installed, with provision for a fourth. The present (stage I) maximum pumping rate is 3450 l/s, that is 8.2 DWF). The sewage is pumped into a screening chamber with two automatically raked, 19-mm spacing, bar screens. A third, manually raked, screen has been provided as a standby. Grit removal is carried out in two spiral-flow grit channels. The grit is pumped into Dorrclone separators which return organic matter to the sewage flow. The flow of sewage is measured and flows below three times design DWF (that is below 1275 l/s) are passed to the primary settlement tanks. Flows in excess of this are passed to storage/storm-flow tanks. These tanks are identical in size to the primary sedimentation tanks and will be used for that purpose after the stage II extensions have been completed.

The primary sedimentation plant consists of four horizontal-flow tanks with a total capacity of 11 197 m^3 giving 7.3 hours retention at design DWF. The desludging system was designed to operate on a timed sequence

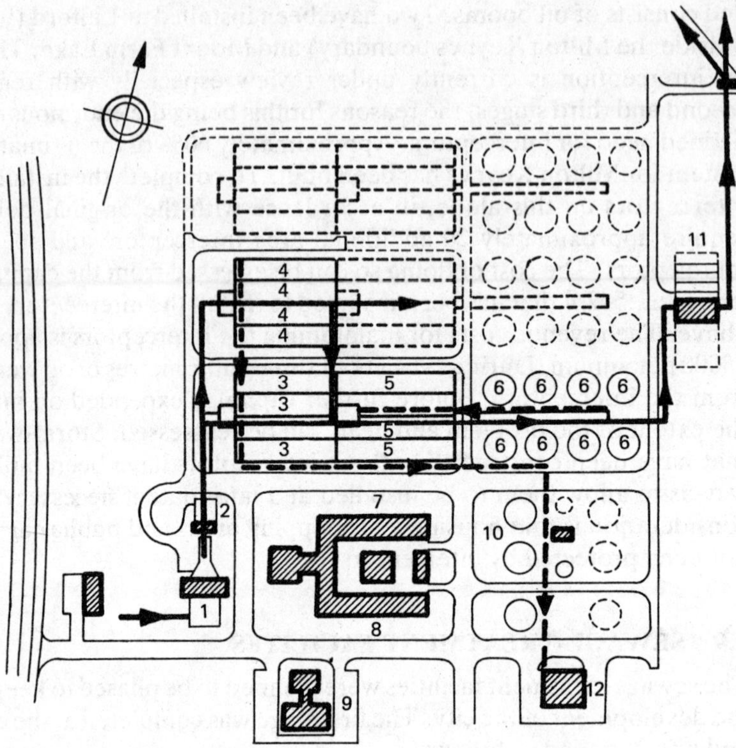

1. Works inlet (PS)
2. Grit channels
3. Primary sedimentation tanks (PST)
4. Storm tanks (stage II PST)
5. Aeration tanks
6. Final settlement tanks
7. Blower house
8. Workshops
9. Administration building
10. Anaerobic digesters
11. Sand filters
12. Incinerator house

– – – Future stages
—— Sewerage flow
— — Sludge flow

Fig. 1.2 — Cotton Valley (Milton Keynes) Sewage-Treatment Works

controlled by nucleonic sludge density meters, sludge being pumped directly into anaerobic digestors to avoid odour problems.

Secondary treatment of the settled sewage is carried out in four parallel aeration tanks with a total capacity of 14 964 m^3, giving a retention time of 10 hours at design DWF. A system of diffused air aeration is used. The domes are arranged in the tanks to give tapered aeration in the ratio 4:2:1:1 in the four sections of each tank. Air is supplied by one of two compressors. Dissolved-oxygen (DO) concentration is controlled by the use of DO probes and meters. The meters are linked by a control computer to the compressors and air output varied accordingly. The mixed liquor from the aeration plant is settled in eight circular final settling tanks. These have a total capacity of 9894 m^3 giving 6.5 hours retention at DWF. Activated sludge is returned by two axial flow pumps and surplus sludge is co-settled in the primary sedimentation tanks.

The effluent from the activated sludge plant receives tertiary treatment by rapid-gravity sand filters and, finally stabilisation in lagoons. The sand filters can treat up to 3.5 DWF or 127 000 m^3/day through a surface area of 418 m^2 and a media size of 1.2–2.4 mm. The filtered effluent is discharged through a 4-km long 2290-mm diameter tunnel to maturation lagoons alongside the River Ouse downstream of Newport Pagnell. The lagoons have a retention time of 2 to 2.5 days.

Sludge treatment on the works consists of heated anaerobic digestion followed by consolidation. Two centrifuges were installed for sludge dewatering, after conditioning with polyelectrolytes. A multi-hearth incinerator was provided for incinceration of the sludge cake if suitable (Department of the Environment/National Water Council, 1981; Anglian Water Authority, 1979) outlets to land could not be found.

The capital cost of the first stage was approximately £8.5m. This included an on-site laboratory, workshops and administration building complete with control room. The current running costs are approximately £175 000 for electricity and £130 000 for labour per annum. The works have now been in operation for ten years. Figure 1.3 shows the build-up of flow to the works

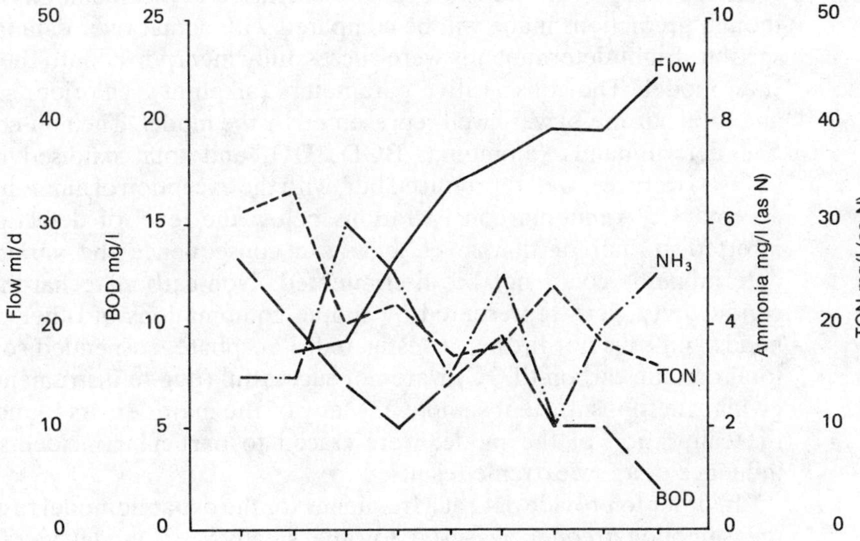

Fig. 1.3 — Cotton Valley final effluent

over that period and the average quality of the final effluent. The second stage is under construction and is expected to be completed in 1986 at a cost of about £6.3m.

1.10 THE BEDFORD OUSE STUDY

This study was concerned with the whole catchment of the Ouse system down to Earith. It is of particular interest with regard to the part of the catchment including and immediately downstream of the new city of Milton Keynes. The study sought to provide mathematical models of the river system in terms of flow and quality. The models were developed and calibrated (and adjusted as necessary) using actual data, and predictions checked against further actual data. Two models were developed, one being a steady-state model to be used for predicting long-term effects of changes in effluent flows, quality, and so on, and the other is a dynamic model which can be used to predict short-term or day-to-day changes in river-water quality.

It was originally evisaged that weekly sampling would be required for the steady-state model. It was not possible to do so, however, with 33 river stretches, 150 sampling points, and 29 significant effluent discharges into the system. It was decided, therefore, to take monthly samples with fortnightly samples at the input points and at gauging stations. Weekly sampling was carried out at certain critical points in the system.

The results of the study are far too extensive to include in this chapter, though predictions made will be compared with actual river data in a later section. Eight determinands were successfully incorporated into the steady-state model. The conservative parameters (alkalinity, chloride, sulphate, and total hardness) were well represented by the model. The non-conservative determinands (ammonia, BOD, DO, and total oxidised nitrogen (TON)) were less well reproduced but, with the exception of ammonia, were acceptable. Ammonia concentrations below the level of detection were inserted at half of that level, but as a consequence the variability of determinands could not be discriminated. Non-carbonate hardness and conductivity were represented by simple combinations of other determinands and thus not further investigated. Phosphate, suspended solids and total organic carbon (TOC) were not successful (due to instrument failure giving rise to insufficient samples). Many of the major errors found during the calibration of the model were traced to particular incidents or the influence of a few extreme results.

In order to provide data at a frequency for the dynamic model to operate, the sampling frequencies used for the steady-state model were clearly insufficient. Thus continuous monitoring was carried out for dissolved oxygen and temperature, and daily sampling for other determinands at four sites. The remaining 40 sites were sampled from daily down to once every 42 days according to their importance. The stretch of river used in the dynamic model was restricted to that between Newport Pagnell and Bedford. This has particular importance with regard to the public water-supply abstraction point at Clapham, immediately upstream of Bedford. Also this stretch is the one most affected by the discharge of sewage effluent from Milton Keynes. The predictive ability of the model was found to be adequate for BOD and DO. It was subsequently applied to chloride and TON, as representatives of conservative and non-conservative substances, with some success.

1.11 APPLICATION OF THE MODELS TO MILTON KEYNES EFFLUENT

The steady-state model has been used to predict average river water quality at various key points on the river system, based on certain assumptions about effluent flow and quality from the sewage-treatment works serving the new city of Milton Keynes. Table 1.1 reproduces some of the assumed data given in the report.

Table 1.1 — Assumed quality of Milton Keynes effluent

Calculation Number†	Date	BOD (mg/l)	Ammonia (mg/l)	TON (mg/l)	Flow (m^3/d)
1	1973	–	–	–	0
3	1975	5	0	30	18 182
4	1981	10	10	20	36 364
5	1981	10	10	10	36 364

†Numbering used in The Bedford Ouse Study-Final Report, 1979

The first case in Table 1.1 (calculation number 1) represents the situation prior to the building of Cotton Valley STW. Number 3 represents the stage immediately after Bletchley, Newport Pagnell, and Wolverton STWs were closed and the flow diverted to Cotton Valley STW. Numbers 4 and 5 represent increased flow at Cotton Valley with a deterioration of the effluent to the design standard. Number 5 assumes that some nitrate removal is being operated at the works. Table 1.1 should be compared with the actual circumstances regarding Cotton Valley effluent and shown in Table 1.3.

The calculated river quality corresponding to the above for sampling points at Sherington, Harrold, and Bedford are given in Table 1.2. A graphical representation is given in Fig. 1.4. Again these calculated data should be compared with the actual data shown in Table 1.4 and Fig. 1.4.

Table 1.2 — The calculated effect of Milton Keynes efluent on the River Ouse

Calculation Number	BOD (mg/l)	Ammonia (mg/l as N)	TON (mg/l as N)	DO (mg/l)
A. *At Sherington (S)*				
1	4.0	0.5	7.0	9.9
3	2.8	0.2	7.9	9.6
4	3.6	1.4	8.1	9.0
5	3.6	1.4	6.8	9.0
B. *At Harrold (H)*				
1	3.9	0.2	6.6	9.5
3	3.0	0.2	7.3	9.6
4	3.5	0.6	7.8	8.7
5	3.5	0.6	6.7	8.7
C. *At Bedford (B)*				
1	3.4	0.1	6.7	10.3
3	2.6	0.1	7.4	10.4
4	3.1	0.2	8.0	9.8
5	3.1	0.2	6.4	9.8

1.12 THE EFFECTS ON FRESH WATER

These will be considered in two ways.

How has the water quality changed since the development of Milton Keynes started?

How does the water quality compare with the predictions made in the Bedford Ouse Study?

Table 1.3 shows the quantity and quality of the effluent from Cotton Valley STW. The effects of the development of Milton Keynes and the increased population are clearly shown in the flow figures. What is perhaps more surprising is the effluent quality. In spite of the significant increase in flow through the works there has been a marked improvement in the BOD and only an insignificant deterioration in ammonia. This has been due to improved operational strategies and correction of mechanical defects since the works was commissioned.

Table 1.3 — Cotton Valley final effluent

Date	BOD (mg/l)	Ammonia (mg/l as N)	TON (mg/l as N)	Flow (m^3/d)
1976/7	10	3	33	17 600
1980/1	9	5	18	36 500
1983/4	2	5	16	44 300

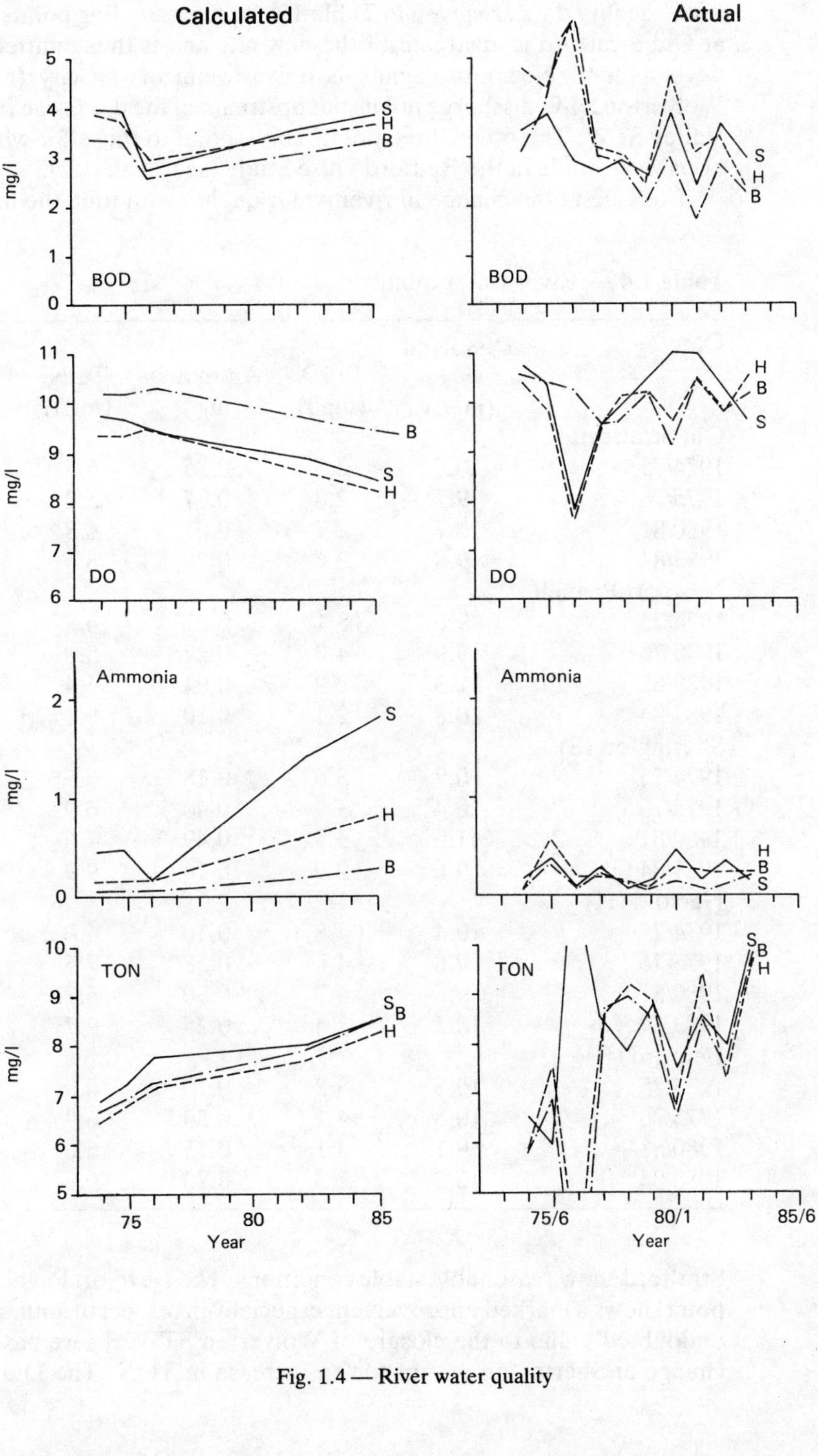

Fig. 1.4 — River water quality

Bletchley and Newport Pagnell STWs closed down in 1975 shortly after Cotton Valley was commissioned; Wolverton STW closed in 1978. River water quality data are given in Table 1.4 for five sampling points. The point at Old Stratford is upstream of the new city and is thus unaffected by the development. Newport Pagnell is downstream of the city (and the old Wolverton STW discharge point) but upstream of the discharge from Cotton Valley STW. The other three points correspond to those for which predictions were made in the Bedford Ouse Study (see Table 1.2).

Looking at the change in river water quality with time the data for Old

Table 1.4 — River water quality

Date	Dissolved oxygen (mg/l)	BOD (mg/l)	Ammonia (mg/l)	TON (mg/l)	Chloride (mg/l)
Old Stratford					
1974/75	10.0	2.5	0.25	6.5	34.5
1975/76	9.5	2.8	0.17	5.2	40.3
1980/81	10.7	2.7	0.07	6.8	36.8
1983/84	9.8	2.6	0.28	8.5	38.3
Newport Pagnell					
1974/75	9.6	3.1	0.22	6.8	40.0
1975/76	9.9	4.0	0.27	5.3	44.3
1980/81	11.3	2.9	0.04	6.4	37.5
1983/84	10.8	2.1	0.10	8.1	41.5
Sherington (S)					
1974/75	10.7	3.6	0.18	6.5	41.0
1975/76	10.4	3.9	0.40	6.0	57.5
1980/81	11.0	3.9	0.49	7.6	47.9
1983/84	9.6	2.9	0.16	9.9	51.1
Harrold (H)					
1974/75	10.4	4.8	0.10	6.0	44.3
1975/76	9.8	4.4	0.58	7.5	59.3
1980/81	9.6	4.7	0.30	6.7	42.0
1983/84	10.5	2.4	0.25	9.7	48.7
Bedford (B)					
1974/75	10.5	3.2	0.10	6.0	43.8
1975/76	10.3	4.1	0.30	6.9	58.0
1980/81	9.1	3.1	0.15	6.6	41.5
1983/84	10.2	2.3	0.20	9.9	47.8

Stratford show reasonably stable conditions. The Newport Pagnell sampling point shows a marked improvement especially in respect of ammonia. This is undoubtedly due to the closure of Wolverton STW. There has been little change at Sherington except for an increase in TON. The DO and BOD levels have improved slightly, but the concentrations of ammonia have deteriorated slightly; (the full sets of data used to produce Fig. 1.4 show these trends more clearly). There has been little change in the quality at Harrold and Bedford except for an improvement in BOD and an increase in TON.

Comparison of the actual quality data with that predicted by the Bedford Ouse Study shows that, with the exception of TON, the predictions were slightly pessimistic. This may in part be due to the better-than-expected effluent quality from Cotton Valley STW. The ammonia values in particular are lower in the river than anticipated. This would not account for the fact that the TON values were higher than predicted. These may well be due to farming practices and an underestimate of runoff nitrate levels in the model.

1.13 CONCLUSIONS

The possible effects of urbanisation on the quality of fresh waters were carefully considered at the planning stage of the new city of Milton Keynes. The importance of maintaining water quality, particularly with regard to the downstream abstraction for the water supply at Bedford, was fully recognised. The expenditure on the Bedford Ouse Study has been fully justified by the success of the predictions made by the river quality models. Likewise the considerable expenditure on sewage-treatment facilities has enabled the river-water quality to be maintained, and in some stretches improved, compared with that which existed prior to the development of the city.

There is a question mark over the need for oil interception on an area basis in the city. Further expenditure has been delayed until the problem has been more clearly quantified. The balancing lakes have prevented any serious flooding and have a considerable amenity value in their own right.

Over all, it is considered that the practices in controlling the effects of urbanisation on fresh waters in the Milton Keynes area have been a success. The expense involved must be considered in the context of the total costs of the development of the city, that is approximately £1500m up to the present.

ACKNOWLEDGEMENT

The author wishes to thank the Director of Operations of Anglian Water for permission to publish this chapter. The opinions expressed are the author's own and do not necessarily accord with those of Anglian Water, the Milton Keynes Development Corporation or any of their employees. The author is particularly grateful to Messrs A. Alcock, A. Beer, P. Cashen, A. Lines, P. Staufert, D. Tester, and D. Wharton without whose help it would have been impossible to assemble the data necessary to produce this chapter.

REFERENCES

Anglian Water Authority (1979) *Manual of good practice for utilisation of sewage sludge as a fertiliser.* (Available from Anglian Water, Ambury Road, Huntingdon, Cambs. PE18 6NZ, England).

Anglian Water Authority/Department of the Environment (1979) *The Bedford Ouse Study — Final Report* of the steering Group to the Anglian Water Authority and the Department of the Environment. (Available from Anglian Water, as above.)

Department of the Environment/National Water Council (1981) Report of the sub-committee on the disposal of sewage to land. Department of the Environment.

Milton Keynes Development Corporation (1970) *The plan for Milton Keynes*. Presented by the Milton Keynes Development Corporation to the Minister of Housing and Local Government. (Available from the Milton Keynes Development Corporation.)

Watson, J. D. & D. M. Technical Supplement No. 10 *Sewerage, Sewage Disposal and Surface Water Drainage.* (Available from J. D. & D. M. Watson and Milton Keynes Development Corporation.)

2

Water quantity and quality in the new town of Lelystad

E. J. B. Uunk and **F. H. M. van de Ven**

2.1 INTRODUCTION

In the Lake IJssel (IJsselmeer) area, four polders have been constructed. Two polders have been completed and two are near completion. In the two most recently reclaimed polders (Eastern and Southern Flevoland with areas of 54 000 and 43 000 ha, respectively) urbanisation, nature preservation and recreation have become important factors in the land use next to the agricultural development. In these polders, two new towns are under construction, named Lelystad and Almere. Their main goal is to accommodate people and industries from the overcrowded towns in the western part of Holland, in particular Amsterdam. Lelystad and Almere are located about 60 and 20 km east of Amsterdam, respectively. The building of Lelystad started in 1966. This town, which had 60 000 inhabitants by the end of 1984, is planned to have about 90 000 inhabitants by the year 2000, on a gross area of 4300 ha. Almere is planned to house 200 000 people by the year 2000 on a gross area of 11 000 ha. The building activities there started in 1974.

2.2 WATER MANAGEMENT OF A TOWN IN A POLDER

Polder areas need an artificial drainage system to discharge the natural water surplus. The water surplus comprises not only precipitation minus evaporation — average precipitation varies from 50 to 70 mm per month, while the potential evaporation ranges from about 120 mm per month in summer to about 5 mm per month in winter — but includes also seepage, entering the low-lying polders from the surrounding areas with higher drainage levels. To maintain an efficient drainage, a network of subsurface drain pipes, ditches, canals and pumping stations is constructed as is shown schematically in Fig. 2.1.

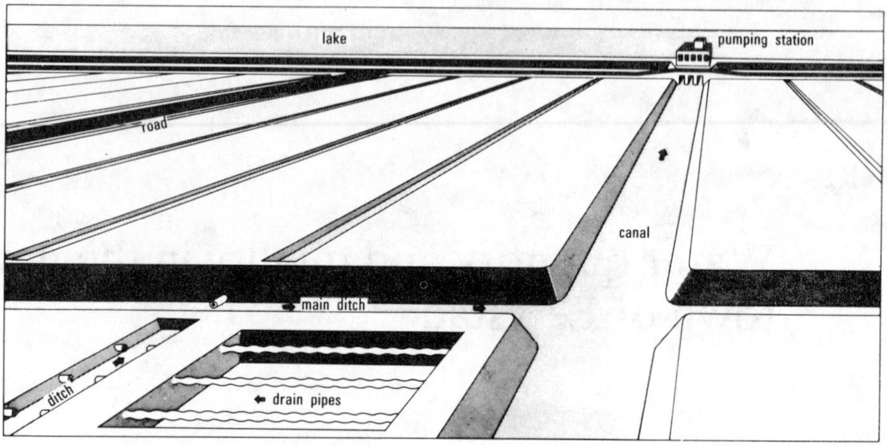

Fig. 2.1 — Schematic view of the water management system of a polder

The water level in the canals near Lelystad is 6.20 m below average sea level. The pumping stations are able to drain a water surplus of 11 mm/d to Lake IJssel. This lake has a water level of 0.2–0.4 m below average sea level. In case there is an extra water surplus, this is temporarily stored in the canals and ditches that cover about 1% of the total polder area. The water management of the urban areas is more or less separated from that of the rural surroundings. In order to have sufficient drainage in these towns, subsurface drains, storm sewers and canals are constructed. In Fig. 2.2 a schematic view of the build-up and the drainage of Lelystad is shown. To facilitate building, the level of building sites in the Lake IJssel polders is moreover raised by one metre of sand.

Lelystad and Almere are provided with a separate sewerage system. The storm sewers discharge into the urban canals. These are modestly sized surface waters, distributed over the urban area. The subsurface drainage system collects the surplus of infiltrated precipitation and most of the seepage and discharges into the canals as well. Finally, the canals receive some precipitation and seepage directly. The urban canals discharge the surplus water of the urban area over weirs to surrounding polder canals. The water level in the urban area of Lelystad is 5.40 m below average sea level. This relatively high water level is mainly due to the metre of sand applied on the building sites. To get the correct drainage depth, the level in the open watercourses had to be raised by 0.80 m. To maintain a stricter water-level regime in the urban canals, the water management in the urban area is

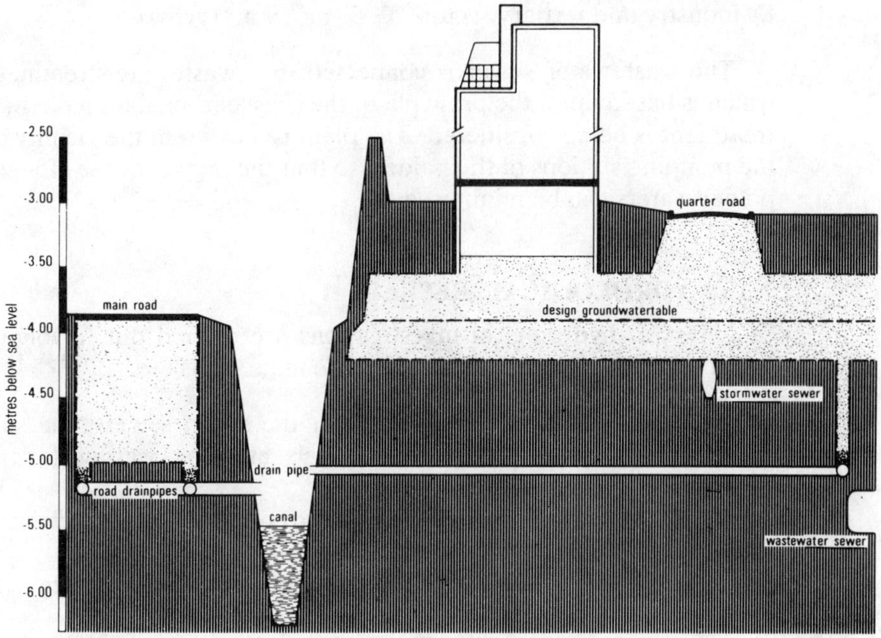

Fig. 2.2 — Scheme of the water management system as constructed in Lelystad (Schultz, 1984)

separated from the rest of the polder — in Lelystad by weirs — and the percentage of open water in the urban area is 3% compared to 1% in the rest of the polder. As a result of the considerable surface area of the canals and of the width of the weirs, the urban canals act as a buffering reservoir for urban runoff control. The urban canals also have a function with respect to town planning and recreation (fishing, boating, ice skating) and form a habitat for aquatic life.

In Lelystad a system has been constructed enabling the municipal canals to be flushed when required. Up to now the system has only been used once (to relieve a calamitous situation), since the quality of the town water generally is better than the polder water available for flushing. Therefore, in Almere a flushing system was not installed.

The subsurface drainage system is designed to maintain a groundwater level of more than 0.70 m below the surface at a discharge of 5 mm/d. The storm sewers are designed to drain the surface runoff. Their design capacity in Lelystad generally is 6 $m^3/s/km^2$. At present 5 $m^3/s/km^2$ is used, carrying the drainage from the paved surfaces.

The wastewater sewer is designed for a discharge of domestic wastewater of 15 l/person/h. Other design discharges for the wastewater sewer are as follows:

- trade and firms in secondary sector 25 $m^3/s/km^2$ (gross)

- industry and tertiary sector 50 $m^3/s/km^2$ (gross).

The wastewater sewer is connected to a wastewater-treatment plant, which is based upon the principle of the 'Pasveer' oxidation ditch. Tertiary treatment is being considered. The plant is located in the vicinity of one of the pumping stations of the polder, so that the impact of the effluent on the polder waters can be minimised.

2.3 HYDROLOGICAL RESEARCH

In Lelystad, hydrological investigations are carried out, involving water quantity and water quality aspects. The main goals are:

(i) to evaluate the design criteria for the storm sewers, the subsurface drainage system, the urban canals and the wastewater-treatment plant(s);
(ii) to control and, if relevant, reduce the pollution of urban surface waters; and
(iii) to evaluate the impact of urban areas on surrounding polder canals and on the management of these waters.

Since 1968 a research project has been conducted on the quantitative aspects of the urban hydrology in two experimental basins. Since 1972 regular investigations have been made on the water quality in the drainage system of Lelystad. In 1982 a third experimental basin was equipped for combined water quantity and quality studies. The main objectives for the measurements in that basin were:

- to determine for various relevant pollutants the loads associated with the storm-water runoff;
- to analyse the impact of the storm-water runoff on the quality of the receiving canal water; and
- to study the behaviour of the storm-water runoff in comparison with the other two existing basins.

The basin is a 4.5 ha large housing area named 'Bastion' and about 66% of the basin area is covered. It contains 60 houses/ha. The buildings are two to four storeys high and the roofs are of corrugated asbestos-cement.

Precipitation and storm-sewer discharge are monitored continuously by a data-logging system. Both in the storm-water and — at three locations — in the canal water, the water temperature, pH, oxygen concentration and conductivity are monitored likewise. The storm sewer discharge and the wet precipitation are sampled automatically. Additional manual sampling takes place in the canal water and in the subsurface drainage discharge. At other sites in Lelystad samples are taken manually to get a reference for the data of the experimental basin. The water samples are analysed on a wide variety of quality variables, including: ammonium-, Kjeldahl- and nitrate-nitrogen (NH_4-N, Kj-N, and NO_3-N); biochemical oxygen demand (BOD) (5 days at

20 °C); chemical oxygen demand (COD); dissolved oxygen (DO); total phosphorus (tot-P); (volatile) suspended solids (SS); cadmium (Cd); chromium (Cr); copper (Cu); lead (Pb); zinc (Zn); mineral oil; polynuclear aromatic hydrocarbons; detergents; major ions, and faecal coliforms and streptococci (faec. col. and faec. strep.).

2.4 WATER QUALITY IN LELYSTAD

Tables 2.1 to 2.4 give an impression of the water quality of precipitation,

Table 2.1
Rain-water quality in Lelystad
Bastion 1984; no. of observations: 7–18

tot-P	0.04	(0.00–0.25)	mg/l	Pb	14	(<5 – 36)	μg/l
Kj-N	1.10	(0.50–1.70)	mg/l	Zn	118	(20 – 415)	μg/l
NH_4-N	1.00	(0.50–1.50)	mg/l	Cu	8	(<5 – 16)	μg/l
NO_3-N	0.70	(0.40–1.20)	mg/l	Cd	0.5	(<0.5– 1.8)	μg/l

Table 2.2
Storm-water quality in Lelystad

Location		Bastion†		Bastion†	Four reference sites‡
Year		83		1982/1983	1982/1983
No. of observations		43–550§		11¶	11¶
SS	(mg/l)	34	(1.2 –250)	1.7 – 38.8	5.6 – 270
BOD	(mg/l)	3.2	(0 – 22)	3 – 4	2 – 13
COD	(mg/l)	23	(5 –230)	7 – 19	11 – 105
tot-P	(mg/l)	0.4	(0.03– 3.5)	0.08– 0.09	0.02– 4
Kj-N	(mg/l)	1.0	(0.3 – 19)	0.8 – 0.9	0.4 – 18
NH_4-N	(mg/l)	1.0	(0.02– 5.2)	0.08– 0.42	0.1 – 13
NO_3-N	(mg/l)	1.1	(0.4 – 11.5)	0.06– 1.2	0.1 – 5.3
Mineral oil	(mg/l)	1.5	(0.14– 7.1)	1.14– 4.3	0.2 – 2.4
Pb	(μg/l)	53	(2 –573)	32.5 –111	14 –1152
Zn	(μg/l)	390	(120 –1900)	196 –938	45 – 800
Cu	(μg/l)	17	(2 – 65)	3 – 34	2.6 – 72
Faec. col.	(No/100 ml)		<95 – >8 10^4	11 10^3–24 10^3	375 – 8 10^6
Faec. strep.	(No/100 ml)		<95 – 2.7 10^5	3.8 10^3–44 10^3	1 10^3 – 2.4 10^6

† Flow-weighted samples and averages
‡ Grab samples; sites consist of 2 housing areas, 1 light industrial area and a large car park
§ Number depends on determinand
¶ Same events; comparable moments of sampling

storm sewer and subsurface drainage discharge and canal water in Lelystad. The data are either average concentrations with ranges in parentheses or merely ranges.

As Table 2.1 indicates, the concentrations of nitrogen and heavy metals

Table 2.3
Drainage water quality in Lelystad

	Bastion 1982/83/84		Four sites 1975				Bastion 1982/83/84
BOD	2	(1 – 3)	1.6	(0.6– 2.8)	mg/l	Pb	<5 – 22 μg/l
COD	24	(19 –30)	27	(19 –38)	mg/l	Zn	19 (9–42) μg/l
tot-P	0.43	(0.08– 2.0)	0.8	(0.1– 2.4)	mg/l	Cu	<2 – 3.6 μg/l
Kj-N	2.1	(1.6 – 2.8)	2.2	(1.2– 4.1)	mg/l	Faec. col.	<95/100 ml
NH_4-N	1.5	(0.95– 2.1)	1.5	(0.9– 3.0)	mg/l	Faec. strep.	475–7500/100 ml
NO_3-N	0.7	(0.0 – 2.3)	0.25	(0.0– 1.2)	mg/l		

Table 2.4
Canal water quality in Lelystad

Location		Bastion		Bastion		Five reference sites
Year		1983		1981		1981
No. of observations		10†		10‡		50‡
Transparency		>30		40–>55		23 –120
BOD	(mg/l)	3.1	(2 – 6)	4.3	(2 – 8)	2 – 9
COD	(mg/l)	24	(11 – 39)	25	(17 – 41)	15 – 50
tot-P	(mg/l)	0.27	(0.07– 0.8)	0.28	(0.1– 0.8)	0.04– 0.6
Kj-N	(mg/l)	1.6	(0.4 – 4.3)	1.1	(0.7– 1.4)	0.7 – 6.8
NA_4-N	(mg/l)	0.7	(0.1 – 2.5)	0.2	(0.0– 0.35)	0.0 – 3.2
NO_3-N	(mg/l)	0.7	(0.2 – 1.1)	0.4	(0.0– 1.3)	0.0 – 2.1
Cl	(mg/l)	116	(67 –237)	129	(47 –223)	113 –344
mineral oil	(mg/l)	0.4	(0.0 – 0.7)	—	—	—
chlorophyll-a	(μg/l)	17	(1 – 89)	33	(4 –104)	2 –128
Pb	(μg/l)	16	(1 – 52)	—		—
Zn	(μg/l)	69	(24 –146)	—		—
Cu	(μg/l)	5	(<2 – 11)	—		—
faec. col.	(No./100 ml)	<95§	(<95–1000)	515§	(<95–12000)	<95–6300

† Sampling at arbitrary times — generally in dry periods
‡ Same data
§ Median

in rain-water are fairly high. As a result, precipitation is a significant polluting source for these constituents. For ammonium- and nitrate-nitrogen, the average concentrations in rain-water are even similar to those in the storm-sewer discharge (Table 2.2). Striking features of storm-water quality are:

- high variability;
- high concentrations of some heavy metals;
- high bacterial numbers;
- high content of mineral oil;

- moderate to high concentrations of nutrients;
- relatively low content of biodegradable organic matter.

The high variability is mainly caused by differences in rain storm intensities and volumes. It appears that the amount of rainfall determines to what extent the sewerage system is flushed, while the maximum rainfall intensity during a rain storm determines the amount and nature of solids and related constituents (among others, P, heavy metals) to be discharged. Apparently, relatively light organic particles are flushed from the surface more easily than inorganic particles. At low rainfall intensities a 'first-flush' phenomenon was observed for soluble constituents.

For some constituents, for example mineral oil, incidents (spills) proved to be significant. The high concentrations of lead are, as elsewhere (Uunk, 1983), attributed to the wet and, even more so, dry deposition of lead, emanating from exhaust emissions. Exhaust emissions must also be responsible for the fairly high concentrations of aromatic hydrocarbons which were found (not shown in Table 2.2). In view of the rather poor rain-water quality throughout Holland, not all of the deposited pollutants originate in the urban area. The zinc concentrations are higher in the Bastion area than in most other sites. This is probably caused by the corrosion of galvanised railings, which are used widely in the area. There is no input of domestic wastewater to the storm-sewerage system of the Bastion. Therefore bacteriological contamination and part of the nutrient and organic-matter content of the storm-sewer discharges must emanate from the faeces deposited by domestic pets, particularly by dogs.

As Table 2.2 shows, storm-water quality in the housing area Bastion is comparable to the quality on other sites in Lelystad. Therefore it can be regarded as an indication of average storm-water quality in Lelystad. The storm-sewer discharges from commercial centres, large parking areas and main roads are somewhat more polluted than the discharges from housing areas and from districts with light industry (not shown in Table 2.2). However, these polluted areas cover only a minor part of the total urban area.

The subsurface drainage water has a fairly good quality, except for its nutrient content (Table 2.3). The high concentrations of phosphate are probably caused by colloidal iron-phosphates. The relatively high chloride concentrations and COD are caused by the contribution of the upward seepage.

As a result of sedimentation and biochemical processes, canal-water quality (Table 2.4) is generally better than the average quality of the storm-water and drainage-water inputs. The high nutrient content of the inputs and the nutrient-rich clayish bottom render the canals eutrophic. In general, this results in an intensive growth of submerged and floating macrophytes. Where residence times are long enough, algal blooms may develop. In the Bastion canal intensive growth of duckweed develops now and then in summer. This may render the water anaerobic, initiating the release of ammonia-nitrogen and ortho-phosphate from the canal bed to the water column.

The bacteriological quality of the canal water is influenced by the bad bacteriological quality of the storm-sewer discharges.

The impact of the storm-sewer discharges on canal water quality can be demonstrated more clearly by comparing the quality of the canal water and the canal sediment at the outfall of the storm-sewer with the quality of water and sediment up- and downstream from that outfall (Table 2.5).

Table 2.5

The quality of water and sediment (sed) in the Bastion canal, 1982/1983

Constituent		Upstream from outfall water	Upstream from outfall sed	At outfall water	At outfall sed	Downstream of outfall Near water	Downstream of outfall Near sed	Downstream of outfall Distant water	Downstream of outfall Distant sed
chloride	(mg/l)	97		44		45		87	
COD	(mg/l–mg/g)	27	45	21	95	22	72	25	47
tot-P	(mg/l–mg/g)	0.32	0.5	0.31	1.1	0.43	1.0	0.36	0.7
Pb	(μg/l–μg/g)	19	27	35	167	33	64	22	26
faec. col.	(counts/100 ml† – counts/g)	215	1000	1000	5000	3950	700	95	700

† Median

In Table 2.5, canal-water quality is based on the results of sampling directly after 19 rainfall events; sediment quality is based on the results of composite samples, taken from the top 5 cm of sediment at each site.

In the water samples, the impact of storm-water is most evident for chloride, lead, and faecal coliforms, while the sediment samples show accumulation near the outfall for all constituents considered. The bacteriological pollution, caused by the storm-water runoff, prohibits swimming in the urban canals.

Investigations on the biotic community in the canals (macrophytes, invertebrates, epiphytic diatoms and filamentous algae) revealed that the canals can generally be classified as eutrophic and β-α-mesosaprobic. On the basis of species composition, a tendency towards the more polluted α-mesosaprobic state could be distinguished in the vicinity of the storm-sewer outfalls (Roos 1986).

2.5 QUANTITATIVE AND QUALITATIVE WATER-BUDGET OF THE CANALS

Table 2.6 shows the average monthly water, chloride, total-phosphorus, total-nitrogen, and BOD budget for a section of the Lelystad area. This section of 1.2 km^2 includes a shopping centre of 0.1 km^2 and 0.025 km^2 of open water.

The total average flow through the canals of 41 mm/month corresponds to an average residence time of the water in the canals of 11 days only. The canal-water output is discharged over a weir into a polder canal.

Fig. 2.3 — The Bastion canal

The water budgets show no retention: canal-water discharge is assumed to equal total inputs minus 'Penman' evaporation.

The seepage, entered in the canal budget, is direct seepage through the canal bed. It is saltier (3000 g Cl/m^3) and richer in nitrogen (25 g N/m^3) than the seepage that enters the subsurface-drainage system, as the latter is influenced by mixing with infiltrated precipitation.

Storm-water runoff and shopping-centre discharge contribute a relatively large part of the input of organic matter to the canals.

A considerable part of the total P- and N-input to the canals is removed from the canal water before it is discharged into the surrounding polder waters. For phosphorus, gravity settling and co-precipitation after the addition of ferrous hydroxide must be the main causes for this removal. For nitrogen, nitrification followed by denitrification plays a rôle in addition to gravity settling, especially at higher water temperatures. The average nitrogen retention is 78% for the period May to October as opposed to 41% for November to April. Budgets for heavy metals and for faecal bacteria show even higher removal efficiencies of the canals. On the other hand, some organic matter is added to the canals as biomass (mainly algae).

2.6 THE ANNUAL WATER BUDGET FOR LELYSTAD

With the previous information and data on the population of the town section studied (5250 people), supply and quality of drinking water (0.145 m^3 per capita per day, 0.01, 0.1 and 0 grammes of P, N and BOD per m^3, respectively), wastewater production (0.12 m^3 of water and 3.0, 12 and

Table 2.6

Average monthly water, chloride, total phosphorus, total nitrogen, and BOD budgets for a study area (all quantities related to the total land area of the town section; all percentages related to total input) (Uunk and Van de Ven, 1984)

	Water (mm)	(%)	Chloride (kg/ha)	(%)	Tot-P (g/ha)	(%)	Total-nitrogen (g/ha)	(%)	BOD (g/ha)	(%)
1. Total monthly inputs to the canals	41	100	117	100	280	100	1660	100	1440	100
Contribution of:										
Subsurface drainage†		54		69		66		43		29
Storm-water runoff†		29		4		20		31		48
Shopping centre discharges		12		5		10		9		21
Precipitation		3		0		0.4		2		1
Seepage		2		22		3.6		15		1
2. Total monthly output from the canals	41	100	117	100	86	31	780	47	1620	113
Contribution of:										
Canal-water discharge		97		100		31		47		113
Evaporation		3		0		0		0		0
3. Retention	0	0	0	0	194	69	880	53	−180	−13

† Excl. the shopping centre

47 grammes of P, N and BOD per capita per day) and treatment plant effluent quality (14.9, 19.5 and 8.5 grammes of P, N and BOD per m^3), an annual water budget for Lelystad could be calculated (Table 2.7).

Table 2.7

Annual water budget for Lelystad (1969 to 1981)
Flows in 10 mm/yr, material fluxes in kg/ha/yr

	Flow	P	N	BOD_5
Drinking water supply	230	neg.†	neg.†	neg.†
Change in urban area	−40	+48	+192	+750
Supply to wastewater treatment plant	190	48	192	750
Removal in wastewater treatment plant	0	14	147	730
Remaining	190	34	45	20

	Flow	P	N	BOD_5
Precipitation	700	0.6	20	7‡
Seepage	120	1.2‡	7‡	1‡
'Natural' supply	820	1.8	27	8
Change in urban area	−327	+1.6	−7	+9
Supply to urban canals	493	3.4	20	17
Removal in urban canals	15	2.3	11	+2
	478	1.1	9	19

	Remaining Flow	P	N	BOD5
Output to receiveing waters from Lelystad	668	35.1	54	39
Output to receiveing waters from rural surrounding of Lelystadt	250	0.3	34	7

† Negligible
‡ Estimate

Table 2.7 illustrates that the urban canals have a similar removal function for runoff and seepage as the wastewater-treatment plant has for the municipal wastewater. Precipitation turns out to be the main input of water into the urban area. The treated wastewater accounts for 28% of the output of water from the urban area, but for 97% of the P-output and 83% of the N-output. As a result of the high removal efficiency of the treatment plant, the BOD output through the effluent is of the same order of magnitude as the BOD output of the canal water.

In order to get an idea of the main sources of water pollution in the Lake IJssel polders, the output from the urban area per unit area is compared with that from the rural area, as established from other investigations of the IJsselmeerpolders Development Authority. It must be noted here that only about 10% of the total polder area is to be urbanised. It turns out that the urban area yields higher outputs than the rural area, especially for phosphorus. At the moment, phosphorus output is of prime interest, because it is a key factor in eutrophication control. The phosphorus budget shows clearly that reduction of phosphorus output can be achieved most readily by phosphorus removal in the wastewater-treatment plant. This measure is in preparation now.

2.7 CONCLUSIONS

On the basis of the experience in Lelystad, the following conclusions can be drawn with respect to the effects of urban land use on the quality of receiving waters.

- Urban storm-water runoff is contaminated with traffic-related pollutants, in particular heavy metals and organic micropollutants. A part of the pollutants is emitted in the urban areas itself, the rest originates from outside but is deposited in the urban area, either wet or dry. Another important source of heavy metals, in particular zinc, may be corrosion.
- Urban storm-water runoff can show considerable bacteriological contamination as a result of faecal deposits in the urban areas by domestic pets (dogs).
- The high variability of storm-water quality is mainly caused by differences between rain-storm events in rainfall volume and intensity. At low rainfall intensities, only soluble constituents and relatively light particles, for example organics, are discharged, while at higher intensities a much higher solids load is picked up, and along with it adsorbed constituents like heavy metals and phosphate.
- The majority of the solids discharged by the storm sewers settles in the vicinity of the outfall, resulting in a local sediment pollution.
- In Lelystad, the urban canals act as a significant sink for many pollutants. As a result, the water that is discharged by the canals into the rural area surrounding the town is much cleaner than the (storm) water entering the canals. Processes responsible for this improvement are not only sedimentation and precipitation, but also biochemical breakdown. Organics and nutrients supplied to the canals are partly converted to biomass (algae).
- The bacteriological pollution, caused by the storm-water runoff, prohi-

bits swimming in the urban canals. Apart from that, canal-water quality is acceptable. Flushing the canals with water from the rural surroundings is not necessary. Besides, this water is of poor quality.

- In terms of saprobic state, the canal water can be classified as α-β-mesosaprobic, except for the vicinity of the storm-water sewer outfalls, where a tendency towards an α-mesosaprobic state was established. In terms of trophic state, the canals are eutrophic.
- The phosphorus output from Lelystad originates primarily from the wastewater-treatment plant. Phosphate removal there is in preparation.
- By separating the water-management system of the urban and rural area and by applying stricter standards for the urban water management system, the impact of the urban area on the rural environment is limited, both for water quantity and water quality.

REFERENCES

Roos, C. (1986) *Biologische gevolgen van de lozingen vanuit regenwater-riolen voor het ontvangende oppervlaktewater.* R.IJ.P.-rapport 1986 — 2 Abw. Ijsselmeerpolders Development Authority, Lelystad.

Roos, C. & Uunk, E. J. B. (1985) *The impact of stormwater runoff on the biota in urban surface waters* (in preparation).

Schultz, E. (1984) *Water management in newly reclaimed land.* R.IJ.P.-rapport 1984–33 Abw. IJsselmeerpolders Development Authority, Lelystad.

Uunk, E. J. B. (1983) Pollution in urban runoff. In: Jolànkai, G. & Roberts, G., *Proc. MAB-5 Workshop on land use impacts on aquatic systems*, VITUKI, Budapest.

Uunk, E. J. B. & Ven, F. H. M. van de (1984) Water budgets for the town of Lelystad. In: Balmér, P., Malmqvist, P-A. & Sjöberg, A. *Proc. 3rd Int. Conf. on Urban Storm Drainage*, **3**, Chalmers, Univ. of Tchnol., Göteborg.

3

Impact of motorway runoff upon surface water quality

R. Perry and **A. E. McIntyre**,

3.1 INTRODUCTION

The development of motorway networks and associated drainage systems has given rise to concern about the possible effects of road-surface runoff on water quality. There has also been increasing control of specific substances of concern within the water cycle, and both the European Community and the World Health Organization are reviewing water-quality standards. Many of the investigations, to date, of this aspect of water quality have been short-term and in atypical locations so that there are insufficient data available with which to assess the magnitude of the pollution problem. For a pervasive appraisal to be undertaken of the problem of motorway runoff, consideration must be given to the sources of the contaminants, means of transformation into and out of the system, taking in to account the complex physical and chemical interactions and their effects on the environment. The loading intensity of the various contaminants at any given site is a function of several variables, and this is reflected in the wide range of values reported in the literature.

Studies of motorway environments, where contamination levels can be characterised and related more closely to vehicular traffic, have shown that drainage water from such areas contains pollutant levels comparable to urban runoff and underlines the importance of vehicle-related pollutants in the urban environment.

Materials present in motorway runoff arise from a variety of sources including:

road-surface degradation;

vehicle lubrication system losses;
vehicle exhaust emissions;
load losses from vehicles;
degradation of vehicle tyres;
road-surface cleaning;
de-icing compounds;
atmospheric deposition/precipitation.

The range of contaminants which have been found to occur in precipitation, road runoff, urban runoff, and combined sewer overflows, are detailed in Table 3.1, together with reported concentration levels.

The construction of a highway through a region extensively modifies the hydrological cycle, with greater volumes of water to be dealt with in a shorter period of time at sites near the road. The effect is twofold; to increase total pollutants by, for example, resuspension of solids from a temporary sink, and to reduce the pollution by dilution. Accordingly, this Chapter aims to consider the hydrological and quality aspects of motorway runoff, and to assess the effectiveness of treatment procedures.

3.2 HYDROLOGICAL ASPECTS

The site used for this work is situated about 56 km north west of London on the M1 motorway. The M1 carries one of the heaviest flows of traffic in the United Kingdom, usually between 25 000 and 40 000 vehicles per 24 hours on the south-bound carriageway, with a high proportion of heavy goods vehicles. The motorway runs in a general south-east to north-westerly direction at the study site.

The drainage area under study (Fig. 3.1) consists of a short (54 m) straight length of three-lane southbound carriageway and hard shoulder of total width 14.8 m, with cross and longitudinal carriagewsay slopes of 2.5% and 0.16% respectively. The annual average rainfall for this area is 648 mm with an annual number of 213 dry days.

Table 3.2 presents a summary of the hydrological characteristics of 24 events sampled in terms of volume, peak flow, mean flow, peak : mean ratio, duration and rise time, this last term being defined as the time in hours from the start of runoff to the peak flow. The data presented cover an 11-month period (November 1980 to September 1981) during which the majority of major seasonal variations in site conditions were experienced. The events sampled constitute approximately 10% and 14% total annual rainfall in winter (November to March) and summer (April to September) respectively. Any such classification of seasons must, to some extent, be arbitrary, but these two periods were chosen since they appear to encompass most satisfactorily the major annual environmental changes at the site.

During the summer period, the change in prevailing weather patterns from that in winter produces a greater number of short, relatively intense

Table 3.1
Summary of available information on runoff quality

	Reported concentration range (mg/l)†			
Determinand	Precipitation	Road/street runoff	Urban runoff	Combined sewer overflow
COD	2.5–322	5–1058	5–3100	93–2636
BOD	1.1	25–165	1–700	15–685
Total solids	18–24	145–21 640	400–15 322	150–2300
Volatile total solids	—	37–1500	12–1600	—
Suspended solids	2–13	11–5500	2–11 300	20–1700
Volatile suspended solids	6–16	100–1500	12–1268	113
Settleable solids	—	—	0.5–5400	—
Total dissolved solids	—	66–3050	9–574	—
Conductance (μmho/cm)	8–395	10 000	5.5–20 000	—
Turbidity (JTU)	4–7	—	3–70	—
Colour (Pt–Co units)	5–10	—	5–160	—
Total organic carbon	1–18	5.3–49	14–120	—
Total inorganic carbon	0–2.8	—	1–17	—
Oils/hydrocarbons	—	28–400	0–110	—
Phenols	—	—	0–10	—
Total nitrogen N	0.5–9.9	0.18–14	1.1–6.2	4.0–63.3
Organic N	0.1–0.32	0.18–3.23	0.1–16	1.5–33.1
Inorganic N	0.69	—	1.0–7.60	—
Ammonia N	0.01–0.4	1–2	0.1–14.0	0.1–12.5
Nitrate N	0.02–5.0	0.31–2.62	0.1–6.9	—
Nitrite N	0.01	—	0–4	—
Total phosphorus	0.001–0.35	0.3–3.5	0.09–4.4	1.0–26.5
Hydrolysable phosphorus	0.8–0.24	—	0.1–10	—
Aldrin	—	—	'trace'	—
Dieldrin	0.003¶	6.8×10^{-6}‡	'trace'	—
p,p′-DDD	—	18.9×10^{-6}‡	—	—
p,p′-DDT	—	17.2×10^{-6}‡	—	—
Heptachlor	0.04¶	—	'trace'	—
Lindane	—	—	'trace'	—
PCB	—	311×10^{-6}	—	—
Bromide	—	0.02–6	5	—
Chloride	0.1–1.1	4–700 000	2–25 000	—
Cadmium	0.013–0.056	0.002–0.4	0.006–0.045	—
Chromium	0.023–0.08	0.018–1.0	0.01–27.0	—
Copper	0.06–0.48	0.007–2.5	0.041–0.45	—
Iron	0–3.05	5–440	0–5.3	—
Lead	0.024–10.4	0.01–113	0.01–14.5	—
Mercury	—	0.001–0.07	—	—
Nickel	—	0.02–1.5	—	—
Zinc	0.02–4.9	1–15	0.01–5.23	—
Total coliform (per ml)	—	<1–79 000	240–991 000	—
Total coliform (organisms per km)	—	15.9×10^{10}	—	—
Faecal coliform (per ml)	—	<1–3000	5500–11 200	—
Faecal coliform (organisms per km)	—	0.9×10^{10}	—	—
Faecal streptococcus (per ml)	—	<1–35 000	120–20 000	—

† Unless otherwise indicated units are mg/l.
‡ kg/km kerb
¶ mg/kg

storms and fewer long-lasting events with steady, low-intensity rainfall. For the data presented, however, only the differences between winter and summer mean peak flows, average mean flows and average peak : mean

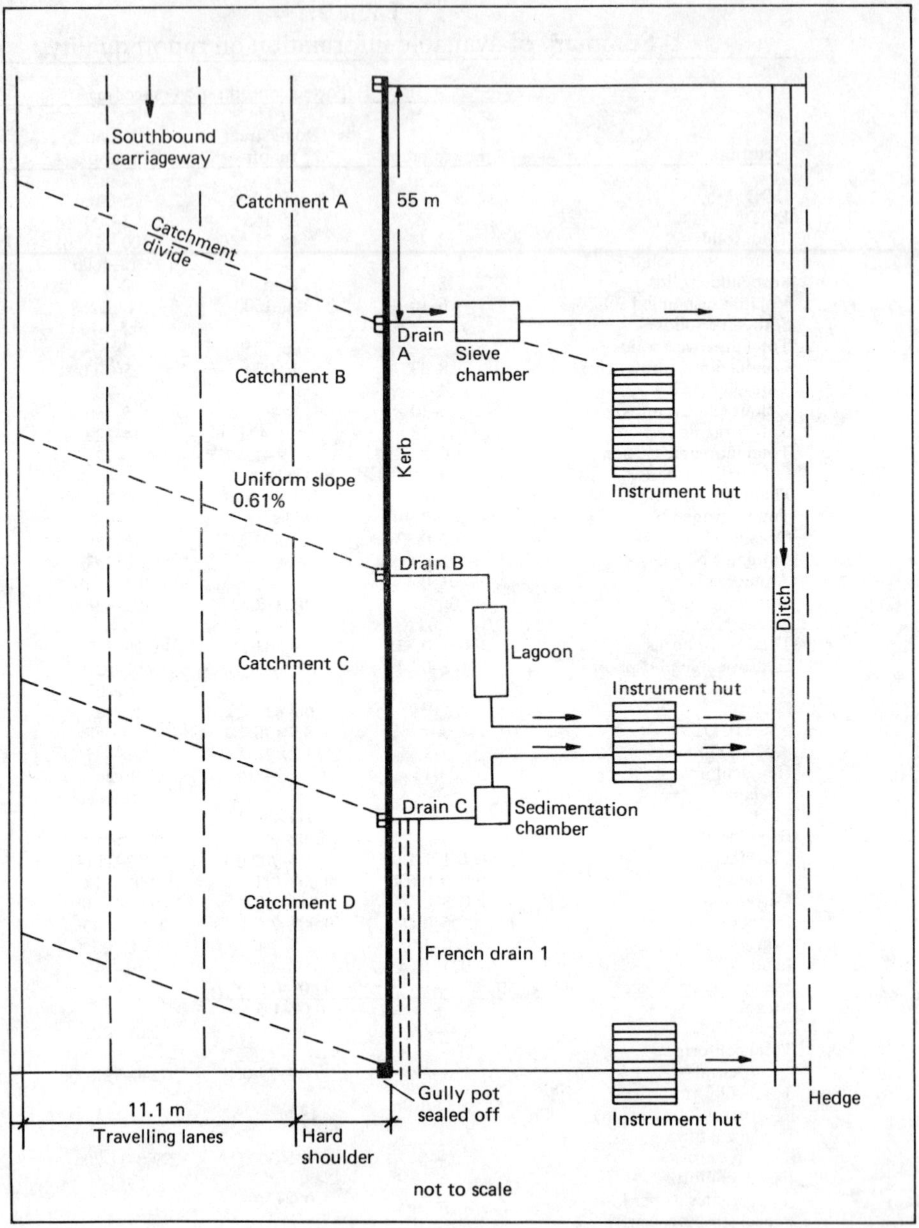

Fig. 3.1 — The layout of the motorway sampling site

ratios are statistically significant at a confidence level of greater than 90%, reflecting the wide spread of data for many of the parameters. Not unexpectedly, the range in values shows that the highest peak flows, mean flows and

Table 3.2
Summary of hydrological data for events sampled

	Volume (litres)‡	Peak flow (l/s)	Mean flow (l/s)	Peak:Mean	Duration (hr)	Rise time† (hr)
Winter						
Mean	20 003	0.55	0.18	2.80	3.29	1.06
RSD¶ (%)	84	93	56	39	71	125
Range	221–6358	0.10–1.87	0.07–0.41	1.30–5.20	0.85–10.20	0.09–4.33
Summer						
Mean	2275	1.22	0.28	3.62	2.37	0.74
RSD (%)	84	125	61	48	67	101
Range	220–6756	0.09–6.43	0.07–0.73	1.29–8.81	0.72–6.86	0.09–2.36
Significance levels of differences between means	<90	>90<95	>95	>90<95	<90	<90

†Time from start of runoff to peak flow
‡To obtain mm, multiply litres by 1.25×10^{-3}
¶RSD = Relative Standard Deviation (%) $(100\sigma_n/\bar{x})$

peak:mean ratios all occurred in summer, while longest durations and rise times occurred in winter.

3.3 RUNOFF WATER QUALITY

Runoff water quality measured over the period November 1980 to October 1981 is summarized in Table 3.3. All the determinands display a wide variation in concentration, and this range is further highlighted by a breakdown of the data into summer and winter months, Table 3.4.

Statistically-significant differences between the mean concentrations for summer and winter periods were observed for total solids (TS), total volatile solids (TVS), chloride (Cl) and bromide (Br). The latter two materials originate from de-icing salt, which is applied in considerable quantities during winter months. The higher concentrations of TS in winter are also largely explained by the contribution from de-icing salt. The reason for increases in TVS during the winter months is not clear, although increased erosion and degradation of road-surface components during salting episodes may be a contributing factor.

There was no statistically-significant difference in total suspended solids (TSS), lead (Pb), zinc (Zn), and cadmium (Cd) levels between the means of winter and summer events. In the case of total suspended solids this lack of significance may be partly accounted for by the compensating effects of greater dilution and increased mobilisation at higher flow rates during summer events. Also important is the contribution to winter total suspended solids by de-icing salt, which has about 6% by weight in the form of insoluble particulate material. It is estimated that salting contributes up to 25% to the total suspended solids recorded in winter.

Table 3.3
Summary of runoff water-quality determinands

Determinand	Winter max	Winter min	Winter n†	Summer max	Summer min	Summer n
pH	7.60	6.10	(13)	7.00	5.70	(11)
Electrical conductivity (μS)	22 000	80	(13)	552	25	(11)
Total solids (mg/l)	12 860	167	(13)	3700	<15	(10)
Total dissolved solids (mg/l)	12 560	1441	(6)	455	12	(6)
Total volatile solids (mg/l)	940	56	(10)	229	<20	(10)
Total suspended solids (mg/l)	717	52	(13)	3430	18	(11)
Oil (mg/l)	40	8	(9)	23	6	(3)
Chemical oxygen demand (mg/l)	362	44	(5)	575	36	(9)
Chloride (mg/l)	6714	12	(13)	40	1.00	(11)
Bromide (μg/l)	6.00	0.2	(13)	2.30	0.05	(11)
Total lead (μg/l)	2874	316	(13)	8015	100	(9)
Total zinc (μg/l)	4047	275	(13)	3425	116	(9)
Total cadmium (μg/l)	102	3	(13)	39	<3	(5)
			max	min		
Total copper (μg/l)	} for winter		30	7		
Total chromium (μg/l)	} and summer		85	—		

†Number of events sampled

Table 3.4
Seasonal water quality parameter means for winter and summer periods

	TS (mg/l)	TVS (mg/l)	TSS (mg/l)	Cl (mg/l)	Br (mg/l)	Pb (μg/l)	Zn (μg/l)	Cd (μg/l)
Winter								
$\bar{x}$†	3003	242	169	1551	2.1	630	1130	15.2
RSD‡ (%)	105	71	40	113	60	47	81	67
Summer								
$\bar{x}$	325	98	230	17	0.58	550	1176	11.4
RSD (%)	24	39	73	56	45	49	25	36
Significance levels of differences between means	>95	>95	<90	>95	>95	<90	<90	<90

†$\bar{x}$ = Seasonal mean
‡RSD = Relative standard deviation

Significant quantities of oil were detected in runoff water (Table 3.3) and polynuclear aromatic hydrocarbons (PAH) tend to be associated pollutants, their concentration in used engine oil being of the order 0.11 mg/g.

The oil fraction found at low initial flows is dependent on the traffic flow during the antecedent dry period. A high soluble proportion is found after shorter dry periods, presumably because association with solid material and oxidation have not taken place. Thus, on average, 70% of the oil is associated with solids, higher proportions being found after long dry periods. Total oil levels generally range between 5 and 30 mg/l, though concentrations in excess of 100 mg/l have been recorded specifically during high intensity events when suspended-solids levels are high. Less than 15% of total PAH is associated with the soluble oil fraction, and the relative proportions of individual compounds associated with road dust and runoff sediments are included in Table 3.5.

Table 3.5
PAH concentrations in road dust and runoff sediment

PAH	Road dust μg/g	%	Runoff sediment μg/g	%
F	2.5	29	9.3	32
B(b)F	2.2	26	6.2	22
B(a)P	1.7	20	5.4	19
B(ghi)P	1.2	14	4.2	14
B(k)F	0.5	6	2.1	7
IP	0.4	5	1.7	6
Total PAH	8.5		28.9	

F — Fluoranthene; B(b)F — Benzo(b)fluoranthene; B(a)P — Benzo(a)-pyrene; B(ghi)P — Benzo(ghi)perylene; B(k)F — Benzo(k)fluoranthene; IP — Indeno(123cd)pyrene

From the overall results of the monitoring and analytical programmes, annual quantities of pollutants contained in runoff water from the motorway have been calculated and are presented in Table 3.6. These are in general agreement with those derived from previous studies (Table 3.7).

3.4 TREATMENT OF MOTORWAY RUNOFF

There is a general paucity of data on the treatment of runoff, particularly with regard to the effectiveness and efficiency of certain treatment processes for the removal of individual polluting substances.

The concept of treating storm water is relatively new and only limited information is available from the literature. Laboratory-based settleability

Table 3.6
Calculated annual loadings of selected pollutants†

		(kg/km/yr)
Total solids	(no salting)	2400
Total suspended solids		1500
Total volatile solids		740
Lead		4
Zinc		8
Oil		126
Polynuclear aromatic hydrocarbons		18×10^{-3}

†Based upon 500 mm runoff per year

Table 3.7
Comparison of mass loadings for motorway runoff (kg/ha per mm of runoff)

Determinand	United Kingdom (This study) M1	Birmingham M6 (Hedley & Lockley, 1975)	Switzerland Zurich N1 (Dauber *et al.*, 1978)	United States† (Federal Highways Administration, 1981)	France‡ A1/A2 (Cathelain *et al.*, 1981)
TSS	2.0	11.8	1.1	2.6	1.3
TVS	1.0	—	—	2.4	—
Pb ($\times10^{-3}$)	5.9	24	3.5	9.7	2.4
Zn ($\times10^{-3}$)	11.5	36	2.7	4	3.8
Oil	0.17	—	—	—	—
PAH ($\times10^{-3}$)	0.025	—	—	—	—

†Mean value from six monitoring sites
‡Estimated upper limits

and filterability tests performed on urban runoff (Bennett *et al.*, 1981; Whipple & Hunter, 1981) have indicated the potential of sedimentation and filtration as means of treatment. The possibility of utilising retention basins is being vigorously pursued, which are a traditional feature in sewerage design for flood mitigation to remove particulate pollutants by settlement (Whipple, 1979; Characklis *et al.*, 1978). The application of more sophisticated treatment processes, such as dissolved-air flotation and microstraining, to treat urban runoff is also being examined (Bratby, 1982; Field & Knowles, 1975), whilst for the treatment of motorway runoff in particular, the use of a lagoon partially overgrown with weeds and a percolating filter consisting of small stones, gravel, and sand have been evaluated (Cathelain *et al.*, 1981; Broch-Jensen, 1978).

Three treatment systems were evaluated for their ability to treat motorway runoff water, a sedimentation tank, a lagoon, and a French drain. These are depicted diagrammatically in Figs 3.2, 3.3 and 3.4. An overall plan of the site is included in Fig. 3.1. Full specifications of these treatment facilities have been given previously (Colwill *et al.*, 1984).

The quality of untreated runoff and the effluents from the three treatment systems, over the period late 1980 to early 1983, is presented in Table 3.8. Significant differences in quality were again evident between summer and winter periods, as has been noted above. It was apparent that all three drainage systems were generally ineffective in removing soluble materials from runoff water.

The drainage systems were found to produce effluents with consistently lower SS levels than untreated runoff, together with lower concentrations of solid-associated pollutants. For the sedimentation tank and French drain, statistically-significant SS removals were evident during both summer and winter periods, with a tendency towards higher removal efficiencies in summer. Removal of SS in the lagoon was consistently greater than 90% with no apparent seasonality.

An accurate assessment of the removal of oil and PAH by the drainage systems is not feasible owing to the lack of data, and it is perhaps best to consider these from the point of view of samples taken with a single storm event. Results of this exercise, for removal of PAH, TSS, and oil in the sedimentation tank and French drain, are contained in Table 3.9. The latter system displayed more efficient removal of both oil and PAH (88% and 84% respectively) (which is perhaps to be expected since the French drain acts as a filter) whilst removal efficiencies for oil and PAH in the sedimentation tank were insignificant.

For heavy-metal pollutants, their association to varying degrees with solid matter in suspension will largely determine removal in the drainage systems. Cadmium was expected to remain relatively unchanged during treatment as it was principally in the soluble form in untreated runoff. However, a considerable reduction in concentrations of Cd was observed, possibly explained by adsorption or transformation in the drainage system. Levels of total lead were found to be strongly associated with suspended solids, and significant removals were evident in all three treatment processes. The high and variable levels of the dissolved-zinc component in untreated runoff produced variable removals during treatment, and these were lower on average than those recorded for lead.

The annual average efficiencies of removal for selected components in runoff by the three treatment processes are included in Table 3.10, where the general efficiency trend of: lagoon$\geqslant$French drain$>$sedimentation tank, is evident. The extent of removal of particular materials will vary considerably between storms, depending upon such characteristics as peak-flow rate and total-volume discharge. A considerable variation in treatment efficiency with season can also be expected, which is particularly true of sedimentation-tank effluents, where, for example, the reduction in lead varied from near zero to 75% from winter to summer. These data contained in Table 3.10 should therefore be regarded as general guides to removal efficiency and not as accurate figures.

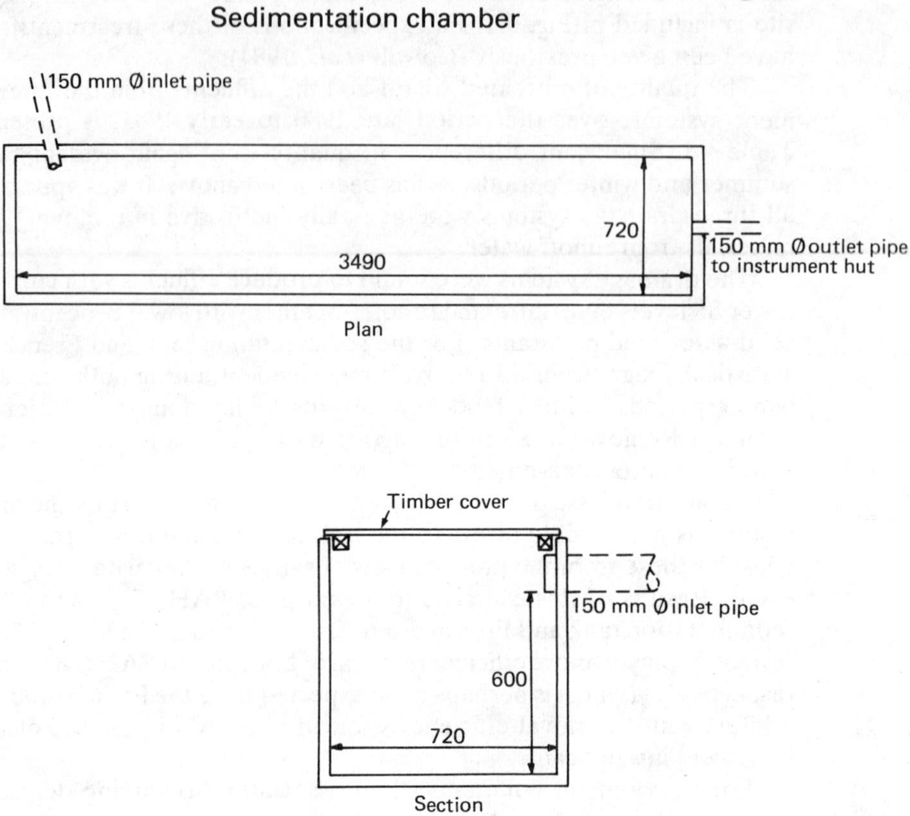

Fig. 3.2 — Schematic diagram of the sedimentation tank (dimensions in mm)

3.5 CONCLUSIONS

Runoff from impervious surfaces, such as heavily-trafficked motorways, has implications for the water cycle, both in hydrological and quality terms. A wide range of organic, inorganic, and metallic pollutants may be removed from the road surface and occur at high concentrations in the runoff water. Discharge of these runoff waters directly to watercourses may adversely affect the quality of the receiving water body, whilst in urban areas, collection in combined sewerage systems may impose increased hydraulic and pollutant loads on conventional wastewater-treatment facilities. In both these cases, where a deterioration in the quality of the ultimate receiving water occurs, the potential reuse for potable-water supply may be limited. Particular seasonal activities associated with the motorway, for example the application of de-icing salts, can also have far-removed effects on water quality. Since de-icing salt contains approximately 220 mg/l by weight of

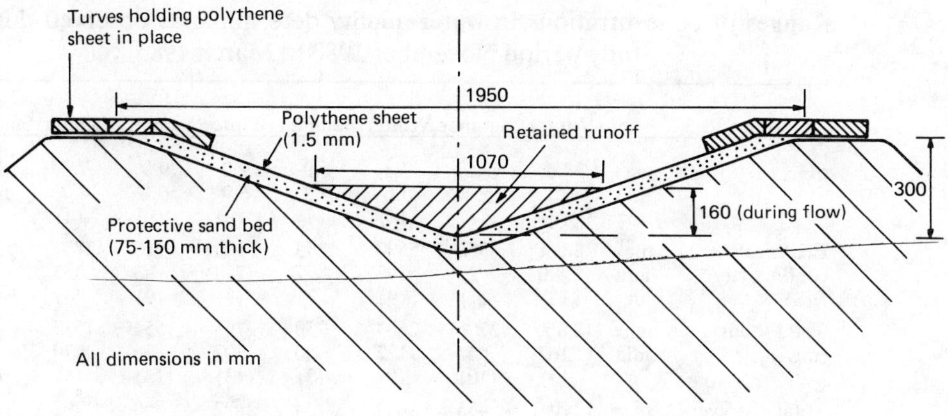

Fig. 3.3 — Schematic diagram of the lagoon

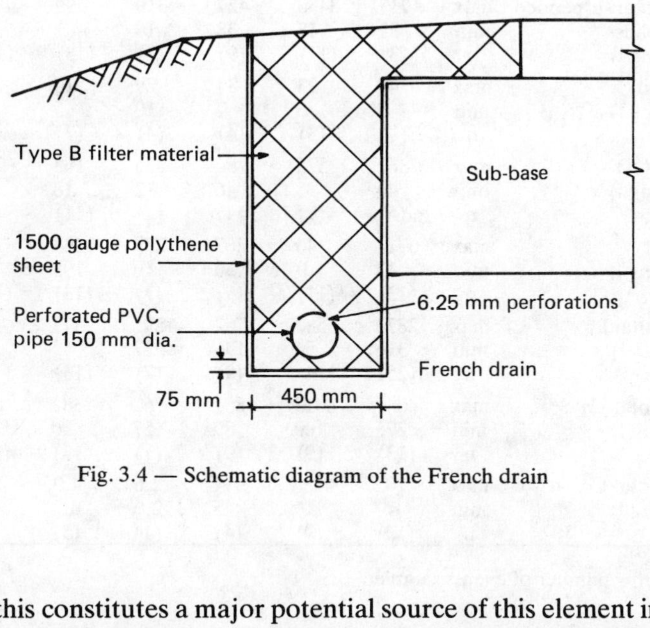

Fig. 3.4 — Schematic diagram of the French drain

bromide, this constitutes a major potential source of this element in surface waters. It has been hypothesised that the occurrence of brominated trihalomethanes in treated lowland river waters arises from the contribution of deicing salts to bromide concentrations in rivers (Luong *et al.*, 1980). However, an additional source of bromide may also arise from car-exhaust emissions, where dibromoethane is used as a lead scavenger in leaded petrol.

An assessment of treatment procedures has indicated that significant removal of solid-associated pollutants from runoff may be achieved, although little effect was observed on the concentrations of soluble compo-

Table 3.8

Ranges of concentrations in water-quality determinands observed during study period November 1980 to March 1983

Determinand		Winter	Summer	Winter	Summer	Winter	Summer	Winter	Summer
pH	max	7.6	7.0	7.3	6.9	9.2	9.9	7.4	8.0
	min	6.1	5.7	6.6	6.2	6.9	6.1	6.7	6.2
	n†	(13)	(11)	(14)	(4)	(8)	(9)	(4)	(6)
Electrical conductivity (μS)	max	22 000	552	5500	1255	5100	910	2750	363
	min	80	25	197	77	778	93	837	56
	n	(13)	(11)	(9)	(7)	(7)	(9)	(4)	(6)
Total solids (mg/l)	max	12 860	3700	2664	750	2796	554	1537	438
	min	167	<15	122	60	307	36	570	25
	n	(13)	(10)	(9)	(7)	(13)	(13)	(4)	(6)
Total dissolved solids (mg/l)	max	12 560	455	2442	734	2692	560	1702	203
	min	1441	12	140	11	254	22	545	68
	n	(6)	(6)	(9)	(6)	(13)	(11)	(4)	(4)
Total volatile solids (mg/l)	max	940	229	216	264	242	208	—	222
	min	56	<20	60	62	30	16	—	3
	n	(10)	(10)	(4)	(4)	(8)	(9)	—	(6)
Total suspended solids (mg/l)	max	717	3430	427	310	99	62	57	59
	min	52	18	38	14	7	8	2	5
	n	(13)	(11)	(9)	(7)	(13)	(13)	(4)	(8)
Oil (mg/l)	max	40	23	34	29	10	20	—	5
	min	8	6	7	10	5	2	—	1
	n	(9)	(3)	(4)	(4)	(7)	(8)	—	(2)
COD (mg/l)	max	362	575	282	236	168	220	128	198
	min	44	36	46	12	18	10	22	18
	n	(5)	(9)	(4)	(4)	(11)	(4)	(7)	
Cl (mg/l)	max	6714	40	1267	273	1444	173	767	26
	min	12	1.0	30	16	198	5	201	14
	n	(13)	(11)	(9)	(7)	(13)	(13)	(4)	(6)
Total Pb (μg/l)	max	2874	8587	1752	1295	373	115	206	141
	min	316	54	136	37	32	7	31	17
	n	(25)	(14)	(9)	(7)	(13)	(12)	(4)	(4)
Total Zn (μg/l)	max	4047	8015	881	762	382	874	289	240
	min	275	100	89	157	36	51	64	43
	n	(13)	(9)	(9)	(7)	(13)	(13)	(4)	(4)
Total Cd (μg/l)	max	102	3542	14	2.6	4.9	2.3	—	—
	min	<3	116	5	2.0	0.7	0.5	—	—
	n	(13)	(9)	(4)	(1)	(5)	(7)	—	—

†n = number of events sampled

nents. Although the average annual removal efficiencies derived are only general guidelines, their application to the figures in Table 3.6 would produce a modified annual discharge loading for a motorway employing treatment of runoff water. These calculated figures are included in Table 3.11, representing the potential discharge contributions to surface waters from runoff-treatment systems.

The removal of suspended solids is obviously a prime factor in the design of a treatment process, owing to the association of pollutants with solid

Table 3.9
Mean concentration of TSS, oil, and total (four†) PAH from samples collected during a single event

Drain	TSS (mg/l)	Oil (mg/l)	PAH (μg/l)	B(a)P (μg/l)
Untreated	92	30	3.70	0.60
Sedimentation tank	90	22	3.30	1.15
French drain	28	3.70	0.59	0.12

†B(a)P : Benzo(a)pyrene
B(ghi)P : Benzo(ghi)perylene
F : Fluoranthene
I : Indeno(123 CD)Pyrene

Table 3.10
Mean annual removal efficiencies (%) of materials for different drainage systems

Determinand	Sedimentation tank	French drain	Lagoon
TSS	52	85	92
TPb	40	83	90
TZn	47	81	76
ZnSA	57	84	84
ZnD	15	56	62
COD	35	59	54
Oil	30†	70†	>70†
PAH	30†	70†	>70†

†Estimated value — not verified statistically

Table 3.11
Calculated discharge of pollutants from a motorway employing runoff treatment

Determinand	Material loads (kg/km/year)†		
	Sedimentation tank	Lagoon	French drain
Total suspended solids (1500)‡	720	120	225
Lead (4)	2.40	0.40	0.68
Zinc (8)	4.24	1.92	1.52
Oil (126)	<88.20	<37.80	37.80
PAH (18×10^{-3})	$<12.6\times10^{-3}$	$<5.4\times10^{-3}$	5.4×10^{-3}

†Based on 500 mm runoff per year
‡Original figures, no treatment

material in runoff. In this respect, the French drain may possess advantages in that its potential lifetime before requiring major renovation may be as long as 20 years. Calculations have indicated that the sedimentatin tank would require de-sludging as frequently as 5-yearly intervals. In terms of requirements for land area, a French-drain system would also be more economic than either a lagoon or a sedimentation tank.

In the foreseeable future, it is unlikely that the deposition of crankcase oil and tyre-related hydrocarbons on the road surface will change radically as a result of developments and improvements in automotive design. Changes in motor fuel (both gasoline and diesel) composition, however, may bring about a change in the quantities and nature of PAH compounds emitted in vehicle exhaust gases, prior to deposition on or in the vicinity of the road surface. In addition, considerable concern over the presence of lead anti-knock additives in gasoline in both the United States and Western Europe has resulted in the promulgation of legislation for either the removal or planned reduction of the lead in motor fuel.

However, in deciding whether to initiate treatment of motorway and urban runoff, it is clearly important to consider not only the possible impact simply upon water quality, but also the likely effects of urbanisation on the hydrological cycle as a whole.

REFERENCES

Bennett, E. R., Kinstedt, K. D., Nilsgard, V., Battaglia, G. M. & Pontius, E. W. (1981) Urban snowmelt: characteristic and treatment. *J. Wat. Pollut. Control Fed.*, **53,** 119–125.

Bratby, J. R. (1982) Treatment of raw wastewater overflows by dissolved air flotation. *J. Wat. Pollut. Control Fed.*, **54,** 1558–1565.

Broch-Jensen, J. E. (1978) Runoff from motorways passing open country or rural areas *Proceedings of the OCED symposium on Road Drainage.* Berne, Switzerland.

Cathelain, H., Friant, G. & Olie, J. L. (1981) Les eaux de Reuissellement de Chaussées autoroutieres: evaluation des charges de pollution. *Bulletin de Liason des Laboratoires des Ponts et Chaussées,* **116,** 9–24.

Characklis, W. G., Bedient, P. D., Gaudet, F. G. & Roe, F. L. (1978) A study of stormwater runoff quality. *Prog. Wt. Tech.,* **10,** 673–699.

Colwill, D. M., Peters, C. J. & Perry, R. (1984) Water quality of motorway runoff. TRRL Supplementary Report 823.

Dauber, L., Novak, B., Zobrist, J. & Zurcher, F. (1978) Pollutants in motorway stormwater runoff. *Proceedings of the OECD Symposium on Road Drainage.* Berne, Switzerland, pp. 33–342.

Federal Highway Administration (1981) *Characteristics of highway runoff from operating highways.* Vol. IV, US PB81/241/937.

Field, R. & Knowles, D. (1975) Urban runoff and combined sewer overflow. *J. Wat. Pollut. Control Fed.*, **47,** 1352–1369.

Hedley, G. & Lockley, J. C. (1975) Quality of water discharged from an urban motorway. *Wat. Pollut. Control,* **74,** 659–674.

Luong, T. V., Peters, C. J. & Perry, R. (1980) Bromide and trihalomethanes in water supplies. *Environ. Technol. Lett.*, **1**, 299–310.

Whipple, W. (1979) Dual purpose detention basins. *J. Wat. Resour. Plann. Man. Div., ASCE,* **103**, 392–403.

Whipple, W. & Hunter, J. V. (1981) Settleability of urban runoff pollution. *J. Wat. Pollut. Control Fed.*, **53**, 1726–1731.

4

Underground water pollution in Milan and the Province by industrial chlorinated-organic compounds

A. Cavallaro, C. Corradi, G. De Felice and **P. Grassi**

4.1 WATER SUPPLY IN THE TOWN AND PROVINCE OF MILAN

The lowland of Lombardy, and in particular the Milan area, has always enjoyed an abundant supply of underground water and a perfectly healthy climate, and the population has thus been able to procure this precious facility with ease. The overall development of both the region and the principal town has been considerably influenced by this abundance of water.

Milan's water system (the second largest in Italy, serving 1 635 000 inhabitants and a surface area of 182 km^2) is directly managed by municipal authorities, and distribution of the water is extended to almost all the dwellings and outlying communities of the metropolitan area.

An annual average of more than 300 000 000 m^3 of water has been distributed, with an average consumption per capita of about 450 to 500 l/d. Milan obtains its water via wells tapping intercommunicating aquifers, the latter partially separated by discontinuous clay strata, situated at depths of 40 to 110 m or more.

The water is raised by submerged electric pumps, and carried to the pumping centres which convey it under pressure to the distribution network, the pipes of which exceed 2000 km in total length.

In 1984, there were 36 principal plants (centres) and four secondary plants (sub-centres), with a total of 590 wells, 510 of which were active. The water, bacteriologically pure for drinking, is distributed as it is; only exceptionally, and for very short periods, is it chlorinated. Water supply for the Province of Milan (2.4 million inhabitants, a surface area of 2762 km^2 and comprising 249 separate municipal authorities) is also effected by tubular wells (about 600) by local municipal water systems, delivering about 200 000 000 m^3 of water per year.

4.2 STRUCTURE AND ORGANISATION OF INTERVENTION TO IDENTIFY AND ELIMINATE SOURCES OF CONTAMINATION

The structure of the soils and sub-soils of the provincial territory of Milan makes them particularly vulnerable to contamination by many different kinds of organic substances, from a variety of sources, sumps, drain hollows in the soil, pits, and dumps. Therefore, local contamination from specific pollutants (phenols, aliphatic and aromatic hydrocarbons, alkali, amines) is found in the underground waters in the more industrialised areas. In any case, the more worrying source of contamination for its size and range in all the provincial territory is pollution by chlorinated organics (trichloroethylene, tetrachloroethylene, chloroform, methyl-chloroform). This kind of pollutant holds an important position because it is used to a very great extent in chemical, pharmaceutical, electronic, and mechanical industries. Analytical tests made since 1975 indicated a serious source of pollution initially in the Milan area, and more recently in all the provincial territory. Provincial and regional sanitary authorities for their own part, conscious of the gravity of the situation, instituted in November 1975 a Working Group of experts in various disciplines: chemistry, geology, hygiene and engineering, to estimate the degree of contamination of the drinking water.

The basic plan of operation which the Working Group originally devised has been refined with the considerable experience gained over the years, to an authentic chronological methodology for the identification and elimination of the sources of pollution in underground water. This chronological methodology draws on techniques and expertise from different disciplines and divides in to four fundamental points:

(i) *Hydrochemical survey.* This is based mainly on geological and chemical techniques and is characterised by a rigorous succession of operating phases:

(a) *Inventory*: entailing census and location (on a map of scale 1:25 000) of all existing wells, public and private, in an area hydrologically capable of circumscribing the point or points of identified pollution, that is the wells serving as water-supply points.

(b) *Hydrological interpretation*: with analysis of stratographic data

(c) *In situ survey*: in which a schedule is rapidly programmed for each well in the census, prescribing the measurement of static levels and the requisite withdrawal of water from the aquifer for chemical analysis, as well as soil observation.

(d) *Hydrochemical interpretation*: in which the hydrological and chemical data acquired enable a map of the piezometric surface of the aquifer to be drawn up, with other maps detailing the distribution of the principal elements of pollution in the underground water. Collation and analysis of the hydrological data leads to the preparation of an isopiezometric map, representing the morphology of the water table, which enabled identification of the principal lines of groundwater flow. On the same map the results of chemical analyses are then detailed, each well being coloured according to a category of concentration of the contaminant in question.

Thus a map of the distribution of each contaminant in the water is obtained. By correlating the points or areas of major pollution along the line of the underground water flow (the movement of contaminants in solvent form depending precisely upon this flow) it is possible to identify the source areas of the contamination.

(ii) *Census of productive activity* This census is carried out in the area identified in the hydrogeochemical survey, focusing on potential users of contaminants and on chemical analyses of industrial wastes.

Experience has indicated the need for particular attention concerning condition of all underground storage tanks containing chemical products, these being potential sources of pollution.

(iii) *Survey of the focal point of soil pollution* Use is made of direct assessment and related surface and/or subsurface sampling for analyses.

(iv) *Reclamation and purification of the aquifer* This is achieved by removing the polluted soil when penetration is limited in extent or in depth, or constructing a drain and barrage well in the hydrogeologically appropriate position.

The results of the periodic water sampling and susequent analysis follow.

4.3 ASSESSMENT AND EVOLUTION OF THE PHENOMENON

4.3.1 Situation in the Province of Milan

Having prepared analytical maps covering the whole province (in order to identify the worst-hit areas), the Working Group then undertook, in close collaboration with local sanitary authorities, a series of surveys specifically related to the identification of the sources of contamination, with a view to the purification of natural conditions of the aquifer. The areas covered by the initial survey and subsequent work corresponded to that occupied by more than two-thirds of the population of the province, in which areas, obviously, the highest degree of industrialisation is to be found.

To demonstrate the location of contamination by chlorinated-organic compounds in the province, a series of maps has been drawn up, detailing the nature and distribution of such contaminants. Fig. 4.1 reflects the overall contamination by the chlorinated-organic compounds; 'global' identification of the pollution permits characterisation of the areas which are worst affected, and which require systematic monitoring as well as practical intervention. It should be borne in mind that this and the following maps represent the current situation, as it stands after the intervention of both programmed and improvised reclamation activity by the Working Group, operating according to the criteria mentioned above.

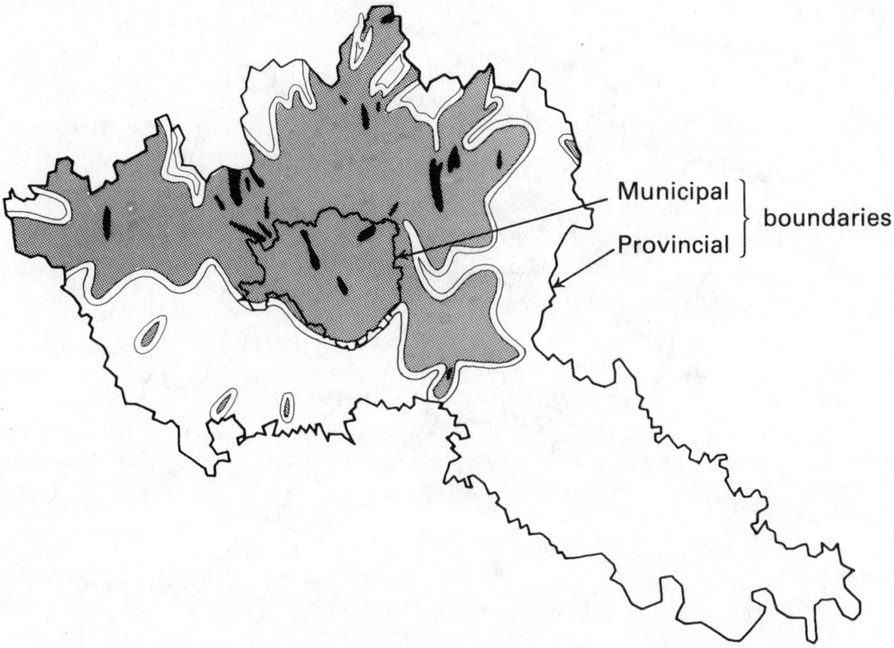

Fig. 4.1 — Distribution† of organic-chlorinated compounds in underground water in the provincial territory

† In all figures the most contaminated areas are heavily shaded.

As may be noted, pollution covers extensive areas, and is particularly acute in the more highly-industrialised zones. Note as well the strips of pollution deriving from outside the privince. There exist, however, large tracts of land to the east, the south and the west of Milan which generally coincide with agriculture or areas of minimal industrialisation, and which may be considered as almost unaffected, where pollution levels are below 10 μg/l, while there are vast areas elsewhere showing levels of between 10 and 50 μg/l.

Appropriate maps, which will be considered later, have been drawn up for the metropolitan area; pollution estimates are related to the average contamination levels to be found in the water of all the wells contributing to the input of the respective centres.

As to the distribution of individual contaminants, it can be seen (Fig. 4.2) that trichloroethylene is the most widespread chlorinated-organic compound, and that it causes the largest rate of acute or sub-acute cases of pollution. Decidedly less diffused, though of similar basic characteristics, is tetrachloroethylene (Fig. 4.3), although chloroform (Fig. 4.4) and methyl-chloroform (Fig. 4.5) are to be found in more localised strips of pollution, which are quite distinct and very difficult, therefore, to correlate.

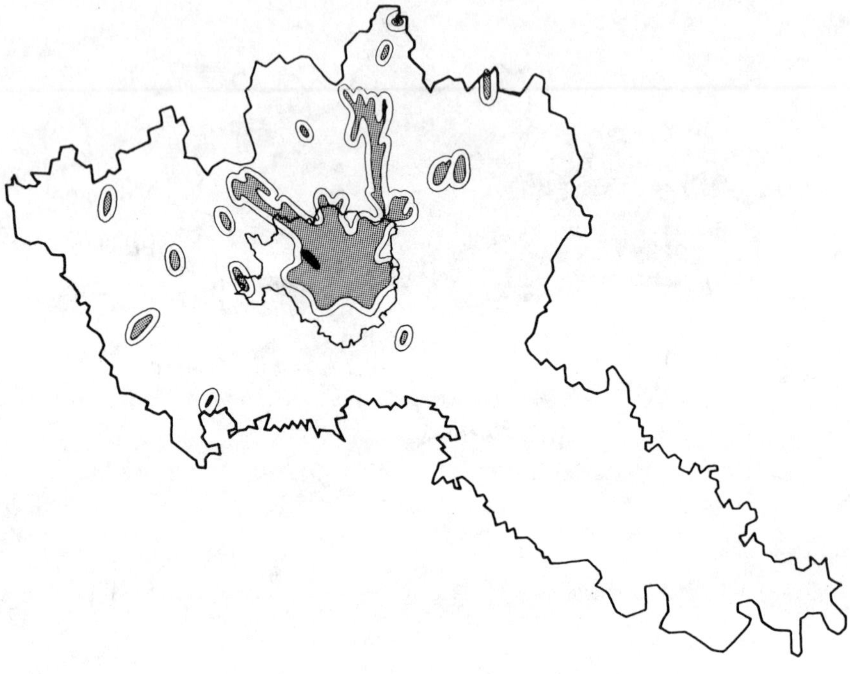

Fig. 4.2 — Distribution of trichloroethylene in underground water in provincial territory

The contaminant which in acute pollution generally demonstrates the greatest intensity is chloroform, while trichloroethylene and tetrachloroethylene often show an intensity which suggests they are the residue of phenomena more active in the past than in the present, although sometimes remaining at a significantly high level.

What the above evidence and the maps show is consistent with the use made in industry of the substances considered. Trichloroethylene and tetrachloroethylene have long been used in industry, the former more than the latter, whilst methyl-chloroform (which is gradually replacing the first two) entered into use relatively recently. Its diffusion in the environment is correspondingly recent, and is therefore notable only in acute cases.

Chloroform presents a different case, in which environmental diffusion is linked to particular production cycles (those of certain pharmaceutical and chemical industries) and is therefore only to be found in those well-defined areas in which such activity exists, or where the soil is subject to the

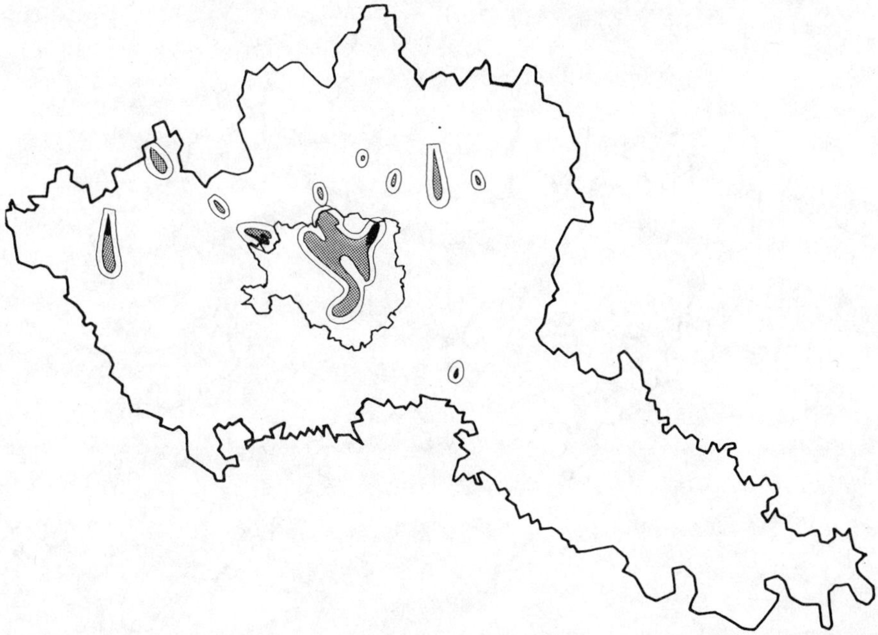

Fig. 4.3 — Distribution of tetrachloroethylene in underground water in the provincial territory

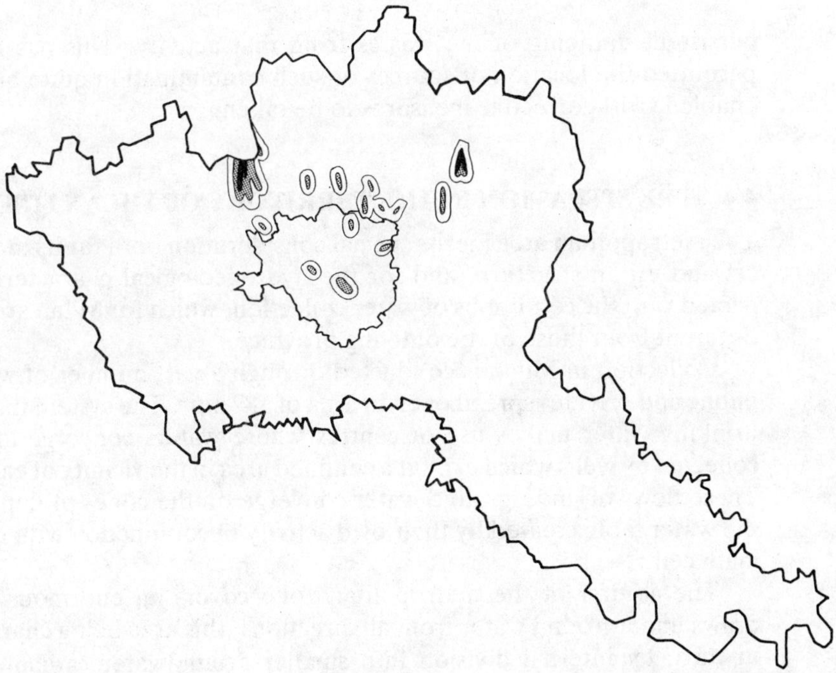

Fig. 4.4 — Distribution of chloroform in underground water in the provincial territory

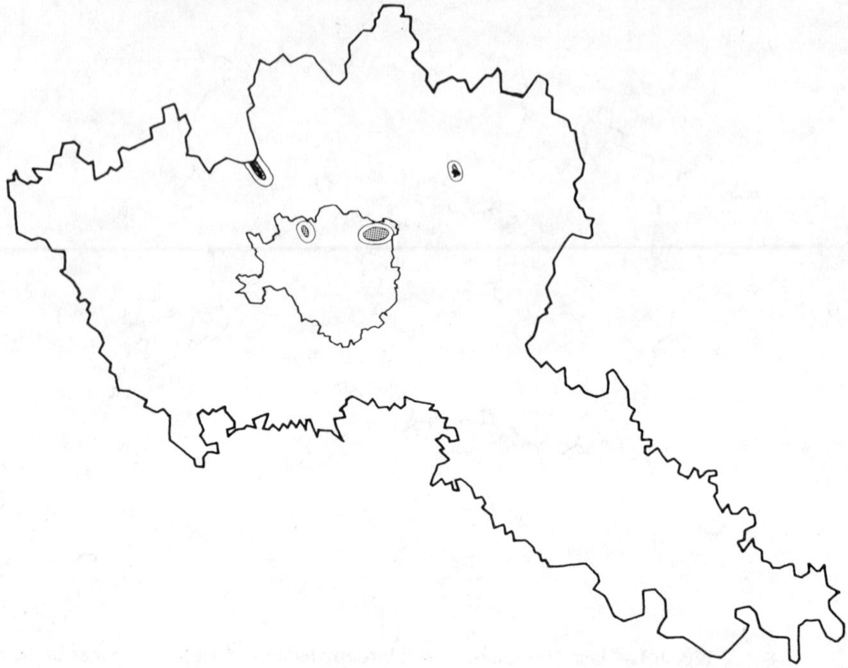

Fig. 4.5 — Distribution of methyl-chloroform in underground water in the provincial territory

persistent dumping of the wastes from that activity. This has frequently permitted the location of sources of such contamination quite quickly and enabled swift corrective measures to be taken.

4.4 THE SITUATION IN THE TERRITORY OF MILAN ITSELF

The metropolitan area merits special consideration both for its morphological and urban structure, and for the hydrogeological characteristics connected with the conditions of water collection, which for Milan are certainly different from those of the outlying province.

Collection in Milan is conducted through a vast number of wells (1200 public and private) spread over an area of 182 km^2. The system that delivers drinking water makes use of centres where waters converge after being collected by wells which exploit a confined area in the vicinity of each centre. These flows of underground water converge on the cones of depression of the water table created by their own activity in conjunction with that of the main centre.

The aquifer of the metropolitan zone covers an enormous area that draws underground water from all directions, this area being characterised, in turn, by internal division into smaller groundwater catchments. This situation, connected with a marked and constant absence of separation between upper and lower aquifers in the area in question, invites 'imported' pollution from external areas; furthermore, it facilitates the infiltration of the deepest levels of the aquifer by the contaminating substances which affect surface strata.

As in the case of the provincial area, maps plotting the presence of solvents have been drawn up, to record both the 'global' conditions of pollution by chlorinated-organic compounds, and the diffusion paths of individual contaminants, with the fundamental aim of tracing the sources of pollution (Fig. 4.6). The map dealing with Milan itself together with the previously-discussed map of the provincial territory, reveals a consistent pattern of pollution in which the northern arc (bordering the more industrial areas of the province) shows greater and wider-spread pollution (with many acute cases locally). In contrast, the southern arc, bordering areas which tend to be residential and agricultural, shows diffused but markedly lower levels of pollution. Indeed, in these areas, which have in any case been urbanised or industrialised only recently, the general level of pollution is below 50 μg/l.

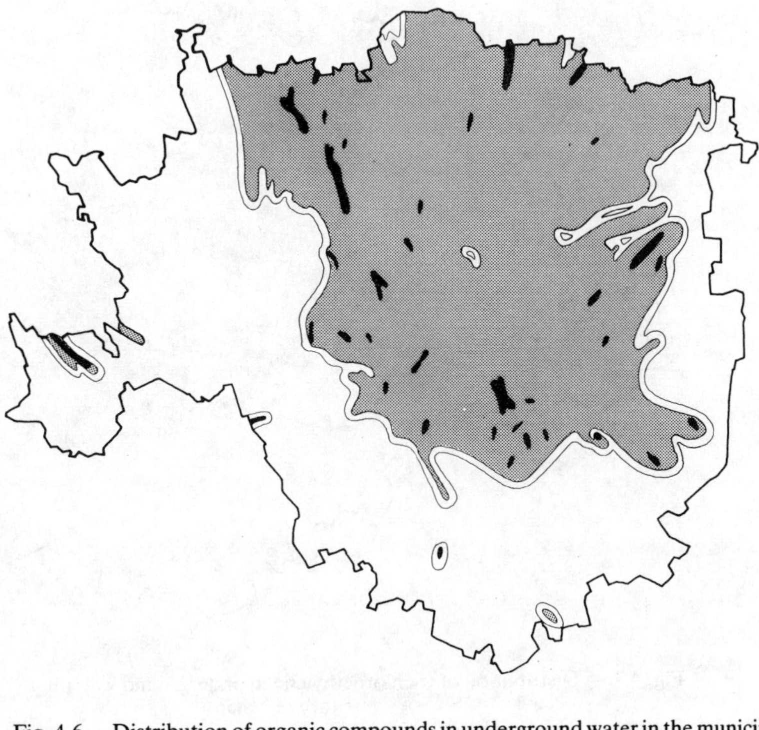

Fig. 4.6 — Distribution of organic compounds in underground water in the municipal territory of Milan

The first observation to make on the different distribution of global pollution is that (except in acute, local cases) levels generally correspond with the historic engagement of the given area in industrial activity; also to be noted are the possible effects of widespread and numerous minor industrial activities, often with instances of use of the contaminants in question. In such zones, disposal systems left much to be desired in the past, widespread use being made of seepage wells deliberately constructed for the dumping of refuse, or of wells originally designed to cater for rain water but adapted, either ignorantly or perniciously, for other purposes. These, then, are the reasons for the profound and diffused state of pollution in the underground water of the urban territory.

Map analysis of specific contaminants reiterates much of what was noted on the provincial maps. Trichloroethylene is again the prevalent and most widespread contaminant (Fig. 4.7), observations on this substance are consistent with those just outlined. Tetrachloroethylene (Fig. 4.8), though considerably less in evidence than trichloroethylene, is nevertheless apparent, whilst methyl-chloroform (Fig. 4.9) is only present at local level, as is the case with chloroform (Fig. 4.10) and 1,2 dichloropropane. The considerations outlined above relating to these substances are equally valid here.

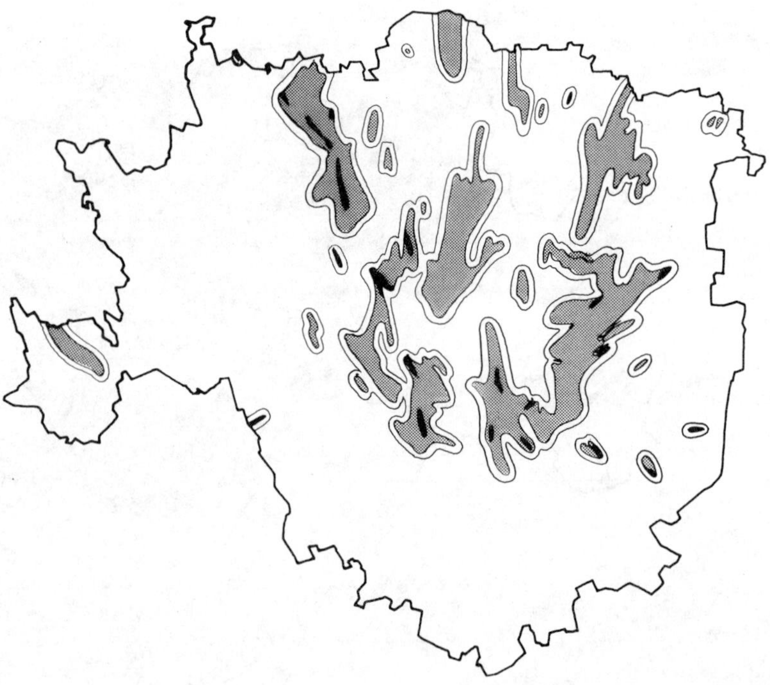

Fig. 4.7 — Distribution of trichloroethylene in underground water in the municipal territory of Milan

4.4.1 Evolution of the phenomenon

From the first assessment, through the gradual progress of survey and intervention, it has been clear that the phenomenon was worrying in terms of both intensity and diffusion. Currently, after the very first phase of the

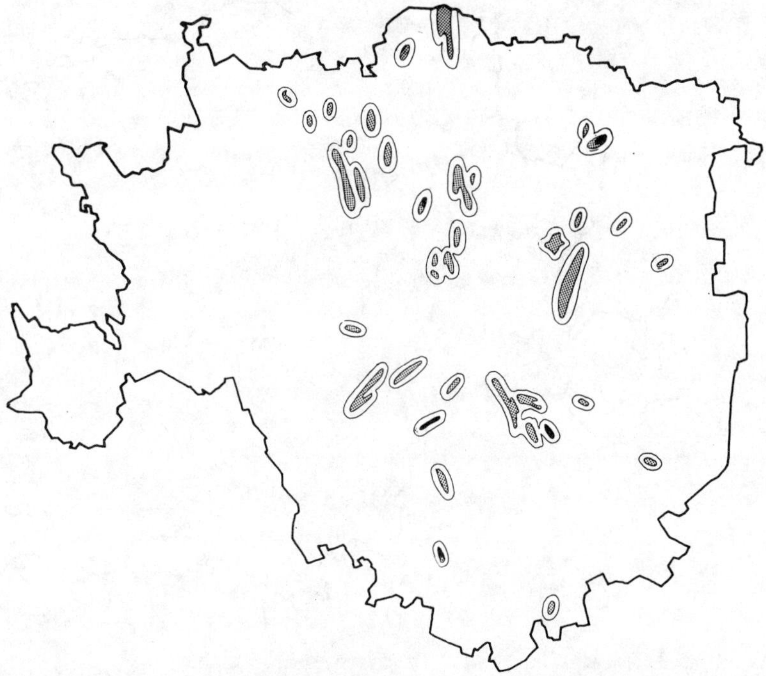

Fig. 4.8 — Distribution of tetrachloroethylene in underground water in the municipal territory of Minan

survey (which occasionally evinced increases in pollution), and as a result of numerous interventions, the situation in the province and in certain zones of Milan itself may be said to be showing a slow but progressive improvement. This is due not only to direct intervention, but also to the energetic activities of local sanitary authorities and the Magistrates, who have implemented deterrents and obliged (*obtortu collo*) industrial operators in particular to assume responsibility. Two reconstructed visual examples are offered of the evolution of acute, localised pollution after identification and elimination of the sources of contamination undertaken by the respective authorities with the technical support of the Working Group. Fig. 4.11 illustrates a localised phenomenon in the north-west zone of Milan, whilst Fig. 4.12 shows a strip of land in the municipality of Sesto San Giovanni. Tables 4.1, 4.2 and 4.3 demonstrate the positive evolution in Milan of chlorinated-organic compound pollution in public wells.

4.5 CONCLUSIONS

From the above and with the benefit of accumulated experiences, pollution deriving from chlorinated-organic compounds, especially in Milan, will certainly not disappear in the near future, even with the present programme of elimination and reclamation, and despite progressive success; pollution is

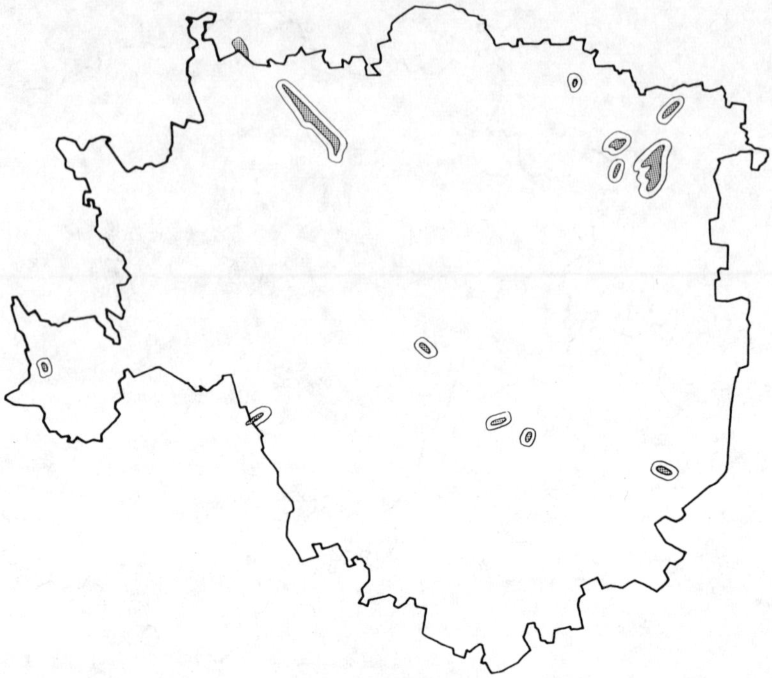

Fig. 4.9 — Distribution of methyl-choroform in underground water in the municipal territory of Milan

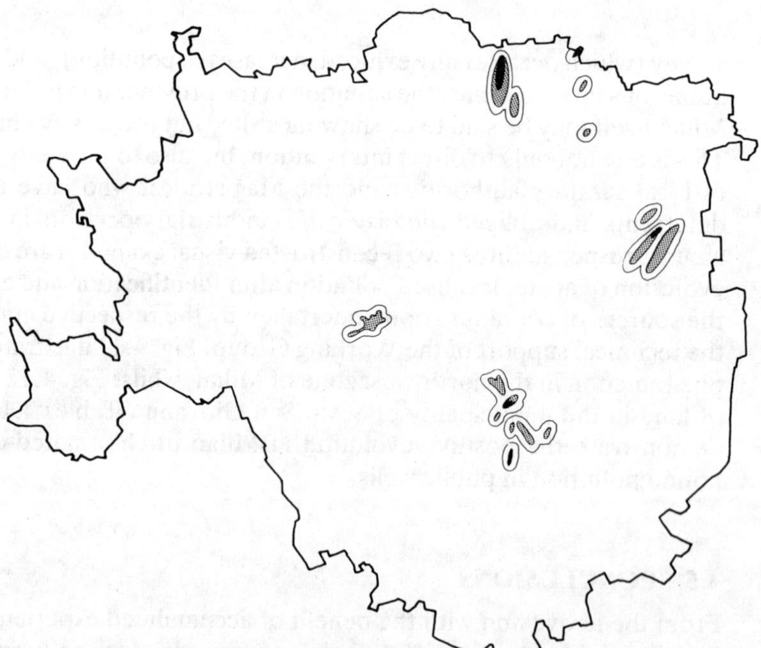

Fig. 4.10 — Distribution of chloroform in underground water in the municipal territory of Milan

likely to remain at significant levels for very many years. It is worth noting, furthermore, that although pollution from chlorinated-organic compounds is today the most widespread phenomenon of its type, it is not unique in its effect on underground drinking water in Lombardy. Hexavalent chromium and other micropollutants should be considered in this respect. Pollution of underground water in Lombardy represents a real, man-made problem which is believed to be unrivalled in other parts of Italy or in Europe for duration or intensity.

The social, hygienic and sanitary conditions of the (approximately) 4 000 000 inhabitants of the area, who depend entirely on underground sources for drinking water, are under tangible threat. Furthermore, pollution represents the major technical problem for the future management of the public water system. Indeed, there are frequent operational breakdowns in collection, feeding pipes, and other highly expensive equipment. In each case, financial losses are aggravated in that significant sums have to be spent on the construction of alternative wells, on the repairing of the network, on survey and on intervention, whether active (reclamation) or passive (treatment of the water in purification plants). But far more important is the very real uncertainty with regard to public health. Pollution in wells can strike anywhere at any time, and recognition of drinking-water contamination always occurs after the event, that is after an indeterminate period during which the populace continues to use the polluted water.

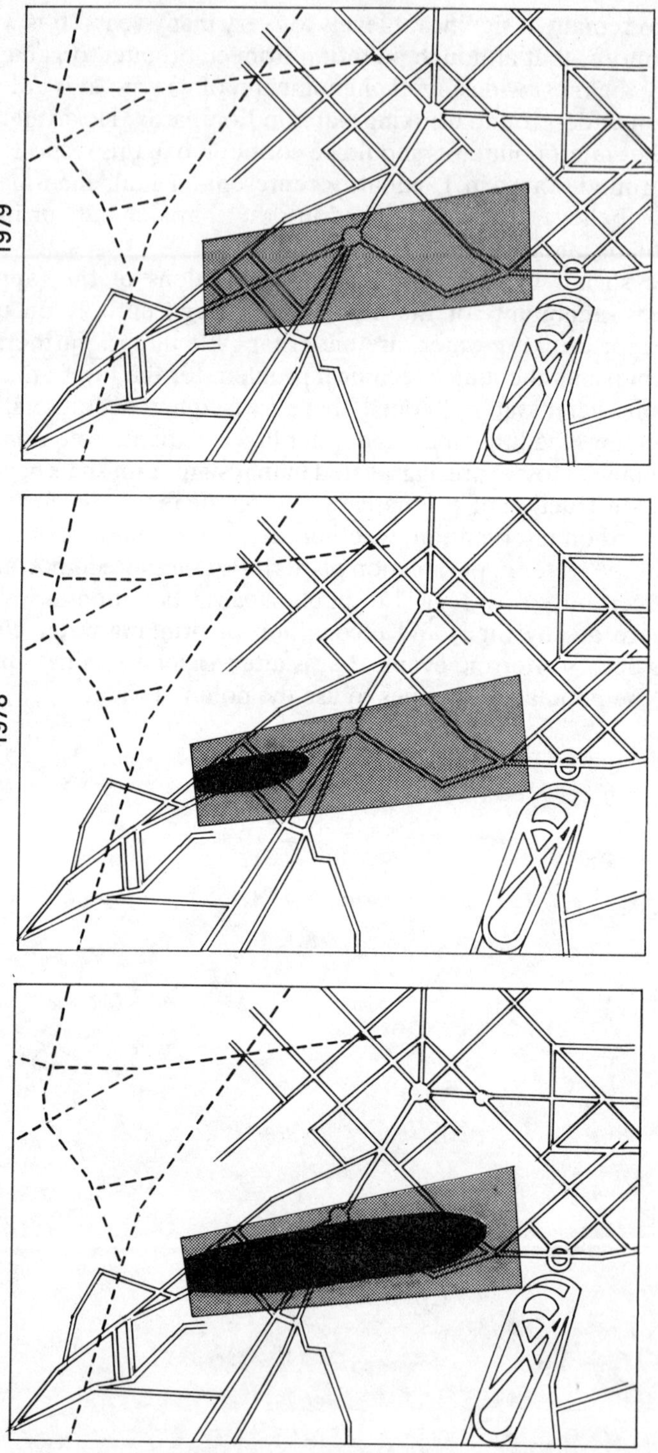

Fig. 4.11 — Evolution of trichloroethylene pollution in the north-west zone of Milan following reclamation intervention

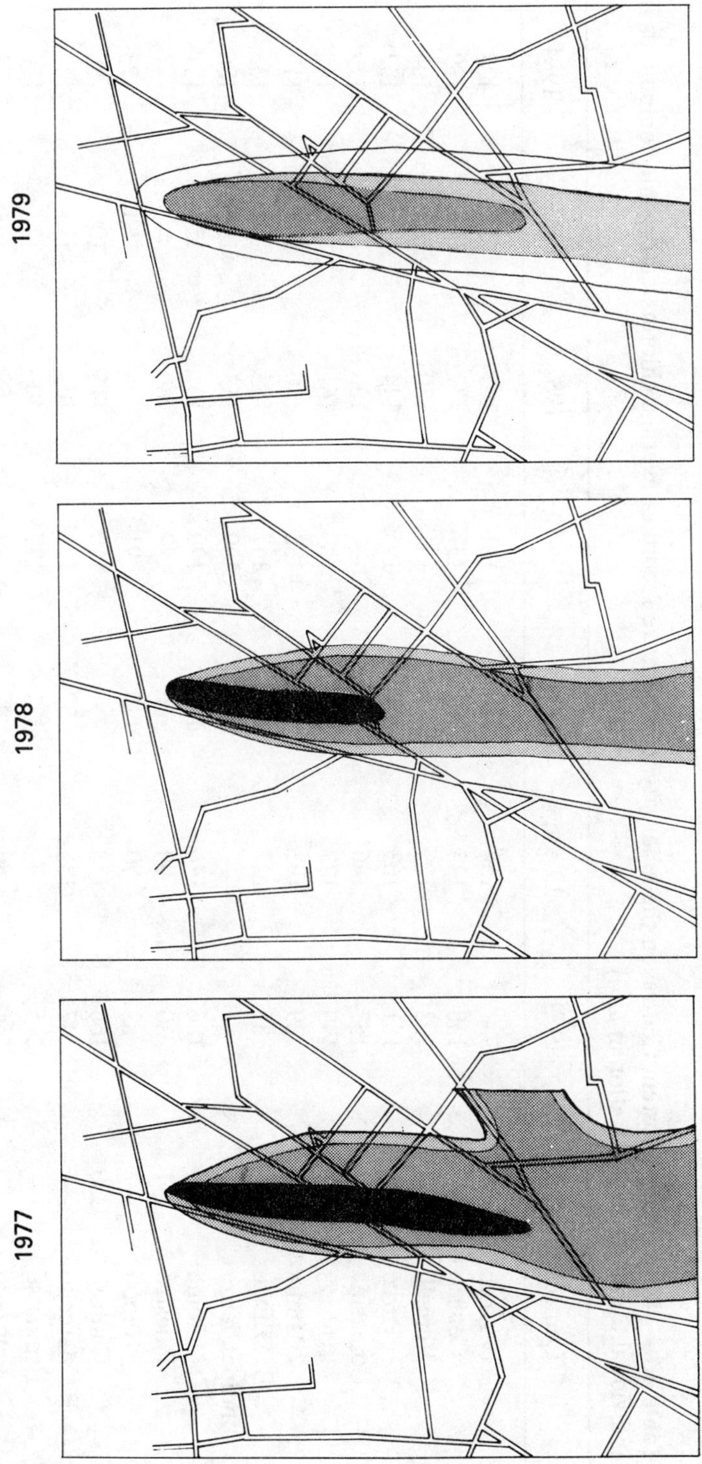

Fig. 4.12—Evolution of trichloroethylene pollution at Sesto San Giovanni following reclamation intervention

Table 4.1 — Milan Water System Centres in the inner city: comparative record of average chlorinated-organic compounds in individual Centres (Concentration in μg/l)

Centres	1977	1978	1979	1980	1981	1982	1983	1984
Armi	—	—	159	158	161	167	155	127
Cenisio	158	158	125	201	137	137	141	139
Vercelli	115	103	123	—	151	130	112	104
Parco	—	181	163	207	195	144	153	132
Comasina	113	157	146	130	—	100	113	119
Anfossi	—	199	177	169	159	161	127	149
Italia	—	229	193	174	183	150	135	115
B. D'Este	—	185	127	147	147	153	139	133
Indipendenza	—	—	—	174	199	153	189	185
Crema	164	186	146	179	179	161	159	151
Palestro	111	80	90	86	87	60	—	—
Napoli	—	82	70	125	91	94	80	73
Cantore	230	166	138	—	110	115	109	88
B. Marcello	—	147	135	137	126	102	115	141
Espinasse	—	245	—	121	127	91	137	139
Crescenzago	30	66	86	92	92	68	63	73
Vialba	143	125	105	112	113	97	105	115

Table 4.2 — Milan Water System Centres in the peripheral zones: comparative record of average chlorinated-organic compounds in individual Centres (Concentration in μg/l)

Centres	1977	1978	1979	1980	1981	1982	1983	1984
Martini	—	—	—	—	—	—	161	93
Tonezza	48	74	74	52	65	48	42	48
Gorla	—	208	172	211	173	191	172	154
Suzzani	105	129	90	116	107	87	85	88
San Siro	8	19	10	13	10	9	11	16
Feltre	—	45	45	69	68	52	48	53
Testi	95	49	65	61	78	61	64	66
Santa Teresa	—	240	363	286	420	223	163	136
Abbiategrasso	37	31	31	43	54	43	50	—
Salemi	13	50	43	41	44	50	45	47
Chiusabella	48	57	57	62	80	65	120	124
Baggio	14	15	14	21	18	17	24	24
Cimabue	64	93	41	45	82	54	53	74
Padova	31	48	43	40	64	71	98	85
Linate	10	11	9	14	11	9	23	56
Novara	—	—	—	26	37	20	47	60
Ovidio	49	20	32	36	38	27	26	28

Table 4.3 — Number and percentage of active wells recorded as contaminated to various degrees

Sampling date	Chlorinated organics (µg/l)	<50	50–100	100–150	150–200	200–250	250–500	500–1000	1000–1500	>1500	Total no. active wells
3.5.1979	No. wells	197	113	76	54	42	60	14	1	—	557
	% wells	35.40	20.30	13.6	9.7	7.50	10.80	2.50	0.20	—	
1.6.1979	No. wells	200	109	83	56	36	60	14	1	—	559
	% wells	35.80	19.50	14.90	10.00	6.40	10.70	2.50	0.20	—	
13.2.1980	No. wells	197	103	93	65	45	52	12	1	—	568
	% wells	34.70	18.10	16.40	11.40	7.90	9.20	2.10	0.20	—	
30.6.1980	No. wells	198	104	96	66	49	49	10	2	1	575
	% wells	34.45	18.08	16.70	11.40	8.53	8.53	1.73	0.34	0.17	
30.6.1981	No. wells	224	112	93	65	54	52	9	1	—	610
	% wells	36.90	18.50	15.20	10.50	8.85	8.40	1.50	0.15	—	
30.6.1982	No. wells	237	123	90	79	65	17	8	—	2	621
	% wells	38.2	19.80	14.50	12.7	10.50	2.70	1.30	—	0.30	

5

Waste disposal from urbanisation and its influence on fresh waters

P. J. Parsons

5.1 INTRODUCTION

In the United Kingdom the management of household and commercial wastes is a procedure that costs more than £1000m a year. The operations of the collection, treatment, and disposal of wastes are undertaken with the objective of being as efficient and economical as possible commensurate with environmental acceptability. The collection and disposal of household waste is the responsibility of municipal authorities but with substantial private-sector involvement; for commercial and industrial wastes, collection is largely arranged inedependently. However, the treatment and disposal of all wastes has to be authorised by the approved municipal licensing authority.

Household waste may be treated by pulverisation, compaction, and baling; it may be sorted with some components retrieved for recycling, or the waste may be incinerated. But whether it is treated or not, the ultimate fate of either the crude waste or the residues after treatment, is burial. Landfills thus accept over 80% of household wastes and so perpetuate a long-established practice.

However, with the passage of time the nature of collected household waste has changed, with large increases in paper and packaging, plastics and putrescible materials and a reduction in the inert components. The overall effect has been an increase in weight and particularly in volume; this change has been most marked in urban areas.

Out of the total household wastes of about 20 million tonnes/yr (Mtpa) (CIPFA, 1984) perhaps 8 Mtpa, originating from large conurbations, is destined for landfilling. Because of the scale of urban waste collection and the need for efficient and reliable handling, disposal has tended to be concentrated on relatively few sites. These are of large capacity; they are supplied by efficient transport systems and have reception arrangements to accommodate the high daily input of waste. They obviously possess the potential for adversely modifying the qualities of surface and groundwater resources.

5.2 THE WASTES

In order that the household waste from conurbations might be seen in perspective, it is noted that this is but part of the 100 Mtpa of waste from households, trade and industry (including demolition), and power stations. The mining and quarrying industries generate at least another 100 Mtpa but these are not subject to the same controls.

As far as the industrial component of waste is concerned about 3 Mtpa is considered potentially hazardous to man or the environment and 75% of this originates in the large urban areas. The most difficult of this material is destroyed by chemical treatment or incineration but 80 to 90% is disposed of to landfill as solids, sludges or liquids, sometimes with household waste and sometimes separately.

The control of landfill in the UK, although comprehensive, does not insist that all sites be lined with an impermeable materials to contain polluting leachate. Instead it acknowledges that some geological settings have characteristics that can absorb and attenuate migrating contaminants so that the residual concentrations do not pose a threat to groundwater resources. The containment of leachate is however required in other geological settings where the landfill is to overlie an aquifer directly and closely or is insufficiently isolated from it to guarantee that its existing quality may be retained. Where lined sites are installed there is always the possibility that rain entering the waste will promote the accumulation of leachate that might then rise and overtop the containment and so contaminate surface waters.

5.3 LANDFILL SITES

Extensive use is made of exhausted mineral workings in which to dump waste. In some parts of the country the numbers of such excavations far outweigh the capacity to fill them with waste but elsewhere there is a shortage. However, very often the strata that have been so exploited for the sand, cement, and aggregate industries also provide aquifers that yield well for water-supply purposes. There has therefore been a strong incentive to examine the characteristics of such strata to determine whether waste disposal into such excavations can be undertaken without endangering these aquifers.

The environmental acceptability of the increasing scale of landfilling was being questioned in the early 1970s following adverse publicity about dubious landfill practices, of odour emissions, and the accidental exhumation of toxic materials. As a result the Department of the Environment initiated research with the objective of placing in perspective the pollution potential of landfill leachate with those of other sources that may pollute water reserves. Thirteen years later this research is still continuing although the direction of investigation has varied much over the intervening period. It is now functioning in an entirely different regime in which the control of landfilling has been tightened by the Control of Pollution Act (1974) and operating standards have improved, considerably.

Initial field investigations concentrated on those landfill sites that accepted toxic wastes with and without other refuse, the emphasis being on sites where pollutant plumes had migrated away from the disposal sites and into the surrounding soils. The resulting document (Department of the Environment, 1978) presented the data by delineating the migration plumes of leachate emanating from 19 landfills chosen for the variety of their wastes and the range of differing geological conditions. The results were accompanied by those of concurrent field and laboratory studies. An important conclusion from this work was that an ultra-cautious approach to landfilling of most wastes was not warranted and that the degree of risk was acceptably low provided that the landfills were responsibly sited and controlled.

Large excavations suitable for waste disposal and close to population centres can be a valuable asset for which a market has developed, the alternative to local disposal being expensive and incurring the costs of transporting wastes to more distant sites. As a result it is usual that a contractor, having secured a licence to operate such a site, will dispose of as much material as possible into the volume available. The typical landfill for the disposal of wastes from larger urban areas thus has a large capacity, it is probably greater than 15m deep and the waste is highly compacted. The compaction of waste is a mode of operation that clearly prolongs the period of usefulness of the site, and reduces the magnitude of waste settlement in subsequent years. Its drawbacks may include problems with the lateral migration of landfill gas into the surrounding strata but this can be largely dealt with.

5.4 REACTIONS WITHIN LANDFILLED WASTE

Experience has shown that the environmental consequences of deep compact landfilling are very different from the shallow filling practised in earlier years, when collection vehicles dumped their loads at the tip and it remained only loosely compacted. Under those conditions air could enter the waste, the materials decomposed aerobically, and the end-products were carbon dioxide and water; sometimes high temperatures were generated (up to 80 °C) causing fires, but water pollution was negligible.

With the denser waste, aerobic conditions may prevail initially for a short time but they then succumb to anaerobic conditions in which the carbohydrates decompose to simple sugars and ferment, forming carboxylic acids that are the major pollutants in the ensuing leachate. Acetic acid accompanied by propionic, butyric and similar acids in lower concentrations are abundant in leachate derived from fresh refuse.

The reactions that occur within a landfill are the product of physical, chemical and biological activities all of which appear to be interdependent. Thus the compaction influences the amount of moisture present, the moisture influences the rate of biological reactions which in turn may influence chemical reactions dependent upon pH, and so on. Landfill temperature is also important and is determined by the comparative rates of the production of heat in a landfill and the loss of heat to the environment. In shallow landfills less than 5 m thick the ambient temperatures exert a control of those temperatures in the landfill, but at greater depths the insulation afforded by the upper layers permits elevated temperatures of 35 to 45 °C to be maintained anaerobically. It has been shown that under such conditions, provided the waste is moist enough, optimum conditions can be achieved in which the carboxylic acids may be broken down by methanogens, producing methane and carbon dioxide as landfill gas (Ham, 1982).

The presence of the organic acids is most pronounced in the early stages of waste decomposition and the buffering capacity of the material influences the limit to which the pH may drop. It is likely to remain within the range 5.5 to 8.0, but if it drops too low conditions become unsuitable for landfill organisms to multiply. As conditions become more favourable for good anaerobic fermentation so the conversion of organic acids to gas will induce the pH to rise and the precipitation of heavy metals will be encouraged. Also the sulphate salts will be reduced biochemically to sulphide and precipitation of metal sulphides will occur. Thus some of the heavy metals from the waste may be mobilised by the organic acids initially only to be immobilised later by precipitation.

5.5 ATTENUATION OF TOXIC WASTES BY REFUSE

In order to examine the capability of domestic refuse to absorb, immobilise or convert selected toxic industrial wastes, trials were run by the Water Research Centre in which four concrete-lined cells were each filled with 200 t of refuse (DoE, 1980). One cell contained a layer of phenolic lime mud with refuse above and below; it was instrumented for temperature and gas sampling and all leachate from rainfall and entrained fluids was collected over a period of four years. Another cell was similarly loaded with a sandwich containing spent iron-oxide waste rich in cyanide. The third was spiked with pesticides, and the fourth, with refuse only, remained as a control. At the end of the experiments the results showed that from all the accumulated leachate and analyses of the exhumed waste more than 95% of phenol and cyanide had been converted to more harmless constituents. Anaerobic conditions had been sustained throughout and the results broadly confirmed experience from monitoring disposal sites that had received these types of wastes.

5.6 WATER CONTENT

It is clear that the water content of a landfill greatly influences the rate and magnitude of fermentation. The moisture content of incoming household waste is typically between 25 and 35%. The moisture content conducive to anaerobic degradation is higher, at 40%. For a granular material such as soil, moisture can be accumulated up to its 'field capacity' before it becomes free-draining for excess water. But for waste, because of its heterogeneity, and perhaps because much of the water is in larger voids rather than in the pores, it begins to drain and create leachate at moisture contents below that of the field capacity (Campbell, 1982).

The density of waste, that is the degree of compaction it receives, influences the amount of water that it can assimilate; a highly compacted waste, like a squeezed sponge, has a lower retention capacity and may be more likely to produce leachate. However, the more tightly it is packed the more difficult it is for water to percolate.

In order to minimise leachate production, it is now common practice to deposit the waste in layers at a rate faster than that at which rainfall can wet it to field capacity. By confining the waste deposition to limited areas landfill cells may be constructed rapidly to their full height and capped with clay, so minimising the entrained moisture, inducing runoff from the surface and deferring the emergence of leachate at the base.

5.7 THE INFLUENCE OF GEOLOGICAL CONDITIONS ON LEACHATE MIGRATION

The main classifications of landfill sites are as follows:

(i) Containment sites — those providing a significant element of containment for wastes and leachates;

and

(ii) Dispersion sites — those allowing the slow release of fluids and significant attenuation of leachates.

(i) Sites that favour a significant containment of waste are those that have soil or rocks that are fine-grained, compact or cohesive and of low hydraulic conductivity (less than 10^{-3} m/d), for example slates, shales, mudstones, siltstones, soft clays and marls.

(ii) Those sites that permit groundwater flow but attenuate the constituents of leachate have the following typical characteristics:

formations exhibiting intergranular flow (sands, sandstones, sandy clays, gravels etc);

and

formations exhibiting marked fissure flow together with an element of intergranular storage, (Chalk, most Triassic sandstones, Coal-measure sandstones).

The extent to which geological characteristics are harnessed to influence lateral migration of contaminants is associated with the groundwater characteristics, and whether the water table lies immediately below or at some depth below the base of the landfill. If an unsaturated zone lies beneath the landfill the migration of leachate will be predominantly vertically downwards and there might well be a capability to attenuate or retain contaminant species within this column. If, however, there are horizontal layers or lenses of impermeable soil interspersed in the column, the vertical migration may be halted and horizontal migration induced until the obstruction is bypassed and the vertical passage resumed.

Once the plume of contaminants intersects the water table the predominant migration will then be horizontal and away from the disposal site. An exception to this may, however, occur if the contaminant fluid is miscible but much denser than the groundwater, in which case the vertical flow continues through the saturated zone until a relatively impermeable layer is reached. At this point its subsequent direction of flow might well be influenced more by the profile of the underlying aquiclude than the general direction of groundwater flow.

Early studies concentrated on the ability of soils to retain heavy metals and particular emphasis was given to the value of any unsaturated zone that may lie beneath a landfill and above the water-table. Studies on the lower Greensand showed that heavy metals were retarded with nickel and cadmium being more mobile than chromium or lead (United Kingdom Atomic Energy Authority, 1981). They also showed how the mobility of each of the metals was influenced by the phase, for example the carbonate phase was more mobile than the sulphide or silicate phases.

As research continued the evidence tended to confirm that the mobilisation and migration of heavy metals was less of a problem than formerly imagined. Attention was expanded to landfills containing household waste, for not only were these more numerous and contained far larger quantities than special or toxic-waste landfills, but the aggregated amount of 'toxic' species deposited as constituents of household refuse may well exceed the total amount deposited as notifiable waste by industry. However, in household wastes the species are highly dispersed and in a form that may be regarded as relatively inert.

5.8 QUALITY OF LANDFILL LEACHATE

Accumulating evidence suggests that the quality of leachate does not vary very much between landfill sites for household waste. Indeed, if hazardous wastes are co-disposed into refuse under properly controlled operations, this leachate also appears indistinguishable from the normal range of constituents (Khan, 1985). These are given in Table 5.1.

Table 5.1 — Range of Leachate Composition (Robinson & Maris, 1979)

pH	6.6—7.6		
	(mg/l)		(mg/l)
COD	66 —11600	Ca	165 —1150
BOD	2 — 8000	Mg	13.3— 480
TOC†	28 — 4440	Fe	1 — 370
Amm-N	0.9— 780	Mn	0.2— 26.5
Na	43 — 2500	Cl	70 —2777
K	20 — 600	SO_4	55 — 456
		Short-chain carboxylic acids (as C)	— up to 3672

† TOC = total organic carbon

5.9 ORGANIC CONSTITUENTS IN LEACHATE

Attention has been focused on the question of how the organic component of leachate might be attenuated. Studies on three landfill site investigations for intergranular aquifers suggest that in the Triassic sandstones the attenuation of organic compounds is minimal in the unsaturated zone, although the composition may have altered (Williams, 1982).

However, the fate of the organics in leachate when they intercept the groundwater is much influenced by the extent to which it is oxygenated. If it is well oxygenated it creates conditions that favour the aerobic breakdown of the organic component; at the same time any ammonia present may be oxidised to nitrate or nitrite and sulphide species similarly may be converted to sulphate. Such reactions clearly reduce the percentage saturation of oxygen — in one case from 80% (background concentration) to 30% beneath the landfill.

At another site, where the groundwater was less oxygenated, and where there was no superimposed unsaturated zone, the groundwater became anaerobic and an organic pollutant plume developed. Nevertheless anaerobic degradation of the organic components occurred close to the source; also, along the length of the plume, analysis revealed an increasing number of organic components of decreasing molecular weight. The progressive breakdown of the large molecular species in this case terminated with methane as the end-product. It may well be that these conditions are most likely to apply only if the groundwater flow is low.

The development of analytical equipment to detect exceptionally low concentrations of contaminants has enabled a large variety of organic micropollutants to be identified, which formerly would have gone unnoted. Such low levels of detection pose problems in determining whether the concentrations migrating from landfills are above or below a threshold that is acceptable. In order to investigate this further the Water Research Centre has studied the background concentrations of organic micropollutants in groundwaters known to be uncontaminated by landfills or other point sources, to examine the variations along four selected aquifer profiles from outcrop to confined conditions (Kenrick *et al.*, 1985). The presence of background levels of specific micropollutants enables groundwater quality near landfills to be assessed and any specific landfill contaminants identified.

The ability of the unsaturated zone to attenuate organic pollution from landfills is proving more difficult to assess. Unsaturated sands from the Triassic sandstone at Ruckley showed that they had negligible attenuating capability for organics. However, sampling of the Chalk for several years beneath a landfill that accepted organic liquid industrial wastes and refuse, showed that the migration of a 'front', while not stationary, is slower than that of the chloride 'front' used as the basis for comparison (since this experiences only dilution).

The most detailed study of this subject has been carried out in a permeable liner of sands and silts placed on the bed of a limestone quarry before waste was deposited (Robinson & Lucas, 1984). With the water table only a metre below the base of the quarry the emplaced material was to be an artificially installed unsaturated zone of homogeneous soil of known permeability and mineralogy. Previous experience had suggested that calcareous soils offered more promise of attenuating organic materials by virtue of their buffering capacity. In this case the opportunity was taken to install over 100 sampling and monitoring instruments within this 6-m thick underblanket of calcareous sands, to obtain water and gas samples from the unsaturated material and to measure soil water pressure, electrical conductivity, and temperature *in situ*. Sampling has been carried out at three-monthly intervals since June 1982 when waste started to be deposited. Now the waste is between 25 and 30 m deep and part of the landfill has been capped and reinstated. The input, all household waste from a large conurbation, was emplaced at rates up to 1000 t/d.

Leachate has appeared at the base of this landfill, though obviously not from rainfall owing to the speed of emplacement; it is thought that this has squeezed from the entrained moisture. The unsaturated material rapidly became anaerobic in advance of the leachate front; and after 15 months elevated concentrations of chloride and organic constituents were measured. The buffering capacity of the soil prevented any reduction in pH. The front of total volatile acids (TVA) advanced to a depth of more than 4 m, but subsequently it has remained stationary and the strength as measured by TVA has declined by an order of magnitude.

Meanwhile there has been a shallow accumulation of perched leachate at the base of the landfill on top of the liner, with a COD up to an order of magnitude greater than in the sand immediately beneath.

This experiment has given confidence in the principle of harnessing the unsaturated zone for leachate attenuation, although it may be some years before the sustained generation of leachate, from the aged waste, reaches the horizon of the underblanket of sand and silt.

5.10 LEACHATE TREATMENT

For the time being the commitment to landfilling as the main route for waste management seems set to continue. Nevertheless the Water Authorities with responsibility for the protection of water resources have to give approval before any site can be licensed for landfilling. In some parts of the country areas with other characteristics favourable for landfilling are therefore rejected on the grounds of aquifer protection.

Accordingly there is a likelihood that an increasing number of sites will be used in future that are lined with an impermeable material or membrane that will intercept and retain any leachate generated. For that reason work has progressed on methods by which leachate might be inexpensively treated so that it may be removed and either discharged as innocuous to surface waters, or transferred to sewage-treatment plants for a final polishing before it too is discharged. Such facilities carry the onus of needing to be operated for many years to handle the inevitable accumulation of leachate. A recent study (County Surveyors Society, 1984) on the treatment of leachates from landfill sites in England showed that the incidence of on-site treatment was small. Of 428 landfill sites only 80 generated sufficient leachate to require collection, treatment and disposal and of these only 11 carried out some form of chemical treatment, and one other that had more sophisticated treatment. Expensive and elaborate treatment plants are, of course, in operation in other countries for the treatment of leachate, so research has been directed towards simplifying both the structures and the mode of operation so that they may be left to function automatically, with only intermittent attention from operators. Present results are encouraging (Robinson and Davies, 1985) and research on the subject will continue.

After landfill sites have been topped off and reinstated they may continue to retain a potential for polluting water resources; periodic monitoring is therefore desirable. At present there is no general opinion about restricting or encouraging moisture in a landfill site. If rainfall is diverted and infiltration minimised the waste remains substantially unchanged; experience has shown from dry, excavated landfills that newspapers can remain readable more than 50 years after burial. Such disposals may be likened to indefinite storage.

If, however, moisture penetrates, the fermentation of the mass will occur with the characteristics of a solid-state reactor, where the chemical, biological and hence physical stabilisation might be accelerated. Present knowledge does not yet enable a prediction of when a landfill will become fully stabilised, for this is dependent upon the time taken for the cellulosic component of waste to decompose completely.

5.11 CONCLUSIONS

The findings of research and experience in the UK substantiate landfill as a valid option for the disposal of domestic and industrial waste, provided that both the site and the type of waste are suitable. Groundwater accounts for about one third of public water supplies, and these resources have been adequately protected from pollutants emanating from waste disposal. As sites become more difficult to identify and less suitable sites have to be considered, so research has been promoted to ensure that if necessary these latter sites may be used and will satisfy the overall objective of preserving water resources.

DISCLAIMER

The views expressed in this chapter are those of the author and are not necessarily those of the Department of the Environment.

REFERENCES

Campbell, D. J. V. (1982) Absorptive capacity of refuse — Harwell Research. *Proc. Landfill Leachate Symposium* Harwell Laboratory.

Chartered Institute of Public Finance and Accountancy (1984) *Waste Collection Statistics*. ISSN 0260–7603.

County Surveyors' Society (1984) *The Treatment of Leachate from Landfill Sites*. Committee 4, Special Activity Group 6. ISBN 0901476242

Department of the Environment/United Kingdom Automic Energy Authority *Uffington Lysimeters Operation and Results*. Parts 1 to 4 published as WLR Technical Notes 36, 40, 42 and 60, DOE London 1976, 1977, 1978, 1979; Part 5, UKAEA, Harwell, May 1981, HMSO.

Department of the Environment (1978) *Cooperative Programme of Research on the Behaviour of Hazardous Wastes in Landfill Sites*. Final Report of the Policy Review Committee. HMSO. 169 pp.

Department of the Environment (1980) *Behaviour of Wastes in Landfill sites. Study of the Leaching of Selected Industrial Wastes in Large-Scale Test Cells, Edmonton, N. London.* WLR Technical Note Series No. 69.

Ham, R. K. (1982) Variables affecting gas generation in landfill samples. *Proc. Int. Symp. on Landfill Gas and Leachate*. Braunschweig.

Kenrick, M. A. P., James, H. A., Gibson, T. M. & Turrell, M. B. (1985) *Trace Organics in British Aquifers — A Base-Line Survey*. Laboratory Report 823M. Water Research Centre.

Khan, A. Q. (1985) Disposal of Toxic Wastes. *Waste Management,* **LXXV** No 2, pp 82–93.

Robinson, H. D. & Davies, J. (1985) Automatic answer to leachate treatment. *Surveyor* **165**, No 4835, 7, 8, 26.

Robinson, H. D. & Lucas, J. L. (1984) Leachate attenuation in the unsaturated zone beneath landfills. Instrumentation and monitoring of a site in Southern England. *Wat. Sci. Tech.* **17**, Amsterdam, pp 477–492.

Robinson, H. D. & Maris, P. J. (1979) *Leachate from domestic waste. Generation, composition and treatment: A review*. Technical Report, TR 108. Water Research Centre. 38 pp.

Williams, G. M. (1982) Studies on Leachate Attenuation in Intergranular Aquifers. *Proc. Landfill Leachate Symposium*. UKAEA, Harwell.

6

Performance of the initial River Tame purification lake

J. M. Hellawell and M. B. Green

6.1 INTRODUCTION

Although the River Tame is a tributary of the River Trent, at their confluence the Tame actually contributes almost 50% more than the Trent to their combined flow. Much of this flow augmentation arises from the import of about 500 Ml/day of potable water into the catchment to meet the needs of a population of approximately 2.3 million, and the demands of the industries concentrated in the West Midlands. The conurbation which includes Birmingham, Britain's second city, differs from other major population centres in that it is situated on the headwaters of what would otherwise be a relatively small river, the flow of which must travel about 225 km before reaching its estuary. The history of the rapid urban growth and industrial expansion of Birmingham and the 'Black Country' in the 19th and 20th centuries, together with the progress in control and improvement of river water quality, have been described by Harkness (1982). Many of the water quality problems caused by rapid growth during the industrial revolution and the early part of this century have been overcome. There are now relatively few direct discharges of industrial effluents, the more polluting areas being controlled and discharged to sewer. However, in low summer flows as much as 80% of the flow in the Upper Tame comprises sewage effluents.

6.1.1 Problems caused by large conurbations

In spite of the concentrated population and industrial activity, the effluents from water reclamation works in the catchment have been of a relatively good standard for the past few years. However the quality of the River Tame remains poor. This is largely caused by the effects of urban surface-water drainage and is particularly acute when, after a dry spell, storms cause rapid runoff from a catchment which has an estimated impermeable area of 20%, and the combined surface and foul water overflows the sewers and enters the watercourses directly. Some 40% of the area is served by a combined drainage system, a further 15% by a partially separated system and the remaining portion, mainly the more recent development, has a separate system of surface and foul drainage. The problem is compounded by the inclusion of considerable quantities of heavy metals deposited by atmospheric fall-out and vehicle emissions, especially from drainage arising from the extensive road and motorway network (Hedley & Lockley, 1975). This material may accumulate in river sediments so that while the subsequent river water quality improves after storms, the benthic fauna continues to be restricted by toxic material in the substratum. Previously, during storms water quality throughout the entire length of the river was usually considerably worse than that of effluents discharged into it (Lester *et al.*, 1971). This event still occurs, but less dramatically. For example, Table 6.1 shows the

Table 6.1 — Mean quality parameters of the River Tame at Water Orton, and of the Minworth WRW effluent discharged at the same location

Site	Date	Determinand (mg/l)				
		BOD	Suspended solids	Ammoniacal nitrogen	Oxidised nitrogen	Dissolved oxygen
R. Tame at Water Orton†	May 1984	10.6	75	2.7	9.3	8.5
	Sep 1984	9.1	62	1.9	6.6	9.0
Minworth ASP Effluent‡	May 1984	5.9	11	2.2	17.3	—
	Sep 1984	2.8	7	0.1	16.4	—

†Based on daily spot samples
‡Based on daily composite samples

monthly mean values of several quality determinands of the effluent discharged from the activated-sludge plant (ASP) at Minworth Water-Reclamation Works (WRW), and of the river at the outfall.

The most serious aspect of the storm event appears to be the creation of a 'surge' of highly polluted water which moves downstream by 'plug' flow, causing rapid re-suspension of sediments which further deoxygenates the river (Fig. 6.1). Summer storms often coincide with the natural reduction phase in the diel cycle of dissolved oxygen caused by the photosynthesis and respiration of plant biomass, and so that problem is exacerbated. These intermittent reductions in water quality threaten survival of coarse fish populations which have become established, although somewhat precariously, in the lower reaches of the Tame, and could, if particularly severe, pose a threat to the important coarse fisheries of the River Trent.

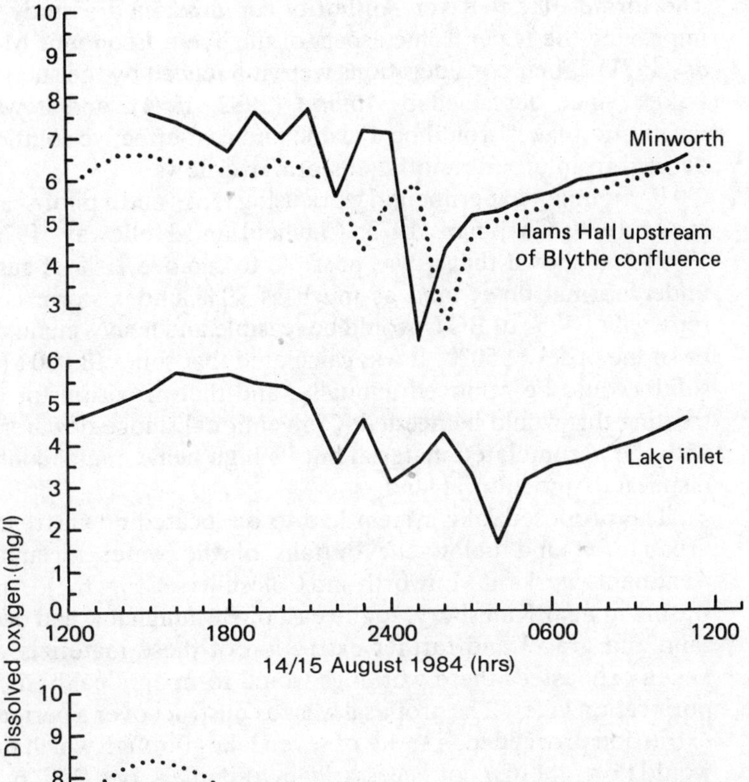

Fig. 6.1 — Variation in dissolved-oxygen concentration at several sites on the River Tame during summer storms (the increased discharge occurred around midnight)

6.2 THE LAKE SYSTEM

The former Trent River Authority commissioned a study of options for improving the River Tame aspect of the Trent Economic Model (Lester *et al.*, 1971). Their considerations were influenced by the success of the Ruhr Lakes, since described by Imhoff (1982, 1984), and it was shown that purification lakes would be a viable and cost-effective solution to pollution derived from urban runoff and storm overflows.

Preliminary experimental work using tanks and a pilot-scale lake (Lester *et al.*, 1971; Chapman, 1973; Clayfield and Holloway, 1976; Thompson, 1979) had shown that it was possible to remove 70% of suspended solids under normal flows, and as much as 90% under storm conditions. The removal of 40% of BOD would be feasible and heavy-metal removal would be of the order of 50%. It was calculated that some 18 000 t (dry weight) of solids would be removed annually, and that provision for extracting and treating this would be needed. Conventional sludge dewatering could deal with the accumulated material but its high heavy-metal content precluded disposal to agricultural land.

The proposed lake system had to be located downstream of the main urban area and below the outfalls of the water reclamation (sewage treatment) works at Minworth and Coleshill (see Fig. 6.2). An area of Tame washland near Kingsbury, to the east of Birmingham, had been worked for sand and gravel and further extraction of these materials was proposed. When exhausted, these workings would form an ideal basis for a series of purification lakes. The proposal was to construct over a period, as the gravel extraction proceeded, a series of seven lakes through which the River Tame would flow, though not necessarily in a simple series (Fig. 6.3). The length would be about 8 km, and over this distance the river would fall some 6 m and so provide the possibility to incorporate low aeration weirs. The first lake in the series was put on-line in August 1980 and the associated bankside work, including the silt consolidation plant, was completed in July 1983, at a total cost of about £7m. This chapter is principally concerned with providing a description of this lake and its performance.

Serious consideration has been given to constructing other lakes further upstream in the catchment in order to provide greater improvement of the river quality through the conurbation, although provision for removal of accumulated sludge would be more difficult in smaller, dispersed units.

6.2.1 Description of Lake 1

The inlet to the lake consists of a bifurcated concrete channel with a gravel trap, and combined floating debris barrier and oil boom immediately above the bifurcation. Measuring weirs (Crump-type) are located at the dividing point of the channel, and the float chambers and housing for flow-measuring equipment are located between the channels. The housing also provides space for the continuous water-quality monitoring station. Each inlet channel is designed to give virtually constant velocity at all discharges and

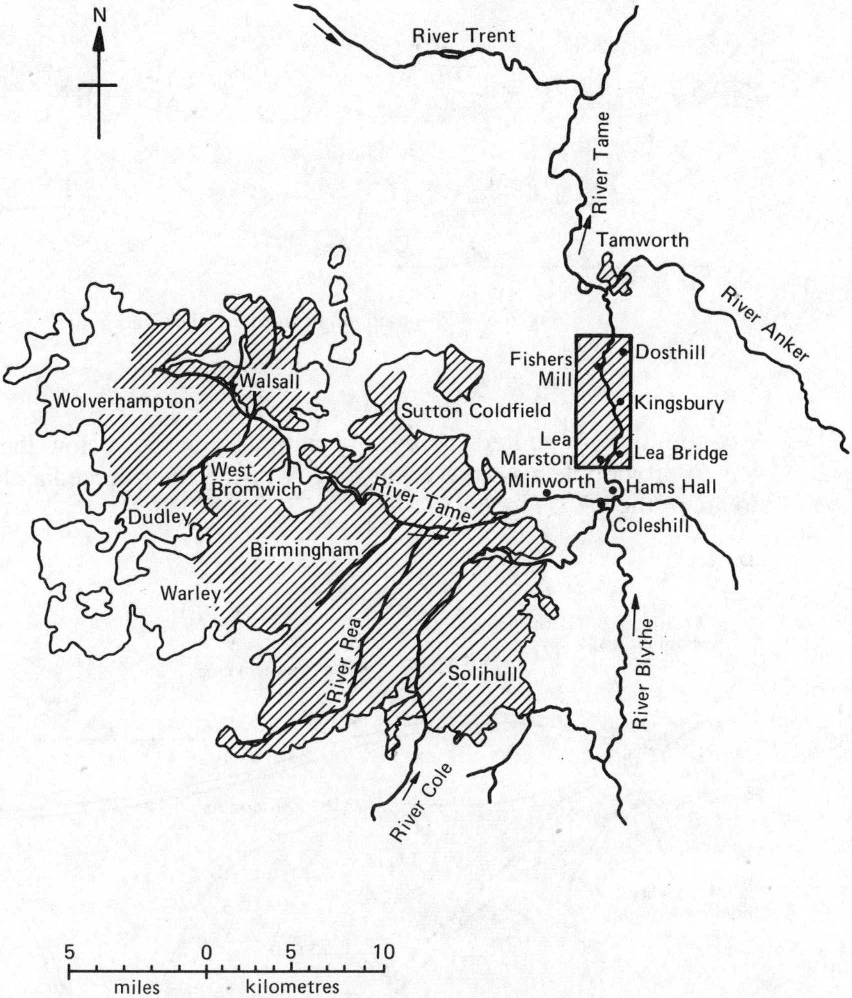

Fig. 6.2 — Map of Birmingham area showing location of rivers, purification lakes and places mentioned in the text

serves to ensure deposition of small gravel and coarse sand, which are also removed periodically by dragline or dredger. The inlet channels then open out into the two halves of the lake, separated for almost their entire length by an isthmus which acts as a central berm. This berm and the lake banks are about 1 m above the normal lake level. The total lake area is 25 ha and the average depth is about 2.5 m, giving a theoretical mean retention time of almost a day at dry-weather flows although in flood discharges the mean retention time would be only two or three hours. The lake discharges through three outflow channels over low weirs which provide some aeration.

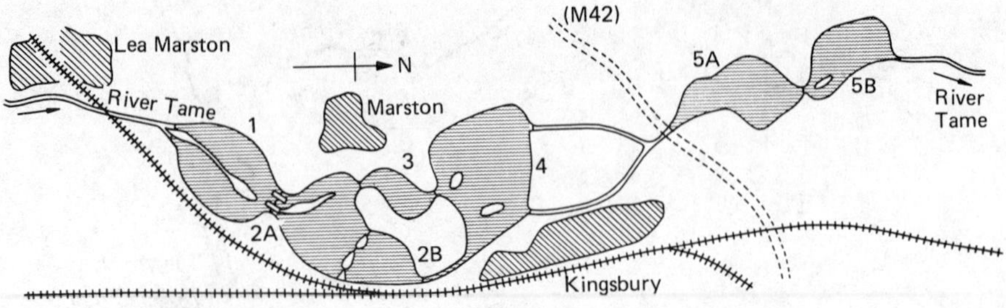

Fig. 6.3 — Proposed arrangement of lake series

A continuously-operated monitoring station is located below the central weir to provide data on the water quality of the river immediately downstream of the lake (Fig. 6.4).

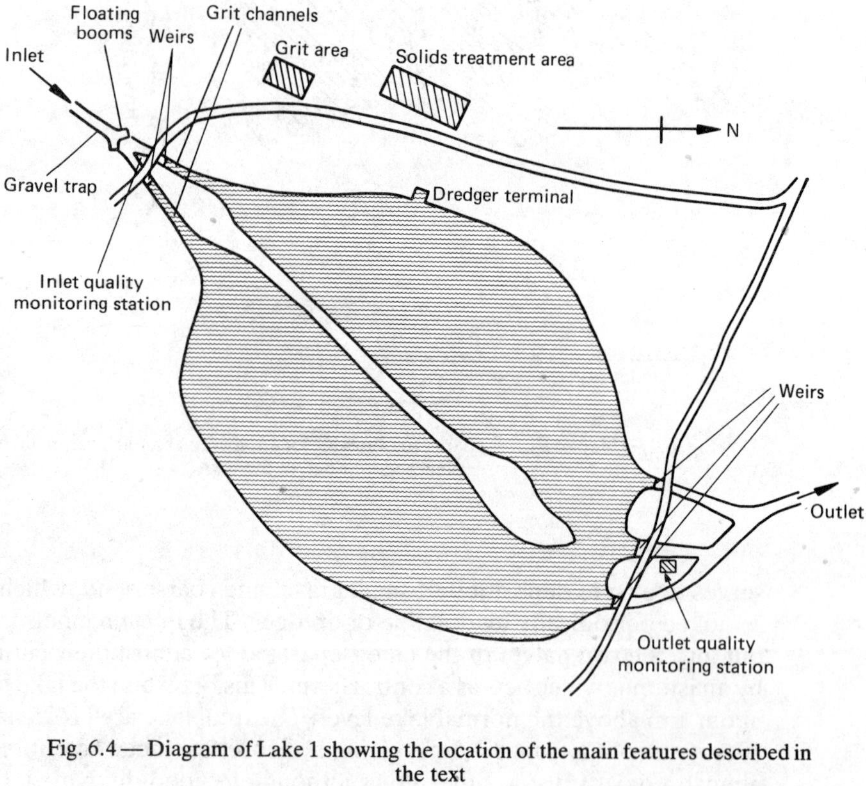

Fig. 6.4 — Diagram of Lake 1 showing the location of the main features described in the text

6.2.2 Operation of Lake 1

As the lake acts as an on-line sedimentation tank, it is therefore necessary to: remove, drain and dispose of the bed load from the gravel trap; remove and dispose, by burning or tipping, the debris deflected by the floating booms, and dredge silt collected in the lake for consolidation, dewatering and disposal.

Solids deposited on the lake bed in the main body of the lake are removed by a suction dredger ('Mudcat') equipped with a horizontal cutting auger and conveyed in a flexible pipeline to the shore for treatment. Sludge is removed in regular 2.4-m wide sweeps across the lake as the dredger hauls itself along cables anchored on the shore and on the central berm.

Since the completion of the sludge-consolidation plant in July 1983, silt dredged by the Mudcat, at about 5% dry solids (ds), has been screened, degritted in hydrocyclones, and consolidated in batch-filled tanks, to 11–14% dry matter (dm), before being pumped 3.3 km to Coleshill Water Reclamation Works. On that site, 9 ha of ash-bottomed drying beds are reserved for the lake silts. The silt is dosed with aluminium chlorohydrate, at about 0.15% w/w as Al_2O_3, and the beds are filled to a depth of 0.46 m. An overall turn-round time of about 14 weeks is achieved, giving a cake with upward of 55% ds. During dry weather the drying time is four to five weeks, but is slower in wet weather.

Annual revenue cost for the operation of the lake and disposal route is about £400 000, which gives a unit cost of about £20/t ds. Included in this are the costs of hauling silt 12 km from Coleshill to the tip, and the cost of tipping heavier material on site; when this latter area is finally reclaimed the overall cost will increase by nearly 10%.

6.3 EFFECT OF LAKE 1 ON RIVER-WATER QUALITY

The success of the Lea Marston lake in improving river-water quality is shown in Table 6.2 which summarises data collected during routine quality surveillance over three years. It achieved a significant reduction in pollutants associated with suspended solids (SS) but, not surprisingly, there was little change in soluble organic carbon or in other soluble determinands. In 1984 there was a reduction of approximately 2 t BOD/d, or about 25%, achieved by passing through the Lake.

The increase in ammoniacal nitrogen of 0.5 mg/l between the inlet and outlet, indicated by these results, is confirmed by more detailed composite sampling at the inlet and outlet, and by continuous monitoring. Ammonia is released by breakdown of nitrogenous compounds, particularly by deamination, within the bottom silts, to give a concentration of about 200 mg/l in the undisturbed sediments, from which diffusion appears to be fairly constant throughout the year.

Biological activity of the sediments is also shown by the release of bubbles of gas, which comprise about 80% methane, during all but the

Table 6.2 — River-water quality at routine sampling points above and below Lake 1 for the years 1981, 1982 and 1984 (1983 data excluded because of the effects of the manual workers' strike)

	Lea Marston (upstream of Lake)			Nether Whitacre (downstream of Lake)		
Determinand	Mean	Range	No	Mean	Range	No
BOD (ATU) (mg/l)	8.0	4.0–44.0	99	5.6	3.0–14.5	89
Soluble organic carbon (mg/l)	13.1	5.2–35.0	59	12.0	5.5–34.0	51
Suspended solids (mg/l)	42	9–460	99	20	7–160	90
Ammonical nitrogen (mg/l)	2.6	0.7–7.0	99	3.1	0.7–7.4	91
Total oxidised nitrogen (mg/l)	9.9	1.8–15.0	99	9.7	3.6–15.0	91
Dissolved oxygen (mg/l)	7.7	0.8–13.0	77	7.7	2.7–12.2	69
Temperature	11.5	2.0–21.1	77	11.5	2.0–21.5	69
pH value	7.5	6.9–8.1	99	7.5	7.1–8.2	91
Conductivity (μS/cm)	750	380–1030	87	750	430–1040	79
Chloride (mg/l)	85	32–198	99	84	30–217	91
Total hardness as $CaCO_3$(mg/l)	263	120–380	60	257	140–315	59
Orthophosphate, as P (mg/l)	3.3	0.4–6.2	75	3.4	0.5–6.8	67
Total cadmium (μg/l)	2.7	0.5–10	56	2.9	0.5–10	34
Total chromium (μg/l)	25	5–80	26	23	5–90	25
Total copper (μg/l)	62	40–130	26	44	30–60	25
Total lead (μg/l)	18	0.5–200	56	10	0.5–22	34
Total nickel (μg/l)	154	90–230	26	137	70–210	25
Total zinc (μg/l)	207	110–350	26	170	100–270	25
Soluble cadmium (μg/l)	1.1	0.3–5	29	0.6	0.1–1.3	9
Soluble chromium (μg/l)	13	5–30	15	13	5–30	14
Soluble copper (μg/l)	30	5–60	17	24	5–60	16
Soluble lead (μg/l)	3	1–8	30	3	1–12	10
Soluble nickel (μg/l)	148	80–230	15	137	80–180	14
Soluble zinc (μg/l)	115	60–180	17	99	60–180	16

coldest periods of the year. Despite the anaerobic condition of much of the depth of the silt, the immediate surface layers are aerobic and harbour a fauna of naidid and tubificid worms and copepods.

6.3.1 Performance in dry weather conditions

Woods *et al.* (1984) showed a diel pattern in dissolved oxygen (DO), at the inlet to the lake, which became more pronounced as the days lengthened and temperatures rose. The cycle is driven by photosynthetic activity of benthic macrophytes and algae upstream. During the summer the diel pattern is reduced between inlet and outlet (Fig. 6.5). Some of this alteration is due to mixing, but there is an overall decrease of about 1 mg/l in the DO

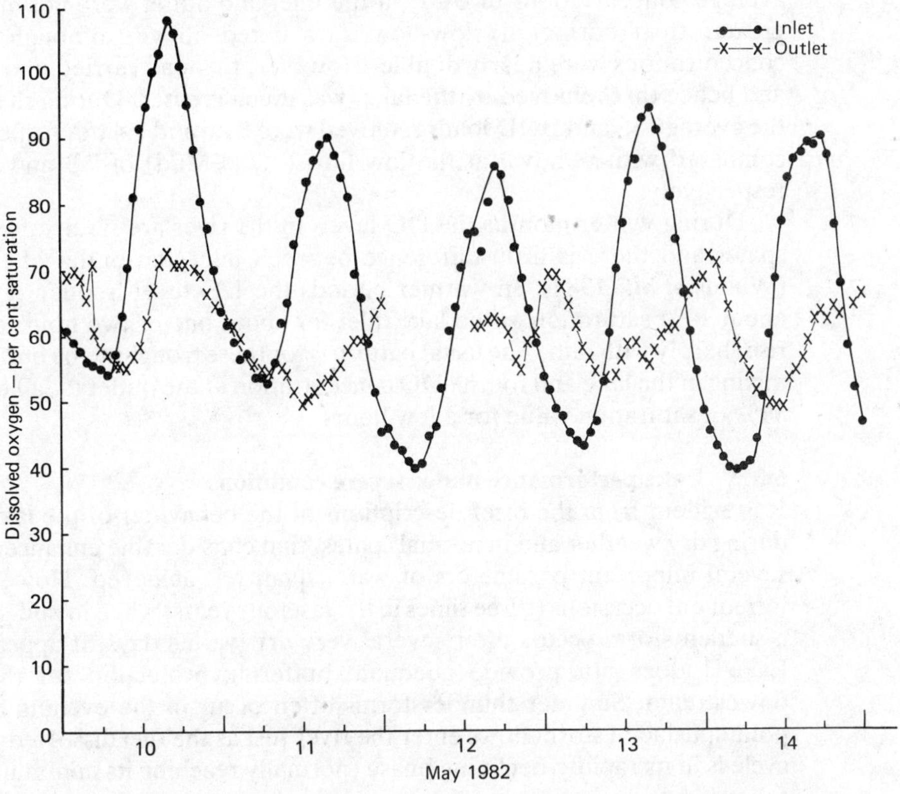

Fig. 6.5 — Diel variation of dissolved-oxygen concentrations measured at inlet and outlet of Lake 1

during this period which amounts to a loss of about 0.8 t O_2/d at summer flows. Respiration rates of the benthic silts vary according to position in the lake and are typically in the range 0.1 to 0.2 g O_2/m^2 h at 18 to 20 °C. On average, this represents an oxygen requirement of 0.9 t O_2/d.

Woods *et al.* (1984) describe a series of dry periods in 1982, totalling 70 days, in which the average SS were reduced from 23.4 mg/l at the inlet to 17.5 mg/l at the outlet, a change of 25%. By contrast, in 1984 for 38 days on which the flow at Lea Marston was less than 700 Ml/d, the SS reduced on average from 21 to 11 mg/l and the BOD from 6.6 to 4.9 mg/l giving reductions in load of 47% and 24.5% respectively.

6.3.2 Performance in wet-weather conditions

The quality of the River Tame upstream of the lake deteriorates sharply with the onset of rain and runoff in the catchment after a dry period (Fig. 6.1), and the improvement afforded by the lake is most evident at such times. Poor river-water quality is not necessarily sustained in periods of high rainfall for in 1984, on 31 days when the flows exceeded 2000 Ml/d, the average concentrations of BOD at the inlet and outlet were not markedly greater than during the low-flow days noted above, although the SS concentrations were nearly double. However, the load carried in the river, and hence that removed by the lake, was much greater. During these days the average SS and BOD loads removed were 81.6 and 7.4 t/d, respectively, compared with removal at the low flow (<700 Ml/d) of 7.3 and 1.1 t/d, respectively.

During winter months the DO levels in the river are rarely affected by spates and there is little difference between inlet and outlet of the lake (Woods *et al.*, 1984). In warmer periods the DO level frequently falls to about 10% saturation at the lake inlet for about one or two hours as flows rise sharply with rain. The usual pattern is for these troughs to be buffered by mixing in the lake and for the DO concentration at the outlet to fall to about 30% of saturation value for a few hours.

6.3.3 Lake performance under severe conditions

It is evident from the brief descriptions of the behaviour of the lake, both during dry weather and in normal spates, that considerable enhancement of several important parameters of water quality is achieved. However, on infrequent occasions (three times in the last four years) when in mid-summer a sudden storm occurs after several very dry warms days, it appears that Lake 1 does not provide adequate buffering protection for the river downstream. Summer thunderstorms often occur in the evening and the pollutant-laden storm flows enter the river just as the diel dissolved-oxygen cycle is in its rapidly-declining phase (normally reaching its minimum in the following early morning). The dissolved oxygen recorded at the lake inlet plummets to zero (or almost so) for about two hours, followed by an equally rapid recovery (Fig. 6.6). At the lake outlet the dissolved oxygen also falls to very low levels; here the duration is greater and the oxygen depression is experienced for a considerable distance downstream. The origin of the material causing these intermittent but very severe events is likely to be the organic solids which have accumulated in the pool sections, at margins, and possibly within macrophyte stands, and which have continued to be digested thereby accumulating a large potential oxygen debt. These, when re-suspended by sudden storm flows, together with the usual material associated with urban runoff, cause severe depletion of dissolved oxygen which cannot be met by surface re-aeration, nor by dilution in the lake which is itself at a trough in its diel cycle. The possibility that surface deposits in the lake are also brought into suspension cannot be ruled out and work on this aspect is continuing.

6.4 PROBLEMS ASSOCIATED WITH LAKE 1

It is abundantly clear from what has been described above, that the performance of Lake 1 has generally satisfied expectations. However, difficulties have been encountered. For example, Wood *et al.* (1984) drew attention to the problem of the design of the combined debris barrier and oil

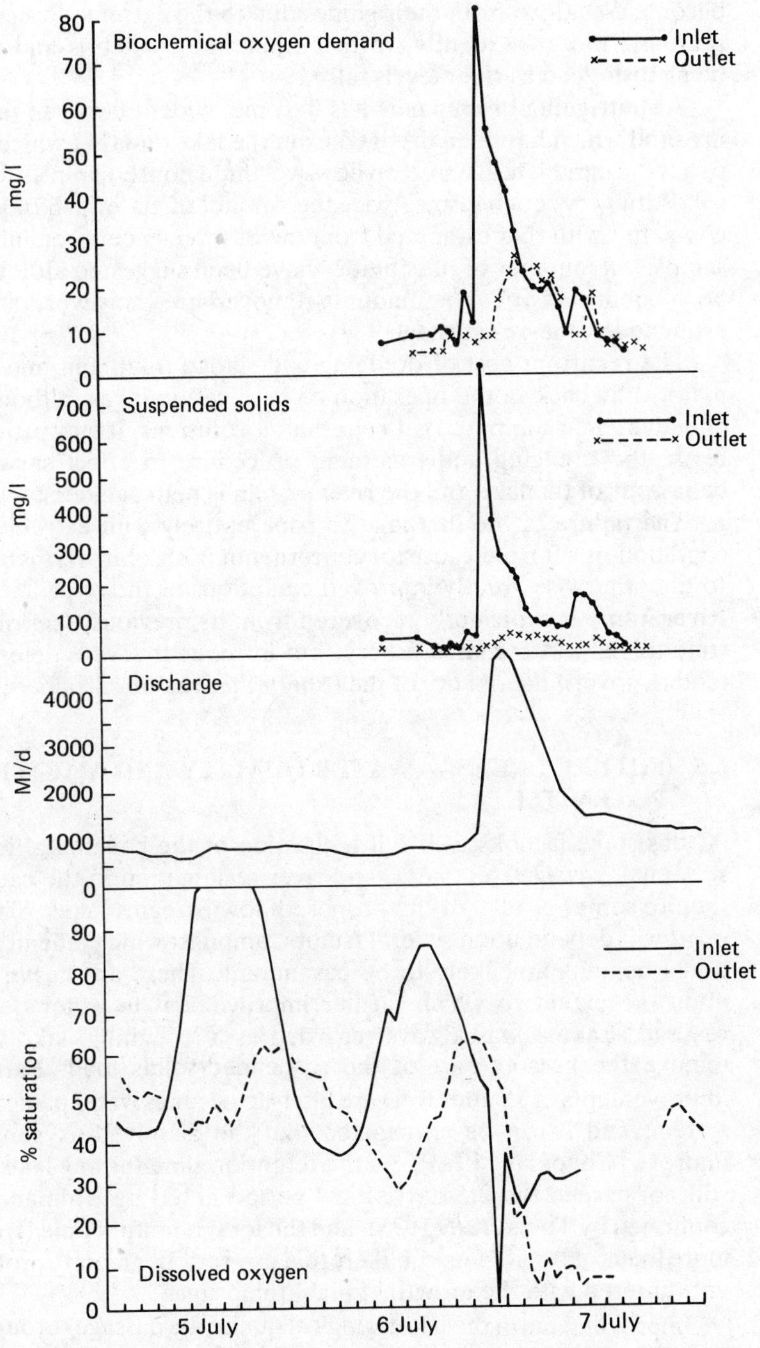

Fig. 6.6 — Variation with flow of BOD, suspended solids, and dissolved oxygen at the inlet and outlet of Lake 1 during a severe summer storm

boom on the inlet to the lake, which has caused the floating booms to become dislodged from their guides due to the extreme forces of a river in spate or, more frequently, to being jammed by debris and subsequently being dislodged as river levels fall.

An intriguing discrepancy has become evident between the amount of material which has been dredged from the lake and that which is estimated to have accumulated, based on flow-weighted hourly samples for suspended solids. In very rough terms, twice the amount of material has been removed compared with that estimated from the difference between inlet and outlet samples. A number of possibilities have been suggested although none has been confirmed yet. The amounts removed are, however, less than those estimated at the design stage.

The recurrent cost of dredging and sludge treatment and disposal is a major drawback in the operation of purification lakes, although it is cost-effective when compared with alternative solutions. It may prove possible to revise the dredging and treatment procedure to effect savings once the behaviour of the lake and the river system is better understood.

The failure of the first lake to cope entirely with extreme low quality conditions is a further cause for concern, not in absolute terms but in relation to the otherwise greatly improved conditions of the river. At present, the River Tame is sufficiently recovered from its previously seriously polluted state to support a coarse fishery, but infrequent severe deterioration will tend to govern the quality of the fishery that can be sustained.

6.5 FUTURE LAKES — WATER QUALITY AND AMENITY POTENTIAL

At this stage it looks as if full realisation of the benefits of investment in sewerage, sewage treatment, and river reclamation in the catchment may require some, or all , of the proposed downstreams lakes. Their development will depend upon several factors, amongst which amenity and recreational benefits are likely to be paramount. These are heavily dependent upon the extent to which further improvement in water quality can be realised. Lake 1 was always regarded as a 'roughing' lake which would remove the greater part of the suspended-solids load. Further smaller improvements in SS and BOD are likely to occur as water passes through the system, and it can be anticipated that the dissolved-oxygen regime will improve (Chapman, 1973). As the retention time for any lake in the series will not exceed the 2.5-day critical period noted by Uhlmann (1971) and confirmed by Toms *et al.* (1975), and the total retention time is unlikely to be more than 5 days, it seems unlikely that any serious amenity problems will be encountered with the growth of planktonic algae.

Improvements in the bateriological quality by passage through a series of lakes could extend recreational usage to include water-contact sports such as water-skiing and wind-surfing. This would greatly increase the amenity value of the scheme in providing facilities which would be within a half-hour's travel of a population of about three million.

6.6 CONCLUSIONS

- Despite considerable expenditure on better sewerage and sewage treatment, and an overall improvement in river water quality, the River Tame is subject to brief but acute deterioration in quality following storms.
- An on-line purification lake has been constructed downstream of the drainage of the greater part of the Birmingham conurbation, and proposals have been made for the construction of a series of lakes.
- The principal function of the first lake is to provide interception and settlement of the suspended-solids load and its associated BOD.
- Facilities have been provided for the removal, treatment, and disposal of deposited solids.
- The first lake works well under dry-weather flows and in normal spates, but on rare occasions it appears to be ineffective in protecting the river downstream from severe depletion of dissolved oxygen. The causes of this apparent failure are not known and are being investigated.
- There is considerable amenity and recreation potential in the construction of further purification lakes.

6.7 REFERENCES

Chapman, B. T. (1973) A mathematical model of a river purification lake, *Wat. Treat. Exam.* **22**, 269–294.

Clayfield, G. W. & Holloway, D. M. (1976) Experimental purification lakes on the River Tame. *Wat. Pollut. Control.* **75**, 341–352.

Harkness, N. (1982) The River Tame - a short history of water pollution and control within an industrial river basin. *Wat. Sci. Techn.* **14**, 153–165.

Hedley, G. & Lockley, J. C. (1975) Quality of water discharged from an urban motorway. *Wat. Pollut. Control.* **74**, 659–674.

Imhoff, K. R. (1982) Experiences with purification lakes and polishing lagoons. *Wat. Sci. Tech.* **14**, 189–204.

Imhoff, K. R. (1984) The design and operation of purification lakes in the Ruhr Valley. *Wat. Pollut. Control.* **83**, 243–253.

Lester, W. F., Woodward, G. M. & Raven, T. W. (1971) The effect and cost of river purification lakes. *The Trent Research Programme*, pp. 74–93. Institute of Water Pollution Control, Maidstone, 120 pp.

Thompson, J. L. (1979) *River Tame purification scheme — development and design.* Paper presented to the Institution of Water Engineers and Scientists, Midland Section, January 1979.

Toms, I. P., Owens, M., Hall, J. A., & Mindenhall, M. J. (1975) Observations on the performance of polishing lagoons at a large regional works. *Wat. Pollut. Control.* **74**, 383–401.

Uhlmann, D. (1971) Influence of dilution, sinking and grazing rate on phytoplankton populations of hyperfertilized ponds and microecosystems. *Mitt. Int. Verein. theor. angew. Limnol.* **19**, 100–124.

Woods, D. R., Green, M. B. & Parish, R. C. (1984) Lea Marston purification lake: operational and river quality aspects. *Wat. Pollut. Control.* **83**, 226–242.

ACKNOWLEDGEMENTS

The authors would like to thank R. Hattersley, Manager, Tame Division, for permission to publish this chapter, and the many colleagues who have helped in experimentation, data gathering and processing, in particular C. Morton, D. McGarr, M. Harley, Dr R. Sedgwick and Miss S. Cooke.

Discussion

Chairman: **F. SANDERSON**
Member, North West Water

For all five discussion sections in this book an attempt has been made to arrange the subjects in a logical order and not necessarily in the sequence in which they arose at Stirling. Thus for urbanisation the headings New Towns, Motorways, Aquifer Pollution, Landfills and Reclamation Lakes have been adopted.

NEW TOWNS

In discussing New Towns, participants were interested in sewer connections, drainage from industrial yards, pollutant sinks, eutrophication, and the interactions between the recreational pressures arising from a new town and the water quality standards required in waters used for recreation.

Sewer connections

Mr D. Hammerton (Clyde River Purification Board) applauded the practice of the IJsselmeerpolders Development Authority of using concrete for storm-water drains and plastic for foul sewers (to avoid confusion and thus wrong connections). Had this idea been used at Milton Keynes or did wrong connections occur? **Mr Coombes** explained that there was no special policy: most drains were of concrete and no systematic use of plastic had been made. Incorrect connections did occur but to an unknown extent.

Mr van de Ven added that wastewater sewers were laid at greater depths than the storm-water sewers. In a recent survey no wrong connections were found in Lelystad and only a few in Almere. Overall, errors were fewer than one per cent.

Industrial yard-drainage

Mr C. D. Bayes (Water Research Centre, Scottish Office) said that contaminated yard-drainage from new-town industrial areas had caused problems in receiving waters in Scotland. Was there experience of such problems in Milton Keynes? If so, since the oil interceptors could not deal with water-miscible pollutants, would consideration be given to diverting yard drainage to foul sewers in sensitive areas? **Mr Coombes** replied that this diversion had happened to some extent. Area interceptors were intended to trap oil-spills from those areas without privately installed interceptors. Spills of miscible pollutants from industrial premises were rare but of course passed through the interceptors and entered the balancing lakes. Another potential problem, not yet causing sufficient nuisance for remedial measures to be taken, was the non-intercepted drainage from certain large car parks.

Pollutant 'sinks'

Following on from this point **Mr J. B. Ellis** (Middlesex Polytechnic) suggested that for a wide variety of pollutants [aromatic hydrocarbons, metals such as cadmium, copper and (perhaps) zinc] structures such as ponds or balancing lakes only delayed the passage of pollutants temporarily and did not act as 'sinks'. Finely-divided organic particles and the substances associated with them were re-suspended and lost from the sink in considerable proportions. **Mr Coombes** confirmed this and emphasised that the primary purpose of the balancing lakes at Milton Keynes was flood control. Any other benefit was a bonus. **Mr van de Ven** said that re-suspension in the canals occurred only in limited areas, near the storm-sewer outfalls. In general the velocity in the canals (rarely more than 5 to 10 cm/s) was not sufficient to transport sediments. The canals were designed to provide storage for accommodating fluctuations in water level. For soluble components however (cadmium and zinc) retention was more limited.

Eutrophication; phosphorus removal

Dr. D. C. Hardwick (Ministry of Agriculture, Fisheries and Food) asked if the Lelystad lakes were as eutrophic as the canals, and whether either supported fish life. **Mr van de Ven**: Nutrient levels were high in the canals and the southern lakes, but despite algal problems in the latter (see below) there was still a good variety of life in them. People fished the canals successfully.

Mr D. R. H. Price (Anglian Water) pursued the subject of eutrophication. What problems had been observed and did they seem to justify the cost of the proposed plant for phosphorus removal? Phosphorus from Lelystad and Almere, **Mr van de Ven** explained, was discharged directly to the large northern lakes but the amounts involved were not sufficient to cause problems. In the southern polders, however, shallow marginal lakes received phosphorus from the sewage-treatment plants alongside the feeder streams to the lakes. Here, in the 1960s, very large inputs of phosphorus had occurred. Eutrophication resulted and the lakes were now dominated by the blue-green alga *Oscillatoria*. Attempts were being made to reduce the problem.

(a) Tertiary-treatment was being installed at every waste water plant; solutions of iron salts were being added to the effluents, at considerable expense.
(b) The residue time of water in the lakes was being reduced to try to flush out the algae.
(c) A feasibility study on the possible removal of the contaminated lake sediments was under way. These sediments still released enough phosphorus to cause algal blooms although the input of nutrients from the rivers was no longer able to do so.

Combining amenity and recreation with the alleviation of pollution: safety aspects

Comments of the authors were sought by **Mr V. W. Howells** (Severn Trent Water) on the health and safety implications of promoting the recreational use of waters which could be subject to pollution incidents. **Mr Coombes**: Boating and fishing took place on the lakes and the normal provisions for safety were the responsibility of the relevant clubs. If a severe spillage in the city resulted in a hazardous situation in the lakes the same emergency procedures would be applied as in any other UK city. Essentially, the public would be kept away from the affected area.

In the Lelystad region, the canals were suitable for boating and fishing but swimming was prohibited because the water quality was not sufficiently high bacteriologically to comply with Dutch standards for bathing waters. **Mr van de Ven** added that the canals at Lelystad could be flushed clean in an emergency, but in fact this has only been required once since 1968, so the system was not installed at Almere, but other measures were available in both places.

MOTORWAYS

Comments were made on the comparability with other situations of the studies reported here, and on the effectiveness and maintenance of the treatment systems.

Relevance of data to other motorways; comparability with other work

Mr C. P. Young (Water Research Centre) wondered whether the runoff quality of motorways was consistent. The drainage systems varied from road to road. For example motorways built on chalk drained straight to the rock so that there was an extended 'linear point-source' of contaminants. **Dr McIntyre** assumed that runoff quality would be broadly similar from road to road, the major differences being due to traffic density. Responding to a question from **Dr R. F. Packham** (Water Research Centre), **Dr McIntyre** referred to similar studies in Switzerland, the USA, and North London and confirmed that the levels of metals he reported were particularly high. They seemed comparable with those arising directly from the metal-plating industry in the UK Black Country 20 years or more ago.

Loading rates

The question of loading rates was taken up by **Mr J. B. Ellis** (Middlesex Polytechnic). He was concerned that re-suspension and deflation†, especially in wet weather, could seriously affect the estimates of loading rates from these small catchments. **Dr McIntyre** agreed. The loading rates quoted were averages, and the variation around these could have been as much as 40 to 50%, depending on the degree of re-mobilisation of deposited material and, for instance, the number of days beforehand without rain.

Treatment of runoff

Efficiency

Mr Price went on to suggest that the efficacy of the various treatment methods was likely to be sensitive to the size of the treatment facility. How were the different systems sized or standarised? He suggested that for practical purposes cost might be the most useful standard to adopt.

Dr McIntyre agreed with this last point but said that, even with the usual policy of providing treatment facilities capable of dealing with the largest flows experienced, the costs were probably of little consequence when compared with the cost of building and maintaining the motorway itself.

Maintenance; life of structures

Dr Packham accepted that the French Drains performed better than the lagoons but asked what happened to the solids in the former. How long would French Drains remain effective? Would it be necessary periodically to dig them out? **Dr McIntyre** suggested that the Drains would last about 27 years. When one metre of the front end of a Drain had been dug out the stones were found to be thickly coated with oil and grease. This must retain other organics and might act as a 'fly paper' for solids. Apart from weeding, the Drains did not require maintenance.

AQUIFER POLLUTION

After brief discussion of the extent of the problem, the need for protection policies was argued.

Occurrence

Mr I. R. Geuze ('De Dommel' Waterboard, Netherlands) questioned whether the pollution of the aquifers below Milan was indeed the worst in Europe, or merely the best-documented example. We know too little about such under-city contamination.

† deflation [Geol.] the sweeping erosive action of the wind over the ground.

Protection polices
Mr R. Aspinwall (Aspinwall & Co) raised the subject of aquifer-protection policies. He knew of only two such policies extant in the UK. Enthusiasm for them seemed to have lessened among Water Authorities, the argument presumably being that the Authorities could make their voice known through the planning process anyway. So, could what has happened and is occurring below Lombardy happen in the UK, as a result of there not being well-documented, proper, aquifer-protection policies which are agreed with the infra-structure planners?

Mr H. Fish (Natural Environmental Research Council) followed up this point. As he saw it, the policy for the protection of groundwater being used or proposed as a source of supply (and clearly that below Milan had immense value) must be that the aquifer was protected in compliance with the WHO Guidelines or the EC Directive for Drinking Water. That must be the basic policy. However, given that every case is different, a generally-applicable protection policy is unlikely to be effective. Each case for industrial development or waste disposal that could affect an aquifer must be argued individually. The question then became: 'How can this proposal go ahead while at the same time the aquifer is protected to the existing drinking-water quality standards?'

LANDFILLS

The monitoring and analysis of leachates and the value of encouraging evapotranspiration from the landfill surface were examined.

Non-biodegradable contaminants
Mr Hammerton: It was comforting to note the considerable reduction in chemical oxygen demand (COD) in the experimental blanket of sand under the landfill, but what was known about non-biodegradable dangerous substances that might be affecting the watercourse? Were these being monitored? **Mr Parsons** said that samples were taken by pressure-vacuum lysimeters established at various horizons in the blanket, and these were analysed for the full spectrum of fatty acids (by WRc). Everything that could be measured was measured, including gas, pH and temperature.

The performance of the monitoring devices had been checked against analyses of core material; good agreement was found. The experiment would go on for another five or ten years. It was a unique study, the first to allow us to understand what was happening in the 'attenuation blanket' or 'unsaturated zone' before the full generation of leachate occurred. What was not understood was how there could be an accumulation of moisture at the base of the landfill when the waste originally only contained 25 to 35% moisture (although rainfall would have added something to this figure before the landfill was capped with clay). It was now covered with topsoil and was being seeded for pasture.

Evapotranspiration

Mr T. G. Brownlie (Department of Agriculture and Fisheries for Scotland) then asked what value Mr Parsons attached to the evapotranspiration of water by grassland overlying landfills; this must help to reduce the influx of rain-water. Although rainfall was greatest in winter and evapotranspiration in summer, a vigorously growing crop of grass must induce a usefully large soil-moisture deficit by the autumn which would help to reduce the penetration of winter rains. **Mr Parsons** suggested that this might be considered as a subject for research.

RECLAMATION LAKES

Questions covered the structure of reclamation lakes, sampling of the solids they trapped and aspects of the metals associated with the solids.

Structural Design

In answer to a question from **Mr A. Malmgren-Hansen** (Water Quality Institute, Denmark), **Mr Green** said that the separation of the inlet into two structures was for practical and operational purposes: it was easier to operate the dredgers under all weather conditions in the lake divided in two parts by the central structure.

Solids

Sampling discrepancies in load

Mr Geuze asked how the suspended solids were sampled at the inlet and outlet. 'De Dommel' Waterboard had found that sediment transport did not completly coincide with water flow so that during periods of high flow very frequent sampling was needed. Quite different results were obtained if the frequency was reduced, particularly at the influent of such lakes. **Dr Hellawell** explained that hourly samples were taken using a multi-sample unit. They were concerned that they were not picking up a bed-load and that the system might be giving a false impression of solids loading, although in calibration exercises the samplers seemed to be behaving satisfactorily. **Mr Green** added that the Flygt pump was situated in such a way that priming was no problem. The rate of pumping was high. A 10-mm strainer might prevent some material entering the pump but this was thought to be a minor source of error. Creeping bed-load could be significant. A new monitoring system was ready to be installed.

Mr Fish pursued the problem of the reclamation lake 'making muck'. **Dr Hellawell**: Ninety-nine per cent of the time the lake trapped solids very satisfactorily but the last one per cent presented the enigma. The lake did not create solids, he thought, and the sediments were not disturbed. Solids might 'skate across the surface' and appear in the outlet samples. 'Fingerprinting' of the material, although difficult, might solve the origin of the solids at the outlet.

Metals; effect on BOD

Mr Ellis referred to Table 6.2 which showed very high levels of soluble metals. Although the lake was described as a 'roughing lake', inevitably there would be a significant passage of metals through the lake. He also asked why COD was not used as an index in monitoring (in view of the well-established inhibitory effect of metals on the BOD test), especially since there must be a considerable delayed oxygen demand in the lake.

Dr Hellawell agreed that the metal levels were high; the initial lake had been designed to remove organic pollutants. The industry of the catchment was thought likely to change to one in which metals featured less prominently. **Mr Green** said that, Divisionally, there was little confidence in the determination of COD at low levels. The levels of metals were not sufficient to inhibit BOD, measurements of which were backed up by analysis for total organic carbon.

Metals; disposal site

Mr Malmgren-Hansen suggested that in removing the metal-rich sediments from the lake and disposing of them on land, another pollution problem was being created, but **Mr Green** could discount this. The material, lifted from the drying beds at about 55% dry matter, (drier in summer), was transported 15 km to a large clay-pit where it was used to cover domestic refuse.

Part 2

Mineral exploitation

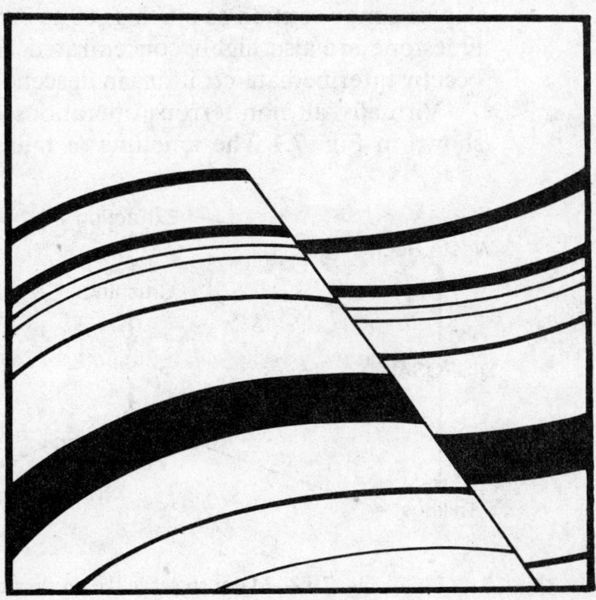

7

The management of liquid effluents from non-ferrous mining and minerals extraction

A. K. Barbour

7.1 INTRODUCTION

The natural concentration of metals and minerals which is economic to extract varies widely. Large-scale open-pit copper operations can be economic where metal concentrations are as low as 0.5%. For uranium and gold mining the concentrations may be even lower. On the other hand, many iron deposits may contain as much as 60 to 70% of the metal; borax, chalk, and limestone are also highly concentrated. Bauxite, zinc/lead, and tin usually occupy intermediate positions in descending order of concentration.

Virtually all non-ferrous operations can be depicted schematically as shown in Fig. 7.1 The amounts of tailings and, often, waste rock to be

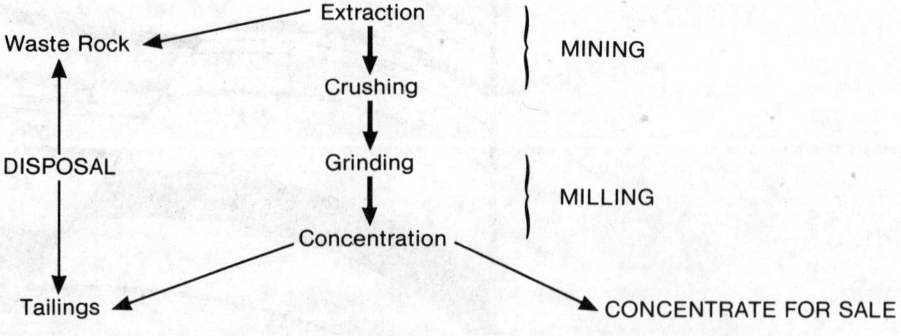

Fig. 7.1 — Major stages in the production of non-ferrous concentrates

disposed of each year can be very large, amounting to many millions of tonnes from an individual copper or uranium mine.

This chapter concentrates on the management of liquid effluents, and their effects on neighbouring surface and groundwaters, arising from:

- rainfall on waste-rock dumps,
- chemical reagents from flotation or leaching operations,
- tailings areas.

Stress is laid upon the chemical aspects of effluent control. For more detailed treatments, and, specifically, analyses of the engineering and hydrogeological aspects, the reader is referred to the published texts of Jones (1975); Williams (1975); Aplin & Argall (1972), and Argall (1978).

7.2. SOURCES OF POLLUTION

7.2.1 Erosion of waste rock dumps

Unlike the chemical and metallurgical processing industries, mines have to be located where economic mineralisation naturally occurs. Because of the tonnages of low-value materials which have to be transported, processing plant associated with the mine also has to be located nearby. Extraction operations naturally break up the terrain and hence increase greatly the surface area of material exposed to rain which, in many parts of the world, falls as intense storms of relatively short duration, giving a high risk of flash-flooding.

In these circumstances, 'wash-out' from waste-rock piles is inevitable. Fortunately, by definition, waste rock contains low concentrations only of the desired elements, which are often relatively toxic, but the clays and silts eroded can cause local streams to become opalescent due to the high burden of suspended solids. Practicable ameliorative measures of general applicability do not exist, though occasionally it is possible to channel runoff streams via the tailings impoundment (see section 7.2.3). Fortunately a corollary of 'spatey' rainfall is that there are often periods of several months of relatively dry weather when erosion is small and stream discoloration is much less marked. Practical problems arise only where streams subject to serious erosion are used for cattle-watering. In these circumstances, provision of alternative supplies of water suitable for the purpose should be arranged by the mine operators. This is usually not a particularly onerous requirement since the area of influence of even large, open-pit mining operations is usually quite small, and clean supplies can be obtained by the provision of relatively small local impoundments, either collecting rainfall or storing the water required in any case for the operations of the milling and processing areas. In the nature of mining operations, areas of unusual mineralisation do not usually occur where intensive agriculture is practised.

Runoff problems can be more serious where sulphuric (pyritic) deposits are being worked, or where high-sulphur coal is being extracted. Acid is generated by oxidation reactions:

$$2FeS_2+2H_2O+7O_2 \rightarrow 2FeSO_4+2H_2SO_4$$
$$4FeSO_4+2H_2SO_4+O_2 \rightarrow 2Fe_2(SO_4)_3+2H_2O$$
$$Fe_2(SO_4)_3+6H_2O \rightarrow 2Fe(OH)_3+3H_2SO_4$$

Many methods have been proposed for dealing with acidic runoff including deep injection, neutralisation with lime, and dilution. None is of general applicability; such treatments can only be applied where runoff follows well-defined channels and, in any case, neutralisation is both expensive and difficult to operate effectively. Reliance usually has to be placed on the natural absorptive powers of local streams and the ameliorative measures outlined in the preceding paragraph.

7.2.2 Liquid effluents from milling

Milling is defined as the comminution of the extracted ore into particles which can be subjected to a recovery process which separates the valuable minerals (concentrate) from the valueless (gangue). The term milling is now usually used to cover the flotation process (or a chemical-treatment process in the cases of gold, uranium, and the production of alumina from bauxite) which is now an essential part of all non-ferrous mining operations.

After primary and secondary crushing and screening, milling operations start with grinding in a multiplicity of ball and rod mills. After classification, the ground material passes to the flotation units where a variety of reagents may be used depending on the chemical composition, density, and so on, of the mineral being concentrated.

Froth flotation, by far the most widely-used concentration method, is based on conferring hydrophobicity to the individual particles, and hence assisting them to become attached to air bubbles. Particles with higher mineral content then rise to the surface of a froth which in turn is skimmed. The remaining barren particles become tailings. The flotation reagents used tend to be specific for particular processes; Table 7.1 shows some general examples (Vick, 1983).

Leaching is the concentration method favoured in some operations, sometimes in conjunction with flotation. The largest-scale example is the separation of alumina from bauxite by the Bayer process, in which caustic soda is used to dissolve out the hydrated aluminium oxide; others are the use of sulphuric acid to acid-leach uranium oxide and some copper-oxide ores, and the use of sodium cyanide in the extraction of gold.

By passage through a multiplicity of cyclones and thickeners the end products of the milling or leaching process are the concentrate of the desired metal or metals together with a slurry containing the discarded process water, unwanted gangue, and the reagents, frothers, collectors, and so on added during the flotation stage. This slurry then passes to the tailings impoundment area, sometimes after chemical treatment immediately after the flotation section to remove by oxidation organic reagents which have a

Table 7.1 — Flotation reagents

Class	Use	Compound
Collectors	To coat particles with water-repellent surface attractive to air bubbles	Water-soluble polar hydrocarbons, such as fatty acids, xanthates
Modifiers		
(a) pH regulators	To change pH to promote flotation, either acid or basic	NaOH CaO Na_2CO_3 H_2SO_4 H_2SO_3
(b) Activators and depressants	Selectivity to modify flotation response of minerals present in combination	Metallic ions Lime Sodium silicate Starch Tannin Phosphates
Frothers	To act as flotation medium	Pine oil Propylene glycol Aliphatic alcohols Cresylic acid
Oils	To modify froth and act as collectors	Kerosene Fuel oils Coal-tar oils

high oxygen demand, and any cyanides which may be present. Overflow water from the thickeners, and so forth, is recycled to process wherever possible.

7.2.3 Liquid effluents from the tailings area

In most non-ferrous mining operations tailings management is an area of major environmental consideration requiring, as it does, acceptable answers to question such as those concerning:

- tailings disposal method;
- if to land, optimum site for the tailings disposal area;
- if to lagoon, dam stability and method of design and construction;
- method of tailings transportation from mill to impoundment or sea;
- tailings stability in seismic areas;

- purity of supernatant water and its disposal route either to recycle or to adjacent streams;
- effects of tailings water on adjacent streams and groundwater;
- revegetation of tailings areas to minimise windage losses and to improve aesthetic appearance.

Tailings are not homogeneous materials, but differ substantially across the spectrum of the non-ferrous industry in relation to:

- particle size (slimes†/sand ratio);
- specific gravity;
- physical characteristics, including abrasiveness;
- chemical composition;
- pH.

The first two parameters influence strongly the flow, settlement and, in seismic areas, liquefaction characteristics; the chemical aspects naturally have a major influence on the levels of toxicity of the tailings water and treatment methods to minimise its effects on receiving streams or other bodies of water. The details of tailings-disposal systems are thus necessarily highly site-specific; the following general outline requires modification and interpretation to suit the details of particular operations in specific locations.

7.2.3.1 Tailings disposal — method and location

Unusually in mining operations, there are sometimes optional courses of action in choosing the method and location for tailings disposal so that the 'best practicable environmental option' (BPEO) can be selected.

Unlike waste rock, tailings can be transported as aqueous slurries, either being pumped or moving under the action of gravity through pipes or culverts. Settlement characteristics can be calculated accurately and this is clearly very important for the avoidance of blockages and breakdowns in operation, as is the determination of the degree of abrasiveness on materials of construction likely to be encountered. Slurry transportation often provides a range of options for the economic disposal of tailings, which is usually not available for waste rock disposal.

Some such choices which may become practical alternatives are considered below.

- *Narrow, deep valleys versus disposal in shallow valleys or plain-land*

  Narrow, deep valleys are usually easier to dam and do not disturb agricultural land, although they may obliterate ecologically valuable areas such as tropical jungle. They are usually visually unobtrusive, partly because they are exposed to the sight of few people. On the other hand, such valley locations are often relatively elevated, thus

† Slimes represent the fraction of total tailings which passes a 200 sieve: sands represent the +200 openings per inch-mesh-size material

increasing pumping costs and often increasing hazard in seismic areas if tailings liquefaction leads to break-out; land at lower elevations is usually more valuable agriculturally.

- *Location to minimize adverse environmental impact on adjacent surface waters and groundwaters*

  For streams and other surface waters, the choice of the BPEO involves weighting and balancing factors such as the degree of treatment (and its cost) required for the disposal of tailings water into a particular stream versus, perhaps, discharging it untreated into a stream more distant from the mill, but which is environmentally and commercially unimportant.

  Although now generally out of favour with many regulatory authorities, tailings disposal to sea can, in the author's view, be a preferred choice in cases where pumping/pipeline costs are not prohibitive. In general, deep outfalls to sea can utilise its enormous absorptive capacity for ions and, usually, the area of serious disturbance to benthic organisms is relatively small. The BPEO studies should be made (but often are not) to assist choice between the various options on the basis of both detailed scientific baseline data of *all relevant ecological aspects, and economics*. Although less visually apparent, any adverse effects on groundwater supplies and purity may be very important indeed. Consequently, BPEO studies based on Environmental Impact Assessments must include hydrogeological assessment of such seepage flows, for the tailings-impoundment area, including the dam, as well as the extraction site.

- *Location for safety*

  The failure of a tailings dam could have disastrous consequences for human beings and activities located nearby. Although the design parameters for tailings dams are now well-developed and incorporate safety factors to accommodate predicted frequencies of earthquake and storm, it remains prudent to locate tailings impoundments away from people and human activity as far as possible. This is another clear benefit of sea disposal where it can be done in an environmentally-acceptable manner.

7.2.3.2 Purity of supernatant water and effects of adjacent streams

At many mine sites water is expensive and recycling of tailings effluent is practised wherever possible. At open-pit operations in arid areas, recycled water is frequently used to spray haul roads, broken rock prior to shovelling and such like, with the object of suppressing dust to the maximum extent possible.

As noted earlier, the aqueous component of tailings slurry from the mill usually contains very low concentrations of surface-active frothers and collectors and, where acid conditions are used in the flotation circuits, relatively high levels of cations if present in the feed to the mill. Examples of the latter are iron, manganese, cadmium, mercury, copper, lead, and zinc. Problems in the tailings area can also arise where pyritic deposits are being worked due to the development of acidity by oxidation in the presence of water. When the impoundment is in active use it is usually saturated with water and access of air is limited; but when the pond level falls, conditions for rapid development of acidity are present, which could pose serious problems with pyritic deposits after operations have formally ceased. Bacterial oxidation with *Thiobacillus ferrooxidans* is thought to be a dominant factor in the development of acidity from tailings containing sulphides.

If cyanide has been used in the extraction circuit, as it is in most gold concentration processes, it may be necessary specifically to convert it to a relatively innocuous cyanate by oxidation immediately upon leaving the flotation circuits.

It is clearly impracticable in this chapter to provide world-wide purity criteria for tailings effluent, but attention has to be focused on the obvious parameters such as: heavy metals and arsenic, suspended solids, pH, chloride, sulphate, occasionally fluoride, and, increasingly, nitrate, together with the flow characteristics and uses of the receiving bodies of water. Dependence is usually placed on utilising the dilution and absorptive powers of the receiving waters. Conventional treatment, for example liming to precipitate heavy metals, pH adjustment, and so on, is used where it is necessary to preserve existing uses of the receiving body. In uranium mining, activity is often reduced by precipitating radium as the double salt $RaBaSO_4$ by treatment with barium chloride (Averill *et al.* 1982).

However, in view of the very large volumes of water involved in most tailings operations, particularly where recycling of supernatant is not practised, sludges and some other wastes from the treatment process usually have to be disposed of separately in small impervious impoundments; this is often not a preferred environmental option compared with dilution into streams as it creates a toxic 'hot-spot' which may be difficult to manage after general operations have ceased. Whatever disposal option is selected, adequate monitoring should be practised to ensure that any significant changes in the quality of the receiving body are quickly detected and assessed.

7.3 REVEGETATION OF TAILINGS AREAS AND WASTE-ROCK DEPOSITS

During the operating life of the mine, the deposited tailings are normally largely covered by the supernatant mill effluent, leaving only the beaches exposed. This is important from the standpoint of minimising wind erosion-which can become a serious problem where prolonged dry seasons are encountered. At 'close out' or cessation of active operations, it is now becoming usual for regulations to require some permanent system for the passive management of tailings and waste-rock areas so that: they are not a health hazard to either human beings or animals; windage nuisance is minimised; continued contamination of watercourses does not occur. Improvement of aesthetics should also be a significant objective — flat sandy areas can be very obstrusive visually in wooded or mountainous terrain.

Where tailings contain major proportions of slimes, the eventual total 'drying-out' process can be very prolonged; but it can be accelerated by evapotranspiration from suitable tree plantations. When tailings areas have adequately dried it is often possible to establish vegetation on this barren and hostile substrate, using techniques which have developed rapidly over the last 10 to 15 years (Bradshaw & Chadwick, 1980). Control of pH by heavy liming is usually a first essential, followed by application of the plant nutrients nitrogen and phosphorus. Plants and grasses indigenous to the area are often the most promising candidates for successful vegetation. Once a limited natural humus cover has been established, legumes can also be incorporated. Where tailings or waste rock is highly pyritic, revegetation is much more difficult due to the generation of acid noted earlier, but progress is being made. Of course, all such areas can be topsoiled before reseeding, but such a procedure is usually inordinately expensive.

Re-vegetation by such techniques has an important part to play in preserving the environment quality of mining areas and their watercourses. In the open-cast extraction of coal, the National Coal Board has demonstrated that such areas, managed carefully and in detail, can be restored to effective agricultural use. Many time-expired quarries are now widely used as centres for water sports and angling.

However, in large-scale open-pit mining for non-ferrous metals, it is usually impossible to disguise the pit itself; historically, the high incidence of local areas of dereliction and (continuing) contamination of waterways are quite unacceptable today. Whilst the scale of major non-ferrous mining is such that some impacts of an adverse environmental nature are inevitable during active operations, these can be controlled by active foresight and planning to acceptable levels *for the lifetime of a mining operation*, say 20 to 40 years. Techniques actively developed over the last 10 to 15 years offer considerable promise that long-term dereliction, and contamination of river systems, can be substantially reduced in the future. Taking science, technology, engineering, and economics together, — and assuming that all continue to develop favourably — the author believes it would be reasonable to claim that the world's essential raw materials can be won in an environmentally-acceptable manner. However, there is some considerable way to go before claiming that non-ferrous metals can be mined with no adverse environmental impact of any kind — perhaps this will require the development of new mining technologies as well as new technologies for environmental protection.

7.3 REFERENCES

Aplin, C. L. & Argall, G. O., Jr., (1972) Tailings Disposal Today. *Proceedings of the 1st International Tailing Symposium*, Tucson, Arizona, Miller Freeman Publications Inc., San Francisco.

Argall, G. O., Jr. (1978) Tailings Disposal Today. *Proceedings of the 2nd International Tailing Symposium*, Denver, Colorado, Miller Freeman Publications Inc., San Francisco.

Averill, D. W., Schmidt, J. W., Moffett, D., Webber, R. T. & Barnes, E., (1982) in: *Management of Wastes from Uranium Mining and Milling*. International Atomic Energy Agency, Vienna, pp. 353-364.

Bradshaw, A. D. & Chadwick, M. J. (1980) *The Restoration of Land*. Blackwells, Oxford.

Jones, M. J. (Ed.(, (1975) *Minerals and the Environment*. Institution of Mining and Metallurgy, London.

Vick, S. G. (1983) *Planning, Design and Analysis of Tailings Dams*. John Wiley and Sons, New York.

Williams, R. E. (1975) *Waste Production and Disposal in Mining, Milling and Metallurgical Industries*. Miller Freeman Publications Inc., San Francisco.

8

Mineral extraction and water quality in Scotland

D. Hammerton

8.1 INTRODUCTION

Mining is generally recognised as being second only to agriculture as the world's oldest and most important industry. The importance of mineral extraction in the early development of mankind is signified by the designation of successive epochs such as the Stone Age, Bronze Age, and Iron Age (Down & Stocks, 1977), from which it can also be deduced that mining was one of the first human activities to have a significant impact on the environment. The degree to which these early mining industries caused recognisable pollution is, of course, a matter for conjecture and the effects must have been very localised.

However, in Europe by the Middle Ages there is clear evidence that mining could result in severe environmental impact. In the first textbook on mining (Agricola, 1556) the impact of mining in Germany is described thus:

> '. . . the strongest argument of the detractors is that the fields are devastated by mining operations . . . the woods and groves are cut down, for there is need for an endless amount of wood for timbers, machines and the smelting of metals. And when the woods and groves are felled, there are exterminated the beasts and birds . . . Further, when the ores are washed, the water which has been used poisons the brooks and streams, and either destroys the fish or drives them away . . . . Thus it is said, it is clear to all that there is greater detriment from mining than the value of the metals which the mining produces.'

This chapter examines the effects of mining on water quality during the past century, the progress which has been made since the River Purification

Boards were established, and, finally, considers whether present legislative controls are adequate.

8.2 HISTORY OF MINING POLLUTION IN SCOTLAND

There is good evidence to suggest that the mining industry did not begin to have a significant impact on the water quality of Scottish rivers until the last two decades of the nineteenth century. The Commission on Rivers Pollution set up in 1868 devoted the whole of the fourth report (Rivers Pollution Commission, 1872) to a painstaking survey of Scottish rivers but, although it showed that many rivers were severely polluted by sewage and industrial wastes, it found relatively little evidence of the direct impact of mining or quarrying. Although mention was made of ferruginous waters discharged from a colliery and from abandoned mine workings to the River Till in the Tweed catchment, and an account was given of pollution of the Glengonnar Water (a tributary of the Clyde) by discharges from the lead mines at Leadhills, these sources of pollution were considered to be of minor importance. The report gave a vivid account of the gross pollution of the River Almond by the newly-established shale-oil industry, but this was due to processes at the various paraffin-oil works and not to the mining of shale rock.

A report in 1881 by the Medical Officer of the County Council of Linlithgowshire showed that pollution of the River Almond by paraffin-oil works was severe, but that the worst problem was ferruginous water from ironstone mining which 'poisons the stream from its fountain-head to its exit in the sea and spoils the whole district through which it flows'. The Medical Officer carried out a series of experiments which proved that the dissolved iron could rapidly be precipitated with slaked lime, though he reported that the cost of this treatment would render the industry unprofitable. In fact mining for ironstone ceased completely in the Almond catchment before he completed his report.

8.2.1 Pollution from coal washeries

The impact of coal mining on water quality only became serious towards the end of the century due to the introduction of coal-washery machines at the pitheads. A report (County Council of Lanark, 1903) showed that in the Clyde basin 'only one or two washers were in operation prior to 1885 and that, since then, there has been a steady increase until 1897 when the number reached 61'. By 1902 the number had reached 68, it being estimated that the amount of coal washed daily ranged from 60 to 300 tons per washer per day, and that the total amount washed daily in the Lanarkshire coal fields was over 15 000 tons. The great development in coal washing was attributed partly to the working of coal seams previously considered unprofitable because of the large amount of earthy and other impurities, and partly to the demand of consumers for cleaner fuel.

The 1903 report referred to complaints from farmers and industrialists and stated that the fine coal solids rendered the water unfit for any purpose.

Cattle either refused to drink it, or suffered afterwards and damages were sometimes recovered for losses due to this cause. One farmer claimed that many sheep had died, and a post-mortem on one sheep carried out in the county laboratory found four ounces of coal and sand particles in the stomach. Many industrialists found that the coal solids made the stream water unusable. One example quoted was that of a Turkey-red dyer whose goods were damaged even after the installation of a filter to remove the coal solids which reached his water intake 2.4 km downstream of a coal washery. The report stated that solids from one coal washer so silted up the bed of the North Calder Water as to constitute an offence under Part I of the Rivers Pollution Prevention Act, 1876, by interfering with the flow of the river.

The report went on to state 'The prevention of pollution seems, on first consideration, such a simple matter that the wonder is it should occur at all. Difficulties, however, soon became apparent when preventive measures are enforced. The difficulties relate to the provision of proper structural arrangements, their upkeep and proper management. A colliery working a poor seam of dirty coal with an old washing plant, can perhaps ill afford to throw away the old plant and provide the new machinery already described'.

A second report published by the County Council of Lanark in 1909 revealed that considerable progress had taken place as a result of prosecutions and continued inspections by the river inspectors. Although the number of coal washeries had increased to 72, 89% of these were now provided with silt-recovery tanks and settlement ponds whereby much of the water was recycled and the volume of dirty water greatly reduced.

8.2.2 Pollution from mine-drainage waters

The 1909 Report, mentioned above, referred for the first time in Scotland to pollution by mine-drainage waters from abandoned mines, and drainage from colliery refuse heaps existing long after mining had ceased. The report also stated that pumped waters from working pits sometimes contained 'a sufficiently quantity of solids to give the water a very turbid, polluted appearance, particularly where the underground roadways were laid on a clay pavement'. At times the suspended solids reached 60 grains per gallon, that is over 850 parts per million (850 mg/l). The amount of suspended matter greatly increased when the sumps at the pit bottom were cleaned out. This report also referred for the first time to iron in pit waters and drainage from refuse heaps which precipitated out in the receiving streams to cover the bed with iron ochre.

The third report of the County Council of Lanark, 1924, stated that, apart from sewage, pollution by coal-dross washings 'probably accounts for most of the pollution affecting the River Clyde and its tributaries'. In 1924 the number of coal washers under supervision was 74, all of which discharged their effluents direct to streams, none having connections with public sewers. The report showed that at almost every colliery satisfactory preventive measures in the form of silt-recovery tanks and settlement ponds were provided but 'difficulty invariably arises in having these measures properly managed and kept in satisfactory order and the absence of simple legal means of enforcing proper attention and adequate maintenance of these preventive measures is greatly felt'.

This report also drew particular attention to the problem that water in the same condition as that in which it had been drained or raised from a mine was, by Section 5 of the Rivers Pollution Prevention Act, 1876, exempted from the operation of the Act, although such waters frequently contained appreciable or even large amounts of suspended solids which rendered them polluting liquids. Only at one colliery in the county, where the pit water was persistently turbid, were settling ponds provided and this could not be insisted upon by the inspectors. A more serious aspect of this problem was the growing number of mines in which the drainage water contained a large quantity of iron, either in solution or suspension, giving rise to severe pollution of the receiving water. Around 1908 pollution of the Dippool Water by ferruginous waters from abandoned mines in the vicinity of Haywood gave rise to complaints, and also at that time water from the nearby Woolfords Colliery was discharged to Cobbinshaw Loch with 'disatrous results to fish life'. The lessees of the fishing rights took proceedings against the colliery owners, recovering damages and obtaining an undertaking that no more water would be sent to the loch. Instead the water was diverted to the Dippool Water causing further serious pollution of that stream! In this connection it is of interest to note that the discharge of a large volume of iron-bearing minewater to the Dippool Water commenced in January 1984 from a pipe serving old abandoned mine workings near the village of Woolfords.

In 1921 the closing of the coal mines for three months due to strike action led to the most severe minewater pollution yet seen in Scotland, and caused the extensive destruction of fish life in the Dippool Water and nearby Douglas Water. At the collieries concerned (Woolfords Colliery and Douglas Colliery) large volumes of water from an ironstone area were normally pumped. During the strike the mine workings filled with water which became acid, took up iron in solution and, when pumping was resumed, caused severe pollution. The polluting character of these discharges can be gauged from the analyses made at the time in the county laboratories (County Council of Lanark, 1924) as shown in Table 8.1.
The analyst commented that the samples from Woolfords Colliery were quite different from any sample previously obtained, and that this was the first time that any free acid had been detected in a pit water.

As a result of these incidents the Public Health Committee drew attention to the difficulty of dealing with pit waters and resolved that the County Council should approach the Secretary for Scotland and urge him to have legislation introduced to amend the law so that pollution of this nature could be effectively dealt with.

Judging from contemporary reports the impact of mining on rivers in the central belt of Scotland remained essentially unchanged until the setting up of the River Purification Boards in the mid-1950s. The best sources of information are to be found in the reports of the Scottish Advisory Committee on Rivers Pollution which was set up in 1923. For example their

Table 8.1 — Analyses of pit waters discharged to Douglas Water and Dippool Water (1921). Results in parts per 100 000

	Douglas Colliery	Woolfords Colliery
Total iron in solution	4.80	128.00
Iron in suspension	4.80	7.80
Total iron	9.60	135.80
Suspended solids	8.90	15.40
Oxygen absorbed	0.685	15.52
Alkalinity as $CaCO_3$	Acid	Nil
Acidity as $CaCO_3$	—	680
Free H_2SO_4	—	22.5

fifth report (1935) showed that the greatest volume of polluting material entering the River Almond was derived from working and abandoned collieries which numbered 26, including a fireclay pit. It was calculated that over 12 million gallons (54.5 Ml) of pit water and effluent from coal washings were discharged daily to the river and its tributaries, and that the ratio of colliery discharges to rain water within the area was for all practical purposes only 1 to 2. A further 2 million gallons per day (9 Ml/d) was derived from shale mines further down the catchment.

The report again stressed the deficiency in the 1876 Act in that 'water in the same condition in which it has been raised or drained from the mines . . . is excluded from the scope of the Act'. The report went on to say 'With regard to pit water, we are convinced that the Almond will continue to be grossly polluted until the law is amended to bring this within its scope. Owing to the clayey nature of the strata in which the collieries are situated the water pumped or raised from the mines is in many instances highly polluted by matter in suspension, although in other cases it is more or less free from pollution and is used for boiler-feed and other purposes. In addition, large volumes of pit water heavily charged with oxide of iron are pumped from the pits or rise from old workings, particularly in the Lanarkshire area and in the Fauldhouse district of West Lothian. This water is in some cases running quite brown on leaving the pit. In others it is perfectly clear, the iron being in solution in the form of salt, but it is soon oxidised on contact with the water in the stream, producing the typical brown oxide. This colour persists right down the Almond, and is only relieved to some extent by the dilution afforded by the stream and by addition of further pollutions. It is a matter of great difficulty whether the onus of purifying should, in view of the exclusion of pit water under the present law, be put on the mine owner or the land owner (if the mine is closed) or the local authority. Meantime we reserve our opinion on this matter'.

The report showed that regular inspections of colliery effluents were carried out, as many as 16 or 17 samples of effluent from each colliery being examined by the Lanarkshire river inspectors during 1931. Where pollution was found immediate representations were made to the company concerned. However, the report noted that the Lanarkshire Coalmasters Association considered that the standard for coal-washery effluents of 4 parts per 100 000 for suspended solids was too stringent and they thought that a figure of 15 might be agreed by the colliery owners.

8.2.3 Pollution from lead mines

The 1924 report of the County Council of Lanark stated that the mines at Leadhills, together with those at nearby Wanlockhead in Dumfriesshire, were the only lead mines working in Scotland. Complaints were received as to the silting up of the Glengonnar Water and inspection showed that although the Leadhills Company had provided catch pits and settling ponds, the effluent was unsatisfactory, depositing silt of a siliceous nature which was difficult to remove. A sample of the stream water below the outfall in April 1924 contained 434 parts per 100 000 of suspended solids, which contained 3.4% of lead. Although the effluent imparted a bluish-grey colour to the stream for some distance downstream, this did not appear to affect fish life.

8.3 WEAKNESSES IN THE RIVERS POLLUTION PREVENTION ACT, 1876

It is evident from the available literature that, despite vigorous efforts by certain local authorities (notably Lanark County Council), pollution of rivers in central Scotland resulting from coal-mining operations was a serious and worsening problem during the first half of this century. There were four main deficiencies in the 1876 Act.

- Enforcement of the legislation was placed in the hands of local authorities and usually carried out, to the best of their abilities, by sanitary inspectors who had many other public-health problems to cope with. Only Lanarkshire appointed specialist river inspectors for this purpose.
- There was no provision for the enforcement of standards, and the standard for suspended solids (4 parts per 100 000) proposed by the Royal Commission on Sewage Disposal was disputed by the coal owners.
- Prosecution of offenders for causing pollution was such a complex operation that such action was very rarely undertaken, thus making enforcement of the law extremely difficult.
- Mine waters as raised or drained from a mine were specifically exempt from the provisions of the Act. Parliament had not foreseen that such waters might provide the most severely polluting of all discharges.

8.4 PROGRESS FOLLOWING THE RIVERS (PREVENTION OF POLLUTION) (SCOTLAND) ACT, 1951

This Act provided a new impetus for the control of pollution by setting up, for the first time in Scotland, independent River Purification Boards, based on river catchments, whose prime function was the enforcement of the Act.

The deficiencies of the preceding legislation were largely overcome by a much-simplified enforcement procedure, and the ability to enforce effluent standards by a consent procedure. Thus no new discharge could be made to the receiving water without the consent of the river purification authority, which could impose such conditions as it thought necessary for the prevention of pollution.

The new River Purification Boards (RPB), most of which were set up between 1954 and 1956, inherited circumstances which had changed little since the pre-war years. In central Scotland, in the areas of the major coalfields, colliery discharges along with sewage effluents were often the main sources of pollution, and other discharges associated with mineral extraction arose from clay mines and washery plants for sand and gravel.

8.4.1 Pollution by solids in suspension

The most common cause of pollution was due to fine coal and clay solids discharged from coal-washery plants and, to a lesser extent, from pit waters. Despite the fact that most collieries had settlement lagoons, the design, operation and maintenance of these was found to be generally extremely poor, resulting in discharges which ranged from thick slurries to waters containing hundreds or thousands of parts per million of solids in suspension. The upper reaches of the Almond and its tributaries were black with coal solids which blanketed the stream bed, thus eliminating much of the normal flora and fauna, and many streams from Ayrshire to Fife were similarly afflicted. However, thanks to regular inspections and sampling by RPB inspectors, the provision of advice to colliery staff, and the willing co-operation of the National Coal Board (NCB), considerable progress in reducing the impact of such wastes on streams was achieved in the early years. At most collieries the NCB was asked to achieve a standard of 60 mg/l for suspended solids and, although their existing discharges were exempt from the consent requirements of the 1951 Act, considerable efforts were made to try and achieve this. In general the collieries were asked to ensure that adequate settlement capacity was provided, that lagoons were regularly cleaned out, that the banks were properly maintained and that discharge pipes were at the correct level. A striking improvement in the average quality of the River Almond throughout its length, between 1957 and 1961, was attributed as much to a reduction in coal solids as to a reduction in suspended solids from sewage works, and contributed significantly to the early return of fish life to this river (Lothians River Purification Board, 1962). However, it is worth noting that during the decade 1955 to 1965 the data gathered from chemical and biological monitoring increasingly suggested that suspended solids, whether derived from coal washeries or sand and gravel washeries, caused damage, even at very low concentrations, due to sedimentation under low flow conditions and consequent blanketing of the stream bed. The opinion of the boards was turning in favour of closed-circuit systems with no discharge to the river. However, prior to 1965 there was no legislative power to compel polluters to stop existing discharges, though the Lothians RPB (Annual Report, 1961) did manage to persuade the Levenseat Sand Quarry to operate on a closed-circuit system from 1961 with very considerable benefit to the Breich Water and River Almond.

8.4.2 Pollution by acid iron-bearing minewaters

During the first decade of the River Purification Boards there was a rapid increase in discharges of acid ferruginous waters which caused severe pollution problems in the areas of the Forth, Lothians, Clyde, and Ayrshire boards. The problem was almost entirely associated with the closure of worked-out mines, but also occurred in open-cast workings and in colliery spoil heaps. The processes by which ferruginous waters are produced have been well summarised by Williamson (Lothians RPB, Annual Report, 1964) who showed that three conditions are essential:

(a) the presence of iron pyrite (FeS) exposed to a supply of oxygen and water;

(b) sufficient water to elute the oxidation products;

(c) insufficient alkaline material in the surrounding strata to neutralise the acidity produced.

Investigations in the USA, cited by Williamson, have shown a complex stepwise process whose rate is determined by the oxidation of dissolved ferrous iron to ferric. An acidophilic iron-oxidising bacterium *Thiobacillus ferrooxidans*, which is found in many mine waters, acts as a catalyst which increases the oxidation rate by five or six orders of magnitude over the abiotic rate. The various chemical and biochemical processes result in the production of ferrous sulphate ($FeSO_4$), ferric sulphate ($Fe_2[SO_4]_3$), sulphuric acid and hydrated ferric oxide (Fe_2O_3 x H_2O).

When ferruginous waters reach a watercourse the rise in pH and presence of atmospheric oxygen brings about a hydrolysis, with the production of the typical orange-yellow precipitate comprising mainly amorphous ferric hydroxide ($Fe[OH]_3$), with variable amounts of Goethite ($FeO[OH]$) and Jarosite ($KFe[SO_4]_2\,[OH]_6$).

Early reports of the Lothians River Purification Board (Annual Reports (1961 & 1962) showed that some damage was caused by such discharges. In November 1959 a major outbreak of ferruginous water from Tynemount Colliery, brought about by the closure and cessation of pumping at an adjacent mine, killed every trace of algae, macroscopic plants and benthic fauna for 5 km downstream, and significantly reduced the abundance of flora and fauna for a further 9 km. A similar discharge utterly ruined the Killandean Burn for a distance of 8 km to its junction with the River Almond which, in turn, was discoloured with iron ochre for a further 5 km. Studies in the Killandean Burn (Lothians RPB, 1961) showed that bacterial action was in part responsible for the deposition process . . . 'Undoubtedly much of the precipitation of iron is brought about by a purely chemical process of oxidation but it is evident on microscopical examination that the deposits on stones and boulders are brought about by the activity of enormous growths of the sheath bacterium *Leptothrix ochraceae*. Both this and *Sphaerotilus natans* have been shown to be growth forms of the same bacterium. *Leptothrix* is stated to be able to obtain energy from the synthesis of organic matter from carbon dioxide and dissolved salts by the biochemical oxidation of ferrous bicarbonate and ferrous sulphate to ferric hydroxide. The ferric hydroxide is deposited in the gelatinous coat of the bacterium and in this way a thick coating is rapidly built up over stones or whatever object the bacterium happens to be growing on'. A remarkable feature of this stream was the appearance in the spring of a brilliant green coating over the whole bed for a distance of several kilometres. On examination this was found to comprise a pure culture of a species of the green flagellate *Euglena*, which flourished at an average pH of 3.0!

8.4.3 NCB policy concerning ferruginous discharges

It is worth noting here that, to begin with, the NCB co-operated extremely well with the RPBs and showed every intention of dealing with ferruginous minewaters, whether these were associated with working or abandoned collieries, and despite the fact that the boards had no legal powers to control them. Remedial measures at two collieries in the Tyne catchment, comprising chemical precipitation and settlement, were comparatively satisfactory and brought a great improvement to the rivers affected (Lothians RPB, 1964). However, with the closing down of increasing numbers of collieries and increasing costs and no doubt acting on legal advice, a change of policy was noted in 1963. During that year both the Lothians RPB and East Lothian County Council became concerned at massive pollution of the Seton Burn and the foreshore at Seton Sands due to the outcropping of ferruginous mine water following the cessation of mining operations in the Tyne catchment. The County Council asked the RPB to insist on adequate treatment for the three discharges, but the NCB replied that any outcropping of water following the closure of a mine and the cessation of pumping must be regarded as surface water and not their responsibility (Lothians RPB, 1964).

8.5 PROGRESS FOLLOWING THE RIVERS (PREVENTION OF POLLUTION) (SCOTLAND) ACT 1965

It has been pointed out elsewhere (Hammerton, 1983) that the 1951 Act provided only the powers necessary to halt the deterioration in river quality by controlling new discharges, while the bulk of existing discharges responsible for virtually all the long-standing pollution problems could not be effectively consented and controlled. The 1965 Act removed this major handicap and provided the RPBs with powers for the control of all existing discharges to inland waters, as well as full or partial control in many areas of tidal waters. During the ensuing decade, which fortunately also coincided with a period of economic growth, substantial progress was made in the restoration of heavily polluted rivers and the RPBs were able to reduce substantially pollution by effluent from mines and quarries.

8.5.1 Control of suspended solids

The majority of these effluents contained only inert solids in suspension derived from coal-preparation plants, a few mine waters and the growing number of sand and gravel washeries. Before 1965 the RPBs had usually imposed a consent limit of 60 mg/l (or occasionally 100 mg/l) on new discharges of this type (Hammerton, 1977), and this limit was generally applied to existing discharges which came under control under the 1965 Act. However, biological surveys were providing increasing evidence that suspended solids at very low concentrations could adversely affect fish spawning sites and benthic fauna by blanketing the stream bed and preventing the free flow of oxygenated water through the underlying gravel. A further problem was that turbidity due to colloidal matter, by interfering with the vision of fish, made the traditional fly-fishing impossible. Moreover, as rivers improved and fish life returned anglers were drawn to the rivers and complained if the water was not clear and sparkling. In the light of these factors, the RPBs developed a long-term policy of requiring closed-circuit operation of treatment plants with full recycling of wash water and no discharge to a watercourse. This is now accepted as the ideal practice throughout Scotland for sand and gravel quarries, open-cast mines and coal preparation plants. However, at such sites provision has to be made for the treatment of contaminated surface water which, following rainfall, usually has to be discharged to a watercourse. Consents for such discharges usually include a limit of 60 mg/l for suspended solids and no visible trace of oil. No difficulty has been experienced with closed-circuit systems provided that proper maintenance is carried out, but large settlement lagoons of more than 100 000 m^3 capacity must be registered with the Inspectorate of Mines and Quarries. In one instance, in the Clyde area, the inspectorate insisted on the installation of a high-level overflow for safety reasons, but this only operates under exceptional conditions.

8.5.2 Control of direct extraction of sand and gravel

A serious weakness in the Rivers (Prevention of Pollution) (Scotland) Acts was discovered when several firms began the direct extraction of sand and gravel from the River Clyde above Lanark, from 1959 onwards. Planning permission having been granted by the County Council, the Clyde RPB found itself powerless to stop or effectively control the operations. By 1965 with as many as four dragline excavators in action severe discoloration of the Clyde was apparent for many kilometres downstream. Even after 1968 when the operations at Lamington were restricted to a single excavator, analyses showed that the suspended solids in the river were unacceptable and caused severe damage to fishery interests. The Board eventually sought and obtained a private Act (Clyde RPB Act, 1972) to control direct mineral extraction and certain other forms of pollution, but the problem was only finally resolved in 1981 after two public enquiries and after the removal of an estimated 3 million tonnes of sand and gravel. This created a lake one mile in length and up to 40 feet in depth, the biggest 'hole' in any river in Britain, the filling of which by natural processes will cause upstream erosion for many years to come.

8.5.3 Problems of ferruginous waters

The increasing problem of acid iron-bearing discharges became so acute that the Lothians RPB advocated national action to combat the problem. The problem was almost entirely associated with the accelerating closure of collieries resulting in two types of discharge:

(a) surface-water drainage from bings (spoil heaps) rich in iron pyrites;
(b) direct gravity discharges from abandoned mines.

The number of seriously polluting discharges increased rapidly from nine in 1969 to 28 in 1973. In the case of working collieries and open-cast mines, where the Board was able to control discharges by means of consents, a general limit of 5 mg/l total iron as Fe was increasingly used (Lothians RPB, 1969). However, with the industry in decline and many mines running at a loss, the NCB was increasingly unwilling to spend money on remedial measures at pits due for closure. For example, in March 1974, the Ayrshire RPB reported the NCB at Dalquharran to the Procurator Fiscal for breach of consent conditions. The NCB was found guilty and fined but subsequently stated that if further proceedings were taken the mine would be closed with the loss of over 150 jobs.

From 1975 onwards increasingly serious problems were reported in Fife, mostly involving discharges from the adits of disused mines, but also from a major new open-cast site at Westfield where the NCB Opencast Executive agreed to provide a full-scale treatment plant (Lothians RPB, 1976; Clyde RPB, 1977). By 1978 surveys by the Forth RPB indicated that Fife had the most serious mine-drainage problem in Britain.

In the area of the Clyde RPB problems of ferruginous waters were also increasing as a result of pit closures. In 1976 the Procurator Fiscal at Ayr Sheriff Court was asked to take proceedings against the NCB in respect of the discharge from the abandoned Kames Colliery at Muirkirk, affecting some 11 km of the River Ayr. It was believed by the Board that the fact that a pipe, some 300 m in length from the pit bottom to the surface, had been left in place at the time of abandonment could be regarded as 'causing' a polluting discharge to enter a watercourse. However, in 1978, after consulting Crown Counsel in Edinburgh, the Procurator Fiscal decided against taking proceedings. He took in to account all the circumstances of the case and, in particular, the provisions of Section 31(2) of the Control of Pollution

Act 1974 which, when implemented, would appear to permit, for the first time in Scotland, the discharge of waters from abandoned mines (Clyde RPB, 1977).

8.5.3.1 Pollution from Dalquharran Colliery

In October 1979, the most serious pollution incident yet recorded in Scotland occurred when acid mine water from the Dalquharran Colliery at Dailly, Ayrshire, broke out from the old mineshaft two years after the mine was closed, and effectively ruined one of the best salmon rivers in Ayrshire. Within 24 hours all fish in the 16 km stretch of the river downstream to the coast at Girvan were killed, including many salmon up to 6.8 kg in weight. The high levels of iron in the river also threatened the closure of Alginate Industries' factory at Girvan which employed nearly 500 persons, while complaints were subsequently received from shipowners in Girvan Harbour claiming that the water was corrosive to ships' hulls. Because of its unique features and the subsequent prosecution of the National Coal Board, which made legal history in Britain, a summary of the case is presented here. A full account of the case will be published elsewhere (Hammerton, in preparation).

The Clyde RPB, anticipating that problems would occur after closure of Dalquharran Colliery, took out a Sheriff's Order in 1976 requiring that any discharge from the land occupied by the NCB at Dalquharran must not, on neutralisation to pH 7 to 8, contain more than 60 mg/l of suspended solids. The NCB did not oppose the Order and also stated that they did not think there was any risk of pollution after closure. However, the Clyde RPB installed equipment in a nearby shaft and monitored the water level which rose steadily after the cessation of pumping. In September of that year the Board's hydrologist forecast that the water would break out around December 1st, but in the event it broke out on October 21st following a period of heavy rainfall. However, the volume and polluting nature of the discharge far exceeded what could have been anticipated. The initial flow at a rate of 2.27 Ml/d increased during 1980 (except during the summer months) to a peak flow of around 6.5 Ml/d in December 1980. Typical chemical analyses of the discharge are shown in Table 8.2.

Immediately following the outbreak the Clyde RPB wrote to the NCB requesting that remedial action should be undertaken as soon as possible but, in reply, the NCB stated that because the mine was closed and they had relinquished the lease of the land, they could not consider themselves in any way responsible for the discharge. Accordingly formal samples were taken and the matter was reported to the Procurator Fiscal. The case was heard in July 1980 when the Sheriff of Ayr found the NCB not guilty. The Crown appealed against this decision and, in February 1981, the Justiciary Appeal Court in Edinburgh reversed this decision and remitted the case to Ayr Sheriff Court with a direction to convict. This decision made legal history, it being the first time that the NCB had been held guilty of polluting a watercourse by means of a discharge from an abandoned mine.

Following the verdict the NCB agreed to undertake remedial measures

Table 8.2 — Typical analyses of the Dalquharran discharge†

	October 1979	October 1980	October 1981
pH	4.0	4.55	5.7
Suspended solids	51	53	30
S.S. at pH 8.0	2334	1876	904
Sulphate	5800	4837	2900
Iron	1050	880	475
Aluminium	125	12.5	5.9
Manganese	49	35.7	27.6
Calcium	373	310	330
Magnesium	425	500	325
Zinc	2.2	1.1	0.52
Nickel	3.8	1.63	1.1
Cobalt	2.7	—	—

†mg/l except for pH

and in October 1981 proposed a completely novel solution comprising three main features:

(*i*) a dam at the mine outlet with control works to regulate the flow from the mine;
(*ii*) injection of lime and sodium hydroxide to precipitate the iron deep inside the old workings, and
(*iii*) diversion of all surface waters which were entering the mine.

The ultimate aim of these measures was to meet a requirement by the Clyde RPB that the total iron concentration in the Water of Girvan should not exceed 1 mg/l (Clyde RPB, 1982). Subsequently, a further modification was proposed and executed by the NCB: it was found that water in the upper mine workings close to the surface at Dowhail was almost unpolluted (pH above 6.0 and iron less than 20 mg/l). Pumping trials, followed by the installation of a siphon which enabled the abstraction of 27 Ml at Dowhail without any diminution in quality proved that here was a means of greatly reducing the polluting discharge at Dalquharran. By means of a massive excavation and drilling through rock a gravity-flow pipeline was installed which now removes 95% of the inflow and discharges it to a tributary of the Girvan.

The works were largely completed by the autumn of 1983 and have fully achieved the desired objective. The average discharge has been reduced to a trickle and the iron content has been reduced from a peak of 4000 kg/d to less than 10 kg/d. Despite the severe drought during 1984, reducing the available dilution, the river has throughout 1984 met the Clyde RPB's requirement and is now in superb condition. The NCB donated £1000 to the Girvan Fishery Board in 1982 for restocking the river with salmon fry (*Salmo salar*) which was carried out in 1983. Subsequently there have also been two good runs of salmon with large specimens being recorded in all parts of the river. A joint working party set up between the Clyde RPB and the NCB in 1981 proved such a valuable means of co-operation that, following the successful conclusion of the work at Dalquharran, a permanent liaison group of officials has been set up (Clyde RPB, 1983).

8.5.4 Impact of other mining operations

No mining operations other than those described above appear to have created significant water-quality problems in Scotland. The closure of a fire-clay mine at Monkcastle, Strathclyde, in 1981 led to a resurgence of acidic water containing 70 mg/l of iron which caused a very localised deposition of ochre in the River Garnock during 1982 (Clyde RPB, 1982). This is another example, frequently encountered in coal mining, where the problem of ferruginous waters has only arisen after closure.

The last remaining lead mines in Scotland, at Leadhills in the headwaters of the Clyde, closed in 1968. Several mine-drainage waters from the area of the old mines discharge into the Glengonnar Water, the flow from the mines being several times that in the stream. A recent study of the stream (Doughty, 1985), shown that elevated levels of lead are present in the stream averaging 148 μg/l (range 80–229) decreasing to 18 μg/l at the junction with the Clyde 7 km downstream. Although trout (*Salmo trutta*) are present throughout, specimens examined near the point of discharge showed typical abnormalities associated with lead poisoning. These included asymmetrical blackening of the caudal region and spinal deformities. Lead concentrations in the fish tissues reflected those in the ambient water.

Three barytes mines are currently working in Scotland (Harris *et al.*, 1984) and others may recommence working to meet the demand for heavy muds for gas- and oil-well drilling. No problems have been encountered and the heavy inert nature of the mineral suggests that pollution problems are unlikely to occur.

8.6 PRESENT-DAY PROBLEMS AND LEGISLATIVE NEEDS

In the 20 years since the implementation of the 1965 Act major progress has been made in restoring Scotland's rivers to a reasonably satisfactory condition after a century of neglect and inadequate legislation. The inert solids from mine waters and coal-preparation plants, and, more recently, from sand and gravel washeries (which feature as one of the two main causes of river pollution in reports covering the first 60 years of this century) are a thing of the past. Indeed it is now increasingly difficult for students at colleges and universities to be shown examples of chronic pollution whether by inert solids, inadequately treated sewage, or oily discharges.

In retrospect, although the methods of controlling pollution of this nature were recognised in the reports of the Rivers Pollution Commission of 1868, and particularly in the work of the Royal Commission on Sewage

Disposal and other reports of the day, the development of legislation and an effective administrative structure for its implementation has been an extraordinarily slow process. Thus only in 1965 were all the necessary ingredients attained for effective control of the main sources of pollution, that is independent river authorities with powers to enforce standards.

The importance of the role of legislation in controlling pollution is well illustrated by the contrast between the ability of river purification authorities to control discharges from working mines and their almost total inability to deal with discharges which are brought about by the cessation of mining. The point is further illustrated by the differences in legislation north and south of the Scottish border. Unlike its English counterpart, the Rivers (Prevention of of Pollution) (Scotland) Act, 1951, did not exempt from control 'minewater in the same condition that it is raised'. The Scottish boards have always had the ability to control discharges of pumped mine water although not always as successfully as they would have wished, particularly where the mines were already uneconomic and moving towards closure.

The problem of ferruginous discharges, as seen in the foregoing account, has become a major problem in central Scotland and this has been recognised by the Scottish Development Department (SDD) and the Commission on Energy and the Environment which have both published reports.

In 1983 the SDD published the results of its survey (SDD, 1983) which showed that in the areas of the Clyde and Forth RPBs there were 62 significant discharges of ferruginous waters (Fig. 8.1). Of these, eight were from active mines, two from open-cast workings, 25 from abondoned mines, 17 indirectly associated with mining, seven from colliery spoil heaps, a further seven from land drainage with no recent mining in the vicinity, and three from rubbish tips. The daily flows ranged from 10 m^3 to 31 400 m^3, and the average iron content from 2 to 500 mg/l. The total flow amounted to 230 000 m^3/d containing 6139 kg of iron as Fe or over 2200 tonnes per annum. The survey showed that some 50 to 60 km of watercourses were moderately or severely affected. The worst pollution was associated with recent outbreaks but problems were recorded for some 15 discharges from mines abandoned 80 to 100 years ago! Since the report was published further outbreaks have occurred and it seems likely that the scale of the problem will increase.

A report published by the Commission on Energy and the Environment (1981) made the following comment on this matter.

> In the meantime no action is being taken by the water authorities or the NCB in remedying problems or in dealing with past or present owners of abandoned mines. The NCB take the view that effects on watercourses are usually localised and unlikely to justify the expense necessary to control any problem. They accept that consents are required where waters are pumped from otherwise unworked mines to protect adjacent workings. There is an urgent need to establish the extent of the pollution and the range of costs of the necessary remedial treatment. The question of responsibility should also be settled. We recommend that discharges from mines abandoned by the predecessors

> of the NCB, or even the NCB some long time ago, should be considered as a form of dereliction comparable to abandoned pit-heaps. We recommend that provision should be made through central government to meet the costs of any essential work, as is the case with derelict land clearance schemes.
>
> Throughout this Report we stress the need to avoid the creation of future dereliction by the proper planning and after-care of new developments. The closure of pits, as we also stress, is an integral part of the mining industry. Thus we recommend that when a pit is to be taken out of operation there should be close co-operation with the water authorities to assess and plan for the effects of the cessation of pumping. The NCB should carry the costs of any necessary preventive or remedial measures. Where appropriate, conditions should be imposed as part of a planning consent. We also recommend that Section 46(6)(b) of the Control of Pollution Act 1974 should be amended to ensure that the powers of the water authorities under Section 46(5) to recoup the costs of remedial works should apply in respect of mines being closed by the NCB.

In the light of these recommendations the government considered the establishment of a working party to look into the question of pollution from abandoned mines, but decided that it would be advisable to establish first the full scope and nature of the problem by undertaking a fact-finding study. Accordingly the SDD requested a more detailed survey by the RPBs to provide accurate data on flow rates, chemical composition, impact on receiving waters and, where possible, estimated costs of remedial measures. The survey was, of course, a national one and parallel surveys in England and Wales were sought by DoE and the Welsh Office.

The Scottish survey was completed early in 1984, with some financial assistance from the Scottish Development Agency, information being supplied by the Clyde (Clyde RPB, 1984), Forth (Forth RPB, 1982), Solway, and Tay RPBs. Altogether about 140 discharges were found, but only about 20 were identified as being of priority for treatment, though it was recognised by the RPBs that if these were effectively dealt with public pressure for others might increase.

At the time of writing there is no sign of further action towards implementing the recommendations of the Commission on Energy and the Environment. No doubt the question of costs weighs heavily with the government and it must be recognised that ferruginous discharges can be particularly costly to treat, while sludge disposal poses a special difficulty.

The Scottish Development Agency funded two studies, at the former Kames and Dalquharran Collieries in Ayrshire, to establish the feasibility and costs of treatment. At Kames the capital cost of a full-scale plant comprising aeration, biological filtration, settlement, and incorporating separate drying beds was estimated at £170 000 with annual operating costs of £1250 (Best & Aikman, 1983). Here the total daily discharge of iron as Fe amounts to only 50 kg. At Dalquharran, where the daily discharge was estimated at 1250 kg, the capital cost was estimated at about £1 million, while a pipeline to the coast into deep water was estimated at over £2 million. These schemes were both rejected by the NCB as being too expensive and the solution adopted (described earlier) has cost about £220 000 and involves minimal running costs. However, this type of solution is not applicable in the majority of cases. The NCB estimated that for most discharges the total costs of treatment by aeration ranged from 10 to 25 p/m^3 mine water at 1980 prices, but for discharges requiring chemical precipitation the costs would be in the range of 10 to 100 p/m^3 (Best & Aikman, 1983).

8.6.1 Control of Pollution Act, 1974, Part II

While generally welcoming the extended powers available under this Act, the Scottish River Purification Boards are concerned at the probable effect of Section 31(2), which would appear to protect for the first time in Scotland the act of permitting water from an abandoned mine to enter controlled waters. In 1982 the Forth RPB and Scottish Anglers Co-operative were received by the Minister of State for the Environment to discuss this problem but in the event no change has been made (Forth RPB, 1982). However, according to the Scottish Development Department, the scope for prosecutions of the Dalquharran type will remain (SDD, 1983). Moreover the recommendation of the Commission on Energy and the Environment that Section 46(6) should be amended [to ensure that water authorities would be able to recoup the costs of remedial works under Section 46(5)] has not been implemented and indeed Sections 46(4) to (7) inclusive were brought into force ahead of schedule. Some boards are therefore concerned that they could be pressurised into carrying out remedial work for which the costs could not be recovered.

8.7 FUTURE OUTLOOK

In conclusion the writer believes that the River Purification Boards are now very well placed in terms of their legal powers to ensure that never again will mining (alongside many other industries) be allowed to damage the aquatic environment. Provided that the boards have the determination to 'conserve in so far as practicable the aquatic resources of their areas' and provided that monitoring and inspection continue to be effective, then there is every likelihood that the rivers will continue to improve. There will undoubtedly be new challenges such as the landward exploitation of oil, but such new industries are not seen as a threat as far as water pollution is concerned. Indeed there is every indication that standards of control will become more stringent in the light of research and monitoring now in progress.

Thus the only area of serious concern is that of pollution from abandoned mines. In the writer's view the government should grasp this nettle firmly and implement as soon as possible the recommendations of its own Commission on Energy and the Environment. It seems unthinkable that this blight on the environment should be allowed to continue and possibly increase in its impact over the next decade. While costs in respect of individual mines may seem excessive, the number of real problem areas in relation to mining as a whole is relatively small. If the whole cost of remedial action was added to the cost of coal (and this has not been suggested by the Commission) the customer would scarcely notice the difference in cost. For example on 1983 production figures a levy of 10 p per tonne would yield an annual income of £11.5 million. The problem is now firmly in the hands of the government which must firstly ensure that funds are made available to carry out remedial work in a phased programme over say ten years, and secondly ensure that the necessary legislative powers are available to Water Authorities to ensure that the same problems do not recur in the future.

REFERENCES

Agricola, Georgius (1556) *De Re Metallica* translated by H. C. Hoover and L. H. Hoover (1950) Dover Publications, New York.

Best, G. A. & Aikman, D. I. (1983) The treatment of ferruginous ground water from an abandoned colliery. *Wat. Pollut. Control,* **82**(4), 557–566.

Clyde River Purification Board (1977) *Annual Report 1977*.

Clyde River Purification Board (1982) *Annual Report 1982*.

Clyde River Purification Board (1983) *Annual Report 1983*.

Clyde River Purification Board (1984) *Pollution from abandoned mines*. A report to the Scottish Development Department.

Commission on Energy and the Environment (1981) *Coal and the Environment.* HMSO, London.

County Council of Lanark (1903) Report on the Administration of the Rivers Pollution Prevention Acts by the County Medical Officer.

County Council of Lanark (1909) Second Report on the Administration of the Rivers Pollution Prevention Acts by the County Medical Officer.

County Council of Lanark (1924) Third Report on the Administration of the Rivers Pollution Prevention Acts by the County Medical Officer.

Doughty, C. R. (1985) In press. Concentration of trace metals in water and brown trout (*Salmo trutta L.*) from a Scottish stream contaminated with mine drainage.

Down, C. G. & Stocks, J. (1977) *Environmental Impact of Mining*. Applied Science Publishers, London.

Forth River Purification Board (1982) *Annual Report 1982*.

Hammerton, D. (1977) Statement to Public Enquiry into Sand and Gravel Extraction from the River Clyde, Lanark, June 1977.

Hammerton, D. (1983) The History of Environmental Water Quality Management in Scotland *J. Instn. Wat. Engrs. & Scientists* **37**, 336–346.

Hammerton, D. (1985) (In preparation) Acid minewater pollution of the River Girvan — A case study.

Harris, P. M., Highley, D. E. & Taskis, D. M. (1984) *Directory of Mines and Quarries*. British Geological Survey.

Lothians River Purification Board (1961) *Annual Report 1961*.

Lothians River Purification Board (1962) *Annual Report 1962*.

Lothians River Purification Board (1964) *Annual Report 1964*.

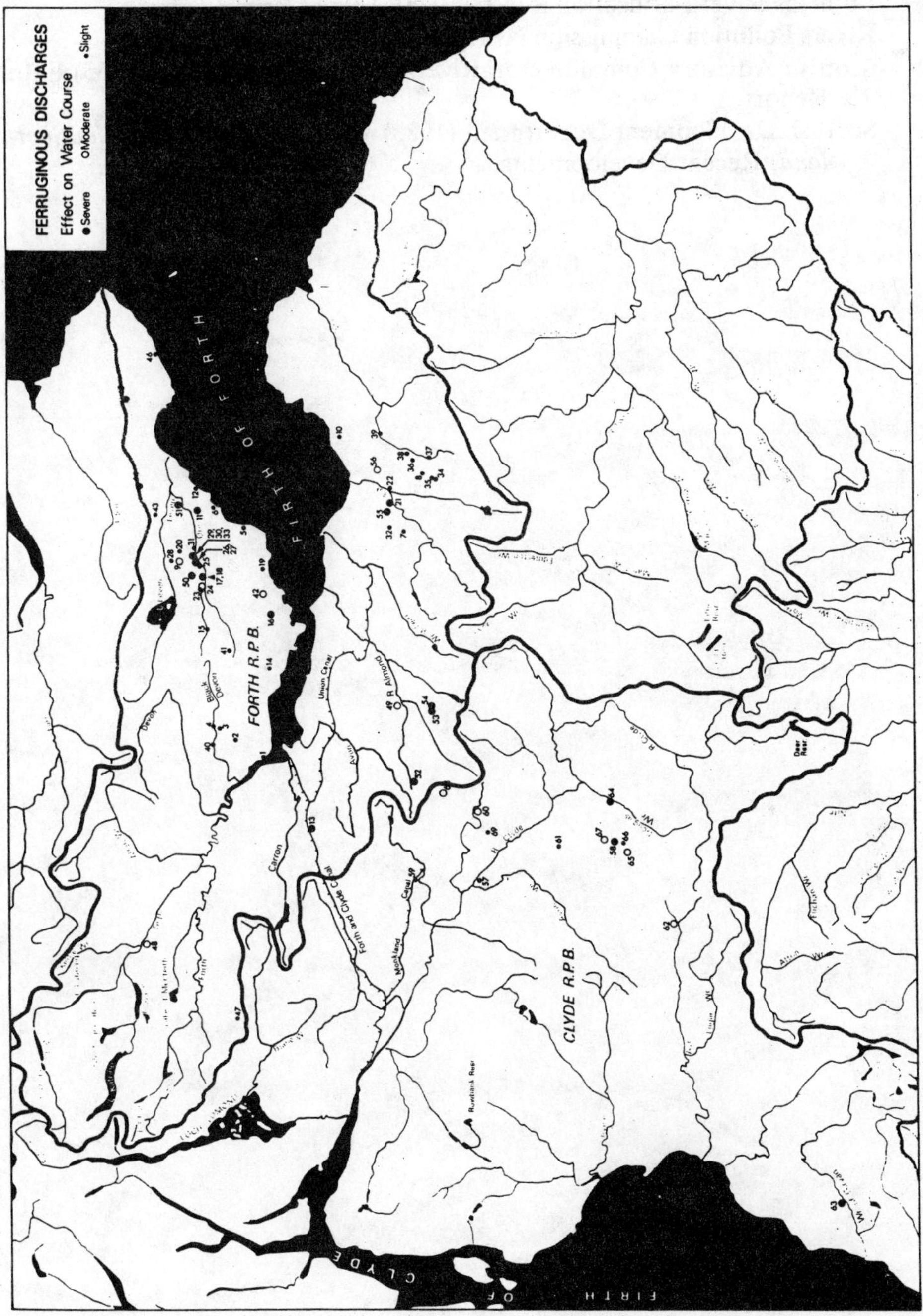

Fig. 8.1 — Ferruginous discharges in Scotland, 1983. (Courtesy of the Scottish Development Department).

Lothians River Purification Board (1969) *Annual Report 1969*.
Lothians River Purification Board (1976) *Annual Report 1976*.
Rivers Pollution Commission (1872) Fourth Report.
Scottish Advisory Committee on Rivers Pollution Prevention (1935) Fifth Report.
Scottish Development Department (1983) *Water pollution control in Scotland:* Recent Developments.

9

Problems of water quality associated with the mining and extractive industries of South West England — some case histories

M. E. Cominetti

9.1 INTRODUCTION

Apart from agriculture, the mineral and extractive industries provide the largest exploitation of land in Cornwall. It is estimated that at least 5500 ha, 1.7% of the total area of the County, is either utilized by active metalliferous mining and china clay production, or has been in the past. Such exploitation can, and often does, have a significant impact upon the environment. This impact can take the form of a visual intrusion or have an effect upon water quality, or both.

This chapter identifies the problems of water pollution in the South West of England and indicates actual and proposed solutions to the problems.

9.2 GEOLOGY

The geology of Cornwall and Devon is such that the mineralisation and kaolinisation is associated with the granitic 'ridge' which runs along the length of the Cornish peninsula. This discordant batholithic intrusion forms five large bosses: Bodmin Moor, Hensbarrow or the St Austell granite, Carnmenellis, the Lands End granite, and Dartmoor. (See Fig. 9.1.) The liberation of super-heated steam and magmatic vapours from the hot granite — hydrothermal activity — gave rise to mineral lodes in fissures in the rock. Kaolinisation is a lower-temperature process in which the feldspars of the granite are attacked, probably by magmatic water and carbon dioxide, and converted into china clay or kaolin. Virtually complete kaolinisation has occurred on a very large scale in the central and western portion of the St

Austell or Hensbarrow granite, although china clay of varying quality is also exploited outside this area and especially on Bodmin Moor, and at Lee Moor, Devon.

The mineralisation in Cornwall can be divided into high and low temperature processes: as the hydrothermal or mineralising solutions or vapours migrated upwards and outwards from the still-molten granite core, the high-temperature minerals (cassiterite, wolfram, copper, and so on) crystallised out where pressure and temperature permitted, whilst the minerals that crystallised at low temperatures (zinc, lead, and so forth) remained in solution until they reached the cooler zones above. The richness of this mineralisation has naturally been, and still is, of considerable economic benefit.

9.3 HISTORY AND ECONOMICS OF MINERALISATION

9.3.1 China clay

Kaolin or china clay ($Al_20_32SiO_22H_2O$) became important during the middle of the eighteenth century, when in a bid to reproduce the quality of Chinese porcelain, William Cookworthy discovered and worked this important product (Derry & Williams, 1960). Thus came the foundation of the china-clay industry in Cornwall which produces one of Britain's most important raw materials for export. Current production exceeds 2.4 million tonnes of which 84% is exported to more than 60 countries. The main users of the product are the paper and ceramic industries which together account for 90% of production. It is also used in the manufacture of rubber, paint and plastics, insecticides and fertilizers. The English China Clays Group employs over 6500 people and is by far the largest single employer in the South West (English China Clays Ltd).

9.3.2 Metalliferous mining

The principal economic mineral of South-West England is tin and the chief ore-mineral is the oxide — cassiterite — commonly known as 'tinstone' (SnO_2). It is estimated that tin oxide has been exploited in Cornwall and Devon since the Bronze Age (1800 BC) when the ore was probably obtained from alluvial gravel deposits (Institute of Geological Science, 1975). More extensive development of the ore desposits took place during the Roman occupation — the Romans already had experience of mining techniques — and it is probable that they introduced underground mining to the region.

Production of tin ranged from about 120 t during the early part of the thirteenth century, to about 2500 t in the early part of the nineteenth century, with a steady increase — probably due to the development of steam-pumping engines permitting mining at deeper levels — to 6000 t in 1850. By 1870, when the tin-plate industry reached a peak, 10 000 t were produced — about half the world yield. Current production from the working mines is over 4000 t per annum as tin concentrate, this material being sent to Capper Pass on Humberside for smelting.

There are also considerable economic benefits and it is estimated that in

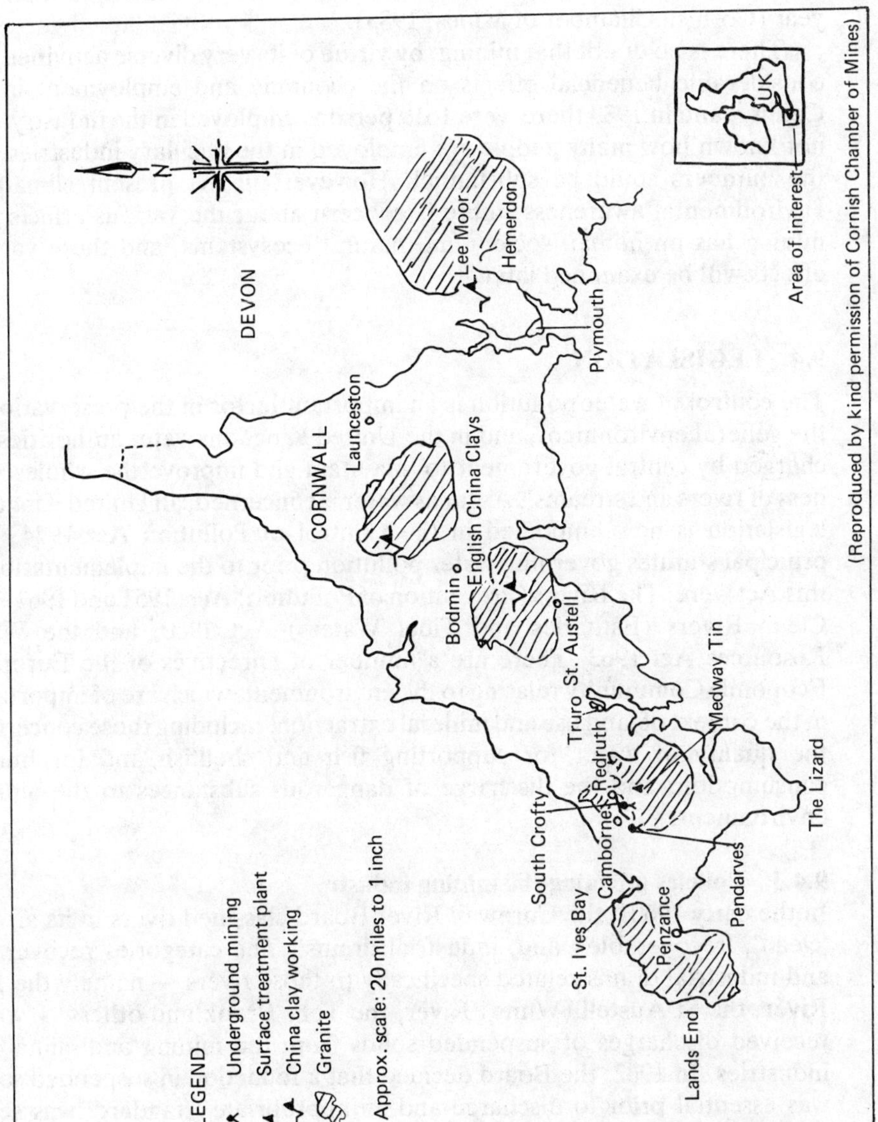

Fig.9.1 — Showing south west granite, china clay and mining areas

1983 the Cornish output of metals, including copper, zinc, silver, and wolfram, saved Britain over £41m in imports of raw materials, and according to the latest figures published by 'Tin International', the Cornish output of tin in 1983 represents about 45% of United Kingdom consumption for one year (Cornish Chamber of Mines, 1983).

There is no doubt that mining, by virtue of its very diverse activities, has considerable beneficial effects on the economy and employment in the County, and in 1983 there were 1619 persons employed in the industry. It is not known how many people are employed in the ancillary industries, but the numbers could be substantial. However, in the present climate of environmental awareness, there is concern about the various effects that mining has on human society and natural ecosystems, and these various effects will be examined later.

9.4 LEGISLATION

The control of water pollution is an important factor in the preservation of the general environment, and in the United Kingdom water authorities are charged by central government to 'maintain and improve the wholesomeness of rivers and streams'. As far as water is concerned, all United Kingdom legislation is now embodied in the Control of Pollution Act 1974. The principal statutes governing water pollution prior to the implementation of this Act were: The Rivers (Prevention of Pollution) Acts 1951 and 1961, The Clean Rivers (Estuaries and Tidal Waters) Act 1960, and the Water Resources Act 1963. There are a number of Directives of the European Economic Community relating to the environment, which are of importance in the context of land use and mineral extraction, including those concerning the quality of water for supporting fish and shellfish and for human consumption, and the discharge of dangerous substances to the aquatic environment.

9.4.1 Policies affecting the mining industry

In the early 1960s, the Cornwall River Board classified rivers in its area as 'clean', 'recoverable', and 'industrial drains'. The categories recoverable and industrial drains related specifically to those rivers — namely the Red River, the St Austell (White) River, the Tory Brook and others — which received discharges of suspended solids from the mining and china clay industries. In 1962, the Board decided that a reduction in suspended solids was essential prior to discharge and an appropriate 'standard' was set at 60 mg/l. However, it was recognised that certain of the industrial drains carried trade effluent and river water in roughly equal proportions with no prospect of recovery, and in the case of the White River no control should be exercised in the quantity of material in suspension so that the industry's expansion could proceed unhindered. Concentrations up to 60% (600 000 mg/l) were permitted (Purves, 1974). In the case of the Red River, discharges of tailings from the mines involved often exceeded 5% (50 000 mg/l).

The Water Resources Act of 1963 created the River Authorities and in 1965 the Cornwall River Authority superseded the Corwall River Board.

Changes in public awareness of environmental matters, complaints of pollution and other factors forced the Cornwall River Authority to re-appraise its rivers policy in 1967. A programme of recovery of all rivers was set and the classification of industrial drains was eliminated. A target of ten years was set to achieve phased recovery, and urgent discussions with the companies concerned ensued. What follows is a review of some case histories.

9.5 IMPLEMENTATION OF POLICY — CHINA CLAY

China clay is 'won' from the host rock by surface-mining techniques, and after stripping overburden, the clay is washed from the pit face — the stope — by powerful remote-controlled water hoses or monitors. The clay stream flows to the bottom of the pit where sand is removed, and the remainder is pumped to the surface for dewatering and refining. The sand forms the waste tips so prominent on the Cornish landscape and the visual intrusion on the environment is self-evident. Further waste material consists of micaceous residues and fines which find their way into rivers and streams. Two most interesting case histories can be outlined.

9.5.1 The Tory Brook

For many years prior to 1967 the Tory Brook was used as a drain for the china-clay working at Lee Moor near Plymouth. The stream was grossly polluted and carried heavy concentrates of clay and sand. Naturally, the stream was entirely fishless. Perhaps of more importance was the fact that the accumulation of silt caused the stream to burst its banks frequently, flooding homes, land and roads leaving everywhere a coating of obnoxious grey slime. Of particular hazard to health was the gradual silting up of the Laira (Plym) Estuary with subsequent failure of surface water and sewage from the City of Plymouth to drain away. Negotiations took several years and the remedy, when agreed, was planned in three phases.

The first phase required the dredging of nearly 4 km of the Laira Estuary from the new Laira road bridge to the confluence of the Tory Brook and the Plym (Fig. 9.2). This was to be conducted in two stages including the reclamation of some 25 ha alongside the river. The second phase involved widening, regrading, and realigning the Plym itself, together with retaining walls, provision of flood embankments, and the diversion of numerous outfalls. The third phase was to regrade the Tory Brook and to provide embankments and new bridges. The work was carried out by the Cornwall River Authority as no satisfactory tender was obtained. Perhaps the dredging operation itself is unique as a specially built dredger was purchased and was one of the first of its kind in the country. Named the 'Sandpiper' she was capable of discharging 200 t of silt an hour through a 25-cm floating pipeline for up to 800m (Fig. 9.3). The first 2.5ha of land had been reclaimed by March 1969.

Naturally if the dredging and other land-drainage operations were to be successful, the discharge of clay residues to the stream had to be stopped, and this operation was undertaken by English Clays Lovering Pochin and Co. Ltd in parallel with the dredging scheme.

The operation, known as the Portworthy Dam scheme, was completed by November 1967 for the settlement of residues from the china clay complex at Lee Moor. Associated with this was the construction of a stilling pool (the Coleland bridge stilling area) for the removal of sand from the Tory Brook washed in from the china clay workings upstream (Fig. 9.4). Soon after completion the mean value for the concentration of suspended solids had reduced from approximately 1700 mg/l to 100 mg/l (Fig. 9.5), and it became commonplace to see the gravel of the stream bed which had been hidden from view!

Ironically, whilst the river was classified as an industrial drain, no complaints were received regarding its quality. However, within a short period of time following improvement, complaints of pollution were made when discoloration occurred following heavy rain! Currently, the discharge from Portworthy Dam is consented at 100 mg/l suspended solids, although the discharge is consistently well below this (Fig. 9.6).

By these means a number of important environmental improvements were achieved and some 12 km of river are flowing free of the heavy concentration of suspended solids. The Tory Brook has been free of the flooding problems encountered in the past.

9.5.2 The St Austell (White) River

Mention has already been made of this river and the original designation of industrial drain was simply recognition by the Cornwall River Board that a considerable tonnage of waste material was being discharged to the river and had been for several hundred years. For example, some 2500 t/d of solid matter was carried in suspension, so that the river channel had become silted to an alarming extent, raising the bed in places almost 2 metres above the original level. Restriction of the channel caused severe flooding in low-lying parts of St Austell and in agricultural areas adjoining the length of the river. Whilst is was no wish of the Board, and later the Authority, to restrict the expansion of the china clay industry, problems occurred at Pentewan Beach (over which the river out-falls) concerning complaints of large areas of beach affected by deposited material. This material was the finer residues and formed dangerous glutinous areas considered to be potentially hazardous to the very young or the older person.

In order to establish the condition of the beach, weekly samples were taken of beach material roughly between the high and low water marks using a grid system. The beach was classified as: *clean* — firm sand; *tacky* — sand with traces of slime, easy to walk on, not dangerous; and *soft* — soft slimes, dangerous. To overcome this last problem, consents were reviewed to restrict the tonnage of 'fines' to that which had been discharged prior to

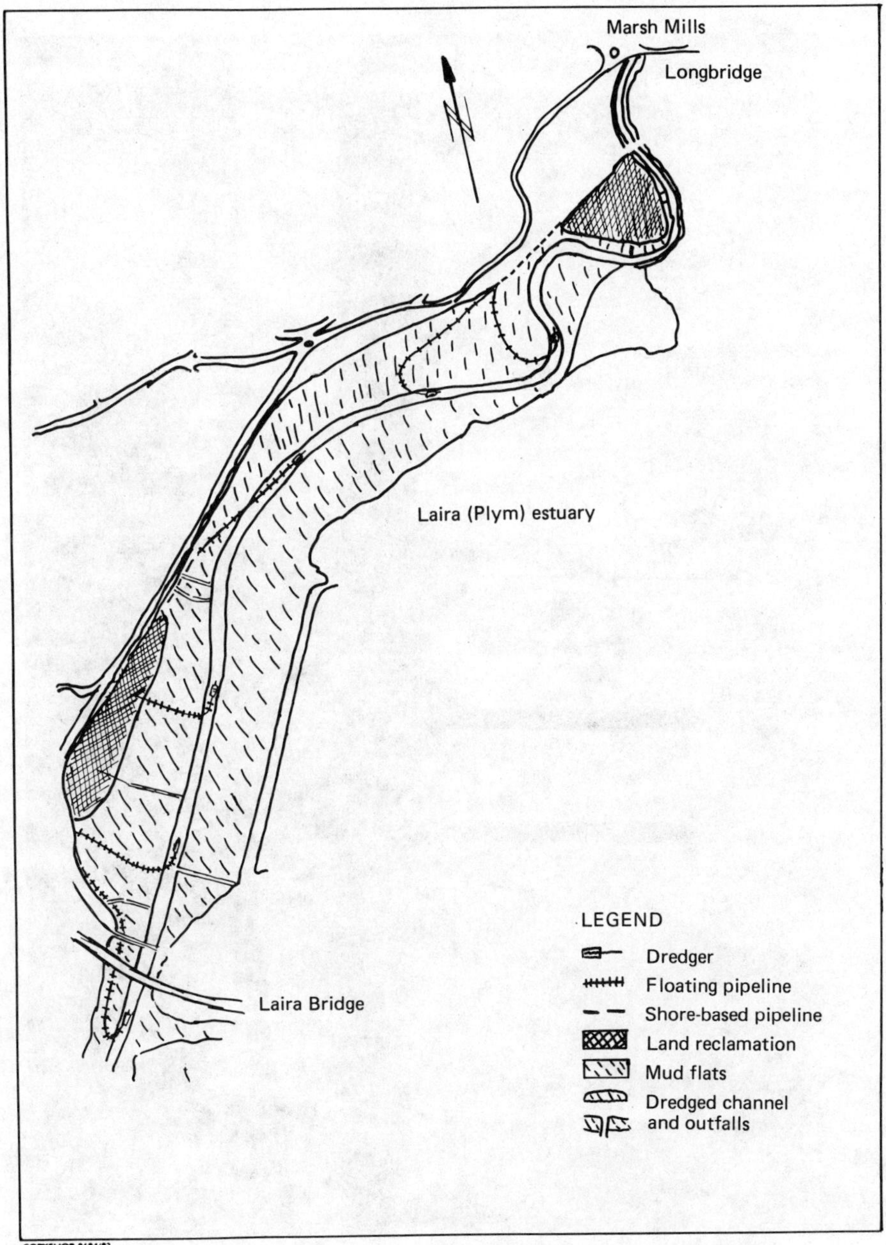

Fig.9.2 — Cornwall River Authority Laira dredging and reclamation

Fig. 9.3 — Cornwall River Authority's new 10″ cutter/suction dredger 'Sandpiper' on the Laira estuary, Plymouth. Silt from the river bed is drawn inside the rotating cutter head in the foreground and is pumped ashore through a floating pipeline to reclaim land at Embankment Road, Plymouth.

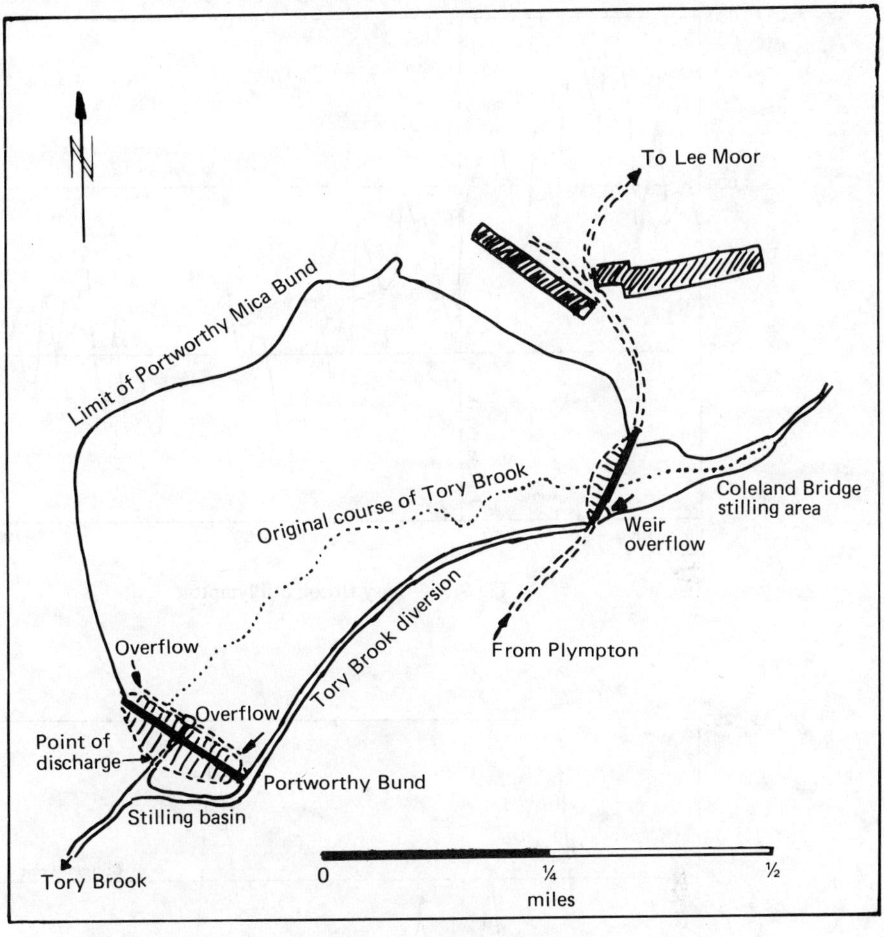

Fig.9.4 — The Portworthy Mica Bund was commissioned in November 1967 as a major scheme to improve control of effluent from the Lee Moor area which discharges to the Tory Brook.

1967. It subsequently emerged that the industry had retained consultants to advise on residue disposal.

The residue scheme that was subsequently brought into operation will be outlined in detail in Chapter 10; suffice it to say that the change was instantaneous and remarkable. Modesty may prevent Mr Comyns from mentioning the fact that the scheme was entered for the Gold Medal Awards to Industry given to encourage the avoidance of water pollution. It was awarded a Silver Medal as a highly-recommended entry and gained an Angling Foundation Certificate of Merit.

There is no doubt that this is an environmental success story — so much so that the word 'white' has been deleted from the name White River and St Austell River substituted. The beach at Pentewan is frequented by thou-

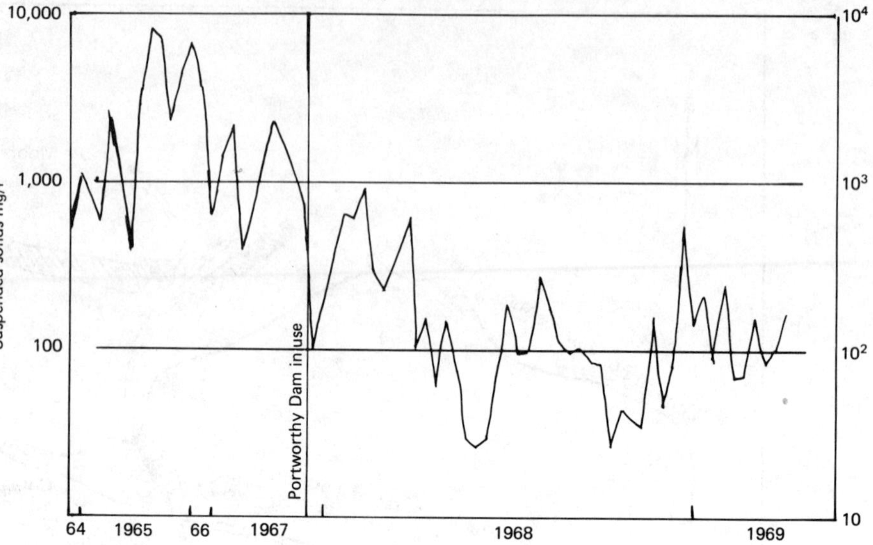

Fig.9.5 — Tory Brook at Plympton.

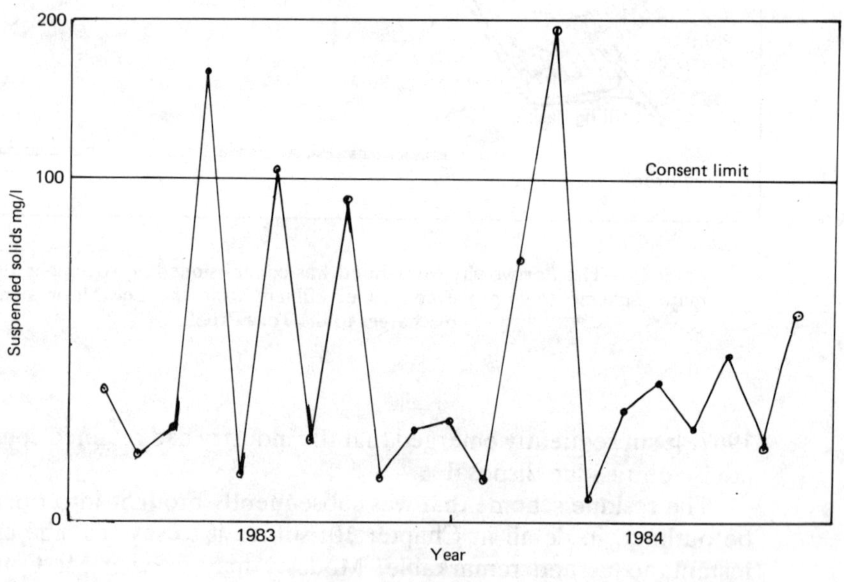

Fig.9.6 — Portworthy Dam discharge control point.

sands of holidaymakers each year in perfect safety, and in fact a large holiday camp has been established by the mouth of the river. However, as with the Tory Brook, the St Austell River suffers from discoloration during wet weather due to polluted surface runoff from the waste tips, china-clay sand roads, and other disposal areas. To overcome this problem of discoloration, work is progressing to provide settlement areas to intercept contaminated runoff and to provide chemical treatment if necessary. Waste tips are now terraced and measures taken to promote growth of vegetation, which prevents erosion and also acts as a filter for contaminated runoff. Whilst much of this control is now legally enforceable by virtue of the Control of Pollution Act, the author is happy to record that agreement was reached with the China Clay Association many years ago regarding such matters on a general 'good-housekeeping policy'. Similarly, a code of practice was produced in 1974 regarding the constructional design of new channels primarily to prevent erosion.

9.6 IMPLEMENTATION OF POLICY — METALLIFEROUS MINING

With the exception of the Hemerdon Tungsten Mine Project near Plymouth, where the ore body can be mined by open-cast methods, the metalliferous mines centred around West Cornwall are traditional underground workings. The grade of tin ore worked is relatively low — between 0.9 and 1.0% Sn, and is processed or dressed on the surface. Ore is dressed for the removal or separation of waste or gangue material, and to increase the concentration of the required mineral. Preparation usually starts with crushing and grinding of the rock; then size classification, separation, and finally smelting and refining.

In Cornish mines, cassiterite is probably the most important mineral and final separation is achieved by a table concentrator (a slightly-inclined plate on which the lighter material is separated from the heavier by a thin stream of water) in which the separating action is augmented by an oscillating motion applied to the table (the shaking table). Further separation is possible by froth-flotation techniques. These processes use large quantities of water and waste material from the shaking tables and flotation — so called tailings — can cause considerable environmental hazard.

Similarly, water drained or pumped from the mines can be contaminated by dissolved toxic metals and can likewise affect the aquatic environment, Examples of unique case histories are given in sections 9.6.1 and 9.6.2 below.

9.6.1 The Red River

The river derives its name from the discharge of tailings material containing high concentrations of iron (as haematite, Fe_2O_3, which often occurs as an accessory mineral in hydrothermal veins). The river rises in the Carnmenellis granite, runs north-east then approximately west and outfalls in St Ives Bay having travelled some 14 km (Fig. 9.7). It is the only example of an industrial drain left in the South West although the river is classified as recoverable in accordance with previous policy. In 1972, the Cornish Chamber of Mines was requested to expedite the completion of the feasibility study for the disposal of tailings from the mining of tin. Consents under the provisions of the Rivers (Prevention of Pollution) Act 1961 were to be issued in respect of the two major discharges, giving a defined period of time for the required improvements. Conditions of Notices were issued limiting the discharge to 2000 mg/l within two years, and then to 60 mg/l suspended solids in line with policy.

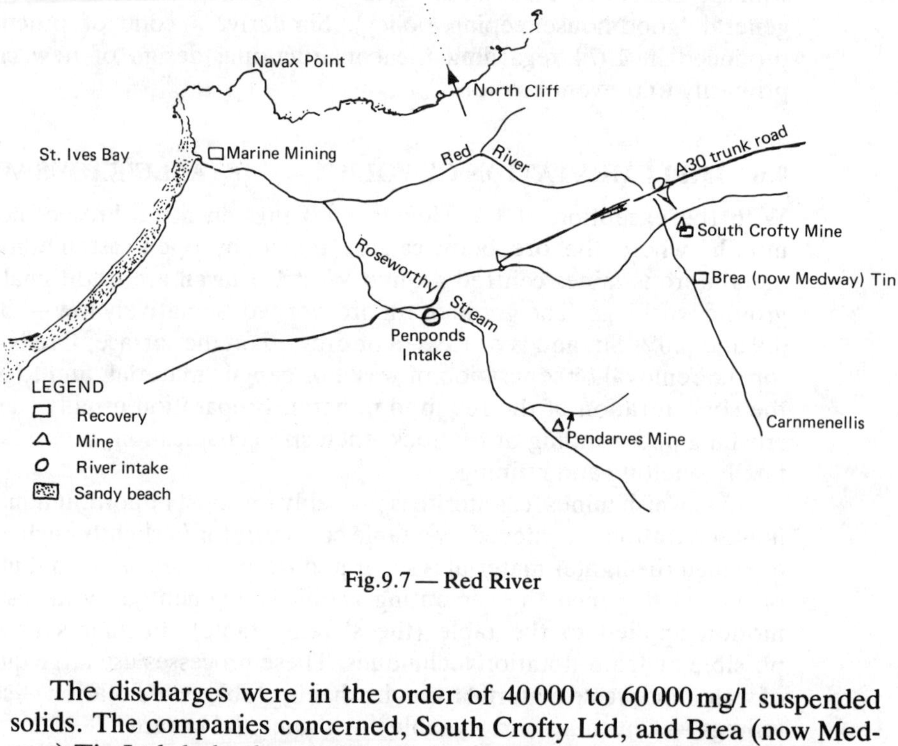

Fig.9.7 — Red River

The discharges were in the order of 40 000 to 60 000 mg/l suspended solids. The companies concerned, South Crofty Ltd, and Brea (now Medway) Tin Ltd, lodged appeals with the Secretary of State for the Environment under Section 6 of the Rivers (Prevention of Pollution) Act 1961 against the Conditions of Notices. The River Authority's case centred on its commitment to 'maintain and improve' the wholesomeness of rivers in its area, and the serious environmental effect the river was having on the beach at Gwithian. The companies were concerned with the cost of a disposal scheme, the construction of a large tailings lagoon being the only available option, and the fact that the financial burden would be such that the mines would probably have to close, with subsequent loss of jobs.

The appeal was heard on 26, 27 and 28 November 1974, and the decision of the Secretary of State made known two years later. The appeal was upheld and discharge from both companies was limited to 80 000 mg/l suspended solids. Taking volume of effluent into account, South Crofty Ltd was, and is, entitled to discharge 750 t/d (as dry solids) and Medway Tin Ltd 30 t. With a maximum permissible discharge of 780 t/d into the Red River, it is no wonder that the river is devoid of life below the two discharges and probably one of Europe's most polluted rivers.

In the last two years, following many complaints from holidaymakers and others regarding the condition of the beach and the staining of the sea for considerable distances, the author, on behalf of the present Authority, is again seeking improvement. Fortunately, ownership of the South Crofty mine passed to Rio Tinto (RTZ) in November 1984; the fact Rio Tinto means coloured river is entirey coincidental! It is pleasing to record that agreement has been reached with RTZ to eliminate the discharge of tailings from the river, and negotiations are proceeding as to the best and most economical method to adopt. Certainly one proposal is to pump the tailings to sea through an existing outfall tunnel at North Cliff (see Fig. 9.7), and hydrographic and other studies are to be conducted to see how far out the outfall needs to be to provide safeguards for local fisheries and the environment generally.

Disposal underground and on land is being evaluated, but it is likely that disposal on land again can be discounted on the grounds of cost and the fact that some 100 ha of lagoon would be required to ensure the productive life of the mine. It is the author's opinion that providing any obstacles can be overcome (including compliance with EEC Directive 76/464, that is, Discharge of Dangerous Substances to the Aquatic Environment) sea disposal is perhaps the one that carries the best hope for the future and could be considered the Best Practicable Environmental Option.

9.6.2 Pendarves Mine (Rio Tinto Zinc)

The existence of this mine near Camborne is perhaps testimony to the fact that metalliferous mining and public water supplies can be compatible. In 1966 an exploratory shaft was sunk and it was intended to extract and process tin ore on site. After treatment in tailings lagoons the effluent would be discharged to a clean tributary of the Red River known as Roseworthy Stream. As the stream was abstracted by the South Cornwall Water Board some 2 km downstream at Penponds (Fig. 9.7), it became clear that strict control of any discharge would be required. It was known that haematite was present in the area which would impart a reddish appearance to any effluent discharge and have similar effects on the watercourse. In May 1973 Pendarves mine was acquired by South Crofty Ltd, and ore was processed in the South Crofty mill. The discharge was restricted to shaft waters contaminated by the activity underground and the Conditions of Consent regarding quality were stipulated appropriate to the production of potable water.

In fact, the limits were the then World Health Organization International Drinking Water Standards, it being felt that with a 5:1 dilution both the stream and the potable- water abstraction would be protected. (These Standards have been met by chemical treatment of the water in five lagoons set in series, using ferric sulphate and a low monomer polyelectrolyte with lime neutralisation.)

Apart from some complaints by riparian owners of red deposits in the stream (re-precipitation of the dissolved iron was believed to be the cause) the Conditions of Consent have invariably been achieved and for 11 years or so mining and water suppy have successfully co-existed. Even greater demands were made on this source of water during the drought of 1984.

9.7 GENERAL DISCUSSION

The preceding examples have illustrated the effect on water quality and the environment generally of active mining in the region. Unfortunately, mining for metals and extraction of other materials has left a legacy of land despoliation which is not only visually displeasing but, when disturbed, can often have very unforseen consequences. In this context, warnings to farmers have been issued by the Ministry of Agriculture, Fisheries and Food regarding soil poisoning as a result of ploughing up grass in old mining areas for arable cropping. Copper and arsenic toxicity have shown the dangers of bringing toxic soil from below grass-root depth. Problems in crops and livestock due to an imbalance of trace elements in soils and pasture are now recognised. Both deficiencies and excesses may give rise to serious problems such as crop failure or death of grazing livestock.

In another instance, hundreds of fish died in a stream and a trout farm through high levels of zinc. It was subsequently discovered that sand purchased from an old lead and zinc mine was used as a base for an impermeable membrane to prevent organic pollution from a refuse tip! Zinc was leached from the sand by rain-water passing through and finding its way to the stream. Removal of the contaminated material cost over £25 000, not including third-party settlement for loss of fish.

However, legacies from the past can be rewarding, especially in the case of tin mining, where losses of tin in tailings averaged 30%. Past mining in the Red River catchment has resulted in a layer of mineral sand some 90- to 150-cm thick in St Ives Bay (Fig. 9.7) containing on average 0.15% Sn and estimated to be in the order of 20 million tonnes. Planning permission has been given to Marine Mining Cornwall Consortium Ltd to suction dredge this mineral for processing ashore.

9.8 CONCLUSIONS

There can be no doubt that the exploitation of land by mining and related industries can have serious impact upon the environment generally, and water quality in particular. However, this impact can be minimised, and often eliminated, by a spirit of co-operation and goodwill between industry and the regulating authorities, and rarely is it necessary to use legal enforcement.

Often there is a complicated interplay between economics, politics, and the environment, and in this context it may be interesting to note the Minutes of Evidence of the Royal Commission appointed in 1860 to inquire into Salmon Fisheries (England and Wales). In the paragraph concerning Cornwall: 'The first question which occurs is, whether it may be possible to reconcile the interests of mines and fisheries, either by diverting the mine water or rendering it by some mechanical or chemical means unnoxious to the fish. Secondly, if this cannot be done, or at least not without an expense which would be unreasonable or prohibiting, and if it became necessary to elect between the conflicting interests, which of the two ought to be preferred'. They concluded: 'That to prefer the salmon rivers of Cornwall, to the great mining interests that form the staple industry of that county, would be to preserve salmon at a preposterous cost'.

In considering the appeals by South Crofty Ltd and Brea (now Medway) Tin Ltd in respect of the Red River in 1976, the Inspector conducting these appeals concluded: 'Curtailment of the Companies' activities would bring about financial hardship and over 600 persons would lose employment and there would be redundancies in those trades associated with mining in the Camborne/Redruth areas. The nation would suffer through lost production of a valuable raw material which would have to be imported at an increased cost in overseas expenditure of something like £14m per year'. He found it difficult to believe that recovery of the Red River to a clean state would increase its value to the public by anything like that amount. It might support fish life and it might become a more pleasant part of the landscape. Matters seem to have come full circle!

REFERENCES

Cornish Chamber of Mines (1983) *Annual Report 1983*.

Derry, T. K. & Williams, T. I. (1970) *A short history of technology*. Oxford University Press, 749pp.

English China Clays Ltd. *The Story of China clay*.

Institute of Geological Sciences (1975) *British regional geology — S.W. England*. Her Majesty's Stationery Office, London.

Purves, J. B. (1974) *Aspects of mining and pollution control in S.W. England — Minerals and the environment*. Institute of Mining and Metallurgy.

10

Remedies to pollution by wastes from china clay workings

R. A. Comyns

10.1 INTRODUCTION

The China Clay Industry started in Cornwall in 1748, and has been in operation ever since, albeit at reduced levels during the two World Wars.

In 1945 at the end of World War II, the total production of China Clay was 280 000 tonnes, of which 164 000 tonnes went to the home market and 116 000 tonnes went for export. By 1950 production had reached 1 000 000 tonnes, and is currently of the order of 2.5 to 3 million tonnes.

Of this production, 80% is exported, 80% of the clay produced goes to the Paper Industry, 15% goes to the Ceramic Industry, and the remainder to the Paint, Rubber, Plastics, and Pharmaceutical Industries and for miscellaneous uses.

10.1.1 Occurrence of china clay

There are five main granite 'bosses' in Cornwall and Devon, (Fig. 10.1) and the extraction of china clay is carried out at four of them, with the main production areas being at Hensbarrow Downs, and the south-west corner of Dartmoor.

The granite consists of quartz crystal together with feldspars, mica and tourmaline. The kaolin within the granite was formed by hydrothermal convective circulation which in turn caused a defluoridation of the feldspars and hence created a china clay.

A well-kaolinised granite is a soft material with a consistency of coarse putty. As the degree of kaolinisation decreases, so the mineral gets harder and more like the original granite.

10.1.2 Production of china clay

The production of china clay (see Fig. 10.2) starts in an open pit, from which, for each tonne of saleable clay produced, some nine tonnes of 'wastes' are also generated. These wastes consist of:

4.5 tonnes of rock and overburden,
3.5 tonnes of sand, which is also known as granular fill, and
1.0 tonne of 'micaceous residue'.

Following the removal of the overburden from the clay deposit, the clay itself is mined hydrualically by the use of high-pressure hoses operating at pressures of 24×10^5 Pa (24 bar). The sand is separated from the clay stream mechanically and the suspension of clay in water is processed to provide the various grades of clay required by the many customers of the Industry.

10.2 DISPOSAL OF BY-PRODUCTS

The rock and overburden are removed from the open pits by dump trucks and deposited on tips. These tips may be close to the current workings, but occasionally the dump trucks can have a round trip of up to 13 km from pit to tip and back again.

Sand is, in general, removed from the pits by conveyor belt and deposited on tips. There are still one or two pits left that remove sand by the old fashioned skip-and-incline system. This old-fashioned method of removal gave the China Clay area of Cornwall and Devon its own characteristic landscape of conical tips. However, the Aberfan disaster of 1966 resulted in the Mines & Quarries (Tips) Act 1969, which altered the method of tipping and so tips now tend to be long and flat, with a maximum height of tipping of 24 m for individual lifts.

Both of these forms of tip have problems with contaminated surface-water runoff, especially in times of high rainfall. However, as part of the requirements for the design and construction of these tips, it is necessary to provide a proper system of tip periphery and under-drains. These can catch the runoff and lead the water into a settlement lagoon prior to its discharge to the river. The newer form of settlement lagoon is usually double, allowing one to be in use settling out solids, whilst the other drains and dries out in the summer, prior to the emptying out of the accumulated solids by mechanical excavators.

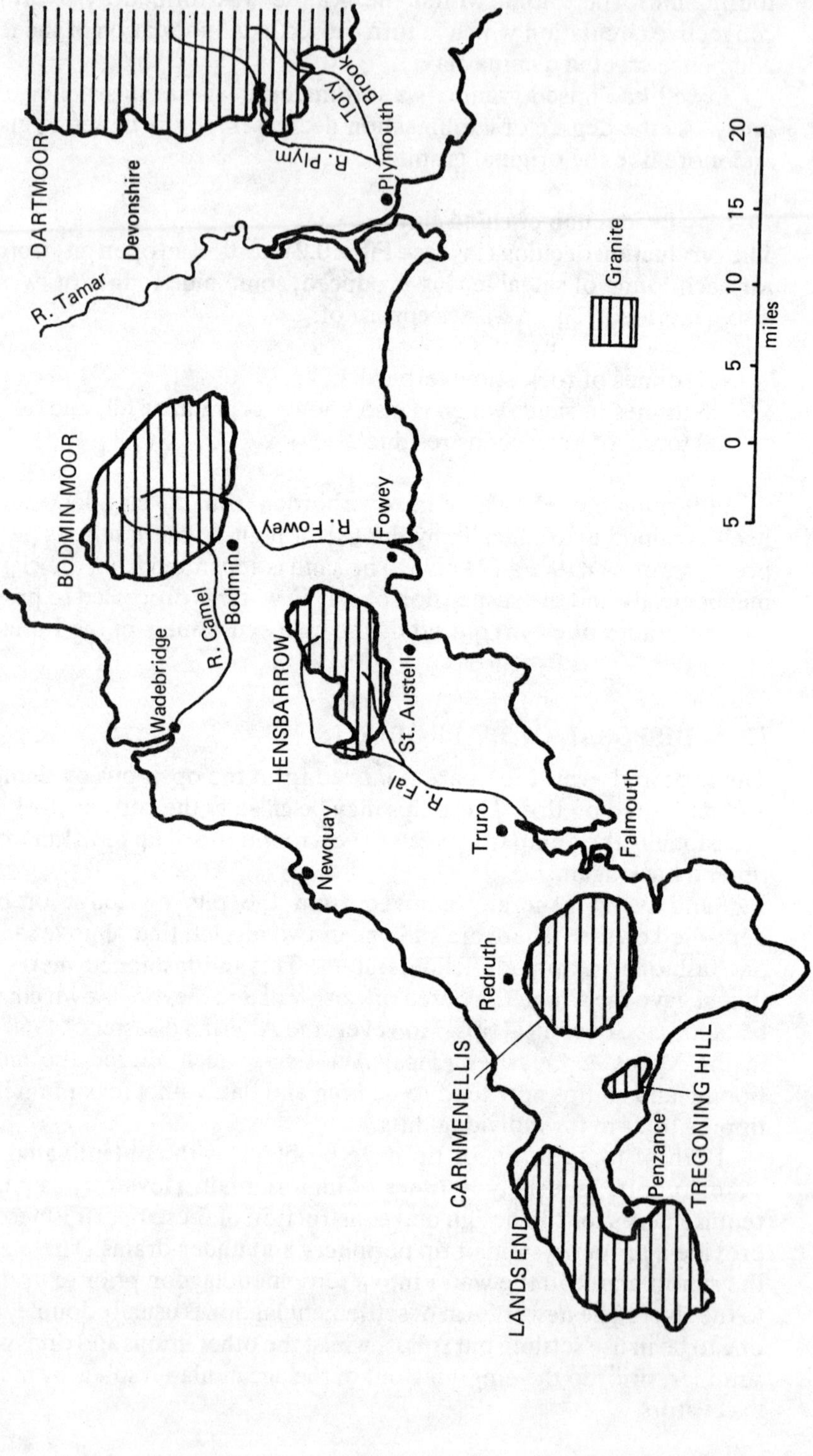

Fig. 10.1 — The principal granite areas of Cornwall and Devon

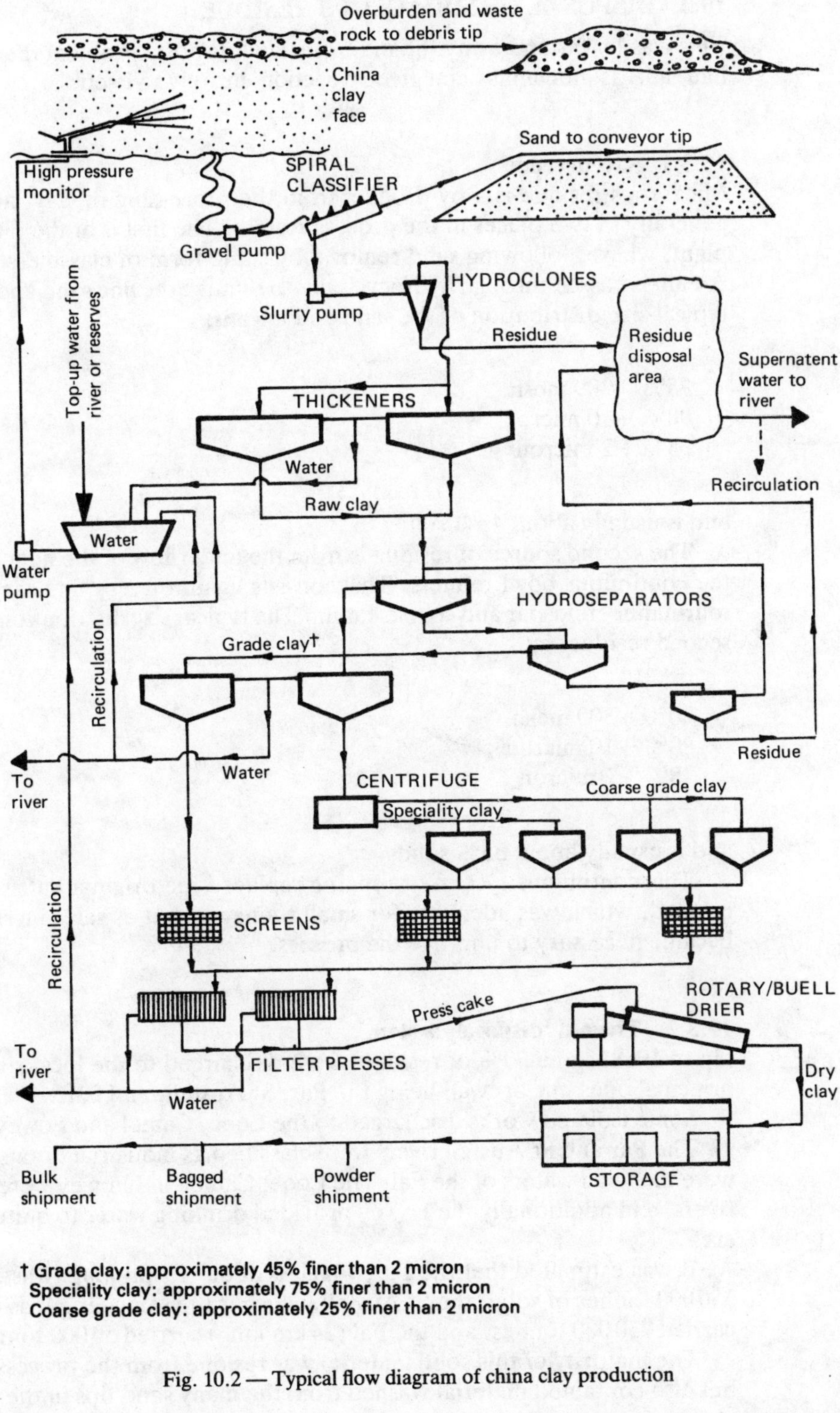

† Grade clay: approximately 45% finer than 2 micron
Speciality clay: approximately 75% finer than 2 micron
Coarse grade clay: approximately 25% finer than 2 micron

Fig. 10.2 — Typical flow diagram of china clay production

10.3 DISPOSAL OF MICACEOUS RESIDUE

This was by far and away the major residue disposal problem the Industry had, and as much has been given a section entirely to itself.

10.3.1 Occurrence

Micaceous residue is a by-product from the processing of clay, and arises generally in two places in the process system. The first is at the pit refining plant, where, following sand removal by some form of classifiers, the clay stream is passed through hydrocyclones to remove the fine sand and mica. A typical size distribution of the sand and mica is:

75% +300 mesh
90% +10 micron
5% −2 micron

and is usually about 45% solids.

The second source of residue is from the underflow of the final stages of the continuous-bowl refiners. This consists again of very fine sand, mica, tourmaline, feldspar and coarse kaolin. The typical size distribution of these second residues is:

40% +300 mesh
80% +10 micron
8% −2 micron

and is usually about 35% solids.

This continuous system of refining replaced the original batch-refining method, which was adequate for small tonnages, but as sales increased, it became necessary to improve the process.

10.3.2 The old 'disposal' system

Up to 1968, some 50% of residues were discharged to the local rivers, the principal ones in Cornwall being the Par, St Austell, and Fal.

Some isolated works discharged to the Looe, Camel and Fowey rivers.

The Par and St Austell rivers were classified as industrial drains, and so were the head waters of the Fal. The Looe, Camel and Fowey were salmon rivers, and additionally the Fowey provided drinking water to quite a large area.

It was estimated that the Par, which is about 13 km long, was carrying 350 000 tonnes of solid, the St Austell, which is the same length as the Par, carried 750 000 tonnes, and the Fal (24 km long) carried 50 000 tonnes.

The majority of this solid material was residue from the process plants, but also contained material washed from the many sand tips in the area.

10.4 CLEANING THE RIVERS

10.4.1 Sampling programme

In order to assess the best method of cleaning the rivers, it was necessary (among other things) to know how much material was entering the rivers and where it was coming from. To do this, a small group was set up under the control of one of the survey staff, with instructions to map the rivers and Company properties, and show on them where all the discharges were taking place, the types of discharge, and to measure the volumetric flow and quantity of solids at each point.

This exercise took four years to complete and was very revealing. In total there were 350 recognised discharge points requiring some form of consent condition. However, many more unofficial discharge points were discovered, and these were stopped and the discharging material diverted to a controlled discharge point.

10.4.2 System design

While the total quantities of residues were being assessed, discussions took place within the Industry with a view to selecting the best practicable method for disposing of residues.

It was agreed to use a civil engineering consultant to assist in the evaluation and selection of an appropriate method.

10.4.3 Options examined

The options considered could be split into two, disposal at sea or disposal on land.

10.4.3.1 Sea disposal

This came to be known as the 'Maenease Point' scheme. All the residues from within the Industry were to be piped to a central collecting station and from there pumped to Maenease Point. From here they would go down a vertical shaft, into a tunnel under the sea bed to a point 800 metres off shore from where the residues would, it was hoped, be naturally dispersed.

A variety of investigations were carried out at the instigation of the civil engineering consultant employed. These covered the currents, tides, and drift in the St Austell Bay, the possible effects on the sea fishing, and the behaviour of the residue at the outfall.

The principal consideration was to determine whether any of the residue would drift back towards the coast. This, from the investigations carried out, did not appear probable but no guarantee could be given that it would not happen. Large-scale tank tests, carried out by consultants in hydrology, indicated that there could well be problems at the outfall in that the residues would not disperse from the end of the pipe.

After an investigation costing, then, over £100 000, it was decided to abandon the concept of the sea outfall.

10.4.3.2 Land disposal

There were several alternative methods of land disposal considered and these are:

(i) *Disposal of the residues on land outside the china clay mining area.*

This would have necessitated acquiring large areas of farmland, outside the Industry's ownership, and would also have had considerable effect on the environment of neighbouring villages.

(ii) *Disposal on land within the Industry's land ownership*

Although feasible, considerable reserves of china clay would have been sterilized.

(iii) *The Five Pit Scheme*

The Local Planning Authority produced a long term plan for the St Austell area which by delineating the boundaries of island settlements to protect the local villages, and by fixing the route of the main roads through the area, which could not be changed, also set physical limits on the areas of ground that could be used for the winning and working of china clay. At the time improvements in clay-processing technology were reducing the amounts of residue that had to be disposed of. This allows the Industry to adopt what became known as 'The Five Pit Scheme'. Five disused pits, where the reserves were sterilized by physical limitations, were to be used for the storage of residues. Whilst the pits themselves provided considerable storage, their capacity could be greatly enlarged by the provision of massive retaining bunds around the perimeter of the pit.

Before the construction commenced, it was realised that there would be considerable engineering problems caused by the nature of the materials being handled. They were known to be mildly corrosive, being at a pH of 3.0 to 4.5, but were very highly abrasive. The particles of fine sand and feldspar are very angular and 'sharp'.

Conventional pumps used within the Industry were worn out in very short periods of time, and pipelines also had a very limited life.

Various pump manufacturers around the world were contacted, and eventually the Mars Pump produced by the Mitsubishi Company of Japan was selected.

With regard to pipelines, a test rig was established and various types of pipe were evaluated against typical residues.

Mild steel had a life of two weeks.

Flame-hardened mild steel lasted six weeks.

Conventional rubber-lined pipe lasted six months.

Pipes lined with a polyurethane rubber lasted over two years.

In order to supply the quantity of pipes required, one of the Companies set up its own production unit to line the pipework with polyurethane rubber. This has proved most successful, and many of the original lined pipes are still in use after a period of ten years or more of handling the very abrasive material.

10.5 THE RESIDUE DISPOSAL SCHEME

This is shown in outline in Fig. 10.3. Residues as they arise are picked up and pumped, either to a central collecting station where they are further pumped to a residue dam, or direct to the dam. In one or two locations it was possible to gravitate the residues to a logoon.

Currently five residue dams are in use, with the largest being at a disused pit known as Kernick. This is being filled at a rate of 500 000 cubic metres of residue per year, and as the level rises in the dam, so the retaining bund is raised. This bund is made from china clay sand, with a crushed and graded rock core. Its construction is controlled by a civil engineering consultant, and when completed will be 100 metres in height from toe to crest. This particular dam has been in use for 12 years and has at least another ten years life left in it.

Supernatant water from the dams will usually overflow to the adjacent river, but in summer time it is pumped back to the clay pits to be re-used in the extraction of the clay. The water that does go to the rivers usually contains suspended solids at levels of less than 100 mg/l, and can be discharging from the Kernick dam at rates up to 600 l/s (8000 gallons per minute).

10.6 PROGRESS ON THE IMPLEMENTATION OF THE SCHEME

Discussions with the relevant Water Authority allowed for a phased improvement in the condition of the rivers classified as 'industrial drains'. By 1968, the Industry had already eliminated about half a million tonnes of residues from the St Austell 'White' river. These residues were disposed of to temporary dumping areas, and so reduced the deposits of residue in the river estuary.

The programme of reduction was such that there were to be no residues in the St Austell river by January 1973, and for the Par river by October 1974. Both of these dates were met, and the change in appearance of the rivers was remarkable. Within a month of the programmed date both rivers had improved in quality status to that of recoverable rivers, and the river-flow concentrations had dropped from 60 000 mg/l to less than 100 mg/l of suspended solids.

10.7 DEVON

The clay pits in the Devon area are sited in the south-west corner of Dartmoor on the edge of the National Park. For many years the residues from these pits and the associated processing plants had been carried by the Tory Brook to the sea. To overcome the problems some small dams were

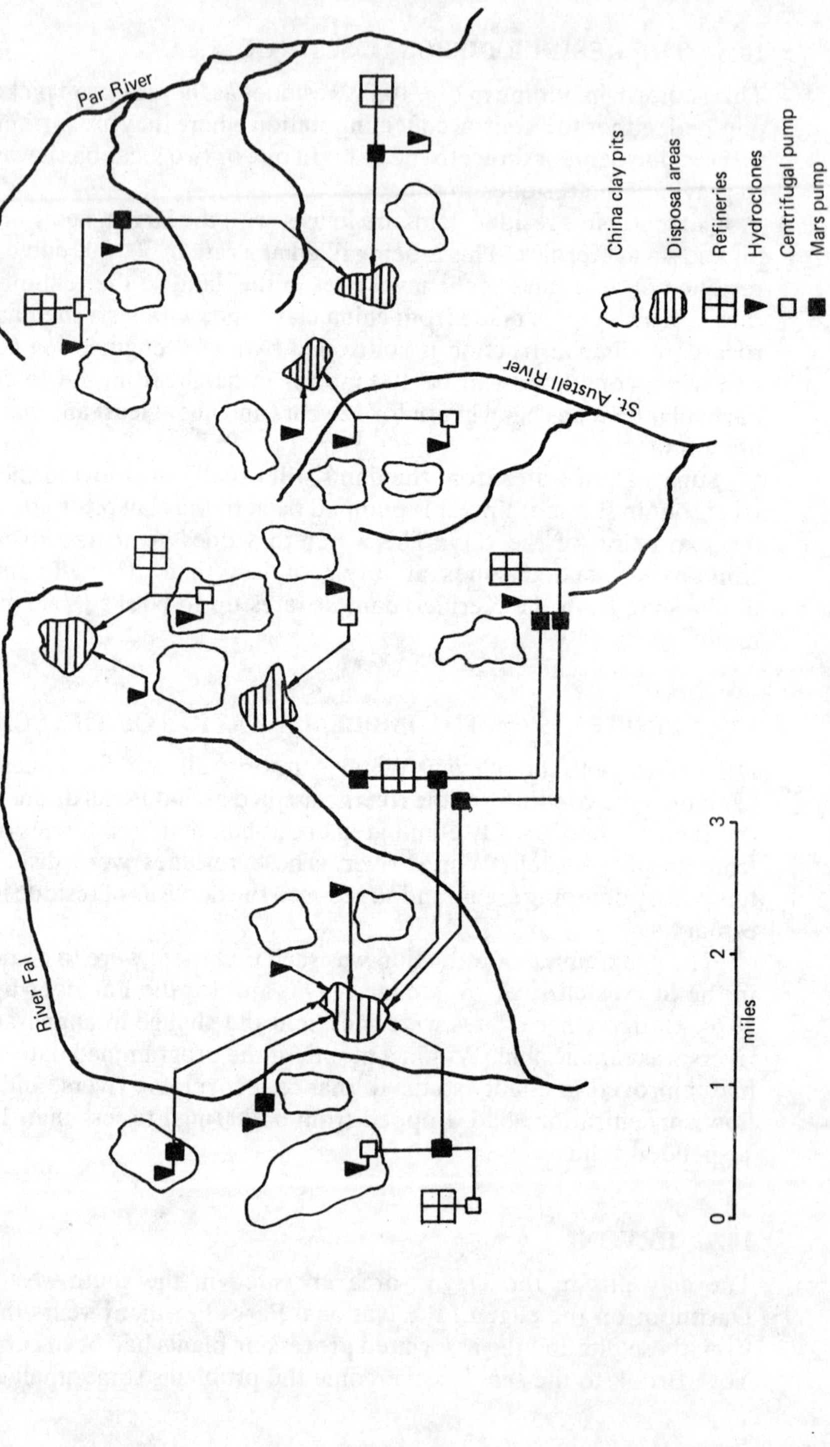

Fig. 10.3 — The Hensbarrow china clay area

replaced by a major dam known as Portworthy, into which all the residues were routed. The overflow from this primary dam went into a second stilling pond to allow a final clean-up before discharging to the river.

One of the smaller rivers was also routed through a stilling basin to arrest the movement of solids entering the stream through the surface runoff from the extensively worked catchment area. This stilling basin is cleaned out every year, the deposited solids being transferred to one of the local tips.

10.8 CURRENT PROBLEMS

The principal problem now remaining is that of surface runoff from the sand and overburden tips. This is being tackled in two ways:

(i) by the provision of catch pits in either the headwaters of the river, or in the drainage leats from the tips before the supernatant water joins the river, and

(ii) by the landscaping and hydro-seeding of the tips. This technique covers the surface with grass, which binds it together, and prevents wash-out by rain-water. This also has the advantage of preventing wind-blown dust, and also makes the tips more attractive visually.

10.9 CONCLUSION

Since the inception of the scheme to prevent residue entering the rivers flowing through the china-clay areas, some £12m has been spent, and annually £750 000 is budgeted to operate the scheme.

This has resulted in rivers, which formerly ran white, in general running clear, and has allowed for the re-establishment of fish life in the streams. Perhaps the most gratifying sight was that of a heron (*Ardea cinerea*) fishing in the St Austell river, now no longer known as the 'White river'.

Discussion

Chairman: **J. I. WADDINGTON**
Director, Environmental Health Service,
World Health Organization

For convenience, the wide-ranging discussion on mineral exploitation has been organised under the sub-titles Suspended Solids, Acid Mine-Drainage and Best Practicable Environmental Options.

SUSPENDED SOLIDS

Questions were raised on the means of separating the solids from the liquid discharges, the quality of water seeping from tailings dams, and the disposal of solids which may be contaminated with metals. Further topics discussed were the criteria to be used in restoring sites and the means of choosing the best option among several methods of disposal.

Solid/liquid separation

Dr R. F. Packham (Water Research Centre, Environment) asked for more information on the methods for achieving such a successful separation of liquids and solids in the industry. **Mr Comyns** said that the wastes containing solids (mica, quartz, feldspar and some coarse kaolinite) were pumped to centralised collecting dams and simply settled. **Mr Cominetti** added that at Pendarves Mine the wastes passed to a tailings lagoon, followed by the addition of flocculating agents — ferric sulphate and a potable-water grade polyelectrolyte.

Seepage from tailings dams

Mr H. Akyar (State Hydrualics Works, Turkey, Visiting Trainee with North West Water) was interested in the quality of seepage water, in particular from borax mines. No more than 3 mg/l boron should be released, because of toxicity to plants, although 50 mg/l might be acceptable in drinking water.

Mr Comyns said that the water emerging at the toe of the china-clay tailings dams was chemically satisfactory for use as a potable supply, but failed bacteriologically. **Mr Hammerton** said that the UK Department of the Environment were likely to specify standards for boron in waters intended for a number of uses. Such standards would be taken in to account by regulatory authorities when giving consents to discharge to various surface waters, in compliance with appropriate EC Directives. Boron was not a problem in Scotland. **Mr H. Fish** (Natural Environment Research Council) recalled that, at least 15 years ago, it was found that the high content of boron had prevented sewage effluents from being used to irrigate certain crops, although the element was beneficial to roses. Perhaps the solution to the seepage problem was to devise an extraction system for borax which did not involve the use of water!

Disposal of waste solids

Metal contamination

Mr E. Geuze ('De Dommel' Waterboard, Netherlands) said that in the Netherlands there was growing concern about metal-contaminated sediments, although the overlying water might be quite satisfactory. He asked for more information on the clean-up of metal-rich sediments. Did the concentrations of metals in the sediments suggest the need for special disposal arrangements? **Mr Cominetti** said that the sediments in the Red River contained high proportions of heavy metals from metalliferous mining. The 'Dangerous Substances Directive' (76/464) indicated the problems to be considered. For disposal to sea the limits were set in terms of soluble metal, but for rivers tighter controls were required using concentrations of total metals, although the metals might be tightly bound to the solids and thus not readily available to the environment. The Ministry of Agriculture, Fisheries and Food specified determination of 'available' metal, defined as that leached out of the material by acetic acid. It was hoped that this would be relevant to the protection of marine life. All these points would be considered when trying to decide the best option for disposal. The important point was not to create a new form of pollution when remedying another. The china clay wastes were sandy materials that did not give rise to metallic pollutants.

Mr Hammerton emphasised the need to dispose of solid wastes containing metals in such a way that leaching and future problems were minimised. He cited cases in Scotland where chromium had leached from landfills which had been reclaimed for industrial uses or as recreational areas. Streams were coloured yellow with chromium for long periods when the leachate broke through.

Criteria and standards in restoration work

Dr S. Ponce (US Department of the Interior) asked what criteria were established at the start of the clean-up of the Red and White Rivers, by which the success of the operation was to be judged. **Mr Comyns** replied that the criterion was that the river should appear visually satisfactory, and to achieve this a standard of <100 mg/l suspended solids (SS) had been set. Other waters in Cornwall (the Looe, Fowey and Camel) were salmon rivers and the standards for these were more stringent, a maximum of 60 mg SS/l. **Mr Cominetti** added that the White River had contained as much as 600 000 mg SS/l but that the only discharges now permitted were from certain drying plants and other processes where the standard was 100 mg/l. On the Red River, depending on the best practicable environmental option finally agreed (and sea disposal appeared to be that option), there could be total elimination of discharge of tailings to the river, so that it could become throughout its length what it already was in its headwaters, an uncontaminated natural stream. **Mr Comyns** also made the point that the number of discharges had been reduced from several hundred in the 1950s and 1960s to less than 100 in 1985.

Disposal options; standards/guidelines, environmental compromise

Mr A. Oxbrow (Napier College, Edinburgh) argued that marine disposal of mining solids was an 'out of sight, out of mind' philosophy not acceptable to many ecologists. The sea's capacity to absorb wastes without problems was not inexhaustible, particularly on continental shelves and in estuaries. Secondly, there were marine seismic disturbances so the sea-bed could not be dismissed as earthquake-free. His third point was to question whether standards of safety and disposal should be different in developed countries from those used in the Third World. In **Dr Barbour's** reply it was pointed out that marine disposal was not often favoured by regulatory authorities and that if large amounts of tailings had to be disposed of, all options should be considered and the best one, environmentally, decided upon. In many circumstances, but depending on each set of local conditions, the marine option might be best. For example, where disposal to sea, at a distance from the shore, was compared with disposal to an environmentally-sensitive mangrove swamp or to a high altitude tropical jungle (a real case), marine disposal seemed the best course. A detailed study would be required before such a decision was made. Seismic submarine disturbances did indeed occur and could even be beneficial if, as in one example in his experience, the result had been to move the deposited tailings further from the shore.

Dr Barbour strongly urged all countries, and particularly Third World countries, to pay attention simultaneously to their regulations controlling discharges and to the proper training of their inspectorates to ensure that regulations were enforced. As far as universally-applicable standards were concerned, he was more used to handling site-specific problems. Careful judgement was needed. He would be more impressed if regulatory authorities in the Third World were as keen on mine safety as they were on universal standards for effluents.

Mr Cominetti endorsed much of what Dr Barbour had said. Although environmental 'compromise' might not seem a suitable phrase to come from a regulatory body, it was necessary. In considering the clean-up of the Red River, options available seemed to be a tailings dam involving an area of land of more than 90 ha (200 acres) (which could bring the regulatory authority into conflict with the planning authority) or a fairly long, deep, sea-outfall. He believed that the latter course would lead to significantly fewer environmental problems. Hydrographic and other surveys would be carried out to confirm whether this choice was the right one. The only other option available seemed to be underground disposal into the 50-km network of mine workings, but this could lead to pollution incidents in the future, whose location, timing, and severity would be impossible to predict. (See also Acid Mine-Drainage, below.)

In summing up this section of the Discussion, **Mr Waddington** said that it was considered that the environment had capacity to receive wastes, if properly managed. In our over-crowded planet there must be untouchable wilderness areas but there must also be environmental compromise. He did not believe in international standards. There were no WHO Drinking Water Standards, only Guidelines. There would shortly be Air-Quality Guidelines. He had visited developing countries that had passed many regulations for protecting the environment but had no trained staff, no inspectorate, without which the laws were merely a 'paper exercise'.

ACID MINE-DRAINAGE

Real examples of solving this major problem were considered, and so was the need to plan to minimise such problems in the future.

Effective solution

As a demonstration of what could be achieved in handling a very difficult environmental problem, **Mr Hammerton** cited the example of Dalquharran Colliery. Here, the conventional options for dealing with the acid, ferruginous, mine-drainage water seemed at first to be either on-site treatment leading to the production of large volumes of iron-ochre sludge, or a very expensive long outfall to the sea. Only by its own research did the National Coal Board (NCB) come up with the excellent solution described in Chapter 8. The need for, and value of, such research and site-specific investigation to tackle the problems of acidic wastes from mining could not be over-emphasised.

Mr H. Goldie (National Coal Board, Edinburgh) argued that the 50 tonnes of lime and 300 tonnes of caustic soda put into the workings at Dalquharran had not been useless. They had given breathing space while a more permanent solution was devised, and he would also assume that stoichiometric quantities of iron were left underground by this early treatment, which would otherwise have entered the river. He asked whether statements of the damaged condition of the river in 1979 and 1980 had been over-emphasised, in view of the opinion, in October 1983, that the river was fully recovered. **Mr Hammerton** responded that the addition of lime and caustic soda before the workings were flooded would have surely been more effective than pumping these materials in after flooding. It was another example where research was needed. The eventual solution was most ingenious, and the NCB were to be congratulated. One of the benefits resulting from the case had been the establishment of a Consultative Party between the River Purification Board and the NCB to consider problems that might arise when mines were closed in the future. As far as the rate of recovery was concerned, **Mr Hammerton** said that without the remedial measures the river would have been written off as a salmon river for the next 30 or 40 years. No one had known what the speed of recovery would be, but it had proved to occur surprisingly rapidly. It was now one of the finest salmon rivers in Ayrshire.

The River Purification Board always tried to achieve its aims without resort to the Courts. Legislation and Government action was required in this field because an industry, faced with many other costs, would not undertake the expensive elimination of pollution unless it was forced to do so.

Dr Packham asked whether, with the value of hindsight, the NCB would have tackled the Dalquharran problem in a different way. **Mr Goldie** said that they would not add lime to the empty workings (as they had often done previously) because it would sit there without dispersing. With 2500 million litres (550 million gallons) of water contaminated they could not afford to treat it all and had to look for other remedies. Chemistry was too expensive; engineering was the answer. Underground walls and dams should have been built to stop the flow of water, the mine entrance should be blocked off and so should any point where the water entered the mine. Now, all shafts were capped and the shaft-bottoms were sealed. This had been a successful strategy. All mines had the potential for trouble but only two of 52 had actually given rise to problems.

Close-out plans

Dr Barbour concluded from his experience that if a 'close-out plan' was prepared and agreed before mining commenced, engineering arrangements could be made from the outset, a far more acceptable situation.

BEST PRACTICABLE ENVIRONMENTAL OPTIONS

The discussion session ended with a consideration of best practicable environmental options (BPEOs), introduced by **Dr Barbour**. In his opinion there was no formal, clear regulatory mechanism for determining a BPEO in the UK or elsewhere, and such a mechanism was needed. **Dr S. C. Warren** (Water Research Centre, Environment) supported the need for methodology but asked how, in the absence of ground rules, an industry should proceed to negotiate a BPEO. Should there be standard governing this or could industry be relied upon to pay due concern to the environment? **Dr Barbour** said that the mechanisms must take into account air, water, land, and occupational health. It must be senior and authoritative, perhaps an applied Royal Commission on Environmental Pollution, headed by senior people from the regulatory authorities. In the interim we should not talk about BPEOs until we have a judicial body to decide what is needed. **Mr Hammerton** reminded the delegates that such a body had been proposed in the 5th Report of the Royal Commission on Environmental Pollution — in the form of Her Majesty's Industrial Pollution Inspectorate. This would discuss and resolve cases where options for waste disposal were in conflict, and was now supposed to be the mechanism to be used. In future a more sophisticated system might be required but not necessarily involving such a massive organisation as the US Environmental Protection Agency, excellent though this was.

Part 3

Agriculture

11

The uses of water in UK agriculture and horticulture

D. C. Hardwick

11.1 INTRODUCTION

Agriculture can affect water in three main ways. Firstly, water is discharged from agricultural land as a result of irrigation and land drainage. Such water can carry with it soil and other matter, and dissolved salts might be deposited if the water evaporates. Secondly, the substances used in farming can be leached into water; fertilizers are quantitatively the most significant but pesticides, because of their biological activity, are also important. Thirdly, some farming (watercress (*Rorippa nasturtium-aquaticum*), fish) occurs in water.

This chapter will outline briefly past trends and future prospects relevant to these topics in the United Kingdom (UK) as a basis for indicating future changes in the potential for water contamination. It must be emphasised, however, that simple extrapolation of past trends is not necessarily the best predictive model at present, when social and economic pressures are changing and the responses of the various sectors of agriculture to such pressures are as yet undefinable. Economic forces, public opinion, and policy at various levels, are all combining to change the course of agriculture to an extent not experienced since wartime when concentration on productivity was the main aim. Now re-evaluation of the need for food in the western world, the protection of the environment, and the public perception of farming, seem likely to slow down or to reverse the rate of expansion. On the one hand, this could reduce the potential of farming to pollute. On the other, any reduction in farm income could reduce the farmer's ability to finance measures to reduce pollution. The balance of forces here depends on political decisions both within the UK and in Brussels, as well as local and world economic forces.

This chapter, therefore, goes somewhat wider than the title implies and deals broadly with the effects agriculture can have on the quality of water, touching here and there on the effects on farming of water of inadequate quality. The term 'quality of water' is used advisedly because water is rarely pure and distinction must be made between 'contamination', where substances are present, and 'pollution', where some demonstrable risk is perceived. The presence of a contaminating substance must always raise questions of acceptability but, until there is evidence that it is harmful, it is not strictly a pollutant.

As a background to the chapter, it is helpful to have in mind the major agricultural causes of pollution and, for England and Wales, these are given in Table 11.1. This shows that animal excreta, liquid effluent from silage, and rain or washing water contaminated with organic materials (for example, faeces, vegetable wastes, milk) are the most frequent pollutants. This is because of their high biological oxidation potential and the consequent effects on river biota. Table 11.2 shows that serious farm pollutions have been increasing.

11.2 IRRIGATION

Irrigation recycles water through land. In some parts of the world, evaporation from irrigation water allows salts dissolved in the water to be left in the soil; up to one tonne of salt per ha/yr could accumulate in this way. In the UK, such conditions rarely, if ever, occur. Of more significance are the dissolved substances or suspended matter carried off in irrigation water. The aim is to keep the soil below field capacity so that runoff does not occur except where unexpected rainfall increases the water load.

The area of land which might be irrigated has varied mainly between 80 and 100 000 ha in England and Wales, 2 to 2.5% of the total arable area (Fig. 11.1). A peak of 130 000 ha in 1977 may reflect the very dry summer of 1976 because the area has subsequently fallen back to 103 000 ha in 1983. The number of holdings practising irrigation remained roughly constant at 7000 until 1977, but 9700 were reported as irrigating in 1983; the basis of the figure may have changed, but, if the increase represents a real situation, it could reflect a potential for expanding the area irrigated and the volume of water required. Figure 11.2 shows that irrigation is predominately in the east and south east.

Prediction of expansion has been optimistic in the past. In 1962, a study predicted a rise of some 6000 ha/yr of irrigated land, and the report 'Water

Table 11.1 — Water pollution incidents caused by farm wastes. Warnings and prosecutions.
(For the year 1 January — 31 December 1983).

Type of waste	Warnings		Prosecutions	
	Verbal	Written	Commenced	Convictions
Overflowing slurry stores	200	298	14	7
Bursting slurry stores	46	40	7	9
Silage effluent	213	362	44	28
Runoff from land	100	86	4	7
Dairy and cattle-yard waste water	275	344	11	3
Piggery and yard waste water	152	164	12	21
Poultry house and yard waste water	12	7	7	3
Runoff from solid manure store	76	48	1	2
Failure of treatment systems	28	29	3	2
Mineral fertilisers	5	8	1	—
Vegetable washing water	19	16	—	1
Pesticides	28	29	—	—
Oil spillages	32	37	2	—
Miscellaneous	88	53	8	4
Totals	1274	1521	114	87

Pollution incident: where a pollutant has reached or appears to have reached a watercourse or underground water.
Source: Regional Water Authorities.

Table 11.2 — Total number of serious farm pollutions in England and Wales.

	1974	1975	1976	1977	1978	1979	1980	1981	1982	1983	1984
Silage	15	50	69	162	41	189	378	600	322	473	663
Cows	3	94	114	339	72	532	690	790	657	874	1640
Pigs	0	6	6	54	65	27	184	267	205	186	186
Other	43	121	84	397	135	412	413	579	668	150	472
Total	61	271	273	952	313	1160	1665	2236	1852	1683	2961

Source: Regional Water Authorities.

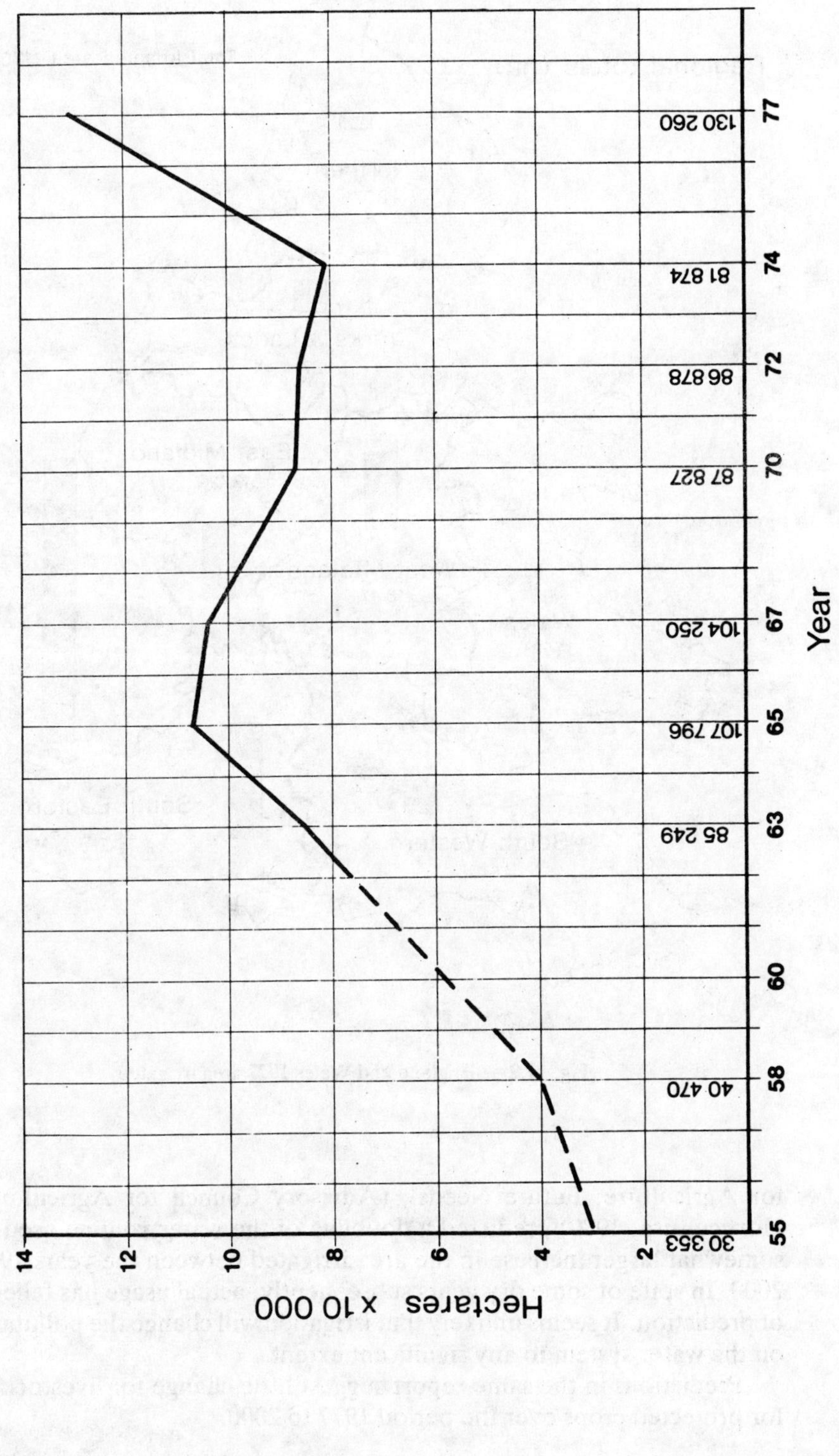

Fig. 11.1 — Area irrigated in a dry season

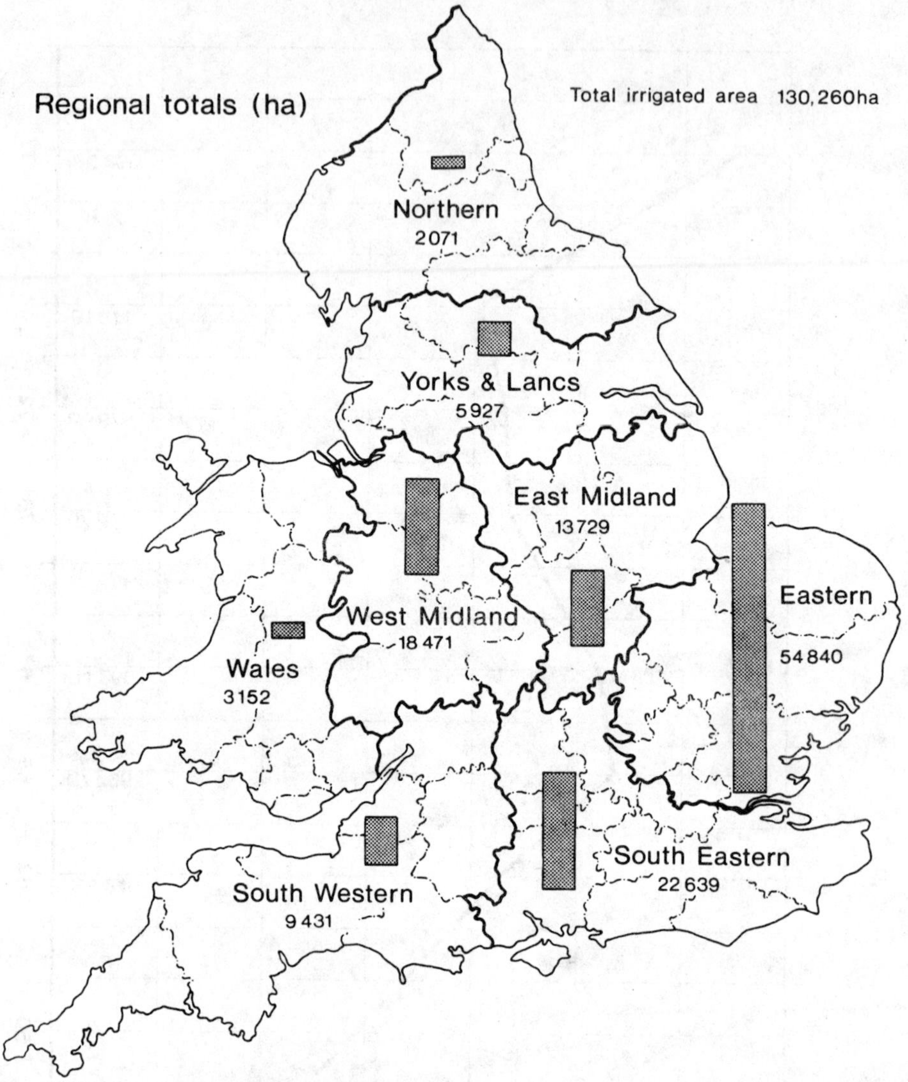

Fig. 11.2 — England and Wales 1977 area irrigated

for Agriculture: Future Needs', (Advisory Council for Agriculture and Horticulture, 1980) predicted a doubling of the water volume used, and a somewhat larger increase in the area irrigated between the years 1977 and 2000. In spite of some dry years subsequently, actual usage has fallen short of prediction. It seems unlikely that irrigation will change the pollutant load on the water system to any significant extent.

Predictions in the same report suggest little change for livestock use or for protected crops over the period 1977 to 2000.

11.3 DRAINAGE

The importance of drainage to efficient agriculture has been recognised for centuries. In the 19th century, drainage proceeded apace and, in the latter half of the century, nearly half the agricultural area of England and Wales was drained at an average rate of 100 000 ha/yr. Figure 11.3 shows the

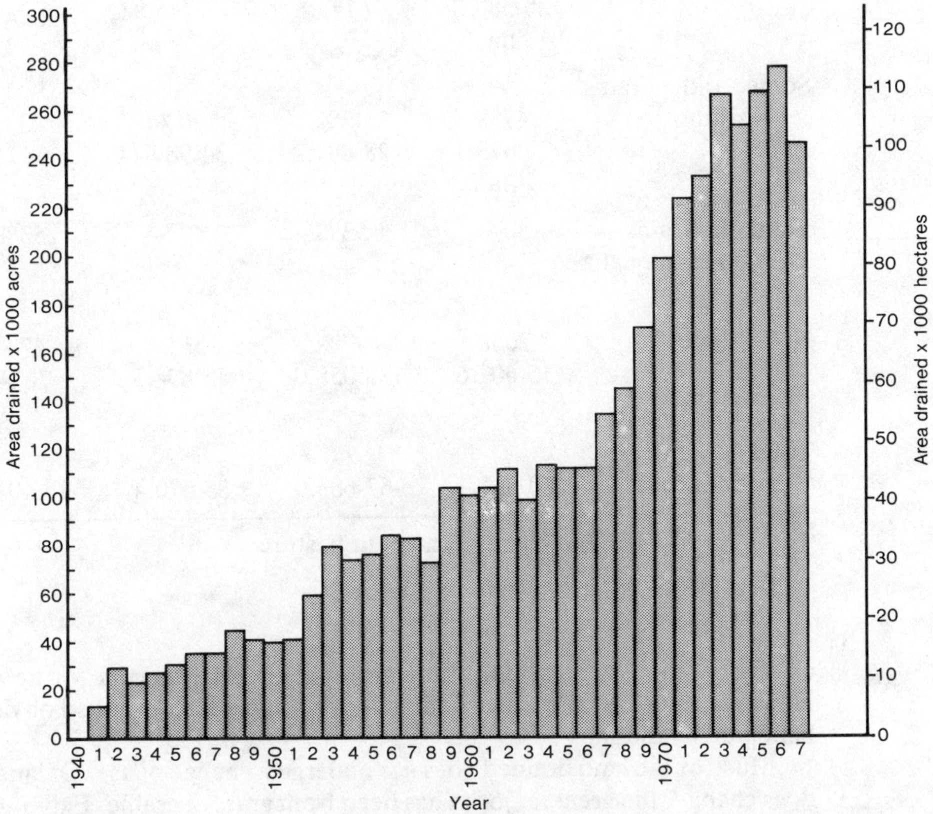

Fig. 11.3 — Underdrainage statistics. Area drained annually, by grant aided underdrainage in England and Wales

annual area drained since the beginning of World War II; only in recent years has the rate been comparable with that of the late 19th century. Currently, 70 000 to 80 000 ha/yr are drained, but precise figures are no longer available because schemes for grant-aiding drainage no longer require approval in advance.

About 50 000 ha need redraining every year to replace old systems (Table 11.3). This leaves only 30 000 to 50 000 ha/yr of land which is newly-drained annually. The maximum requiring drainage is of the order of two to three million ha. Figure 11.4 shows the drainage activity for 1978 to 1979, a pattern similar to that of earlier years. It shows that drainage activity is

Table 11.3 — Old drains in area of drainage benefit

England and Wales Source: MAFF Underdrainage Statistics

	1971	1975	1979	1983
Old drains present but not working	33% 26 580.8 ha	10% 6743.8	9% 7746.8	10% 1534.4
Some old drains working	47% 37 873.7 ha	43% 28 881.2	40% 34 989.1	43% 6892.2
Total old drains (% of benefit area)	80%	53%	49%	53%
No evidence	20% 15 602.6 ha	47% 32 058.0	51% 45 834.5	47% 7674.6
Total area of benefit (ha)	80 057.1	67 683.0	88 570.4	16 101,2

Total 915,514 ha arable and permanent pasture

chiefly on the eastern side, although higher rainfall occurs to the west. However, much of the west country is grassland which is not so obviously benefited by drainage.

Much of the land drained does not undergo a change of use. Of land that does change, the great majority has been from grass to arable. Experiments have indicated that change in use may be more significant in relation to the quantity of nitrate leached than drainage itself. (See Chapter 15.)

With the reduction in grant aid for drainage operations in the UK, activity is likely to lessen. Overall, the area currently drained is unlikely to increase dramatically. Therefore, any contribution of drained water to water contamination is unlikely to increase.

The function of drainage is to reduce the water content in the topsoil and to keep the soil below field capacity. One consequence of this is that water is in contact with the topsoil for a shorter time and substances with a low rate of solution (generally those with a low solubility) may be taken up less. Drainage reduces the opportunity for evaporation, but the volume of water leaving the land may not be greatly changed, although the route as well as the speed of movement is altered parallel to the lowering of water tables. In a few cases, the greater availability of oxygen allows bacterial oxygenation

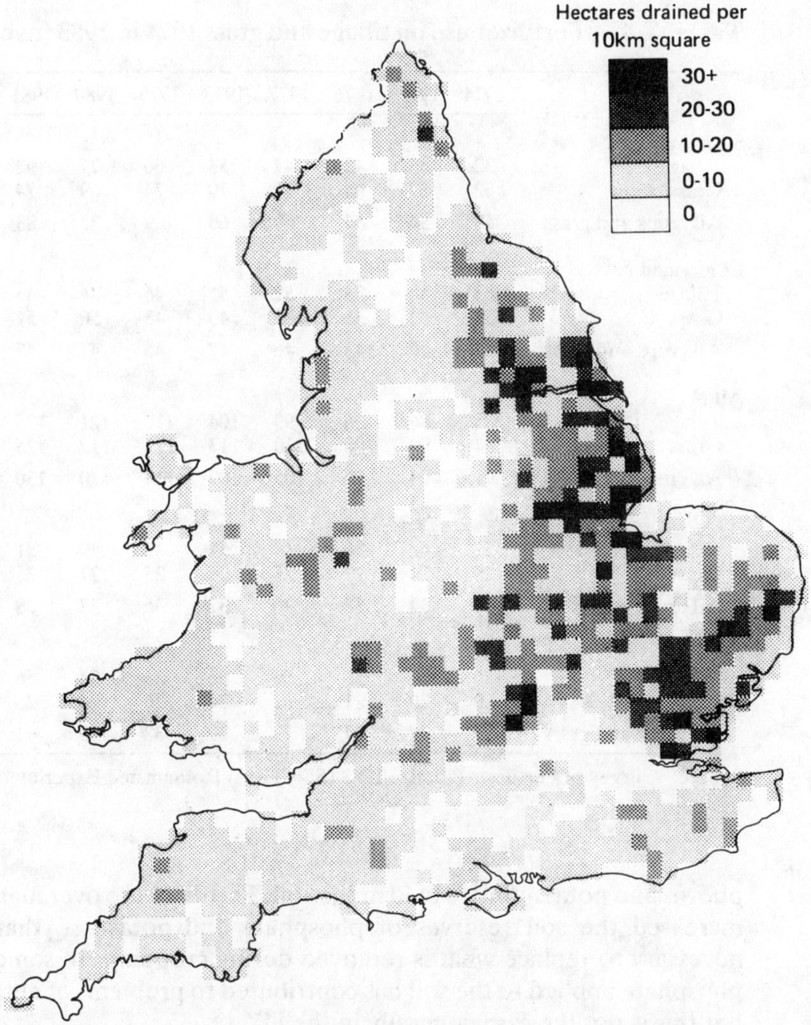

Fig. 11.4 — Location of Drainage activity 1.4.78 to 31.3.79

of iron to produce ochre which looks unsightly, but probably has little or no polluting effect.

Drainage in itself, therefore, causes little or no increase in pollution but may change the distribution of contaminants already present into surface or groundwater. Little change is envisaged.

11.4 FERTILIZERS

The increase in agricultural production in the last few decades has depended on a number of factors, of which an important one has been the increased use of nitrogen fertilizers (Table 11.4). Crops need a number of elements in considerable quantity, notably nitrogen, phosphorus, and potassium. Phos-

Table 11.4 — Fertilizer use on tillage and grass 1974 to 1983 (overall kg/ha)

	1974	1975	1976	1977	1978	1979	1980	1981	1982	1983
Straight N										
Tillage	32	36	42	47	53	66	77	92	99	117
Grass	62	67	64	71	70	74	69	74	71	68
All crops and grass	48	54	54	59	60	69	73	83	85	92
Compound N										
Tillage	53	50	54	52	51	46	44	43	42	37
Grass	32	32	35	39	43	43	50	51	52	57
All crops and grass	43	40	43	46	47	45	47	47	47	47
All N										
Tillage	85	86	96	99	104	112	121	135	141	154
Grass	94	99	98	110	113	117	119	125	123	125
All crops and grass	91	94	97	105	107	114	120	130	132	139
P_2O_5										
Tillage	51	46	50	49	51	49	49	51	55	54
Grass	28	24	27	25	26	25	27	25	24	26
All crops and grass	40	34	38	36	37	36	37	38	39	39
K_2O										
Tillage	56	51	56	56	56	53	54	56	61	60
Grass	20	20	21	22	24	26	26	26	28	28
All crops and grass	37	34	37	37	39	38	40	41	44	44

Source: Survey of Fertilizer Practice 1983 (SS/CH/15) Rothamsted Experimental Station

phorus and potassium are held in the soil. Fertilizer use over many years has increased the soil reserves of phosphate and potash so that it is only necessary to replace what is removed during cropping. In some countries, phosphate applied to the soil has contributed to problems of eutrophication but this is not the case generally in the UK.

Unlike phosphorus and potassium, nitrogen has an immediate and obvious effect on crop yields. The value of the increased yield is several times the cost of the fertilizer (Williams, 1984; England, 1984) and so there is a strong incentive to use nitrogen to the maximum effect. There is a reservoir of nitrogen in the soil in the form of organic compounds, some of it as constituents of soil organisms. In arable soils about 7 t/ha can be present (about 29 t/ha in rough grazing) and some plant growth is supported by the slow conversion of this reservoir to nitrate (The Royal Society, 1983). Cultivation stimulates the conversion of soil nitrogen to nitrate and the greater supply of nitrate leads to better crops in cultivated land. Roughly speaking, crops obtain half their nitrogen from fertilizers and half from soil reserves, which then are replenished by nitrogen from fertilizers. It is wrong to assume that any nitrogen from fertilizers which is not taken up by plants is therefore leached into water.

The concentration of nitrate in water leaching from the soil depends not only on the quantity of nitrate present in the soil, but also on the quantity of water. In dry years, peaks of nitrate concentration occur when water flow starts again and the small volume of water carries a large quantity of accumulated nitrate.

Efficiency demands that the right amount of nitogen is applied at the right time as fertilizer. The right time is when the crop is actively growing and can make use of the nitrate formed from whatever compound of nitrogen is used as fertilizer and from the soil reservoir. For a long time the Agricultural Development and Advisory Service (ADAS) has advised the farmer not to apply nitrogen in late autumn and winter when cultivation supplies all that is necessary.

For the future, it seems likely that the application of nitrogen per ha would be at about the present level, or slightly above, for most crops; for grass there is considerable scope for a greater increase. However, the total usage of nitrogen fertilizer in the UK will depend on the policy developed for regulating production in the EEC and it is possible that the total nitrogen would be reduced parallel to any reduction in total production, although the application per ha is likely to remain about the same.

The concern about leaching of nitrate stems from the EEC Directive that water in supply should not exceed 50 mg/l nitrate. There has been considerable discussion about the medical significance of the intake of nitrate by humans and it is not obvious that nitrate in water below, or even somewhat higher than, 100 mg/l poses any risk, certainly of cancer. Nevertheless, the EEC Directive must be complied with and, in any case, it is wise to act prudently pending better medical evidence. Within the UK, the Nitrate Coordination Group, chaired by the Department of the Environment, is laying the scientific and technical basis for long-term decisions. In regard to agriculture, it is necessarily a long-term matter because current levels of nitrate in groundwater reflect inputs up to several decades ago — wartime ploughing effects may only now be becoming apparent in some cases. Effects on surface waters can be more immediate, although here, too, there is a notable groundwater contribution and again these effects will be delayed.

11.5 ANIMAL MANURES AND AGRICULTURAL WASTES

Until this century, nitrogen fertilization of agricultural land depended almost entirely on animal and human excreta. Changes in agriculture as well as the availability of cheap synthetic nitrogen compounds have meant that, increasingly, animal excreta are viewed as an adjunct to synthetic fertilizers and less important for fertilization than they were. One factor in their use is their polluting potential which is several hundred times greater than human sewage. The ADAS advice on manures emphasises that good practice in storing and spreading manures must take account of this polluting power. Stores must be adequate in size, not only for the amount of manure produced, but also for the probable rainfall on the store or any rain collected and run into it. Also, liquid from stores must not be allowed to leak into the soil. A notable number of water pollution incidents each year are caused by inadequate storage. Fewer incidents arise due to spreading operations.

Another significant pollutant is the liquid produced when silage is made. If the grass is well-wilted with only 75% or less water content, little liquid is produced. Above 75% water, however, an increasing volume of a very polluting liquid is produced — up to 50 l/d per tonne of silage. This must be collected in a suitable tank and spread back onto the land, an additional load of work when the farmer is hurrying to make silage before the weather turns wet again.

For these materials reported water-pollution incidents have been on the increase in recent years. With the coming into force of the UK Control of Pollution Act the opportunity is taken to impress on farmers the necessity of behaving responsibly and according to good agricultural practice. The Water Authorities havc been active in some regions in persuading farmers to take steps to prevent pollution.

Undoubtedly, some problems will remain, but a reverse should be possible in the increase of complaints (Table 11.2) recently experienced. In this connection, EEC-grant systems should take in to account the considerable cost to farmers of upgrading slurry and silage systems to avoid pollution.

11.6 PESTICIDES

Since the war, the use of pesticides (including herbicides) has increased (Table 11.5) and has played an important part in the increasing efficiency of

Table 11.5 — Estimated annual usage of pesticides on crops in agriculture and horticulture in England and Wales, 1971 to 1974, 1975 to 1979 and 1980 to 1983 (treated hectares and tonnes of active ingredients)

Pesticide	1971 to 1974		1975 to 1979		1980 to 1983	
	Treated hectares	Tonnes	Treated hectares	Tonnes	Treated hectares	Tonnes
Organochlorines, insecticides, acaricides	148 000	131	146 000	166	160 000	130
Organophosphate, insecticides	845 800	430	975 000	534	923 000	591
Pyrethroids	1 000	1	41 000	2	172 000	8
Other insecticides, acaricides, molluscicides	922 000	1 286	556 000	905	956 000	664
Seed treatments	371 800	565	3 753 000	591	3 883 000	301
Fungicides	1 895 000	2 400	2 253 000	2 336	6 715 000	4 341
Herbicides, defoliants†	6 003 000	15 250	7 868 000	19 925	12 402 000	3 138
Area of crops and grass	5 631 000‡		9 322 000		10 511 000	

†including chemicals for burning-off. ‡permanent grass was not included in 1971 to 1974.

the industry at all stages. Pesticides are, by their very nature, biologically active substances, with possible side-effects which are difficult to predict. These problems have been recognised for many years and, side by side with increased use, increased selectivity, decreased application rates and more precise placement have all progressed and will continue to do so. For example, the withdrawal of persistent substances in favour of those more easily broken down reduces the chances of long-term problems arising by the build-up of concentrations to polluting levels.

Monitoring the environment for possible effects is also developing as the study of ecology and related disciplines expands. But, in a complex and changing situation, it is not easy to relate the occurrence of a substance in the environment to the observed ecological or other changes which occur at the same time or later. The potential for erroneous conclusions is great. Keeping the contamination of the environment (including water supplied for human and other uses) to a minimum is the only responsible aim, but the balance between benefits (including cheap food) and risks is the essential area of discussion in arriving at that balance. Current moves to control pesticide use in order to reduce contamination are likely to ensure a reduction in risks but, with the continuing introduction of new substances and new methods of application, continued vigilance is essential.

11.7 HORTICULTURE

Water is used in horticulture during growth in fields and glasshouses (see section 11.2), for hydroponics, and for preparation of produce for market.

Hydroponics (nutrient film techniques (NFT) or rockwool) have been available to the industry for many years and have an established use which is relatively small nationally (130 ha, some 9% of the total heated glass). The water is generally recycled for NFT and discharged only when unacceptable for one reason or another. Provided the discharge is subject to appropriate Water-Authority scrutiny, pollution should be minimal. The NFT use is not increasing but some expansion of rockwool use is expected. Here the nutrient fluid is not recycled and some drainage from the rockwool slabs occurs. Attempts to use hydroponic techniques for growing animal fodder seem unlikely to increase rapidly.

Of greater significance is the use of water for washing vegetables and removing unwanted parts of the crop. The disposal of spent wash-water can pose problems if it is too polluting for discharge to a stream. Spreading on land while avoiding runoff to water would be the obvious course, but there is the danger of recycling disease in this procedure.

Statistics for this sector are scanty. The report 'Water for Agriculture: Future Needs' (Advisory Council for Agriculture and Horticulture in England and Wales, 1980) indicated that water demand for protected crops would be constant at 12 000 000 m^3. For washing and processing it indicated that 1 million m^3 were used in 1977, 3 million m^3 were projected for 1985, and 7 million m^3 for the year 2000. Given the move towards a greater consumer demand for cleaned vegetables, this forecast seems reasonable. The potential for pollution is therefore likely to increase and, like other sectors, give rise to problems that can be considerable locally.

11.8 FISH FARMING

In the UK, the most important fish farmed in fresh water is trout (*Salmo gairdneri*), although in Scotland it would be unwise not to mention the production of young salmon. There are minor quantities of other species such as carp (*Cyprinus carpio*) and eel (*Anguilla anguilla*), and this balance reflects also the situation in other northern European countries. During the 1970s, trout production increased sharply. In the UK, the growth was five-fold in seven years. Estimates have been made of future production which suggests rather slower growth in the 1980s, but it is recognised that market forces which are at present unpredictable are likely to have a larger effect on trout production than on the staples of agriculture — cereals, potatoes, milk and meat.

The potential for pollution by fish farms is therefore likely to increase somewhat but, as in other sectors, the pressures to reduce pollution should provide a balancing force and some improvement of the present situation should result, provided profit margins are adequate. One factor here is the possibility of reducing the amount of faecal and other suspended matter discharged, which is a potential cause of pollution. Reduction in the use of cereal binders and fibre in the feed is a possible way forward. With total production at 7000 to 12 000 t/yr, the problem is essentially a local rather than a national one, although the local effects can be considerable. In a few cases, the proportion of stream water abstracted can be high, and small increases in contaminants in the effluent could have a significant effect on the stream below the discharge point.

Trout farms require well-oxygenated water of a high standard of purity. In one survey of producers in the UK, over 10% had substantial losses due to pollution. Although this does not excuse any pollution caused by fish farms, it is a factor in the industry's costs and therefore in their ability to spend money reducing pollution.

11.9 WATERCRESS

Watercress production has remained relatively stable over the years and is unlikely to change dramatically. The rate of use of molluscicides, for example, is therefore not likely to change greatly and, given the current climate of opinion, consideration of new chemicals will involve pollution reduction as a major factor.

11.10 CONCLUSIONS

This chapter has laid out the changes in recent years and indicated, if only in vague outline, possible future changes. Insofar as these can be predicted, major factors arising from agriculture which would increase contamination or pollution seem limited. The two topics of most concern in relation to water supply for consumption are the inputs of nitrogen to agriculture and the use of pesticides. In neither case is the contamination demonstrably causing pollution but prudence, as well as legal constraints, counsel continuing vigilance and the search for an appropriate balance of risks and benefits.

In regard to the wider environment, not only pesticides but general organic materials on the farm caused notable contamination and pollution, The UK Government, under the requirements of the EEC Directive 80/778, has brought into force the Control of Pollution Act (Pollution of Water). Under this, the Minister for Agriculture, Fisheries and Food has issued a Code of Good Agricultural Practice to minimise the risks of pollution by farming operations. Very few, if any, of the pollution incidents recorded in Table 11.1 would haved occurred if good agricultural practice had been followed. Under the new Act, too, Water Authorities have greater powers to prosecute and, if necessary, to propose limitations of agricultural activities on specific areas.

The operation of the Act is supervised by the Department of the Environment. The Water Authorities and the Ministry of Agriculture, Fisheries and Food are currently conducting a publicity campaign to ensure that the agricultural industry is aware not only of the problem, but of the ways in which it can improve its operations — often neighbouring farmers are the first to suffer and there is a widespread appreciation in the industry of the problems of water pollution.

But we cannot afford to be complacent. Increased knowledge of the environment and of human health will undoubtedly reveal new problems which will have to be evaluated. But, on the other hand, increased knowledge will also help to evaluate better factors which are at present of general concern. Perhaps some, maybe nitrate, will be seen to be less fearful than is thought at present. It must be ensured that both legal and operational frameworks are sensitive to new knowledge, whether to tighten or to relax constraints.

REFERENCES

Advisory Council for Agriculture and Horticulture in England and Wales (1980) *Water for Agriculture: future needs*. Ministry of Agriculture, Fisheries and Food, London.

England, R. A. (1984) *Reducing the nitrogen input on arable farms: price effects at farm level*. Divisional Note DN 1254, National Institute of Agricultural Engineering, Silsoe, England.

Royal Society (1983) *The nitrogen cycle of the United Kingdom*. The Royal Society.

Williams, N. T. (1984) Assessing the cost of environmental legislation to the farmer. In: *Impact of environmental legislation on agriculture*. Wye College, Ashford, Kent, England.

12

Leaching of methyl bromide and bromide ions into surface water after fumigation of glasshouse soils

H. de Heer, Ph. Hamaker, L. G. M. Th. Tuinstra, and A. M. M. van der Burg

12.1 INTRODUCTION

In glasshouses in the Netherlands, methyl bromide is commonly used as a soil fumigant to control fungi and nematodes. In 1980, methyl bromide was used on approximately 2500 ha at an average dose of 100 g/m^2.

Until June 1981, low density polyethene (LDPE) films (4 m wide and 0.03mm thick) were used in glasshouses to cover the soil during the fumigation period. It has been estimated (de Heer *et al.*, 1984) that when using an LDPE film, approximately half of the dose is lost into the atmosphere during the first day after application. The emission of methyl bromide into the air caused increasing public concern about the consequences for people living in the vicinity of the fumigated glasshouses and led to demands to reduce the concentrations of methyl bromide in the air. As a result, in 1981 the use of methyl bromide was prohibited, except under special authorisation. In addition, research was initiated to improve the method of applying methyl bromide in horticulture, in order to reduce its emission to the air.

In the Netherlands, experiments were set up to study the efficacy of using gas-tight films to reduce losses of methyl bromide into the air. These trials showed that gas-tight films greatly reduced the emission. To reap the benefits of gas-tight film, the prescribed cover-time of the soil was extended from two days (for LDPE) to ten days for the gas-tight films (de Heer *et al.*, 1984).

It is well known that transformation of methyl bromide during fumigation of glasshouse soils leads to the formation of bromide ions in the soil profile, particularly in the upper part of the root zone. As bromide ions are easily taken up by crops grown on these soils, the acceptable bromide-ion content in the crop may be exceeded, especially for leafy vegetables. Therefore, in the Netherlands it is common practice to leach glasshouse soils with up to 600 mm of water after fumigation.

It was found (Wegman *et al.*, 1981) that when leaching started directly after the films had been removed (that is one week after the start of the fumigation) the concentrations of methyl bromide leached into the drainage water were high (up to 9.3 g/m^3). These concentrations caused the death of small fish in the surface water near the drainage discharge point.

This chapter reports on the effects of using gas-tight films at different doses of methyl bromide, on the concentrations of methyl bromide and bromide ions leached into the drainage water.

12.2 DESCRIPTION OF THE TRIALS

12.2.1 Lay-out of the two trials

The first trial was carried out in February 1981 in two compartments of a glasshouse in Honselersdijk, on a loam soil. Methyl bromide was applied in both compartments using the hot-gas method, at 103 g/m^2. One compartment (1360 m^2) was covered with LDPE. In the other (1900 m^2) a gas-tight film, Saranex 11, was used. The cover time was five days in both cases.

Soil temperature, averaged over a period of three weeks, ranged from 9 °C (at 0.15 m depth) to 11 °C (at 1.05 m depth).

At the end of the cover time (five days), the films were opened up. After three weeks from the start of the fumigation, the films were removed and leaching was started at the rate shown in Fig 12.1.

In the second trial, a dose of 20 g methyl bromide per m^2 was applied to a glasshouse soil in Naaldwijk. The soil was a clay loam. The soil surface was covered with Saranex 11 film and the cover time was 10 days. The average soil temperature during the fumigation period ranged from 14 °C at the surface, to 15 °C at 1.05 m depth.

Three weeks after the start of the fumigation, leaching was started on part of the treated area (4060 m^2).

Drainage water from the tile-drains was sampled in glass bottles (250 cm^3) which were closed with a polyethene cap. The samples were transported and stored upside down (at 4 °C), to minimise losses of methyl bromide through the gas phase.

12.2.2 Analysis for methyl bromide and bromide ions

The samples of drainage water were analysed for methyl bromide within 24 hours. A subsample was taken from each bottle and transferred to a vial (30 cm^3) and stored at room temperature. Then, 500 μl of the top gas phase were injected into a gas chromatograph, equipped with a ^{63}Ni electron-capture detector. The glass column (180 cm long, 0.2 cm inner diameter) was packed with Teenax GC (60 to 80 mesh). The carrier gas was argon/methane (90:10) at a flow rate of 30 ml/min. The temperatures of the injection port, the column, and the detector were 150 °C, 80 °C, and 330 °C, respectively.

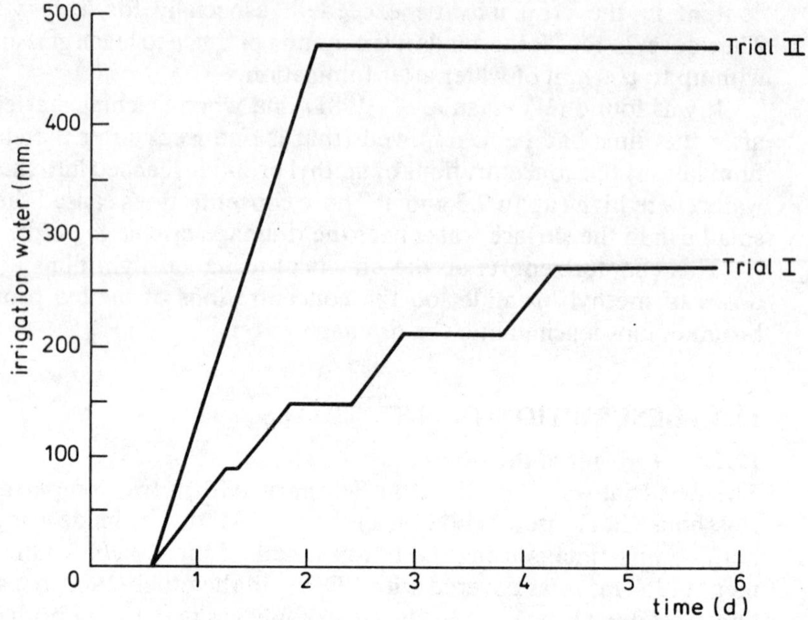

Fig. 12.1 — Cumulative amount of irrigation water in the leaching period of Trials I and II

The average recovery of methyl bromide from distilled water was 100%, with a coefficient of variation of 2%. In the laboratory, four drainage-water samples were spiked with different concentrations of methyl bromide: the average recovery was 108%, with a coefficient of variation of 12%. The decrease in the methyl bromide concentration in four samples of drainage water stored in a refrigerator (dark, 4 °C) was measured. A relatively short half-life of three to four days was found. This suggests that some methyl bromide was lost during the short period of storage, before the top gas concentrations were measured. No corrections were made for recovery or for losses during the period of storage in the refrigerator. The limit of detection of methyl bromide in drainage water was approximately 0.4 mg/m^3.

The concentrations of bromide ions in the drainage-water samples (250 cm^3) were measured using an ion-specific electrode (Orion, type 94–35). A correction for the interference caused by chloride ions was made according to the procedure described by de Bes (1982).

12.3 RESULTS OF THE LEACHING EXPERIMENTS

The concentrations of methyl bromide in the drainage water during leaching in Trial 1 are presented in Fig. 12.2. When using the gas-tight cover (Saranex), relatively high concentrations of methyl bromide (maximum: 4.2 g/m^3) were measured in the drainage water during the first few days. The

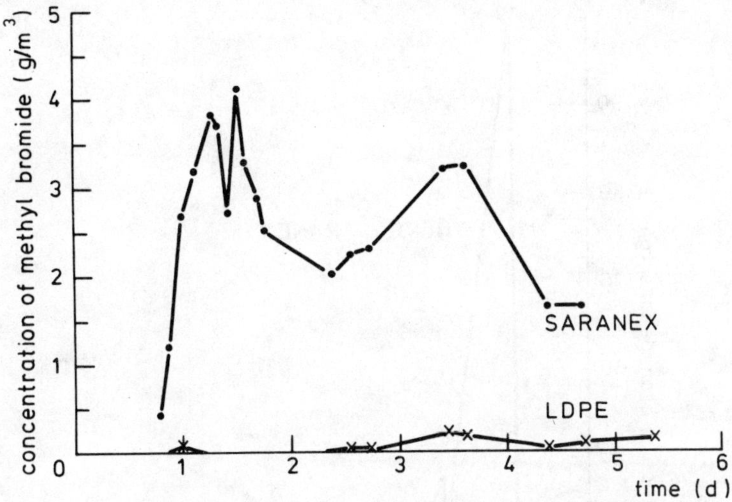

Fig. 12.2 — Concentration of methyl bromide in drainage water during the leaching period of Trial I

concentrations fluctuated considerably, because of the irregular irrigation pattern (see Fig. 12.1). When the LDPE cover was used, the concentrations of methyl bromide in the drainage water were much lower (Maximum 0.22 g/m^3) than with the Saranex cover.

In the second trial, the methyl bromide concentration in the drainage water ranged from nil to 0.014 g/m^3. In half of the samples of the second trial, no methyl bromide was found (the detection limit being 0.0004 g/m^3). The low concentrations in the second trial were clearly associated both with the low dose and with the relatively-high soil temperature.

The concentrations of bromide ions measured in the drainage water during both trials are shown in Fig. 12.3. At the start of the leaching period, the bromide-ion concentrations increased sharply. The concentrations in the second trial ranged from a maximum of 101 g/m^3 on the first day of the leaching, to 25 g/m^3 after four days.

In the first trial, the intake point for the irrigation water was located near a wide canal. Fairly low concentrations of bromide ions (1.4 g/m^3) were measured at the intake point in the leaching period.

In the second trial, the concentration of bromide ions in the irrigation water increased from 4.0 g/m^3 to 8.6 g/m^3 during two days of continuous leaching. This increase was partly caused by water flowing from the drain outlets to the intake point, via small watercourses.

12.4 DISCUSSION

The decrease in methyl bromide in the first 1.2 m of the soil profile during both trials has been reported by de Heer *et al.* (1984): 21 days after the start of the fumigation in Trial 1 (just before leaching) the residual amounts of

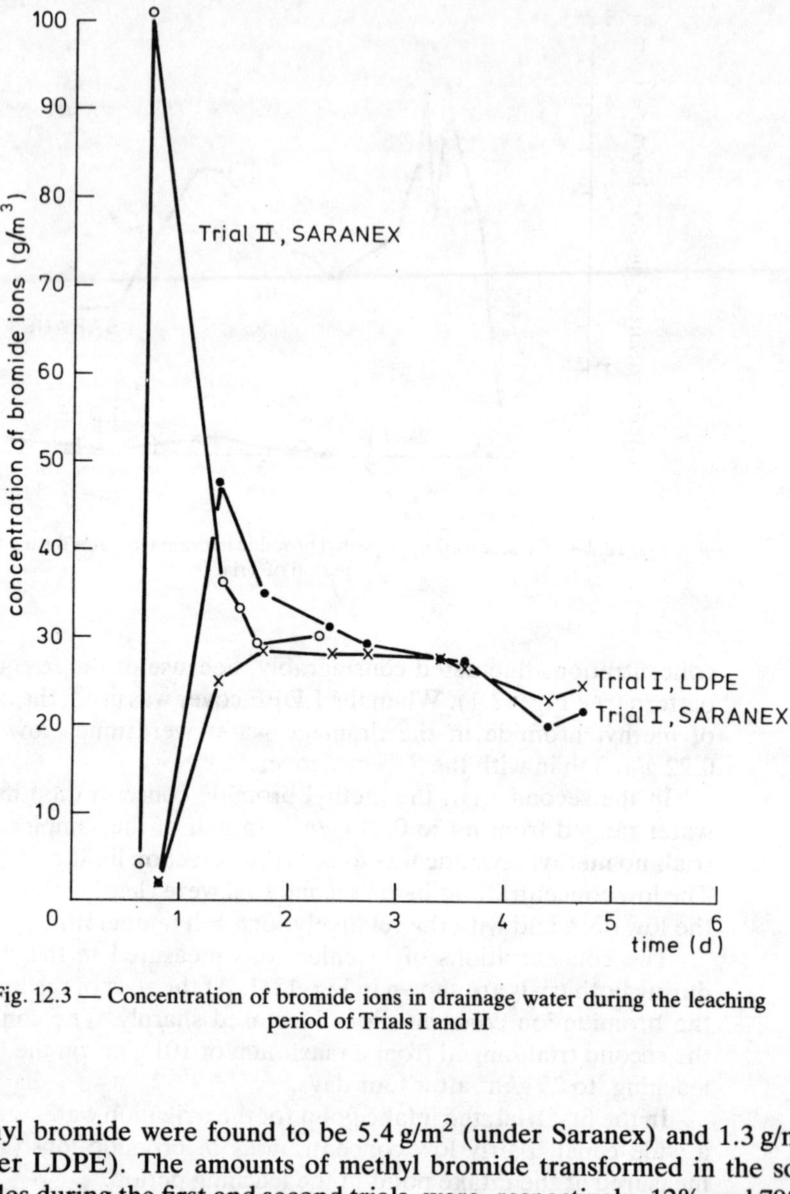

Fig. 12.3 — Concentration of bromide ions in drainage water during the leaching period of Trials I and II

methyl bromide were found to be 5.4 g/m^2 (under Saranex) and 1.3 g/m^2 (under LDPE). The amounts of methyl bromide transformed in the soil profiles during the first and second trials, were, respectively, 12% and 79% of the applied methyl bromide dose.

The amounts of methyl bromide discharged via the drains can be estimated from the methyl bromide concentrations found in the water, and the volume of water discharged. The latter was assumed to be equal to the amount of irrigation water supplied.

In the case of the Saranex cover in the first trial, approximately 13% of the residue of methyl bromide was leached. In the case of the LDPE cover, only some 2% of the residual methyl bromide was leached.

In the second trial the residue of methyl bromide in the soil just before leaching was less than 0.04 g/m^2 (de Heer *et al.*, 1984). The amount of methyl bromide leached was less than 5% of the residue. In fact, this leached amount was less than 0.01% of the applied dose.

Monitoring studies carried out since 1979 in the glasshouse district of the Netherlands (Wegman, 1983) show that the problem of high concentrations of methyl bromide has been greatly reduced since 1981. In autumn 1982 Wegman found that the maximum concentration of methyl bromide in surface water in the vicinity of treated glasshouses was about half of the limit of 0.06 g/m^3 provisionally set by the National Institute of Public Health. Clearly, residues of methyl bromide in drainage water were minimised when leaching of the soil was postponed until after the prescribed waiting period of three weeks, especially if soil temperature was low.

Before leaching in the first trial, the total amounts of bromide ions in the first 1.2 m of the soil profile were 14 g/m^2 under Saranex, and 13 g/m^2 under LDPE. The corresponding value before leaching in the second trial was 21 g/m^2. Under the same assumptions adopted for the leaching of methyl bromide, the amount of bromide ions leached in the first trial was estimated to be 53% of the amount present in the soil profile before leaching. (The figures for under the Saranex and LDPE covers were virtually identical.) The corresponding figure for the second trial was much higher: 83%.

To ascertain how the use of methyl bromide as a soil fumigant contributes to the concentration of bromide ions in surface watrer, a bromide mass balance was set up by Hamaker *et al.* (1984) for a polder district in the west Netherlands for a 12-month period in 1982 to 1983. They found that the total bromide input from precipitation, from water pumped into the canal system of the district to control water levels, and from seepage, was in the order of 1.0×10^5 kg. The total output, which primarily resulted from the discharge of excess precipitation by pumping stations, was in the order of 2.7×10^5 kg. The discrepancy of 1.7×10^5 kg between output and input was attributed to the transformation of methyl bromide. This indicated that approximately 40% of the total use of 4.9×10^5 kg of methyl bromide during the twelve-month period was transformed in the soil. From this it was concluded that the use of methyl bromide for soil fumigation accounted for 65% of the total input of bromide ions into the polder district. The other important input factor was seepage: this accounted for approximately 20% of the total input in the district.

To prevent high concentrations of bromide ions in the surface water, Water Boards nowadays refresh the main watercourses more frequently than in the past. Clearly, leaching with irrigation water that contains high concentrations of bromide ions is not very effective in reducing the concentration of bromide ions in the soil.

12.5 CONCLUSIONS

The leaching of methyl bromide and bromide ions into surface water after the fumigation of glasshouse soils was studied as part of a programme aimed at reducing the environmental load of methyl bromide. In the first trial, the dose of methyl bromide was 103 g/m^2 under gas-tight plastic film (Saranex 11) and under low density polyethene film. The cover time was five days. Three weeks after the fumigation commenced, leaching was started. The maximum concentrations of methyl bromide leached into the drainage water were 4.2 g/m^3 for the gas-tight cover, and 0.22 g/m^3 for the LDPE cover. In the second trial, the methyl bromide dose was reduced to 20 g/m^2, and the cover time with Saranex extended to 10 days. The methyl bromide concentration in the drainage water ranged from nil to 0.014 g/m^3. Concentrations of bromide ions in drainage water ranged from 101 g/m^3 on the first day of leaching to 25 g/m^3 after four days. Residues of methyl bromide in drainage water can be prevented by waiting sufficiently before leaching. The extent of bromide-ion contamination leached into the surface water depends on the dose of methyl bromide, the transformation rate, soil temperature, and the irrigation rate.

REFERENCES

de Bes, S. S. (1982) Storingen bij de ionselectieve bromidebepaling in grond en water en onderzoek naar een spectrofotometrische methode. Stichting Proefstation voor Tuinbouw onder Glas, Naaldwijk, Intern verslag 45, 41 p.

Hamaker, Ph., de Heer, H. & van der Burg, A. M. M. (1984). Use of gas-tight plastic films during fumigation of glasshouse soils with methyl bromide II: Effects on the bromide-ion mass balance for a polder district. *Acta Horticulturea.* **152**, 127–135.

de Heer, H., Hamaker, Ph., Tuinstra, L. G. M. Th. & van der Burg, A. M. M. (1984) Use of gas-tight plastic films during fumigation of glasshouse soils with methyl bromide I: Significance of permeation and leakage for the emission into the outside air. *Acta Horticulturea.* **152**, 109–126.

Wegman, R. C. C., Greve, P. A., de Heer, H. & Hamaker, Ph. (1981) Methyl bromide and bromide ion in drainage water after leaching of glasshouse soils. *Soil, Water, Air, and Soil Pollution,* **16**, 3–11.

Wegman, R. C. C. (1983) Personal communication. Unpublished data. National Institute of Public Health, P.O. Box 1, 3720 BA Bilthoven, The Netherlands.

13

The effects of the agricultural use of herbicides on fresh waters

B. T. Croll

13.1 INTRODUCTION

Relatively little information has been published regarding residues of herbicides in water resulting from application in arable agriculture, despite the very large quantities used. 2,4-D (Frank, 1972; Junk *et al.*, 1976); 2,4,5-T (Junk, 1976); Silvex (Frank, 1972) and Atrazine (Junk, 1976) have been found in rivers, and to a much lesser extent in underground waters. A wider range has been observed in drainage streams below treated areas and in irrigation outflows (Wauchope, 1978). The use of herbicides to control weeds in or near water has been researched much more widely (Frank, 1972), but although their use for clearing land-drainage channels could arguably be called agricultural, this aspect will not be considered in this chapter.

The potential hazard to waters arising from the use of a herbicide will depend upon the quantity used, its solubility in water, persistence, and toxicity. In assessing these factors it must be remembered that chronic toxicity resulting from accumulation in the body can be more important than high acute toxicity, and that persistence in soil can be different from water. For example 2,4-D, which has a persistence in soil of only a few days, has been recorded as undergoing no degradation in stream water after six months (Robson, pers. comm.). Willis and McDowell (1982) have assessed the relative hazards to waters of a wide range of pesticides. Although they show insecticides as having a generaly higher hazard rating than herbicides, quantities used were not taken in to account. Tonnages of herbicides used are generally very much greater than insecticides and therefore herbicides could present a serious hazard to the aquatic environment.

13.2 QUANTITIES OF PESTICIDES USED IN THE UNITED KINGDOM

A survey performed by the Ministry of Agriculture, Fisheries and Food during the years 1971 to 1974 showed the general picture of quantities of synthetic pesticides used in the United Kingdom (UK) (Table 13.1).

Table 13.1
Quantities of pesticides used in the UK 1971 to 1974

Pesticide	Annual usage (tonnes active ingredients)
Insecticides	11 to 95
Fungicides	17 to 390
Herbicides	
MCPA	1660
Mecoprop	1550
Dichloroprop	490
2,4-D	240
Others	26 to 680

The figures shown in Table 13.1 deal only with formulations having a single active ingredient; in addition probably equally large quantities of herbicides were used in admixtures with other pesticides, mainly other herbicides. Mixtures of other types of pesticides were generally much less important.

It can be seen from Table 13.1 that the quantity of herbicides used, particularly phenoxyalkanoic acids, is far greater than any other type of pesticide.

13.3 PERMISSIBLE LEVELS OF HERBICIDES IN WATERS

The toxicity of herbicides to mammals and fish is generally low and the critical concentrations in waters are often determined by organoleptic considerations, with respect to potable water supply or toxicity to plants during overhead irrigation. The lower permissible levels are typically in the range 0.01 to 0.1 mg/l, although in specific instances effects have been recorded at much lower concentrations. For example 2,3,6-trichlorobenzoic acid originating from an industrial effluent affected tomatoes at levels as low as 0.5 μg/l (Williams *et al.*, 1977). A selection of published allowable levels is given in Table 13.2.

It will be seen that the EEC 'Drinking Water' Directive (80/778/EEC) limit is a blanket number. This limit also includes all other types of pesticide, and therefore makes no provision for the very different toxicities of the

Table 13.2
Permissible levels of herbicides in water

Herbicide	Authority	Permissible level µg/l	Max. level detected µg/l
2,4-D	USEPA†¶	87.5	40
	WHO‡‖	100	
2,4,5-T	USEPA	700	
MCPA	USEPA	8.75	
Alachlor	USEPA	700	2.9
Butachlor	USEPA	70	0.06
Paraquat	USEPA	59.5	
Atrazine	USEPA	150	5.0
Propazine	USEPA	325	
Simazine	USEPA	1505	
All	EEC	0.1	

†United States Environmental Protection Agency
‡World Health Organization
¶Ref. National Academy of Sciences, 1977
‖Ref. World Health Organization, 1984

individual compounds. The other limits are derived from toxicological considerations of each compound and show much higher figures, which must be more realistic.

13.4 ANGLIAN WATER SURVEY

In 1982 Anglian Water initiated a survey of all its water sources for a wide range of chemicals, in order to obtain a data base for assessing monitoring requirements under the EEC Drinking Water Directive. The area covered by the Authority is largely used for arable agriculture and therefore a range of pesticides, including herbicides, formed part of the list of determinands. The analytical work was performed by the Huntingdon Research Centre. Grain is the predominant crop in the Anglian region and the herbicides most frequently used are the phenoxyalkanoic acids (2,4-D, MCPA, Mecoprop, MCPB); the tonnage used far exceeds that of any other type of pesticide, as shown in Table 13.1. This group was therefore included in the initial survey list. In mid-1983 gas chromatography-mass spectrometry (GC-MS) work at the Water Research Centre (WRc) and other UK Water Authorities showed that Atrazine, and to a lesser extent Simazine were common contaminants of surface waters, these compounds were therefore added to the determinand suite following development of the analytical methodology in January 1984.

13.5 ANALYTICAL METHODS

13.5.1 2,4-D

The initial method was solvent extraction, butylation and electron-capture gas chromatography (McKone & Hance, 1972), however, following the development of the pentafluorobenzylation method for the MCPA group, this method was used.

13.5.2 MCPA, Mecoprop, MCPB

The analysis of these compounds at the low and sub-microgram per litre levels in waters is difficult. Because the molecules contain only one chlorine atom they give insufficiently low limits of detection to electron-capture gas chromatography unless a more electron-capturing derivative is formed. Several such derivatives including nitro, trichloroacetyl, nitrobenzyl and pentafluorobenzyl have been used, all suffer from interference due to reaction with other carboxylic acids present in the water and in some cases impure reagents. Gas chromatographic methods using halogen-specific detectors, such as Hall conductivity and microcoulometry, and also high pressure liquid chromatography (HPLC), suffer from an inability to achieve sub-microgram per litre limits of detection without excessive manipulation.

Previous analyses performed by the WRc had shown that GC-MS using selective ion monitoring was prone to interference and that careful selection of ions, including comparison of intensity, was necessary in order to obtain reliable results. Regular use of this technique was not available for the Anglian Water survey.

Of the techniques available for analysis of the MCPA group of herbicides electron-capture gas chromatography of the pentafluorobenzyl esters appeared the most attractive, and had proved successful for the analysis of crops (Chau & Terry, 1976). This technique was used for the survey after considerable initial development. The procedure was as follows.

(i) solvent extraction from the acidified sample;
(ii) extraction of the solvent using alkali;
(iii) acidification of the alkaline extract and re-extraction into solvent;
(iv) evaporation and formation of pentafluorobenzyl esters;
(v) electron-capture gas chromatography using a capillary column.

13.5.3 Triazines

Solvent extraction followed by evaporative concentration and determination by gas chromatography using a nitrogen-selective flame thermionic detector (Pressley & Longbottom, 1982), proved successful. Atrazine and Simazine were separated by using a Carbowax 20M column. No interference with the Atrazine response was observed but significant interference occurred with Simazine. This interferant could be removed by clean-up using a de-activated alumina column; however, as Simazine is invariably present at a much smaller concentration than Atrazine, only Atrazine was determined for the survey in order to reduce analytical effort.

13.6 RESULTS

The survey was not complete at the time of writing (completion due in March 1985) but the results to September 1984 are presented.

13.6.1 Phenoxyalkanoic acids

Approximately 600 samples were examined over the period June 1982 to September 1984. The sites where phenoxyalkanoic acids were detected are shown in Fig. 13.1 at concentrations of 0.2 to 2.7 μg/l in surface waters, and 0.2 to 0.4 μg/l in underground waters.

Fifty-nine surface-water sampling points were analysed quarterly or six-monthly over the period of the survey, the more frequent analyses being connected with the larger water-supply schemes. Each borehole was analysed only once during the survey period. Each site marked in Fig. 13.1 has at least three bores, in most cases all the bores at each site contained detectable concentrations of Mecoprop. The limit of detection of the analytical method was taken as 0.2 μg/l for the purposes of this chapter, as this gave responses which were readily distinguishable from the blank. This is regarded as a conservative figure.

Four surface-water samples containing higher levels of herbicides were examined by GC-MS at WRc's Stevenage Laboratory and the presence of Mecoprop confirmed. Confirmation of herbicide identity in samples containing less than 1 μg/l Mecoprop did not prove possible using this technique on the extracts prepared for analysis by GC only.

MCPA was detected less frequently than Mecoprop in surface waters (0.2 to 2.3 μg/l) only. The presence of MCPA was confirmed in the samples examined by GC-MS.

2,4-D was detected less frequently than either MCPA or Mecoprop in both surface and underground waters at levels of 0.2 to 2.5 μg/l. These results are unconfirmed.

13.6.2 Triazines

The sites at which Atrazine was detected are shown in Fig. 13.2. It was found in almost all the surface waters examined at concentrations of 0.2 to 1.4 μg/l. Detection was less frequent in underground waters (0.2 to 0.5 μg/l) where its presence corresponded to unconfined aquifers with light soils and shallow unsaturated zones. Analysis for Atrazine was begun early in 1984 and some 300 samples have been examined. The sampling programme was identical to that described for phenoxyalkanoic acids, but fewer underground water sites have been examined due to the shorter sampling period. The presence of Atrazine has been confirmed in only one sample, by GC-MS.

The results confirm the GC-MS findings of North West Water, Severn-Trent Water, Thames Water, and the WRc, in UK surface and underground waters.

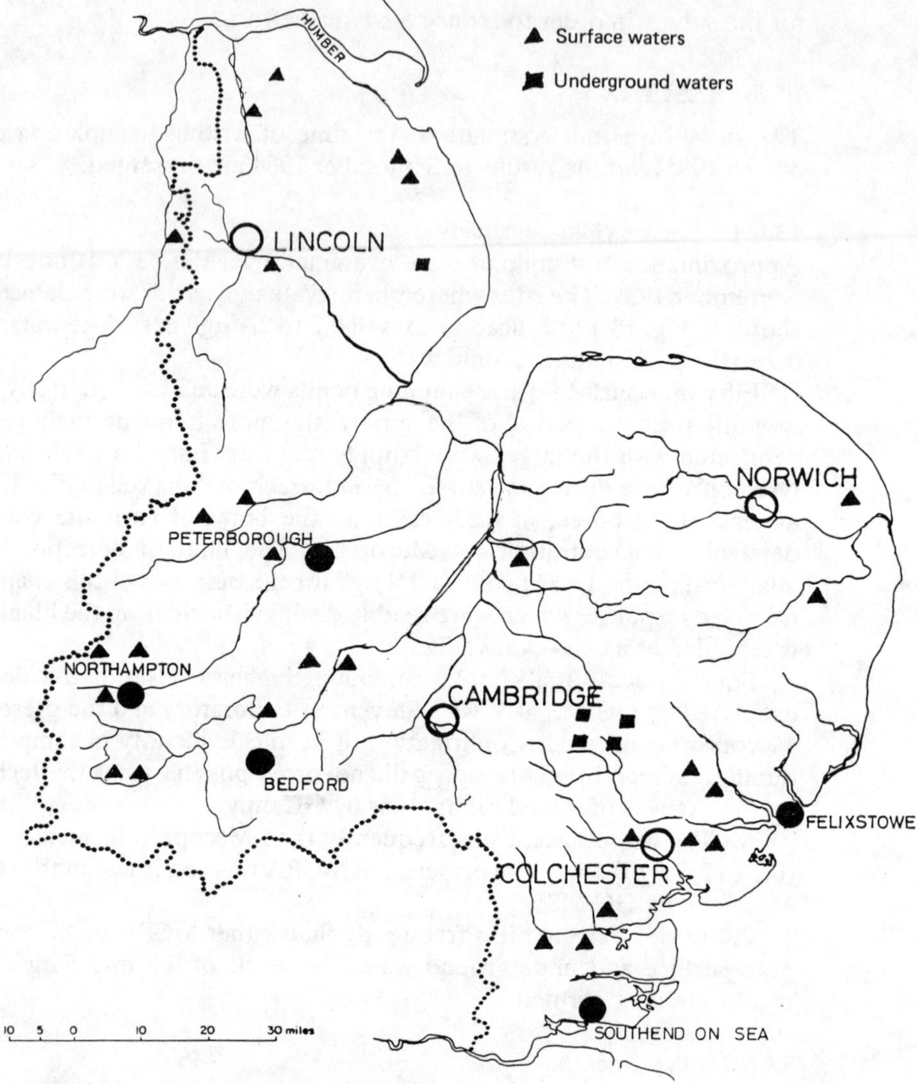

Fig. 13.1 — Sampling points at which phenoxyalkanoic acids have been detected

The widespread presence of Atrazine is unlikely to be due to agricultural use as this is largely confined to sweetcorn and maize, which are minor crops in the UK. It is more likely that its presence at most sampling points arises from total weed control on railway embankments, roadsides, and industrial areas.

As indicated in the section dealing with analytical methods, the presence

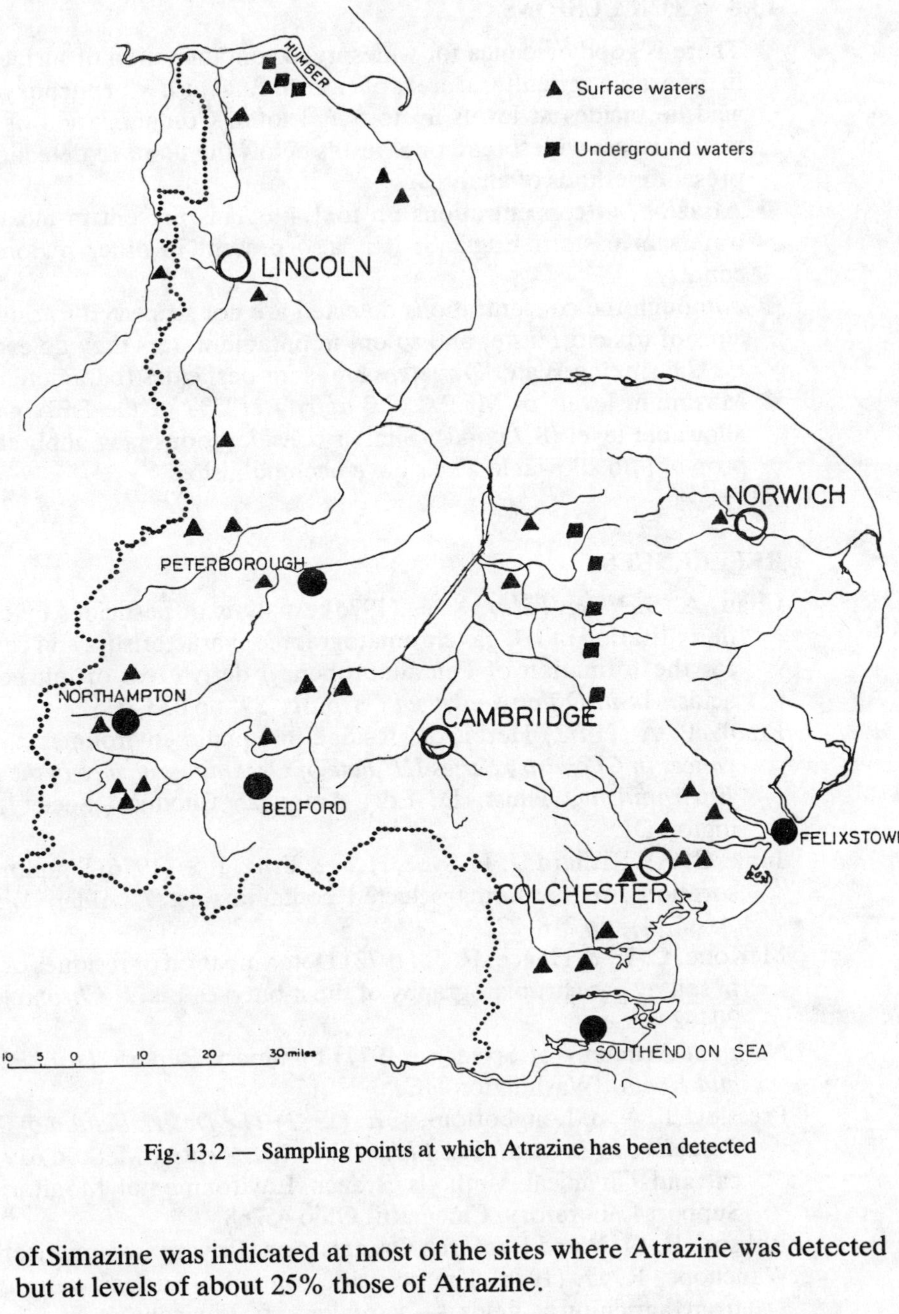

Fig. 13.2 — Sampling points at which Atrazine has been detected

of Simazine was indicated at most of the sites where Atrazine was detected but at levels of about 25% those of Atrazine.

13.7 FURTHER WORK

On completion of the present survey, regular monitoring will be initiated at fewer samping points. Wherever possible, all positive results will be confirmed by an alternative method of analysis.

13.8 CONCLUSIONS

- There is good evidence for widespread contamination of surface waters in the arable agricultural areas of eastern England, with phenoxyalkanoic acid herbicides at levels up to 5 μg/l total. Contamination of groundwaters is less widespread or at levels below the limits of detection of the present methods of analysis.
- Atrazine, at concentrations up to 1.4 μg/l, is present in most surface waters in eastern England; it is also present in other regions of the country.
- Although the concentrations detected are not suspected to cause problems of toxicity, taste, and so on, in potable waters they do exceed the EEC Drinking Water Directive limits for pesticides (parameter 55).
- Maximum levels of MCPA (2.3 μg/l) are 25% of the EPA maximum allowable level (8.75 μg/l). Similar considerations may apply to Mecoprop but no allowable levels have been published.

REFERENCES

Chau, A. S. Y. & Terry, K. J. (1976) Analysis of pesticides by chemical derivitisation pt III, gas chromatographic characteristics and conditions for the formation of Pentafluorobenzyl derivatives of ten herbicidal acids. *Assn. Offic. Analytical Chemists,* **59**, pp 633–636.

Frank, P. A. (1972) Herbicidal residue in aquatic environments, in: *Advances in Chemistry Series III, Fate of Organic Pesticides in the Aquatic Environment*. (Faust, D. Ed.) American Chemical Society, Washington DC.

Junk, G. A., Richard, J. J., Svec, H. J. & Fritz, J. S. (1976) Simplified resin sorption for measuring selected contaminants. *J. Amer. Wat. Wks Assn.* (4), pp 218–222.

McKone, C. E. & Hance, R. J. (1972) Determination of residues of 2,4,5-T in soil by gas chromatography of the n-butyl esters. *J. Chromatogr.* **69**, 204–206.

National Academy of Sciences (1977) Summary Report: *Drinking Water and Health*. Washington DC.

Pressley, T. A. & Longbottom, J. E. (1982) *The Determination of Triazine Pesticides in Industrial and Municipal Wastewater*. Method 619. Physical and Chemical Methods Branch Environmental Monitoring and Support Laboratory, Cincinatti, Ohio 45268.

Robson, R. O. Weed Research Organisation. Private Communication.

Wauchope, R. D. (1978) The pesticide content of surface water draining from agricultural fields — a review. *J. Environ. Qual.,* **7** (4), pp 459–472.

Williams, J. H., Kingham, H. G., Cooper, B. J. & Eagle, D. J. (1977) Growth regulator injury to tomatoes in Essex, England. *Environmental Pollution,* **12**, pp 149–157.

Willis, G. H. & McDowell, L. L. (1982) Pesticides in agricultural run-off and their effects on downstream water quality. *Environ. Toxicol. and Chemistry,* **1** (4), pp 267–279.

World Health Organization (1984) *Guidelines for Drinking-Water Quality,* Vol 1. Geneva.

14

Water quality—the French problem

S Henin

14.1 INTRODUCTION

In line with recommendations made in France by various administrations, the Ministries of Agriculture and the Environment decided to set up an *ad hoc* Working Group in which all representatives of activities concerned with water quality were to be represented. This 'Farming and Water Quality' Working Group consisted of three committees concerned respectively with plant production, livestock production, and crop-protection products. This subdivision was justified from the public services standpoint owing to the fact that it concerned three areas on which the authorities could take specific action.

Since this chapter is necessarily comparatively brief, it will deal mainly with the problem of nitrate pollution. In view of the nature of this conference, however, it would be worthwhile remembering that the Working Group concluded that phosphate pollution was mainly of urban origin (from detergents) and that only 1% of the crop-protection products found their way to water. This last finding naturally raises problems such as the unknown fate of these substances and the analytical difficulties that arise if the chemicals undergo changes after application.

14.2 CHANGE IN THE NITRATE CONTENT OF WATER

When deciding what approach to adopt, the first step of the Working Group was to identify the source of the nitrates present in the waters under investigation.

The French Service responsible for monitoring surface water used to have 711 sampling points over the entire national territory; there are now 1250. Comparision of the 1971 and 1976 surveys produced the results given in Table 14.1. Therefore throughout the whole of France there has been a

Table 14.1 — Changes in the concentration of nitrate in French surface waters between 1971 and 1976.

Concentration of nitrate observed (mg/l as NO_3)	Number of sampling points					
	<3	10	20	50	>50	Total
1971	232	317	130	30	2	711
1976						
— no change	114	203	96	18	1	432
— improvement	40 ←	21 ←	9 ←			
	3 ← (from 20)					75
		1 ← (from 50)				
		1 ← (from >50)				
deterioration		→ 112	→ 73	→ 10	→ 2	
			→ 6 (from <3)			204
				→ 1 (from 10)		
Nett situation in 1976:	157	338	184	29	3	711

distinct improvement in 75 cases and deterioration in 204. Furthermore the picture varies considerably from region to region and, overall, the greatest deteriorations occurred in the north-western area of France, while in the south and south-eastern regions, the improvements and deteriorations were evenly matched.

Results from the period 1976 to 1981 have not yet been fully analysed, but the tendency appears to be much the same. Unfortunately, the data for aquifers are not so systematic but, as an example, the situation in the Department of the Yonne is shown in Table 14.2. This study involved

Table 14.2 — Department of the Yonne — nitrate content in the areas of the public supply networks

Nitrate (mg/l as NO_3)	Percentage of catchment areas 1969	1979
<22	98	34.5
22 — 44	2	44.4
>44	0	21.1

analysing the waters of catchment areas supplying the distribution networks. The number of sites studied was roughly constant (310 to 350) and the comparison deals with the years 1969 and 1979.

There is a marked and rapid deterioration in quality which unfortunately confirms the analyses performed on water of the same origin in other Departments. Generally, the increase in nitrate concentration is of the order of 1 to 5 mg/l/yr. However, in the case of this Department, if one examines the small natural regions in the east where livestock are raised, the nitrate concentration of the water remains constant, whereas it has increased in areas where intensive agriculture is practised.

The Ministry of Health in turn is monitoring the quality of potable supplies. The tendency to deterioration is less marked but the overall result is that approximately 2% of the French population is supplied with water which has a content in excess of 50 mg NO_3/l. On closer examination there is a definite relationship between the areas where the population is supplied with water having a comparatively high nitrate content and zones where intensive farming is practised. The possible sources of these nitrates therefore have to be assessed.

14.3 SOURCES OF NITROGEN

The three committees of the Farming and Water Quality Working Group have assembled the data given in Table 14.3. Not all these figures are of the

Table 14.3 — Quantities of nitrogen involved annually as a function of their origin

Origin of nitrogen	Quantities released or contributed in France (millions of tonnes of nitrogen per annum)
Mineralisation of organic matter in the soil	3.0
Inorganic fertilizers	2.0
Atmospheric contributions	0.5
Domestic sewage	0.2 — 0.25
Farm wastes (livestock rearing)	2.0
Symbiotic and non-symbiotic fixation	1.3
Nitrogen of an industrial origin (excluding inorganic fertilizers)	0.1
'Import-export' balance of food products	0.1

same precision and although the quantity of industrial fertilizers is well known, the figures for mineralisation and fixation of nitrogen are naturally much less certain. Nonetheless it is true that the orders of magnitude of these figures are fairly reasonable in the sense that they agree fairly well with data obtained for small surface areas.

Overall, the quantity of fertilizing components therefore only represents a little over 20% of the quantity of nitrogen available over 50 000 000 ha of French countryside. It will subsequently become apparent that since the use of nitrogen fertilizers of industrial origin is comparatively localised, their regional importance can be greater. These estimates are comparatively useless and to make use of them one must place them in the more general perspective of the nitrogen cycle.

14.4 THE GLOBAL NITROGEN CYCLE

Recent work on the global nitrogen cycle (Söderland & Svenson, 1976) suggests that losses of nitrogen by the gaseous route are of the samc order of magnitude as contributions from the atmosphere. It should be pointed out, however, that on a world scale, nitrogen contributions of inorganic fertilizers are of the order of one third of the fixation associated with natural processes, whereas in developed countries, particularly in France, these quantities are of an equal order of magnitude.

14.4.1 France

In fact, all natural inputs represent 1.3m tonnes (biological fixation) + 0.5m tonnes (precipitation) an amount bordering on the 2 000 000 t inorganic fertilizers. It can therefore be asked if the intensity of the process leading to losses by the gaseous route is sufficient to bring about the loss of the supplementary inputs. This leads to the calculation of the balance on a national scale. The input data shown in Table 14.4 have been drawn from Table 14.3. For the outputs, the nitrate content of surface water has been calculated from the measurements at the 711 sampling points on rivers (section 12.4): these data (0.8×10^6 tonnes) are therefore comparatively reliable. The losses to the atmosphere by denitrification from the water are of the order of 0.4×10^6 t, so the amount of nitrates leached from the soil is about $(0.8 + 0.4)$ 1.2×10^6 t.Losses to the atmosphere (from animal wastes) are as ammonia and other forms of nitrogen, their estimates, based on experimental balances are of a fairly reliable order of magnitude: about 10^6 tonnes. On the other hand, the loss of 10^6 t ascribed to denitrifaction from the soil is calculated from more questionable data.The analyses reflect a difference of 0.5×10^6 t; but this could be attributable to uncertainties in the data.

However, a series of observations has been carried out both in France and in other countries and in fact, it has been found that beneath farmed land, waves of nitrates each representing an annual leaching are moving downwards at speeds of 0.5 to 1 m/a. Consequently, the overall quantity of mobile nitrates stored at a depth of 1 to 3 m below certain crops can be of the order of as much as 0.5 t/hectare.

Therefore qualitatively, the deficit appearing in the balance analysis is confirmed and it appears that above our aquifers there may be a mass of nitrates that present a risk of pollution sooner or later. It is also interesting to

Table 14.4 — Balance of the inputs and outputs of nitrogen in France on a national scale (millions of tonnes)

Inputs		Outputs	
Inorganic fertilizers	2.0		
		Nitrates in surface water	1.2
Atmospheric contribution	0.5		
		Losses by the gaseous route (ammonia — oxides of nitrogen and nitrogen)	2.0
Biological fixation	1.3	Import-export balance	0.1
Total	3.8	Total	3.3

Δ (Inputs — outputs) = 0.5

consider the ranges of nitrate content in the water in the aquifer (the first beneath the groundwater) as a function of land use. The data in Table 14.5 have been obtained from the region under the control of the Agence de Bassin Seine-Normandie. This result corresponds to very simple logic: since

Table 14.5 — Concentrations of nitrate in some aquifers of the Bassin Seine-Normandie

Land use	Number of Observation Points	Range of concentrations recorded (mg NO_3m/l)
Ancient forest	30	0 — 8
Mixed woodland and pastureland	80	1.5 — 15
Mixed farming	30	3 — 19
Intensive farming	200	15 — 130
Semi-urbanised farming area	50	20 — 150
Industrial and urban area	20	25 — 150

nitrogen is vital to life, the increase in the quantities of nitrogen used and the risks of attendant losses are rising at much the same pace. The problem of the protection of water is therefore to break this pattern, which leads one to consider how the nitrogen losses manifest themselves based on timely and hence more accurate observations.

14.5 RESEARCH INTO METHODS FOR PREVENTING WATER POLLUTION

Table 14.6 reproduces the observations made in clay/limestone soil using a lysimeter in the Châlons/Marne region. It is clear that, between bare ground

Table 14.6 — Nitrogen losses as a function of plant cover (Chalons-sur-Marne) (Rainfall 630 mm; Clay/lime soil on limestone)

	Crop		
	Bare Ground	Grassland	Arable/crop/ sugarbeet/wheat
Drainage in mm	266	144	164
Fertilization (kg N/ha/yr)	0	0	129
Leaching losses (kg N/ha/yr)	116	11	35
Concentration in drainage water (mg/l as NO_3–N)	43.6	7.6	21.7

and grassland, there is a tenfold difference in nitrogen losses, while under cultivation losses are only about three times greater than from grassland.

Table 14.7 deals with crops growing on sandy soil. Once again, figures demonstrate the low rate of leaching beneath permanent grassland compared with an annual maize crop. The effect of a catch crop (Italian rye-grass) between two crops of maize is to reduce loss of nitrogen while there is a considerable increase in losses where soya is grown. Old experiments performed at Versailles in lysimeters, unfortunately fairly small-sized, had shown that below a crop of lucerne, the nitrate losses were very low. On the other hand, after the land had been worked, a very sizeable increase in losses occurred over the next two years representing 150 kg N/ha.

In lysimeters installed at Theix, for comparing soils of various depths covered either by permanent vegetation or by annual forage crops, the results shown in Table 14.8 have been recorded. Below permanent grass-land, with soil up to 40 cm deep, the annual losses per hectare were 24 kg after the addition of 373 kg of fertilizer, but under the same conditions, in

Table 14.7 — Nitrogen losses as a function of plant cover (Landes region) (Rainfall: 880 mm + 410 mm irrigation; sandy soil)

	Crop			
	Grassland (Fescue)	Maize (permanent)	Maize (permanent + Rye-grass in winter)	Soya
Drainage (mm)	573	707	703	648
Fertilization (kg N/ha/yr)	122	207	207	30
Leaching losses (kg N/ha/yr)	4.1	44.3	23.8	80.4
Concentration (mg/l as NO_3–N)	0.7	6.3	3.4	12.4

soil 80 cm deep, the losses were reduced to 3 kg. Beneath an annual forage crop, although the fertilizer application was only 152 kg/ha, the drainage losses range between 57 kg/ha/yr for soil 40 cm deep to 32 kg for a 160 cm depth of soil available to roots. The effect of depth is still evident but much attenuated compared with the permanent grassland and it should be noted that the average concentration in the drainage water equalled or exceeded the admissible thresholds. However, the estimated total exports through drainage and harvesting are above the dose of fertilizer applied. On the other hand, these lysimeters have only recently been installed and although the tests have been running for six years, release of nitrogen from reserves in the soil may still be playing an appreciable role.

Thus, there are situations where the nitrogen exports considerably exceed the imports over a number of years despite fairly abundant harvests.

Finally, the reader's attention is drawn to the work of Kolenbrander, whose work is referred to in the report of the Working Group (Ministry of Agriculture and the Environment, 1980), who has collated data from the Netherlands and shown that beneath permanent grassland, the nitrogen losses are roughly identical irrespective of the type of soil up to a certain rate of nitrogen fertilization, but this is of the order of 350 kg for sandy soils and 600 kg for clayey soils.

Beneath arable land, all other things being equal, a clayey soil loses seven to eight times less nitrogen than a sandy soil and the losses increase slowly with the rate of application of nitrogen fertilizer up to doses of the order of 200 kg after which they increase very quickly.

This collection of findings can be summarised by drawing up the following list.

Below permanent vegetation, losses remain low until high rates of nitrogen fertilizer (from 350 to 600 kg/ha) are applied.

Table 14.8 — The influence of ground cover and depth of soil on the losses of nitrate (Theix).

	Permanent grassland			Annual forage crops			
Depth of Soil (cm)	20	40	80	40	80	120	160
Fertilization (kg/ha/yr)	373	373	373	152	152	152	152
Export of nitrogen by harvesting (kg/ha/yr)	272	310	352	119	152	178	187
Nitrogen losses by drainage (kg/ha/yr)	24	24	3	57	47	47	32
Concentration in drainage water (mg/l as NO_3)	31	31	4	66	57	62	44

The ploughing up of permanent vegetation results in very marked leaching of nitrates which may continue for five or ten years and perhaps even longer depending upon the type and quantity of organic matter present. This effect persists even when the soil is covered with vegetation although it is lessened under these circumstances.

The presence of legumes, for several years at least, does not cause any additonal losses of nitrogen in water. But the amounts accumulated are released rapidly after destruction of plant cover as can be seen even after only an annual crop such as soya — at least in sandy soil, and after three years of lucerne.

The type of soil plays an important part in terms of (a) depth, and (b) structure.

(a) The depth of the soil must exceed 80 cm for there to be a definite protective effect even below permanent vegetation.

(b) The less clayey the soil the stronger the leaching process. For a given quantity of leachable nitrates the amount leached would seem to vary in the ratio:

$$\frac{Dr}{Cr+Dr}$$

where Cr is the soil's water-retention capacity over the depth used by the roots, and Dr is the quantity drained, both, expressed in millimetres.

One of the problems currently facing researchers is to understand the mechanisms by which, below vegetation-covered soil or for soil of a certain depth, nitrate losses remain low up to substantial nitrogen concentrations. However, below permanent grazed pasture the export of nitrogen is rarely known to exceed the order of 50 units by this means while fertilizer applications occasionally exceed 300 kg N/ha.

14.6 PREVENTIVE MEASURES

From these findings, a certain number of techniques can be envisaged for limiting the leaching of nitrates.

Since this occurs particularly beneath bare soil, one idea might be to maintain plant cover between crops by introducing green-manure crops or catch crops. The experimental data referred to earlier (Table 14.8) show that the process is effective but the practical conditions of use are limited by the following constraints.

Soil and climatic conditions must enable the protective crop to be sown at the time of harvesting the main crop and at the time of ploughing-in, for the sake of the production crop to be sown next. From the few tests performed in France, only half of the country seems capable of being treated by this type of water protection. As far as perennial crops, fruit trees, and vineyards are concerned, the cultivation of cover plants could be considered. However, the handling of this vegetation must be studied to prevent it from competing with the main crop without losing ability to retain nitrate. On the other hand, this ground cover may cause increased spring frosts, so the use of these techniques will probably be limited to certain pedoclimatic zones and some systematic experimentation must be undertaken particularly since the fight against erosion is also involved.

For annual plants, the problem is to adjust the use of nitrogen fertilizers both to the crops' general and immediate needs. The method of forecasting the nitrogen balance is used for this purpose.

The quantity of fertilizing elements required to supplement the crop's needs are calculated by first assessing the soil's supplies (mineralisation from humus, organic wastes and mineral nitrogen content, and atmospheric contributions). The total crop requirements are determined by multiplying the anticipated yields by the nitrogen content and adding an additional quantity of nitrogen to cover the non-utilised margin, this last estimated parameter being arrived at experimentally from a series of tests. The dates of fertilizer application are determined by the stages in physiological development of the crop. Use of this method has yielded fairly satisfactory results for some winter crops. The anticipated yield is taken as the average of the two best yields for the last five years for the variety grown. Only few data on the pollution reduction resulting from application of this idea are available as yet. Some experiments are in progress which will seek answers to this question. (Obviously, the crop must be protected by pesticide treatment as necessary.) It should be pointed out that hopes of using rational methods of determining fertilizer use are necessarily limited because the margin of acceptable error is comparatively narrow, given that 10 mg/l of nitrogen in water represent 20 kg of nitrogen per hectare for 200-mm drainage. This also means that the risks of pollution are very closely linked with climatic conditions.

Other possibilities will be covered by experimental reports such as the effect of reducing tillage, thanks to direct sowing, the effect of soil structure on the nitrogen utilisation factor and the inhibition of nitrification at some period by chemical treatment.

Lastly, the ploughing-in of stubble, particularly when this has a high C:N ratio, might offer a partial solution inasmuch as the microbial growth which it induces would either modify the inorganic nitrogen present in solution or cause denitrification. Research is in hand to define the conditions ensuring the efficiency of these processes. More references are needed providing information on the required levels of nitrogen fertilization for orchards and the various horticultural crops.

14.7 REMEDIAL MEASURES

The flux of pollutants may be channelled by field drainage to prevent them from penetrating down to the aquifers from which potable supplies are drawn. In some zones, and for a considerable length of time, soakaways have been constructed to get rid of drainage water and these are often used nowadays for the disposal of wastewater. This is a source of pollution which has been spotlighted by measuring the abundance of the nitrogen isotope ^{15}N. However, effects of the same order may result from poorly protected boreholes resulting in communication between deep aquifers. These are known to have nitrate contents which vary considerably according to their origin and although not a general fact, the water from deep aquifers is often of better quality than that of groundwater nearer the surface.

Some sufficiently-detailed hydrological surveys provide knowledge of zones supplying the aquifers and the communications that exist between them; these are vital data for planning the rational utilisation of water resources. They could serve to rationalise the planning of land use in such a way that the quality of the aquifers is protected by maintaining permanent vegetation in the infiltration zones which supply them.

It is also possible that in confined aquifers denitrification may be the purifying factor, but the mechanisms involved have yet to be defined.

As regards surface water, pollution can occur where there are subsurface discharges. Then again, the distribution of crops over slopes, and the maintenance of permanent farmed vegetation on flood plains could also attenuate the concentration of nitrates in these waters.

Certainly, denitrification is a phenomenon which can be put to good use.

Pilot plants are available to purify water intended to supply major towns. At present, adapted techniques are being sought for treating the small volumes of water required for less important uses or for small communities.

14.8 CONCLUSIONS

Deterioration of water, particularly as a result of increases in concentrations of nitrate, is a fact of civilised life associated with the increase in human activities and with the intensity of land use. The authorities have thought to combat this phenomenon by setting up a system of taxes but in view of the complexity of the phenomena involved, no rational basis can be found for such a structure at the moment, and so no positive result can be expected from the point of view of water quality.

On the other hand, there are a large number of possibilities, the main ones of which have just been mentioned, which should help to improve — and perhaps even master — the current situation. Their use entails carefully coordinated technical experimentation and research effort. This is the attitude adopted in France where the government has set up a Standing Committee for the Study of Nitrate Pollution on which all interested parties are represented. Its programme involves monitoring water quality with a view to assessing any advances which might be made by modifying farming techniques. It is also up to this Committee to set out the rules for protecting groundwaters and even more importantly perhaps to ensure their rational administration.

At the moment research work is concentrating on acquiring a better understanding of denitrification, the mechanisms of nitrification and reorganisation of nitrate in soils using the latest agricultural techniques.

REFERENCES

Minsitry of Agriculture and the Environment (1980) *Farming activities and water quality*. Working Group Report (2 vols), Paris.

Söderlund, R. & Svenson, B. H. (1976) Nitrogen, phosphorus and sulphur global cycles. SCOPE Report 7. *Ecol. Bull.* Stockholm, pp. 22.

15

Nitrate in groundwater and the effects of ploughing on release of nitrate

C. P. Young

15.1 INTRODUCTION

Increases in the concentration of nitrate in groundwaters abstracted from certain aquifers in the United Kingdom have been reported since the mid 1970s, principally from the Chalk of East Anglia, Yorkshire and southern England (Greene, 1980; Forster & Crease, 1974; Greene & Walker, 1970) the Lincolnshire Limestone (Greene, 1980; Oakes *et al.*, 1981) and the Triassic sandstone of Nottinghamshire and the West Midlands (Oakes *et al*, 1981; Satchell & Edworthy, 1972; Severn-Trent Water Authority, 1976). A recent review of the situation (Standing Technical Advisory Committee on Water Quality, 1984) has confirmed the persistent upward trends in many of those areas and has recorded that about 100 groundwater supplies, with a total output of 500 Ml/d, currently produce water which consistently or intermittently exceeds the WHO Guideline for nitrate in drinking water of 11.3 mg N/l.

Detailed field investigations of the problem were initiated by the Water Research Centre in 1974 and by the British Geological Survey (formerly the Institute of Geological Sciences) in 1975. These studies, including those undertaken by certain Water Authorities, were co-ordinated through a Working Group convened by the Central Water Planning Unit of the Department of the Environment.

The results of the studies have been extensively reported (Oakes, *et al.*, 1981; Foster & Young, 1981; Young *et al.*, 1976; Young & Gray, 1978; Young *et al.*, 1983; Young, 1981) the principal conclusion being that a strong positive correlation exists between high leaching losses of nitrate from agricultural soil on the outcrops of the principal aquifers, and arable cultivation. By contrast, leaching losses beneath long-term grassland or woodland were found to be generally small. The presence of markedly sinuous profiles of nitrate with depth was found to be characteristic of the unsaturated zones of the Chalk, and to a less marked extent, the Triassic Sandstone aquifers (Young & Gray, 1978) in areas of arable/ley rotations. These observations, together with the presence of distinctively peaked depth profiles in the concentrations of thermonuclear tritium in interstitial water of the unsaturated zone (Smith *et al.*, 1970) lead to the hypothesis that the rate of release and leaching of nitrate from agricultural soils could be related to antecedent farming practices, and that a high proportion of the leached nitrate moves relatively slowly downwards through the unsaturated zones of the Chalk and Triassic Sandstone (Young *et al.*, 1976; Young & Gray, 1978; Young *et al.*, 1983; Young, 1981).

Mathematical models have been developed by the Water Research Centre (WRc) to describe the leaching of nitrates from agricultural soil and its transport and dispersion in groundwater systems (Foster & Young, 1981; Young, *et al.*, 1976; Young, *et al.*, 1983; Oakes, 1982). The models have been successfully applied to groundwater catchments in the Chalk, Triassic Sandstone and Lincolnshire Limestone aquifers, and validated by comparison with historical groundwater quality data. The models initially (Young *et al.*, 1976) relied on empirical relationships between the type of crop, of fertilizer rate of application, and other factors, derived from published information, to estimate probable annual leaching losses. Because of the importance of the use of such predictive models in assisting the planning of future supply strategies, field-scale experiments were carried out over the period 1978 to 1982 to provide additional data for inputs to the models.

15.2 SOURCES OF NITRATE LEACHED FROM AGRICULTURAL SOILS

A recent review (Central Water Planning Unit, 1977) of the relative contributions of potential sources of nitrogen to agricultural land in the United Kingdom, indicates that the proportion of inputs supplied by inorganic fertilizers has increased substantially during the last 45 years (Table 15.1), so that in areas dominated by arable regimes inorganic fertilizers form the main input.

Studies of stable cultivated and uncultivated soils (Smith & Young, 1975) have suggested that their nitrogen pools remain essentially constant, reflecting a dynamic balance between nitrogen inputs and removals in crops, in

Table 15.1 — Comparison of ranking of sources of nitrogen in arable, arable/ley, and dairying catchments (data from Central Water Planning Unit, 1977)

Catchment area	Farming type	Period	Source (as percentage of total)				
			Fixation	Rainfall	Animal waste	Human waste	Inorganic fertilizer
Witham and Steeping	Arable	1938–1939	55	11	24	4	6
		1970–1972	35	7	15	5	102
Thames	Arable/ley	1938–1939	49	11	20	19	1
		1970–1972	35	6	18	17	23
Bristol Avon	Dairying	1938–1939	53	11	28	6	2
		1970–1972	40	7	25	5	23

gaseous loss and by leaching. Significant changes in the nitrogen pool have been found to accompany the change from one stable regime to another (Meintz et al., 1977), with the cultivation of land generally being accompanied by the loss of significant quantities of nitrogen (Reinhorn & Avnimelech, 1974).

The principal soil types developed on the outcrops of the recharge zones of the aquifers studied are classified under the Rendzina (Chalk and Lincolnshire Limestone) and Brown Earth and Podzol (Triassic Sandstone) groups. In all cases the soils are characteristically thin and free-draining, from which losses by denitrification would be expected to be minimal. Under normal climatic conditions in the United Kingdom it is probable that a high proportion of applied inorganic fertilizer is assimilated by crops, weeds and microflora during the growing season. It was, therefore, recognised early in the WRc's research programme (Young *et al.*, 1976) that direct leaching losses of inorganic fertilizer to aquifers were unlikely to account directly for the observed increase in groundwater nitrate. Instead, it was postulated that the nitrogen in the increased crop residues and biomass resulting from the use of fertilizers on arable crops is converted to soluble inorganic forms by microbial mineralisation following harvest and cultivation. The residue from a crop consists principally of decomposable plant material and increased soil biomass with decay half lives of less than two years (Jenkinson & Rayner, 1977). Experimental studies of the effects of the incorporation of fresh crop residues in soils (Atkins, 1976) have suggested that a high proportion of the added nitrogen is lost within a 3-year period, from which empirical control rates for nitrate release in the mathematical models were deduced (Young, *et al.*, 1976) (Table 15.2).

Table 15.2 — Nitrogen release from crop residues (as percentage of total N available)

Crop type	Year 1	Year 2	Year 3
Roots and brassicas	0.70	0.20	0.10
Cereals	0.50	0.30	0.20

The initial estimates of the relationships between crop types, fertilizer application rates, and leaching losses, used to generate inputs to the predictive mathematical model were derived from published information on lysimeter experiments and crop uptake trials. A thorough review of 57 lysimeter experiments, covering a range of crop and soil types, (Atkins, 1976) demonstrated that the rates of loss of nitrate leaching from grass-covered, arable cropped and fallow soils were in the approximate ratio 1:5:30 respectively. Crop uptake experiments have been conducted in which comparisons were made between the nitrogen content of specific crops at specified rates of fertilizer addition and crops of the same varieties grown on unfertilized control plots. As a first approximation it was assumed that the unaccounted for portion of applied nitrogen was incorporated into the crop residue and soil biomass, and would form the pool of readily biodegradable material from which nitrate would be released by mineralisation. The principal evidence employed was from field trials at Rothamsted and Woburn (Johnston, 1976), although data from other sources were incorporated (Dampney, 1979; Cameron *et al.*, 1978; Herbet, 1977). Based on the available information, and to comply with the requirement that when no fertilizer was applied there should not be an unaccountable loss, relationships were expressed in the form y=mx, where

y = unaccountable N (kg N/ha)
x = rate of fertilizer application (kg N/ha).

The results are summarised in Table 15.3, which indicate that for a range of commonly-grown arable crop types, a mass of nitrogen equivalent to approximately 50% of the applied fertlizer remains unaccounted for, to form a pool from which losses by leaching or denitrification may subsequently occur.

Table 5.3 — Summary of empirical relationships between applied fertilizer N and unaccounted N, for various crops

Crop type	m (fraction unaccounted for)	r (product movement correlation coefficient)	Sy (standard error of estimate)	Size of sample	Source of data
Winter wheat	0.49	0.99	5.3	10	Johnson, 1976. Dampney, 1979
Spring barley	0.53	0.99	5.1	11	Jonhson, 1976 Dampney, 1979
Potatoes	0.55	0.99	3.7	6	Johnson, 1976 Cameron *et al.*, 1976
Sugar beet, whole crop removed	0.41	0.99	8.9	5	Johnson, 1976 Herbert, 1977
Sugar beet, tops returned	0.74	0.99	5.0	5	Johnson, 1976

The high percentage for beet crops in which the foliage has been returned to the field, has been attributed by Widdowson (1974) to the presence in the leafy part of the major portion of the nitrogen taken up by the crop, and its rapid degradation and leaching during the autumn and winter.

Lysimeter trials (Atkins, 1976; Garwood & Tyson, 1973) and field-drainage monitoring (Johnston, 1976; Williams, 1976) had indicated that beneath cut grass swards situated on freely-draining soils, nitrate losses were low (generally less than 5 kg N/ha yr) even at fertilizer application rates of up to 250 kg N/ha. However, lysimeter trials on fallow soils, and longer-term experiments on the exhaustion of bare soil (Reinhorn & Arnimelech, 1974; Atkins, 1976; Russel & Richards, 1920), had suggested that very substantial losses could occur from recently cultivated long-term grassland. In order to assess the effect of the conversion of pasture to arable land, simulations using a mathematical model were forced to match nitrate depth profiles measured beneath land on which grass of known ages had been ploughed in (Young *et al.*, 1983; Oakes, 1982), giving the ranges of values in Table 15.4.

Table 15.4 — Chronology of the release of nitrogen from grassland of known age

	N(kg/ha) release by ploughing			
Years in grass prior to ploughing	Total	Year of ploughing	Following year	Next following year
1	100–120	60– 72	30– 36	10–12
2	180–190	108–114	54– 57	18–19
3	240–310	144–186	72– 93	24–31
4 or more	280–380	168–228	84–114	28–38

15.3 WATER RESEARCH CENTRE FIELD EXPERIMENTS, DEEP DEAN, SUSSEX

Much of the evidence used to provide initial estimates of nitrate-leaching losses for incorporation in the predictive mathematical models was of an indirect nature. In order to gain direct evidence of the rates of loss of nitrate from agricultural soils on the Chalk, a series of experimental plots were maintained during the 5-year period 1978 to 1982 inclusive. Five plots, each 12 metres square, were located along the base of a dry valley in the Chalk close to Eastbourne Waterworks Company's Deep Dean pumping station, 1.5 km east-north-east of the village of Litlington (NGR TQ 520 530). Ground level was at about 45 m above Ordnance datum and the water table lay approximately 43 m below ground level. The soil present was a typical Chalk rendzina, approximately 650-mm thick, overlying between 2 and 5 m of disturbed Chalk (Coombe Deposits) which graded downwards into undisturbed Middle Chalk. Prior to the start of the experiment the valley floor had not been cultivated for at least 50 years. The plots were first cultivated in March 1978, prior to which in March 1975 and December 1977, profiles of nitrate had been measured using standard drilling and pore-water extraction techniques (Gray *et al.*, 1977). Both of the deep profiles exhibited the low concentrations of nitrate in interstitial water, characteristic of unfertilized permanent grass. From March 1978 until October 1982 the experimental plots followed the regimes summarised in Table 15.5. Soil sampling and analysis was undertaken by ADAS South-East Region Sub-centre, Wye, Ashford, Kent, at approximately ten-weekly intervals from May 1978 until June 1980; samples were taken from four depth intervals (0–150, 150–300, 300–450 and 450–600 mm) in each plot. Cored boreholes to a depth of eight metres were drilled at regular intervals through each plot by WRc personnel, to provide sequential records of the evolution of nitrate from the soils and its downwards migration. A total of 35 boreholes were sunk and analysed over the period October 1978 to October 1982.

Throughout the period of the experiment the concentrations of nitrate in the soils and underlying Coombe deposits and Chalk in Plots 0 and 1 remained essentially similar, whilst the evolution of nitrate from plot 4 followed a similar trend to that of Plot 3. The progressive changes in soil nitrate in Plots 0 (control), 2 (fallow), and 3 (cereals) are shown in Fig. 15.1. The initially higher nitrate content of soils in Plots 2 and 3 compared with Plot 0 reflects the extent of mineralisation between ploughing in March 1978 and the first soil sampling in May 1978, plus the addition of fertilizer to Plot 3. Available nitrate continued to increase in both Plots 2 and 3 until July, when uptake by the maize began to remove considerable quantities from Plot 3. Following the harvest in September further mineralisation appears to have generated nitrate in both plots, but by January 1979 leaching losses had reduced concentrations to levels comparable with those of the control plot. Approximately 200 kg N/ha appeared to have been leached from the fallow plot, and about 70 kg N/ha from the cereal plot. During 1979, a wet spring followed by a dry summer gave rise to poor crop performance, but a substantial mass of nitrate was again evolved from the soils of Plots 2 and 3 following the harvest and apparently removed by leaching during the winter period.

Sequential sampling to a depth of eight metres beneath the three plots (Figs. 15.2, 15.3, 15.4) confirmed the differences in leaching rates beneath the three examples of land use. The profiles beneath the control plot (Fig. 15.2) showed consistently low nitrate concentrations below about 1.5m. Detailed studies by the Institute of Hydrology (Wellings & Bell, 1980) of moisture and solute fluxes in the upper few metres of the Chalk have shown that upward flows, in response to evapotranspirative losses, from depths of up to three of four metres, are characteristic of chalk soils during summer months, and the higher concentrations recorded from the upper 1.5 m in borehole samples during the autumn may reflect this process. Similarly, the very high concentrations of nitrate in the soil recorded in all plots in September 1979 may reflect concentration effects due to an upward flux during the preceding dry period.

Table 15.5 — Summary of agricultural regimes on experimental plots

Plot No. Description Year	0 Control	1 Fertilized Grass	2 Fallow	3 Cereals	4 Cereal/ley
1978	Unfertilized, uncut	Fertilized, 63 kg N/ha, May. Cut 4 times; grass removed	First cultivated March 1978. Rotavated and harrowed 4 times. No fertilizer	First cultivated March 1978. Maize, 126 kg N/ha, May	First cultivated March 1978. Barley, 70 kg N/ha May
1979	As 1978	As 1978	Rotavated and harrowed 4 times. No fertilizer	Wheat, 70 kg N/ha April	Barley, 70 kg N/ha
1980	As 1978	As 1978	As 1979	Maize, 126 kg N/ha May	Grass, 63 kg N/ha, cut 4 times and removed
1981	As 1978	As 1978	As 1979	Wheat, 70 kg N/ha April	Grass, 63 kg N/ha, cut 4 times and removed
1982	As 1978	Revert to grass	Revert to grass	Revert to grass	Revert to unfertilized grass

Note: Harvest in late August/early September during experimental period

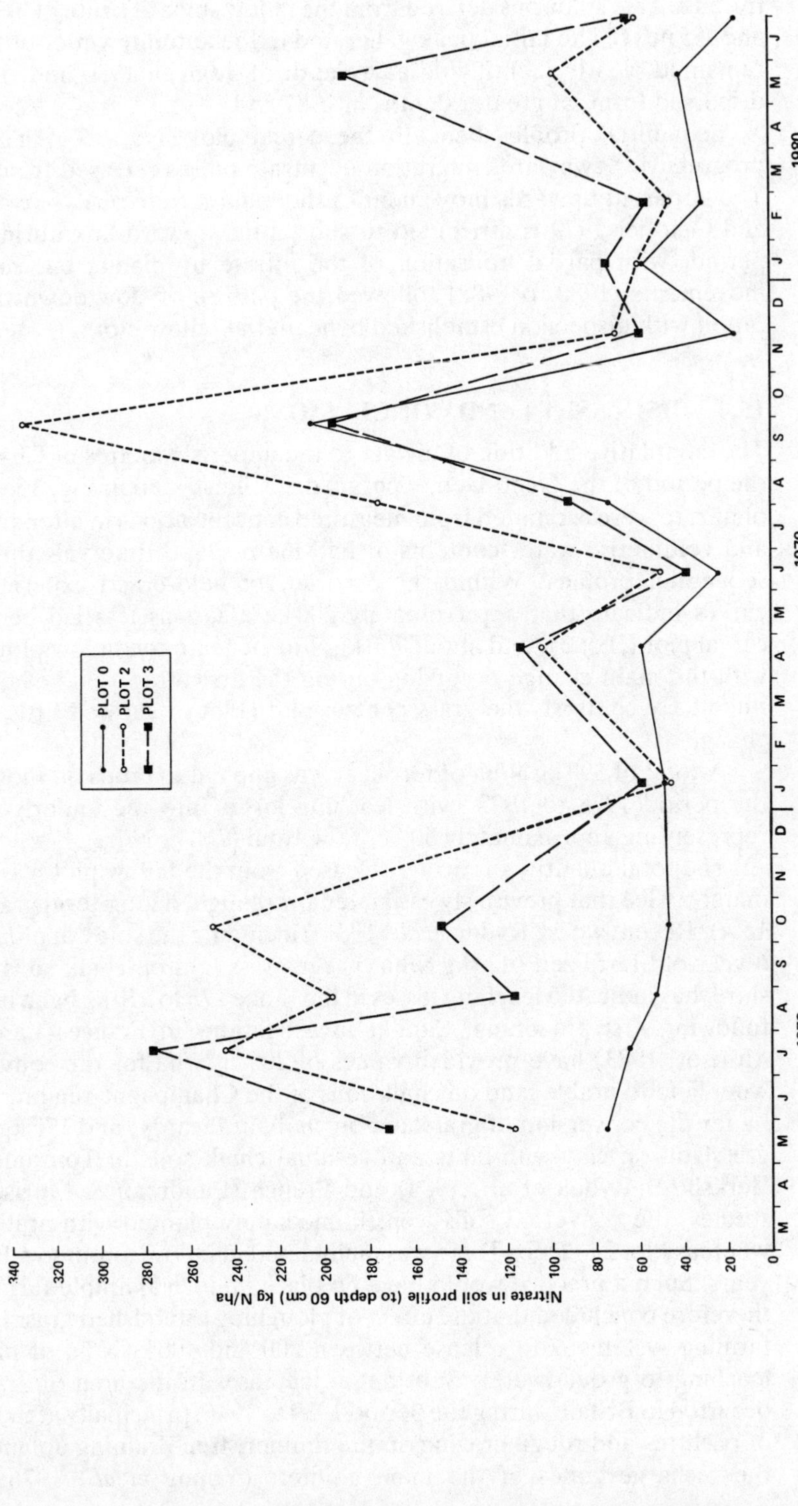

Fig. 15.1 — Changes in nitrate masses in soil, May 1978 — July 1980

By comparison, the profiles beneath the fallow plot (Fig. 15.3) recorded the progressive downwards movement, with some dispersion, of concentrated nitrate solutions derived from the fallow soils, at a rate of between 1.0 and 1.5 m/yr. The initial peak generated in the autumn/winter of 1978/1979 remained clearly identifiable at a depth of 3.0m in 1980 and, in a more dispersed form, at greater depths in 1981 and 1982.

Sequential profiles beneath the cereal plot (Fig. 15.4) also exhibit progressive downwards migration of nitrate pulses released from the soil. The apparent upwards movement of the main nitrate peak between April and October 1979 is atrributed to substantial upward flux during the dry period, with partial utilisation of the nitrate by plants, but subsequent movements (1980 to 1982) followed the pattern of slow downwards migration with dispersion established beneath the fallow plot.

15.4 DISCUSSION AND CONCLUSIONS

The cumulative addition of nitrate to the upper six metres of Chalk during the period of the Deep Den experiments is illustrated in Fig. 15.5. Masses of nitrate were estimated from measured concentrations in interstitial water and volumetric water contents at half-metre depth intervals through the sequential profiles. Within the accuracy of field-based experiments the curves indicate that approximately 200 kg N/ha was leached beneath the cereal plot (Plot 3), and about 700 kg N/ha beneath the fallow plot (Plot 2), with the main change occurring during the first three years of the experiment. By contrast, the grass control plot (Plot 0) showed little apparent change.

A total of 390 kg N/ha of fertilizer was applied to crops on Plot 3 during the period 1978 to 1982, with leaching losses into the underlying Chalk representing approximately 50% of the total (200 kg N/ha).

The total quantity of nitrate released from the fallow plot was approximately twice that previously estimated for ploughed long-term grass (Table 15.4). Recent work (Ryden *et al.*, 1984), including the study of ploughing up 6-year-old fertilized (420 kg N/ha yr) rye-grass leys on chalk soils in Berkshire, has indicated leaching losses in the range 275 to 310 kg N/ha in the year following first ploughing. Similar investigations in France (Landreau & Morisot, 1983) have provided values of 205 kg N/ha for the conversion of woodland to arable land on chalk soils of the Champagne region, 235 kg N/ha for the conversion of grassland on chalk in Picardy, and 170 kg N/ha for grassland on clay-with-flints and residual chalk soils in Touraine. In the Berkshire (Ryden et al., 1984) and French (Landreau & Morisot, 1983) studies, the converted soils were immediately planted with arable crops, whereas Plot 2 in Deep Dean was maintained in fallow continuously for four years. Such a practice would have no place in normal arable farming. It is therefore concluded that the effect of ploughing established grass in normal farming systems is to release between 200 and 400 kg N/ha of nitrate for leaching to groundwater. Substantial increases in the area of arable land occurred in Britain during the period 1939 to 1946, principally at the expense of pastures and rough grazing on the thinner, free-draining upland soils of the recharge zones of the main aquifers (Young *et al.*, 1976; Young,

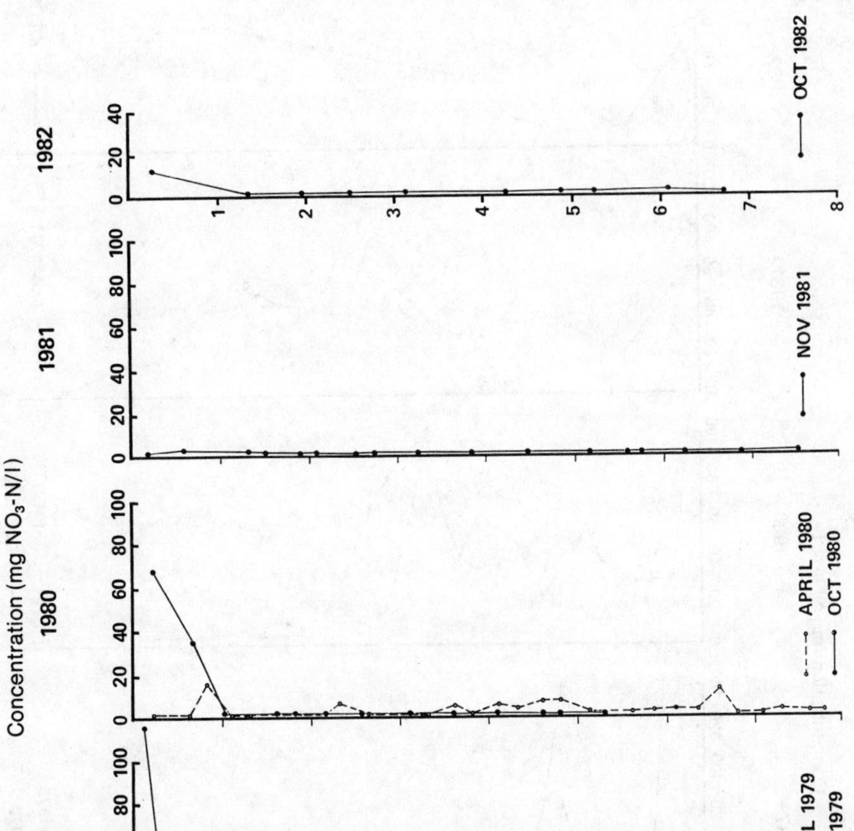

Fig. 15.2 — Nitrate profiles below Plot 0

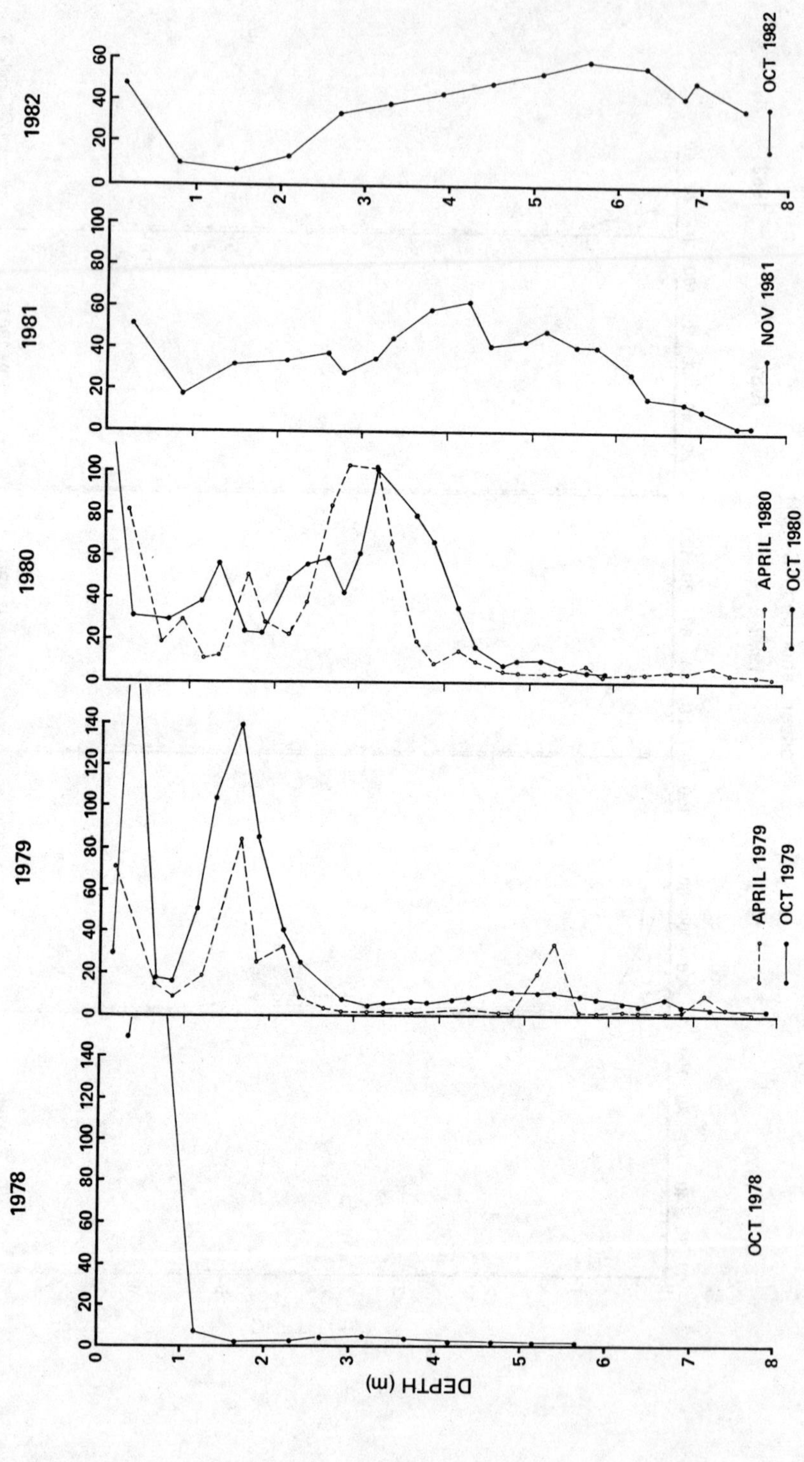

Fig. 15.3 — Nitrate profiles below Plot 2

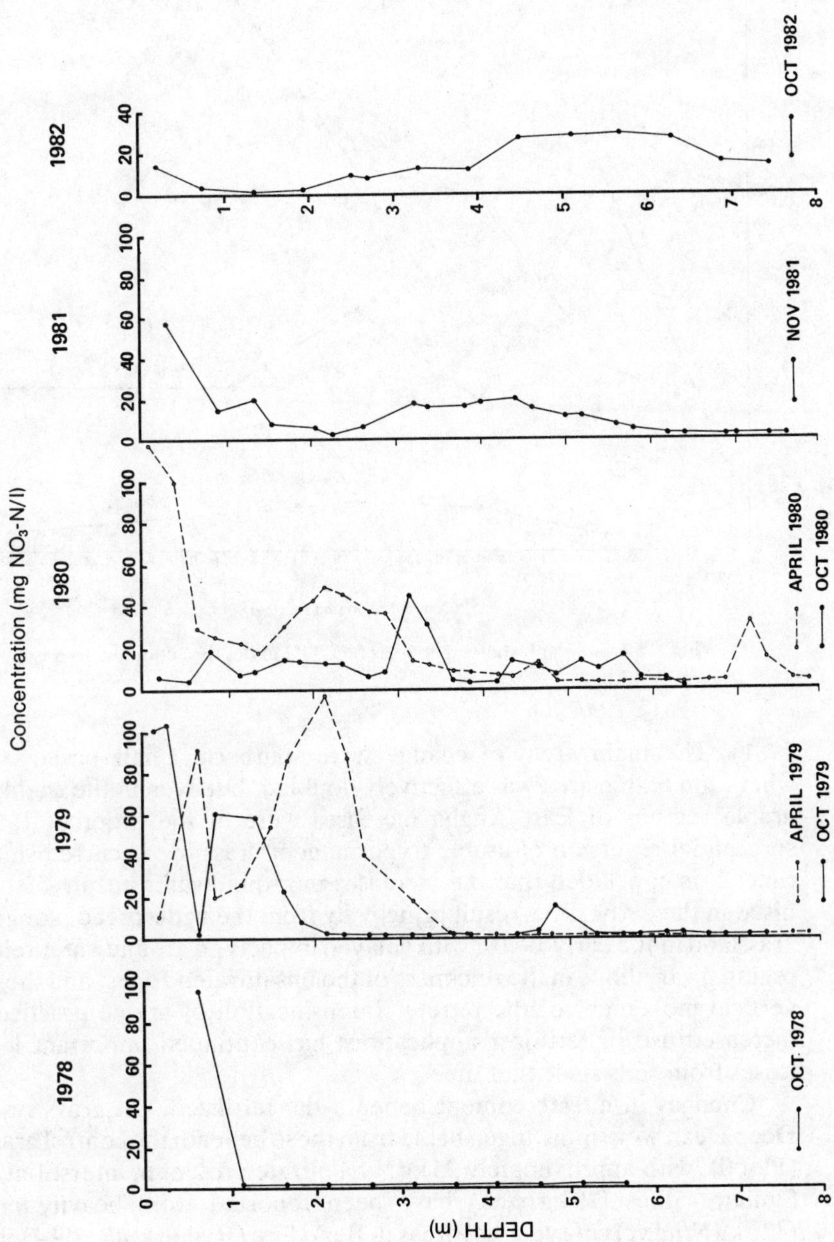

Fig. 15.4 — Nitrate profiles below Plot 3.

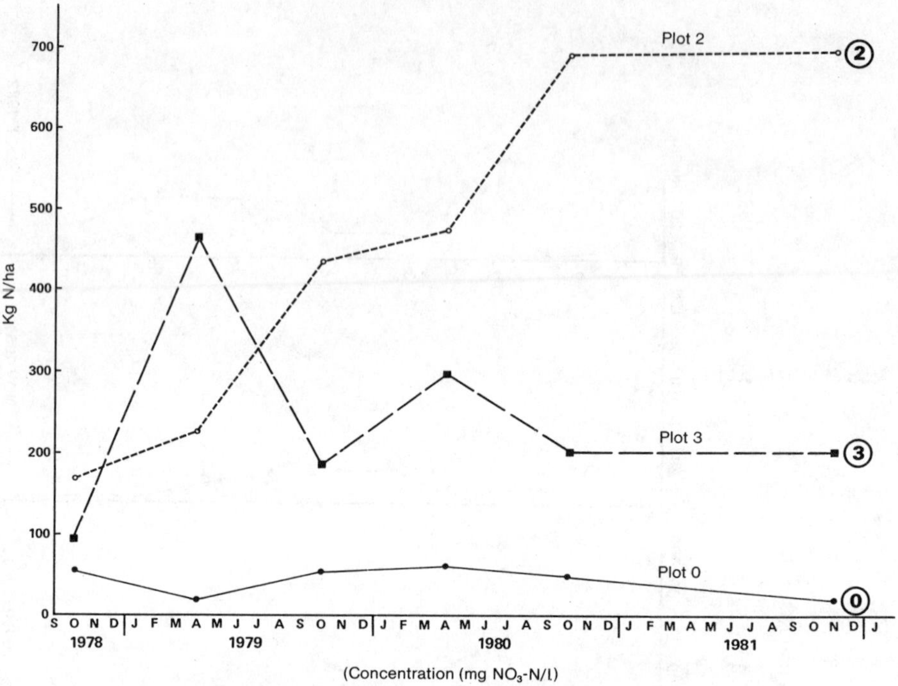

Fig. 15.5 — Graph showing masses NO_3—N in depth section 0.6 to 6.0 m

1981). The main areas of change were southern, Chalk-based counties where the arable area was effectively doubled, but even in the traditionally arable regions of East Anglia increases were of the order of 15%. No substantial reversion of arable to permanent grass has occurred since that time. It is concluded that the increases in groundwater nitrate first recognised in the early 1970s result principally from the widespread ploughing of grassland in the early 1940s, with delayed impacts on groundwater related to regional variations in the thickness of the unsaturated zones, and the rate of vertical movement of the nitrate. Intensification of arable practices with increased use of fertilizer applications has continued important leaching losses from soils since that time.

Changes in nitrate content beneath the fertilized, cut grass swards at Deep Dean were indistinguishable from those beneath the control grass area (Plot 0), with approximately 25 kg N/ha nitrate present in interstitial water. Similar values (29 kg N/ha) have been reported from heavily-fertilized (420 kg N/ha yr) cut rye-grass areas in Berkshire (Ryden *et al.*, 1984), but the same studies indicated important leaching losses (162 kg N/ha yr) from the same grassland when grazed rotationally by steers at a stocking rate of 9.3 head/ha. Losses in the range 10 to 25 kg N/ha yr and 65 to 70 kg N/ha yr have also been reported from non-irrigated and irrigated pastures respectively grazed by sheep, on alluvial soils in New Zealand (Burden 1982). In both cases the losses were ascribed to leaching from areas of concentrated dung and urine, resulting from intensive stocking, and it is concluded that although low losses are characteristic of even highly-fertilized cut grass swards (Young *et al.*, 1983; Young, 1981; Garwood & Tyson, 1973), important leaching losses may occur from intensively-stocked grazing.

REFERENCES

Atkins, S. F. (1976) *Nitrogen Leaching from Fertilizers: Lysimeter Trials Published Results from Europe and USA*. ICI Agric. Div. Report No A128, 607. 76 pp.

Burden, R. J. (1982) Elevated levels of nitrates in groundwater beneath intensively grazed pastureland in New Zealand. In: *Memoires International Association Hydrogeologists*. XVI (3), proceedings of Prague Congress 1982. pp. 197–206.

Cameron, D. R., De Jong, R. & Chang, C. (1978) Nitrogen inputs and losses in Tobacco, Bean and Potato fields in a sandy loam watershed. *J. Environ. Qual.*, **7**(4), 545–550.

Central Water Planning Unit (1977) *Nitrate and Water Resources, with Particular Reference to Groundwater*. The Unit, Reading, Berks. 64 pp.

Dampney, P. M. R. (1979) *Lysimeter studies at Gleadthorpe E.H.F. — Nitrogen Leaching*. Paper to ADAS Soil Scientists Training Visit 1979, Berkshire College of Agriculture.

Foster, S. S. D. & Crease, R. I. (1974) Nitrate pollution of Chalk groundwater in East Yorkshire: A hydrogeological appraisal. *J. Inst. Wat. Engrs*. **28**, 178–194.

Foster, S. S. D. & Young, C. P. (1981) Effects of agricultural land-use on groundwater quality with special reference to nitrate. In: *A Survey of British Hydrogeology 1980*. Royal Society, London, pp 47–60.

Garwood, A. E. & Tyson, K. C. (1973) Losses of nitrogen and other plant nutrients to drainage from soil under glass. *J. Agric. Sci. Cambs.*, **80**, 303–312.

Gray, E. M. Holland, J., Breach, J. & Rowland, C. D. (1977) Nitrate in groundwater, procedures in the collection and preparation of rock and water samples. In: *Proceedings Water Research Centre Conference, Groundwater Quality — Measurement, Prediction and Protection*. Reading, September 1976. Water Research Centre. pp 373–384.

Greene, L. A. (1980) Nitrate in groundwater in the Anglian Region. *Proc. Inst. Civ. Engrs*. **69** (2), 73–86.

Greene, L. A. & Walker, P. (1970) Nitrate pollution of Chalk waters. *Wat. Treat. Exam.* **19**(2), 169–182.

Herbet, J. (1977) Control of the nitrogen fertilization in intensive farming for minimizing water pollution. *Proceedings of International Seminar on Soil Environment and Fertility Management in Intensive Agriculture*. Tokyo. pp 325–337.

Jenkinson, D. S. & Rayner, J. H. (1977) The turnover of soil organic matter in some Rothamsted classical experiments. *Soil Sci.* **123**(5), 298–305.

Johnston, A. E. (1976) *Additions and Removals of Nitrogen and Phosphorus in Long Term Experiments at Rothamsted and Woburn and the Effect of Residues on Total Soil Nitrogen and Phosphorus.* MAFF Tech. Bull. 32, pp 111–114. HMSO, London.

Landreau, A. & Morisot, A. (1983) Evaluation de la vulnérabilité des aquifères libres aux nitrates d'origine agricole. Report No BRGM 83 SGN 026 EAU. 20 pp. Bureau de Recherches Geologique et Miniere, BP 6009, 45060 Orleans, France.

Meintz, V. W., Kurtz, L. T., Melsted, S. W. & Peck, T. R. (1977) Long term trends in total soil N as influenced by certain management practices. *Soil Sci.* **124**(2), 110–116.

Oakes, D. B. (1982) Nitrate pollution of groundwater resources — mechanisms and modelling. In: K-H Zwirnmann Ed. *Non-Point Nitrate Pollution of Municipal Water Supply Sources; Issues of Analysis and Control.* International Institute for Applied Systems Analysis. Collaborative Proceedings Series CP-82-S4, pp 207–230. The Institute, Laxenburg, Austria.

Oakes, D. B., Young, C. P. & Foster, S. S. D. (1981) The effects of farming practices on groundwater quality in the United Kingdom. *Sci. Total Environ.* **21**, 17–30.

Reinhorn, T. & Avnimelech, Y. (1974) Nitrogen releases associated with the decrease in soil organic matter on newly cultivated soils. *J. Environ. Qual.* **3**(2), 118–121.

Russel, E. J. & Richards, E. M. (1920) The washing out of nitrate by drainage water from uncropped and unmanured land. *J. Agric. Sci. Cambs.,* **10**(1), 22–43.

Ryden, J. C., Ball, P. R. & Garwood, E. A. (1984) Nitrate leaching from grassland. *Nature,* **311** (5981), 6–10 September, 50–53.

Satchell, R. L. H. & Edworthy, K. J. (1972) *Artificial Recharge: Bunter Sandstone.* Trent Research Programme, No 7. Water Resources Board, Reading, Berks.

Severn-Trent Water Authority (1976) *Water Quality 1974/75.* 453 pp. The Authority, Sheldon.

Smith, D. B., Wearn, P. L., Richards, H. J. & Rowe, P. C. (1970) Water movement in the unsaturated zone of high and low permeability strata by measuring natural tritium. *Proceedings of Symposium of Isotope Hydrology.* IAEA Vienna. pp 73–87.

Smith, S. J. & Young, L. B. (1975) Distribution of nitrogen forms in virgin and cultivated soilds. *Soil Sci.* **120**(5), 354–360.

Standing Technical Advisory Committee on Water Quality. (1984) *Fourth Biennial Report; February 1981 — March 1983.* 111 pp. HMSO London.

Wellings, S. R. & Bell, J. R. (1980) Movement of water and nitrate in the unsaturated zone of Upper Chalk, near Winchester, Hants. England. *J. Hydrol.*, **48**, 119–136.

Widdowson, F. V. (1974) Results from experiments measuring the residues of nitrogen fertilizer given for sugar beet, and of ploughed-in sugar beet tops, to the yield of following barley. *J. Agric. Sci. Cambs.* **83**, 415–421.

Williams, R. J. B. (1976) *The Chemical Composition of Rain, Land Drainage and Borehole Water from Rothamsted, Broom's Barn Saxmundham and Woburn Experimental Station.* MAFF Tech. Bull. 32., 174–200. HMSO, London.

Young, C. P. (1981) The distribution and movement of solutes derived from aquifers of the United Kingdom, with particular reference to nitrate. *Wat. Sci. Tech.* **13**, 1137–1152.

Young, C. P. & Gray, E. M. (1978) *Nitrate in Groundwater — the Distribution of Nitrate in the Chalk and Triassic Sandstone Aquifers.* Technical Report No. TR 69. 66 pp. Water Research Centre, Medmenham, Bucks.

Young, C. P., Hall, E. S. & Oakes, D. B. (1976) *Nitrate in groundwater: studies on the Chalk near Winchester, Hampshire.* Technical Report No. TR 31. 67 pp. Water Research Centre, Medmenham, Bucks.

Young, C. P., Oakes, D. B. & Wilkinson, W. B. (1983) The impact of agricultural practices on the nitrate content of groundwater in the principal United Kingdom aquifers. In: G. Golubev Ed. *Environmental Mnagement of Agricultural Watersheds.* 11ASA Collaborative Proceedings Series CP-83-S1. International Institute for Applied Systems Analysis, Laxenburg, Austria. pp 165–197.

16

Groundwater pollution in Hungary with special regard to nitrates

L. Alföldi and **A. Homonnay**

16.1 LOCALITY AND HYDROGEOLOGICAL SITUATION

Hungary is situated within the Carpathian basin. The range of mountains surrounding the basin consists of aged, folded, consolidated and metamorphic formations. Similarly the bedrock of the basin consists of aged formations (see Fig. 16.1).

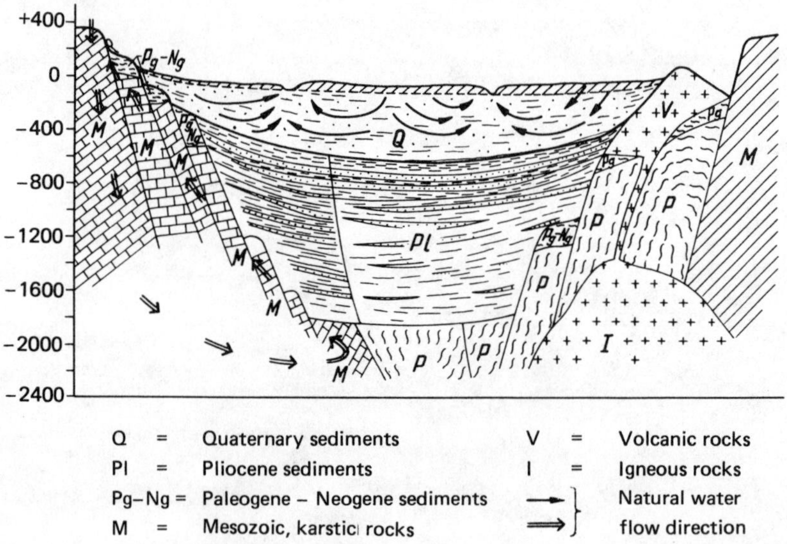

Fig. 16.1 — Geological scheme of great Hungarian basin

Streams heading from the Carpathian Mountains toward the basin have, ever since the Pliocene, filled up the basin with a loose and deep deposit consisting of gravel, sand and clay. The largest depth of this deposit exceeds 400 m in some places.

On the site of the former land-locked sea the bedrock is covered by marly clay several hundred metres thick, on which a series of confined artesian aquifers are situated to a depth of several hundred metres. After the Carpathian Mountains had emerged the land-locked sea was rapidly filled up by alluvial deposits, and gradually converted into a freshwater lake. This has filled up with sandy deposits providing a series of aquifers with excellent water-yielding capacities. Wash-out of debris and sediments from the Carpathian Mountains has been continuing, creating gravel cones along the foot of the mountains, that is along the border of the basin, with alluvial gravel fields along certain river reaches which have a close connection with the river itself. Finally, in the west the Pannonian basin is divided in to two by a transverse chain of mountians consisting of carbonate-karstic formations of some 3000 m depth, which also provide the means of interconnection with the deep confined aquifers of the basement complex.

Due to the hydrogeological situation just outlined, the aquifers of Hungary are closely interconnected and consists of several subsystems which are hydrodynamically balanced. These subsystems have different characteristics according to the degree of contamination. In other words their degree of natural protection varies with the local hydrogeological situation.

The following subsystems can be considered fairly independent.

- *The karstic ranges in the western part of the country, to which the extended karstic thermal aquifers of the basement complex are also connected*
  Under natural conditions the geothermic flow systems, controlled by the infiltrating cold karstic water from the surface, are exposed to pollution only in the zone of infiltration (see Fig. 16.2).

- *Sediment cones and bank-filtered systems: sand-gravel formations along the rivers of 15 to 20 m depth in average*
  Locally the depth of these formations may reach 300 m, for example in the Danube region. Due to the close proximity of these formations with the river water, there is a natural exchange of water between the parent river and the aquifer. Generally this gravel aquifer is open to the surface with no obstacle against vertical infiltration.

- *There are aquifers of diverse character, consisting of holocene flood-silts, sediment-cones along the foot of mountains, and valley-basement deposits.*
  Their near-surface zone is usually open or confined with a thin impermeable cover and the transitional upper zone of the Quaternary deposits is generally 10 m deep, with a maximum depth of 30 m (see Fig. 16.3).
- *Quaternary alluvial aquifers of layered structure with lenses*
  These are characterised by a natural water circulation extending to considerable depths and mainly with indirect communication to the

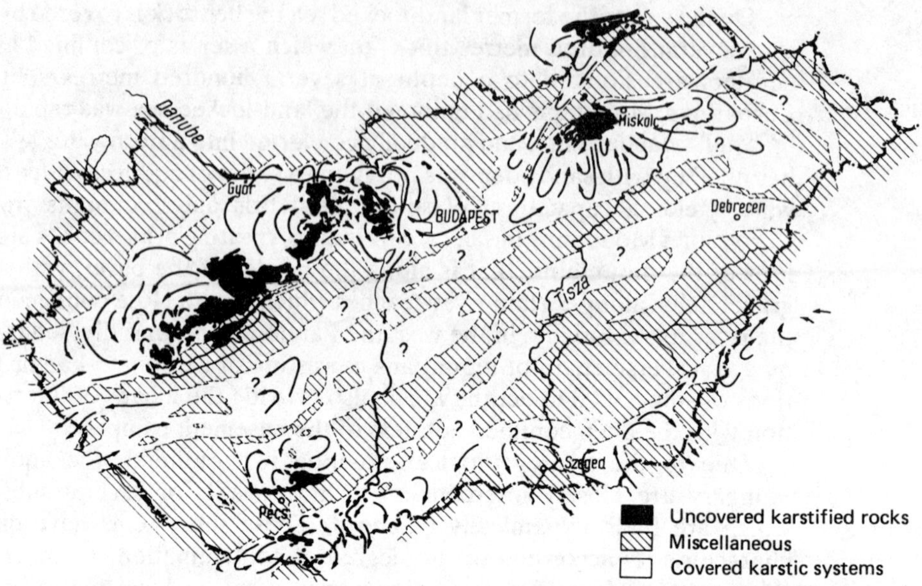

Fig. 16.2 — Mesozoic karstic water-flow system

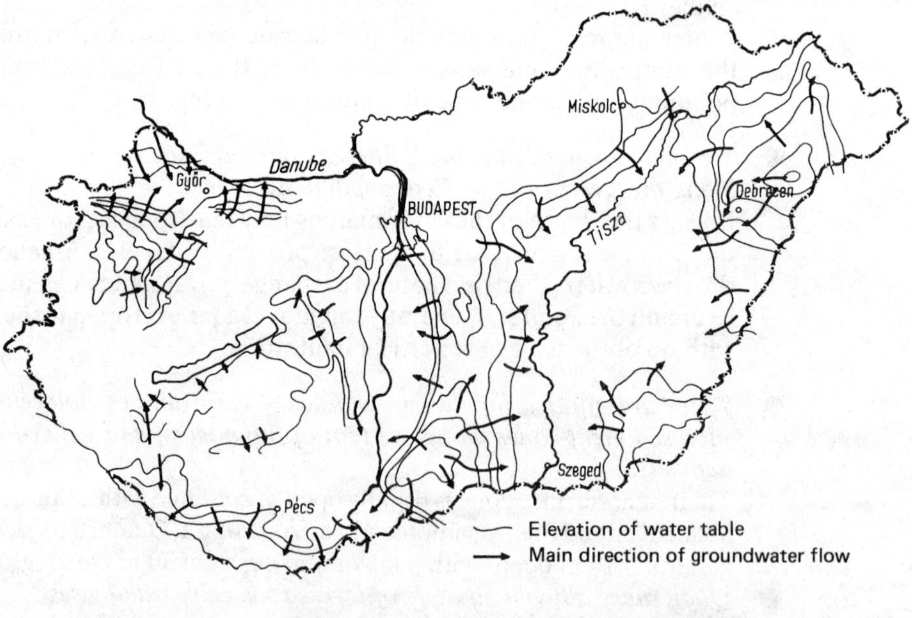

Fig. 16.3 — Natural shallow groundwater flow directions (potential transport route of pollutants)

surface. On the basis of head differences of the artesian aquifers, infiltration zones and upward-flow zones can be clearly distinguished.

- *The upper Pannonian aquifer system*

  This only has distinct indirect surface connections, yielding cold water along the basin's border and thermal water in the middle.

16.2 NATURAL PROTECTION AGAINST AND EXPOSURE TO CONTAMINATION

An aquifer can be considered naturally protected only if infiltration, vertical or otherwise, cannot occur to contaminate the water. An obstacle preventing the contamination of the aquifer could be an impermeable upper confining layer, or another upper aquifer.

In Hungary the surface karstic systems and the near-river 'bank-filtered' systems are practically unprotected and open. The early Holocene aquifer systems are partly open and partly confined; however, due to their shallowness they are *a priori* subjected to contamination, since the upper confining layers, if any, are frequently cut through by foundation works, trenching operations, or well-drilling, thus opening a pathway for direct access from the surface.

The Quaternary and Pannonian artesian systems are generally protected, and the deeper aquifers can be considered sterile.

In the systems outlined, flow velocities are of an order of magnitude of decimetres per year and 100 m per year in the deep and near-surface zones, respectively; thus no contamination of the deep, non-utilised, aquifers will occur within the 'human time scale'. As a consequence of abstracting from the aquifers, flow velocities are accelerated, changing the flow pattern completely and resulting in the acceleration of the transport of contaminants.

16.3 CONDITIONS AND TENDENCIES

The data of more than 60 000 boreholes are available, in the Hungarian Well Register, for the characterisation of the condition of the aquifers. In addition to this there were in 1962 about 12 500 dug wells, but recently their number has decreased to 3000.

The distribution of the 60 000 drilled wells according to their depth is as follows.

Depth	Per cent of the total number
<50 m	50
50–100 m	21
100–200 m	20
200–500 m	8
>500 m	1

The data of the Well Register indicate that the quality of the water of 50% of the wells is endangered due to their relatively shallow depths. The results of a survey made in 1950 on the quality of dug wells of detached ranches indicated that 77% of the wells had good-quality water. However, according to the results of a similar survey made in 1978 only 15 to 30% of the wells were able to meet public-health standards. Today, in the vicinity of settlements the water of near-surface aquifers is highly contaminated and the nitrate content frequently exceeds 100 mg/l, reaching up to 1000 mg/l in certain locations.

According to the results of a survey in 1977 the nitrate content of the water of public or private wells of 720 small settlements exceeded the 40 mg/l value, specified as allowable for human consumption in the Hungarian Standards. Consequently for these settlements bottled or plastic-bag packed drinking water had to be provided (see Fig. 16.4).

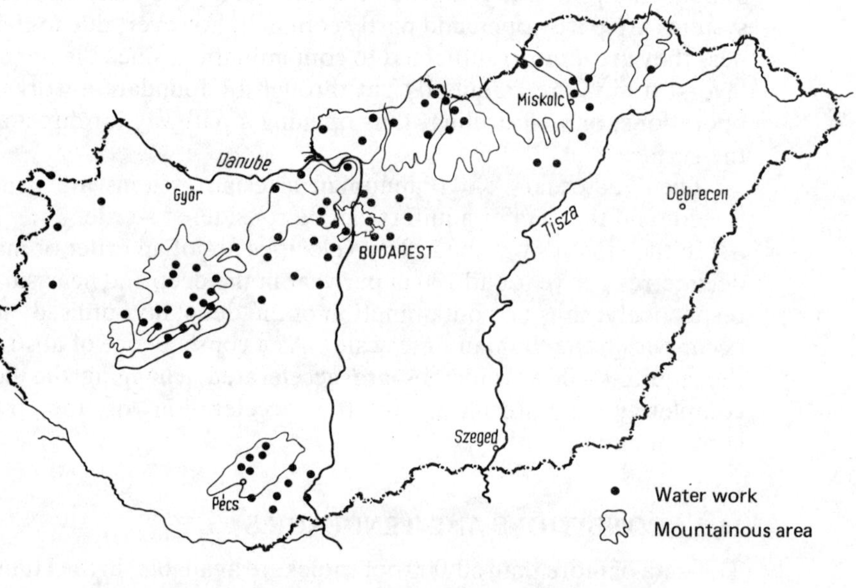

Fig. 16.4 — Nitrate affected water works in 1981

In 1981 a survey of the nitrate content of the water of 1000 water works resulted in the following figures.

Nitrate level	Number of plants
near to 40 mg/l	17
40–50 mg/l	15
< 50 mg/l	39

The total capacity of the water works investigated was 2×10^6 m^3 per day.

The nitrogen contamination of the water of subsurface resources utilised by these water works consists mainly of nitrate, and to a lesser extent of nitrite and ammonia. Nitrate pollution is almost entirely restricted to the near-surface aquifers, with the exception of karstic regions where nitrate enrichment can be observed in depths of 155 to 200 m. Nevertheless, even in such cases, only the upper zone of the water body is contaminated, and then only when the first water-yielding layer is the one which is at considerable depth. Nitrate enrichment is most severe in small agricultural settlements situated in river valleys of mountains regions, and also in bank-wells that are near to high-rise settlements. According to past experiences no nitrate pollution of wells deeper than 50 m has been unequivocally proven.

Considering the trends of water-quality deterioration and nitrate enrichment, an additional 140 to 150 water-production sites will have to be abandoned by the year 2000. A substitution of these with new resources will have to be provided in due time.

16.4 SOURCES AND CAUSES OF NITRATE CONTAMINATION

In Hungary the nitrate contamination of subsurface waters originates from communal sewage and from the intensified use of fertilizers. Industrial wastes of high nitrogen content are not discharged regularly and directly into subsurface recipients, but rather into surface waters.

Hungarian rivers and streams are mainly contaminated by nitrates via surface runoff, or through leaching. Pollutants, and thus also nitrate, do not accumulate in the streams; contaminants are transported downstream, and with the elimination of pollutant inputs the water quality of the streams will recover without any further water-pollution control actions. Pollutants will reach their final destinations in the sea and ocean. Nevertheless with the increase of nitrogen pollution from non-point sources, via leaching and land-runoff, the ammonia and nitrate concentrations of streams are steadily increasing. In summertime concentrations of nitrate in the waters of the Rivers Danube and Tisza are reaching 8 mg/l. These values are still significantly lower than the allowable drinking-water standard, and will not affect the quality of near-river subsurface water resources utilised by bank-well filtration. According to the results of investigations by the Research Centre for Water Resource Development, Hungary, a biologically-active 'filter layer' of several centimetres thick develops on the river beds and retains nitrogenous contaminants via biochemical processes and physical filtration effects, preventing the contamination from intruding into the adjacent aquifers. The results of the investigations indicate that the water of bank-wells is not deteriorated by the quality of the parent river, but rather by pollution originating from the off-river background zone.

The development of the situation outlined was demonstrated by actual concrete control problems, such as the case of a city on the River Danube where pollution by nitrate of the bank-wells used for drinking-water abstraction could only be prevented by constructing a long cut-off wall into the alluvial deposit between the city and the wells, thus forming a barrier against the subsurface flow from the adjacent 'background' aquifers.

The nitrate enrichment of near-river subsurface water resources, utilised by bank-well filtration, was not due to the intensified fertilization, as the primary source of nitrate, but to the fact that the degree of sewerage of the near-river settlements is much behind that of the communal piped supply of drinking water. Due to this lag between sewerage and water supply large quantities of organic pollutants had been discharged into subsurface zones and, with the subsequent decomposition of this organic matter, nitrogenous materials have reached the aquifers below the settlements. The water works were located at a distance from the settlements that was sufficient to prevent the intrusion of pollutants into the production wells via natural subsurface flows. Nevertheless, owing to the increasing abstraction of water, subsurface flow velocities were accelerated and the flow pattern in the non-homogeneous media changed also in an unpredictable manner. This resulted in certain cases, in the intrusion of pollutants originating from the settlements in the vicinity of the water-abstraction site, causing significant enrichment with nitrates.

This situation has been worsened by the ever-expanding recreational use of the river banks during the past two decades, because no sewerage for the rapidly-increasing number of summer cottages was provided. Sewage from individual septic tanks was infiltrating the soil and the aquifers, causing locally excessive loads of organic matter.

16.5 CONTAMINATION OF THE VALLEY-BOTTOM ALLUVIAL DEPOSITS IN HILLY REGIONS

Presumably the organic wastes of the small settlements hiding in the alluvial valley bottoms of the hilly regions had been polluting these formations for several centuries. However, with the vast development of piped-water supply, the quantity of communal sewage has been increased by orders of magnitude, and an additional input from the liquid manure of large-scale livestock breeding farms worsened this problem. Nitrate pollution caused in these alluvial deposits by the increasing input of organic matter has, simultaneously, moved downstream with the subsurface flow to emerge in the alluvial-cones of the valley entrances also. Figure 16.4 clearly demonstrates this situation by the fact that the largest number of water works that had to be shut down occurred among the small settlements of the hilly region. This was a natural consequence of the practice of utilising first the shallow, easily and inexpensively exploitable, near-surface aquifers, even in regions such as the karstic hilly-range of western Hungary, where good-quality water resources were available in the aquifer of 100 to 200-m depth, although with much higher costs of production.

16.5.1 Lowland and Alluvial Plains

The organic manure generated in Hungary would be sufficient only for the fertilization of about 10% of the land utilised by agriculture; consequently the use of mineral fertilizers is a natural prerequisite.

Today in Hungary 53 000 t of nitrogen is used as fertilizer annually, which means a dosage rate of 225 kg/ha to agricultural land. In the view of international tendencies in this respect, a further increase of nitrogen fertilizer usage has to be taken in to consideration.

According to the results of Hungarian experiments the quality of near-surface groundwaters will deteriorate if fertilizer dosages exceed 100 kg N/ha. In the case of excessive use of nitrogen fertilizers, a loss of 20 kg N/ha might occur which, on a national scale, would amount to 115 to 130 $\times 10^3$ tonnes of nitrogen contamination. Calculating with an increase of nitrate levels from 10 mg/l to 40 mg/l, one may conclude that the nitrate quantity lost would be sufficient for the deterioration of about 17 600 million m^3 of water annually. If only one tenth of this quantity got into the groundwater this would be sufficient for the deterioration of about the half of the total quantity of water produced by the water works of Hungary, annually.

The extent of loss, however, depends also on the quality of the fertilizer and also on the technology of application. At present the use of carbamide-type fertilizers dominate which, considering their ready solubility in water is unfavourable. According to the results of a survey made in 1976 to 1977 the loss of fertilizers due to incorrect storage was more than 10%. Today, however, this value is estimated at 3 to 4%.

This loss of nitrogen fertilizers outlined above results, primarily, in an excessive burden on the water and nitrogen budgets of the large lowland plains of Hungary. In these regions the near-surface aquifers, the first to be used to yield water, were already contaminated, thus the water-supply systems had to be based on the artesian aquifers, which have not yet deteriorated. In the Quaternary artesian aquifers the nitrate content is practically zero.

In the first water-yielding layer the natural nitrate content is about 10 mg/l, however in artesian aquifers with depths exceeding 50 m there is practically no nitrate, even in such regions where downward filtration is a proven fact. One likely explanation could be that the water infiltrating downward reaches a reducing zone and looses its nitrate content. The upper limit of this reducing zone is, in the Hungarian basins, at about 20 m depth, where the alluvial formations contain abundant quantities of organic matter or other reducing agents. However, the amount of reducing substances, and thus also the reducing capacity of this zone, is given once and for all, which means that this capacity is exhaustible when severely loaded. This also explains the fact that in areas with high nitrate loads, the lower limit of nitrate-enriched waters is found at greater depths of 40 to 50 metres.

The location of this natural reducing zone, the type, quantity and distribution of the materials present, and the processes prevailing in this environment, are not yet fully known and understood, thus creating an important task for future research.

16.6 PREVENTION OF THE CONSEQUENCES OF NITRATE CONTAMINATION

The most effective means of preventing the undesirable consequences of nitrate contamination would be the controlling of pollutant emissions. The discharge of nitrogenous organic matter into the ground could be eliminated by sewerage of settlements and recreational regions, and by the construction of proper treatment and storage facilities for individual detached homes. Consequently this is 'only' a question of economics, although, in Hungary, some 15 to 20 years will probably be needed to overtake the arrears of work in this field.

As far as the agricultural-production technologies are concerned any decrease in use of nitrogen fertilizers to the turn of the century must not be counted on, but rather the increasing use of fertilizers should be taken into consideration. Nevertheless, decreasing losses of fertilizers due to the development of fertilization technologies and fertilizer qualities, as well as to the switch-over to more reliable nitrogen 'carriers', may be expected, the more so, since this also meets the economic interests of agriculture. Actually, interest in this respect was also promoted by the new decree on sewage penalties of 1984 which also provides for the charging of penalties on the use of solid fertilizers with increased water solubility. Although the overdosage of fertilizers can hardly be fined due to the uncertainties in providing proof on such violations, the improper and unlawful storage of fertilizers, creating potential sources of pollution, has already been subjected to such fines.

By assigning protective zones in the vicinity of water resources and/or water intakes of special importance, there is a possibility for limiting agricultural activities too. However, the economic consequences of the loss of production due to such restrictions may be the sources of the development of conflicting interests (and stresses.)

Considering the orders of magnitude of fertilizer uses outlined, and the non-point source character of the processes of contamination with nitrate, one may not expect the elimination or even a significant (orders of magnitude) decrease of the nitrate pollution of subsurface waters. Consequently, research into nitrate removal from water has been intensified. The well-known processes of reverse osmosis and electrodialysis are too expensive and too energy demanding, thus no wide-scale use of such techniques in Hungary is foreseen. Experiments, with promising results, have been carried out with ion-exchange resins of the chlorine and hydrocarbonate cycle in pilot-scale equipment. It appears that the use of these techniques be appropriate only in the case of smaller water works.

In the Research Centre for Water Resources Development, VITUKI, experiments with several novel nitrate-removal techniques have been carried out, as described in somewhat more detail in section 16.8.

Potential techniques for the *in situ* subsurface denitrification by chemical methods are also the subject of research.

In Hungary, more than 90% of the total piped water supply relies on subsurface waters. Only 2 to 3% of the total communal water use of more than 10^9 m^3/yr is used for direct human consumption; the rest is utilised for other purposes such as washing, bathing, flushing, and so on. Consequently there is good reason to promote research into nitrate-removal equipment which can be fitted to the water taps of individual homes.

The complex problem area of nitrogen and nitrate elimination, and the decreasing of relevant pollution discharges raises several questions of legal, administrative, and public health natures. Consequently there is a need to establish and improve the related legal-administrative framework and to carry out research in this respect.

16.7 NITRATE ELIMINATION TECHNIQUES INVESTIGATED AND USED IN HUNGARY

The appearance of relatively high levels of nitrate contamination in groundwater is mainly due to the increasing use of agricultural nutrients, the lack of adequate sewerage in settlements, and to the development of industrial processes releasing nitrate-contaminated wastes. Safe drinking-water supplies rely upon high-quality water resources and on reliable water-treatment technologies. Thus the rapid development of methods for the elimination of nitrate is an urgent and indispensable task. The Research Centre for Water Resources Development conducted a co-ordinated research programme in to possible methods of economic and efficient elimination of nitrate. Some of these techniques rely on principles other than the well-known ion-exchange technique.

16.7.1 Drinking water treatment: nitrate elimination with fungi

The nitrate-contaminated water to which other nutrients are added, flows through an aerobic biological reactor where lower-fungi species have been grown on the surface of the granulated filter media. The fungi, through their metabolic activity, take up the nitrate and the effluent from this system is treated by conventional methods, that is, clarification, filtration, activated-carbon adsorption, and chlorination. A pilot-plant with a capacity of 1 m^3/d is in operation.

16.7.2 The BIONIT method

The equipment consists of two fluidised-bed reactors where denitrifying bacteria decompose nitrate compounds. The main advantage of this method is that the nutrition of bacteria and the decomposition of nitrate are split in to a two-phase process, separated both in space and time. Of these two reactors of similar character, one is used for feeding the bacteria and the other for the denitrification. The bacterial mass attached to the granulated media is continuously removed and recirculated between the reactors. Thus the treated water will not be enriched by the biomass of high organic carbon content, and the final effluent is a nitrate-free, good quality water needing only adequate disinfection. A pilot plant with a capacity of 100 m^3/d is under construction.

16.7.3 Chemical denitrification

The nitrite and nitrate content of the water will be reduced to nitrogen in a single technological step by a strong reducing agent, in the presence of a catalyst.

16.7.4 Nitrate elimination by UV irradiation

Ammonium salts, fit for human consumption, are added to the water to be denitrified and subsequently irradiated with UV rays, resulting in the decomposition of nitrates to nitrogen. The patented UV irradiation equipment produces 0.4 to 0.8 l/hr nitrate-free water. An appliance which can be fitted to water taps is under development with the aim of producting potable water with a capacity equivalent to the daily water consumption of a family.

16.7.5 The NITREX method

This method has been developed for nitrate elimination from waste waters. The outflow of a totally nitrifying activated-sludge process flows to a basin maintained under anoxic conditions which biologically decomposes nitrate compounds. A plant using this process is in operation.

16.7.6 *In situ* denitrification

Special chemical and biological agents are injected into the aquifer to enhance the building up of a microflora for the decomposition of ammonia and nitrate compounds. Pilot-scale experiments in the vicinity of the metropolitan water works of Budapest will be started in 1985.

16.7.7 Continuous nitrate monitoring of surface and drinking waters

An instrument for the continuous measurement of nitrate has been developed in the Research Centre for Water Resources Development, VITUKI. Nitrate in the water is reduced electrically to ammonia in a cell where the ammonia content of the cell-space is measured by a cooled conductivity electrode.

16.8 LEGAL-ADMINISTRATIVE PROTECTION OF GROUNDWATER RESOURCES

The limited space of this chapter does not allow a detailed description of the relevant legislation. A brief summary of the principles governing water pollution control legislation in Hungary was presented recently (Homonnay, 1984).

Comprehensive legislation on the protection of groundwater against pollution does not exist at present; however, effective elements of such protection can be found in the Water Act, and in the economic incentive system for water-pollution control.

General prohibition of water pollution as contained in the Water Act relates to both surface and subsurface waters. Discharge limits under the economic incentive system are also taken as a legal basis of fining when groundwater quality is endangered.

Thus the enforcement of the general provisions of the Water Act, in respect to water pollution, is guaranteed through a legal/economic regulation. When a wastewater discharged into a river or groundwater does not comply with the prescribed limits, the enterprise pays a fine, the magnitude of which depends on the quantity of the waste discharged and also on the level of danger caused by the pollution concerned. The effluents are regularly checked by the regional water authorities in the framework of a sampling programme, the samples are analysed, and the quantity of the discharged polluting agent is calculated for periods of a year or a season. The resulting annual load of the polluting material is then multiplied by the 'fine lot', for example 1 kg NO_3^-= 1 Forint,† and also by different modifying factors reflecting *inter alia* the 'importance' of the recipient resource.

A categorisatin of surface water and groundwater resources reflects the order of importance. Greatest protection is given to the open karstic aquifers and to bank-filtered water resources. Thus the fine levied on discharges let into water resources of this category might amount to ten times the 'normal' fine.

Discharge limits vary according to the degree of protection of receiving waters. The effluent standard for nitrate when discharged into waters of the highest category is 40 mg/l. No fine is levied for nitrate when the receiving water is a watercourse of less importance, however ammonia discharges are levied in any circumstances. Effluent standards for ammonia vary between 2 mg/l and 30 mg/l according to the degree of protection.

The economic incentive system is not the only legal means for the protection of subsurface waters. There is a system of water-use licensing, which, through the conditions of discharges prescribed in the licence granted, represents a legal prevention of groundwater pollution.

In areas where drinking-water intakes are in operation or foreseen, so-called protection zones are established in order to prevent contaminants from leaching into the aquifers.

Protection against non-point source pollution is legally unregulated. This kind of pollution originates mainly from the use of fertilizers, and also from other production technologies of the agricultural sector. In Hungary one of the major sources of nitrate pollution is agriculture. In the framework of a United Nations Development Programme/Food and Agriculture Organisation project of three-years duration, methods for the prevention of non-point source pollution in agricultural areas are investigated, also including possible measures for legal protection. This project will be completed by 1986, and it is hoped that it will result in the elaboration of different methods to control this kind of pollution.

BIBLIOGRAPHY

Alföldi, L. (1982) Hydrogeological aspects of groundwater nitrification. IIASD Collaboration Proc. Series CP. 82–45.

Alföldy, L. (1982) General report on movement and interaction of nitrates and pesticides in the vegetation cover-soil-groundwater-rock system. International Symposium IAH, Prague, Czechoslovakia.

Almássy, E. & Deák, B. (1981) Variations in the quantity and quality of the sources of public water supply. *Hidrológiai Közlöny*. No. 10, Budapest, Hungary.

† Forint is approximately US $0.017; 90 Forints approximately equal £1.00 sterling (May '86).

Csáki, F. & Endrédi, I. (1981) Pollution by nitrates of the subsurface waters in Hungary. *Studies in Environment Science,* No. 17, Elsevier, Amsterdam.

Homonnay, A. (1984) Water quality standards as a basis of an economic incentive system. *Wat.Sci.Tech.* **16,** 21–31.

17

Acidification of groundwaters and land use

J. Forslund

17.1 INTRODUCTION

In the last ten years many studies have been made of acidification of the environment. Nearly every month the journals bring out articles about acidification, with special focus on the acids arising from incineration of coals and fuels. Other explanations of the problem of acidification have been characterised as red herrings in an article by Havas *et al.* (1984).

This chapter describes one of these red herrings, that of land use as an important factor in the acidification of groundwaters, and acidification in other areas. Focus will be especially on the acid production in the soil in connection with cultivation. Also described is the possible acid production in the atmosphere arising from ultra-violet radiation of ammonia evaporated from cultivated fields.

17.2 BACKGROUND

The results and measurements used in this chapter are from studies made in Denmark, the smallest of the Scandinavian countries. The total area of the country is 4.3 million ha. On 2.9 million ha barley, wheat, rye, turnips, sugar beet, and grass is grown. Agriculture is very intensive with high productivity, and extensive use of fertilizers. The animal production is considerable and, unfortunately, distributed very unevenly, with a high concentration of cattle and pigs and manure in some areas. Perhaps because of the limited size of the country accurate statistics (Agricultural Statistics: 1960–1982) have been able to be compiled about production, the use of fertilizers, and the types and amount of animals and their production of manure. Fifteen state farms in Denmark spread all over the country measure the amount and the composition of precipitation, and have done so since 1955 (Jensen, 1962; Jørgensen, 1974 and 1979). Further, daily measurements are made in two European monitoring evaluation programme (EMEP) air and rain monitoring stations (Heidam 1981/82).

17.3 REASONS FOR ACIDIFICATION OF GROUNDWATER AND SURFACE WATERS

In the following a number of reasons for acidification will be mentioned.

17.3.1 Natural acidification

When percolating through the soil, rain dissolves soluble parts. By such leaching processes, the alkaline parts such as calcium carbonate will be removed from the upper layers of the soil. In some uncultivated areas in Denmark the upper 0.5 m contains very small amounts of lime, while in other areas, up to a depth of several metres, the content of lime is very small (Christensen, 1962). In some areas degradation and dissolution of the rocks may compensate somewhat for the dissolved alkaline parts. In other areas, however, the leaching process is so far advanced that the quality of the rain-water passing through the soil will not change significantly. So the upper groundwater and the water in lakes and watercourses will resemble rain-water. In such areas acidification of rain-water will be the main reason for the acidification of groundwater and surface waters. This situation seems to be characteristic of Northern Scandinavia and Scotland.

17.3.2 Acidification of precipitation

The atmosphere and the rain contain significant amounts of nitrogen and acid compounds. Table 17.1 shows the concentrations in rain, in particles, and in air. The calculation as total deposition is made on the specified deposition velocities and is given as an average from 1977 to 1981.

The total yearly deposition of nitrogen compounds is calculated as being 17 kg N/ha yr, of which the deposition from gases represents about 25%. Some doubt can be raised about the calculation based on model studies of acid deposition from gases, which seems unbelievably large. Hydrogen-ion deposition from gases is, in this table, three times higher than from rain.

17.3.2.1 Changes in rainwater quality

The concentration of nitrogen compounds in the air fluctuates over the year, with peaks normally in April, May, and September, as shown in Fig. 17.1. Very often the concentrations of ammonium and nitrate go together, though the sources should differ completely. Ammonium derives mainly from evaporation from fields, manure, and degradation of organic matter, whereas nitrate is said to derive from combustion of oils, petrols, gas, and from industrial activities. The relationship between ammonium and nitrate can also be seen in Fig. 17.2, where the deposition of ammonium and nitrate

Table 17.1 — Nitrogen deposition from atmosphere in Denmark (National Agency of Environmental Protection (NAEP, 1984a)

	Concen-tration	Deposition Velocity	Deposition kg N/ha yr	Deposition kg H^+/ha yr
Rain	mg/l	660 mm/yr		
NH_4-N	1.0		6.6	
NO_3-N	0.7		4.5	
H^+	44×10^{-3}	660 mm/yr		0.29
Particles	$\mu g/m^3$	m/sec.		
NH_4-N	3.1	2×10^{-3}	2.0	
NO_3-N	0.5		0.3	
H^+	42×10^{-3}			0.03
Gas				
NO_2-N	2.2	4×10^{-3}	2.8	0.8
NO_3-N	0.5	8×10^{-3}	1.2	
Total			17.4	1.1

is plotted over a 20-year period. The slopes of the line coincide remarkably and may suggest another explanation.

It is known that ammonia in water will be converted to nitrite and nitrate under ultraviolet radiation. Perhaps the same conversion can take place in the atmosphere when the sun shines on evaporated ammonia. The transformation will lead to an acid formation of 1 mol H^+/mol NH_3.

Attempts have been made to estimate the total deposition of NO_x-N from combustion of oils, petrols, gas, and so on, and an estimate for Denmark has been set at about 6 kg N/ha yr (NAEP, 1984b). As the total nitrate N is estimated at about 9 kg N/ha yr, it cannot be excluded that part of the nitrate originates from oxidation of ammonia. If so, the increasing evaporation of ammonia may contribute to the acidification of the rain.

17.3.2.2 Nitrification of precipitation

As the precipitation still contains ammonia salts, it can be oxidised to nitrate by nitrifying bacteria under acid production at the soil surface or in the lakes and streams. The acid production from 8 kg NH_4-N/ha/yr will be 1.1kg H^+/ha/yr, or about four times more acid than the rain already contains.

17.3.3 Acidification of cultivated fields

The processes occurring in cultivation of fields are mainly acid producing. The main acid processes are the microbiological oxidation of ammonia and ammonium fertilizers, and the degradation of residues of plants, roots, and so on, after harvesting. Denitrification processes take place too, but under

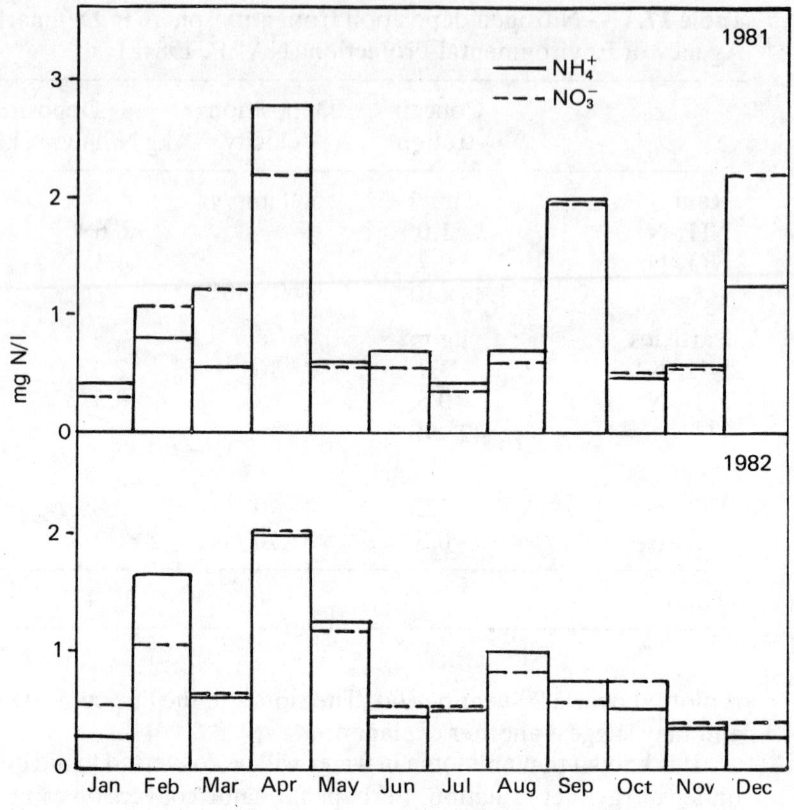

Fig. 17.1 — Ammonium and nitrate concentration in rain at EMEP station 9005, Keldsnor (Heidam, 1983 & 1985)

acid consumption. In order to neutralise the produced acid, lime is frequently added to the soil. Figure 17.3 shows the use in Denmark of different types of fertilizers over the last 20 years (Agricultural Statistics: 1960–1982).

Before calculations can be made of the acid production and consumption, values must be set up for the different nitrogen compounds included in the process.

17.3.3.1 Models for the nitrogen cycle in fields

In recent years a number of attempts have been made to calculate the nitrogen cycle and nitrogen requirements of agriculture. The agricultural industry wishes to work out optimal and economic production models, while environmentalists wish to establish a model to minimise the nitrogen leaching from the fields to air and groundwater. Table 17.2 shows some of the models (Aslyng, 1976; Tjell 1983; Schrøder & Simonsen, 1985).

The basic condition for setting up a model is a constant level of nitrogen in the ground. This will not be the case where shifts between cultivation of crops and grass occur. But for reasons of simplification, the vast amount of nitrogen bound in the soil is not taken into account.

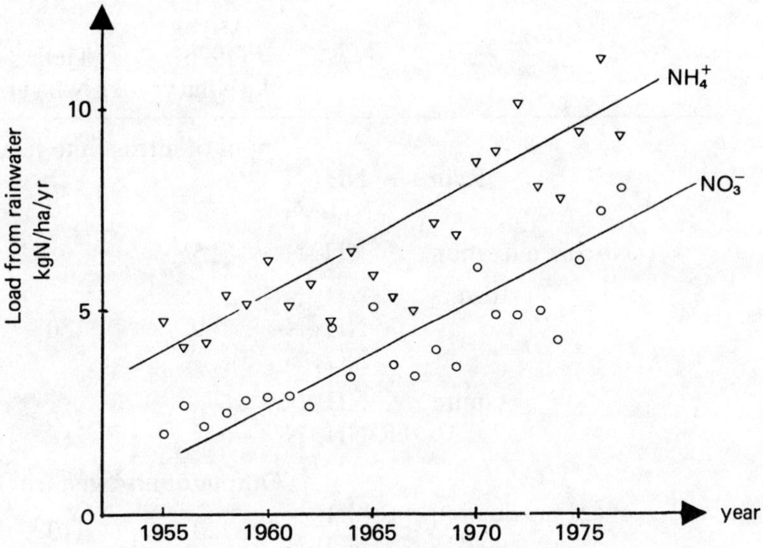

Fig. 17.2 — Ammonium and nitrate load from rainwater in Askov, Denmark (1955 to 1977)

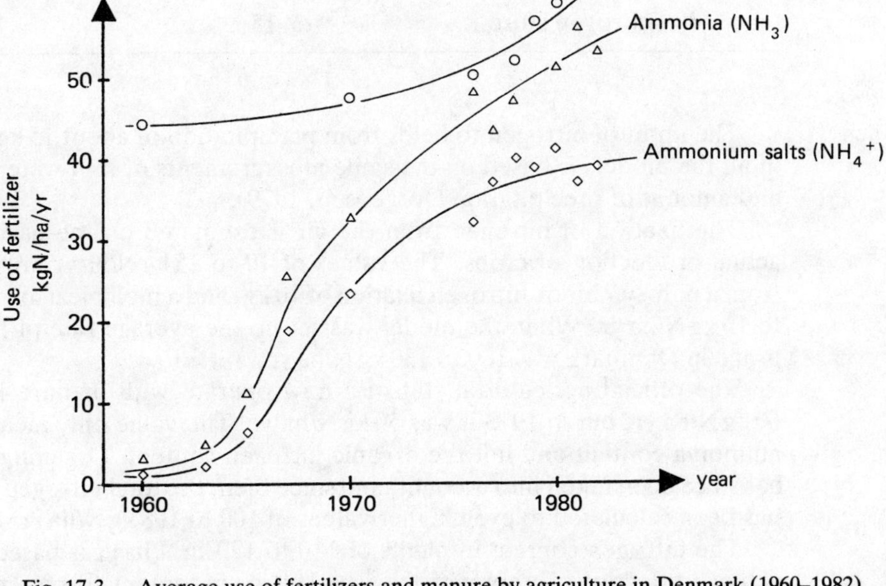

Fig. 17.3 — Average use of fertilizers and manure by agriculture in Denmark (1960–1982)

Table 17.2 — Nitrogen models for cultivated fields

		Aslyng (1976) kg N/ha yr	Tjell kg N/ha yr	Schrødeer kg N/ha yr
		Input of nitrogen to field		
Rain	NH_4-N	15	30	8
	NO_3-N			8
Nitrogen fixation	N_2-N	25		10
Fertilizers	NH_4-N			40
	NH_3-N	100	130	50
	NO_3-N			40
Manure	NH_4-N	50	100	108
	R-NH_2-N			
		Output of nitrogen from field		
Plant production	R-NH_2-N	120	110	111
	NO_3-N			
Loss to groundwater	NO_3-N	20		80
			150	
Loss to air:		35		
Denitrification	N_2-N			30
Evaporation	NH_3-N			38
Combustion	NO_x-N			5
To nitrogen store		15		

The input of nitrogen to fields from precipitation of about 15 kg N/ha yr in all the models is based on the same measurements of rain-water quality and amount of precipitation (Jørgensen, 1979).

The fixation of nitrogen from the air is estimated on the basis of the actual production of crops. The values of 10 to 25 kg N/ha yr are derived from a non-symbiotic nitrogen fixation of 10 kg, and a biological fixation of 5 to 15 kg N/ha yr. When the model was set up the average rate of fertilizer usage in Denmark was 100 to 130 kg N/ha yr.

The official agricultural statistics now operate with manure loads of 60 kg N/ha yr, but in 1973 it was 50 kg N/ha yr. The value only includes the ammonia content and not the organic nitrogen content. Organic nitrogen has since been taken into account, and since then, the total nitrogen content has been calculated to even higher values of 100 to 108 kg N/ha yr.

The nitrogen content in plants of 110 to 120 kg N/ha yr is based on the actual production of crops with the actual nitrogen content. Typical nitrogen uptake in plants appears from Table 17.3.

The loss of nitrate to groundwater mainly depends on the fertilizer

Table 17.3 — Nitrogen uptake in plants in relation to use of fertilizer (Jørgensen & Lassen, 1984)

	Fertilizer applied kg N/ha	Uptake in per cent of fertilizer applied
Barley	120	57
Wheat	120	62
Wheat	180	49
Rye	120	39
Potatoes	120	90
Grass	240	92

applied, the type of crop, and the type of soil. In one of the first models set up by Aslyng (1976) the loss was 20 kg N/ha yr, calculated from analysis of drainage water in clay areas, and therefore was limited. Later Aslyng & Hansen (1984) in lengthy investigations on different soil types, found the results shown in Fig. 17.4. The values in sandy soils are not far from the loss to groundwater of 80 kg N/ha yr estimated by Schrøder & Simonsen (1985).

The loss of nitrogen compounds to the air is of significant size in Schrøder's model. The ammonia evaporation has also been estimated by others (NAEP, 1984c) who have come up with values rather similar to the proposed value.

The estimation of the denitrification contribution is very uncertain as it cannot be measured directly. Aslyng & Hansen (1984) predicted denitrification values of 0 to 60 kg N/ha yr in differents crops and soils.

Another type of model, shown in Table 17.4, has been established as a total balance for the country (NAEP, 1984c). This model cannot be used as a basis for calculating the acid production in the fields, but the various parts seem to support some of the values in Schrøder's model, especially if the main part of the remainder is added to the leaching into the soil. A conservative estimate of nitrate leaching is 70 to 90 kg N/ha yr.

Though much work has been done to make a 'true' nitrogen-cycle model, there is still uncertainty about some of its elements, especially evaporation of nitrogen compounds, the denitrification process, and the organic nitrogen level in the ground.

17.3.3.2 Acid production in the field

Ammonium and ammonia fertilizers applied to the soil will in a few weeks be oxidised to nitrate by nitrifying bacteria. The reactions are:

$$NH_4^+ + 2O_2 \rightarrow NO_3^- + 2H^+ + H_2O$$
$$NH_3^+ + 2O_2 \rightarrow NO_3^- + H^+ + H_2O.$$

So together with one mole of nitrate, one (by NH_3) or two (by NH_4+) moles

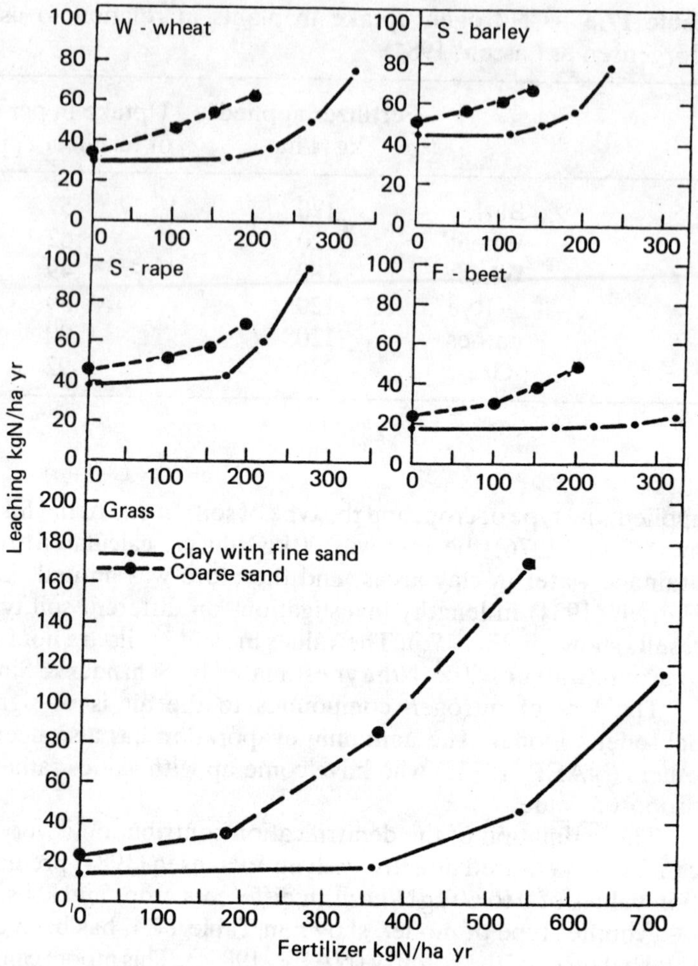

Fig. 17.4 — Average yearly leaching 1st April to 31st March (1968–1971) from clay with fine sand and coarse sand by different application of fertilizers

of hydrogen ion will appear in the soil. The same will happen to the ammonium salts from rain and manure. The organic nitrogen compounds in manure could presumably be compared with the ammonia reaction. According to Dam Kofoed (1983) the content of ammonium in manure varies from approximately 50% (cattle) to 70% (pig). The following calculation is based upon the average of 60% in manure. The evaporated ammonia does not form part of the acid balance of the soil.

Using the Schrøder model the total ammonium and ammonia input to the field can be calcuated as in Table 17.5. The growing of crops contributes to the acid production/acid use. Uptake of nitrate from the soil into plants will lead to an ion exchange with hydrogen carbonate ions or hydroxyl ions from the plants to the soil. On the other hand uptake of cations as calcium

Table 17.4 — Nitrogen balance for Denmark

Input	kg N/ha/yr	Output	kg N/ha yr
Fertilizers	130	Nett sale of crops	10
Imported feed + fish products	60	Nett sale of animals	25
Precipitation + dry deposition	20	Combustion of straw	2
Biological N fixation	10	Denitrification	30
		Evaporation of NH_3	45
		Infiltration and discharge of manure etc.	20
		Leaching of the ground	53
		Remainder	35

Table 17.5 — Calculated input of ammonia and ammonium to the fields, and loss of ammonia to the atmosphere

Ammonia	kgN/ha yr	Ammonium	kg N/ha yr
from fertilizers	50	from rain	8
from manure	43	from fertilizers	40
to the air	−38	from manure	65
Total	55	Total	113

and potassium will be exchanged with hydrogen ions. The uptake of cations is calculated to be about 80 meqv/gN (NAEP, 1984a) or an exchange of 1.1 mol H^+/mol N. If so, every uptake in the plant of cations and anions will lead to a small surplus of hydrogen ions, where the amount depends especially on the relation between the release of hydroxyl ion and hydrogen carbonate ion from the plant. In the following calculation a possible hydrogen-ion production from the plants will not be taken into account.

Denitrification of the nitrate produced will lead to an acid consumption from the soil. One mole of hydrogen ion will be used for the denitrification of one mole of nitrate.

In addition to the reaction between the growing plants and the soil, dead vegetation, residues of plants, roots, straw, and so on, will influence the acid balance when degraded to carbon dioxide and nitrate.

The Schrøder model which is based upon a constant organic nitrogen level in the soil, estimates the annual decomposition of nitrogen at 45 kg N/ha. The nitrogen content bound as protein can be degraded to ammonia and further nitrified to nitrate by formation of one mole hydrogen per mole nitrogen. Degradation of nitrate in roots and in plant residues will affect the acid balance. An estimate for the acid production by nitrification of dead vegetation with a content of 10 to 25% NO_3 of the total nitrogen content will be 2.4–3.0 kg H^+/ha. Carbon-dioxide production has been estimated at 4.5 kg H^+/ha (NAEP, 1984a).

In Table 17.6 Schrødder model is used for calculation of the hydrogen-ion balance on cultivated fields to which there has been no application of lime.

Table 17.6 — Hydrogen-ion balance for cultivated fields without application of lime

		kg H^+/ha yr
Hydrogen-ion production		
From ammonia	$\frac{55}{14}$†	4.0
From ammonium	$\frac{113}{7}$†	16.2
Degradation of nitrogen compounds	$\frac{3\times45}{4\times14}$	2.4
Degradation of organic matter		4.5
Hydrogen-ion deposition from air		
Rain, particles		0.3
Gas deposition from NO_2+SO_2		0.8
Hydrogen-ion use		
Denitrification	$\frac{30}{14}$	− 2.2
Hydrogen-ion excess		~26

† Values from Table 17.5

When using the nitrogen models for calculation of the excess of acid in connection with cultivation, there will always be some uncertainty about the processes in the soil and plants; the evaporation and use of the carbon dioxide produced is also difficult to determine.

17.4 LIME NEEDED FOR NEUTRALISATION OF ACID PRODUCTION

The nitrification process is closely connected to acid formation, and one should expect to find a low pH in drainage water or groundwater if no neutralisation takes place in the ground. The neutralisation medium would be calcium carbonate, which exists naturally in the ground, or as lime applied to the soil. The neutralisation process will be as follows:

$$CaCO_3+2H^+ \rightarrow CA^{++}+CO_2+H_2O$$
$$CaCO_3+CO_2+H_2O \rightarrow Ca^{++}+2\,HCO_3^- \,.$$

The amount of lime required for neutralisation of the hydrogen ions will depend mainly upon the possibility for evaporation of the carbon dioxide, and upon the required pH in the soil.

For neutralisation of the first step a lime consumption of 1300 kg $CaCO_3$/ha yr is needed for an acid amount of 26 kg H^+/ha yr. The second step will double the amount. Practical tests have shown a demand for 750 to 1800 kg $CaCO_3$/ha yr for neutralisation of an acid production of 26 kg H^+/ha yr resulting from nitrification of ammonia (NAEP, 1984a). In Fig. 17.5 is shown the average yearly amounts of lime applied to cultivated fields since 1960 (Personal information from Danish Centre for Cultivation of Plants).

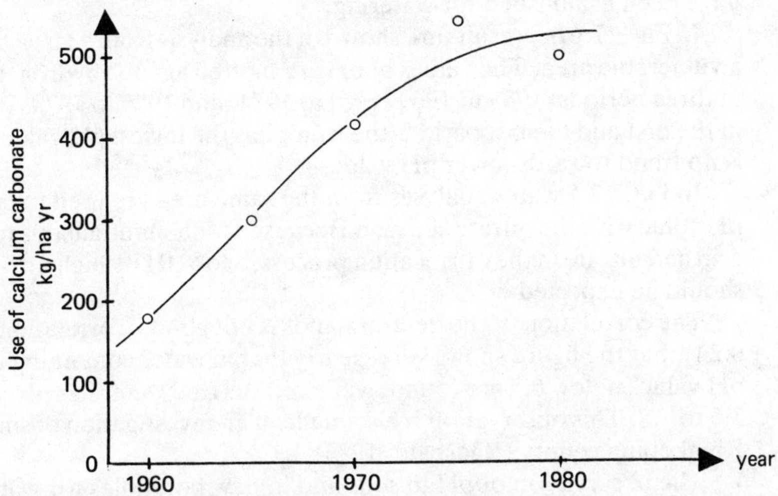

Fig. 17.5 — Average use of calcium carbonate in agriculture in Denmark (1960–1980)

If the average acid production based on the Schrøder model is correct, more lime is needed in the field than is used at present. The result will be acidification of the soil and/or groundwater if no neutralisation takes place in the subsoil.

17.5 ACIDIFICATION OF SOIL AND GROUDWATER

As shown in the calculation of acid production in cultivated fields based on Schrøder's nitrogen model and the addition of lime, an excess of acid should still occur.

Especially in Western Jutland a descreasing pH in soils would be expected. In Table 17.7 pH measurements in the upper soil are shown for the last 10 years. No sign can be seen of an increasing acidification. On the contrary the upper soils seem to have an increased pH compared with 1974. Uncertainty in these measurements prevails especially for soils with pH values below 5.5. It cannot be seen if the acid soils still have the same pH values, or whether a distribution of lower pH values around the value 4 has occurred.

If acidification is not likely to appear in the soil, an excess of acid should be found in shallow groundwaters.

An attempt has been made to show if there is an increase in acidity in particularly vulnerable sandy areas in Western Jutland. The data required to demonstrate such a relation could come from raw-water wells in existing waterworks, monitored over the years, or from new water borings established in different periods. Unfortunately, very few waterworks monitor the raw-water quality, so a measured pH decrease can not be representative of the degree of acidification. The number of new boreholes registered at the data base in the Geological Survey of Denmark is very small in those areas before 1965. But in the period 1965 to 1980, a number of shallow boreholes have been established for watering.

In Fig. 17.6 the results are shown of the analysis from 93 new boreholes in a vulnerable area. The values of pH are plotted against hydrogen carbonate in three periods: 1965 to 1969, 1970 to 1974, and 1975 to 1979. It seems as if in the first and the last periods the water has the lowest pH values, but there is no trend towards lower pH values.

In Fig. 17.7 water analyses from the same area are used to compare the pH value with the nitrate content. Because of the simultaneous formation of acid and nitrate in the nitrification process, a low pH by high nitrate contents should be expected.

The correlation of the determinands is not good, (correlation coefficient 0.24), but the figure shows very clearly that in water containing nitrate only pH values below 6.5 are found, while nitrate-free water has pH values from 3.5 to 7.8. This observation is also made in an investigation of single wells in Ringkøbing county (Bådsgåld, 1983).

The observation of pH in soils and in new boreholes for water gives no strong indication of an acidification in connection with cultivation of the soil. An increasing nitrate concentration is shown, and an increasing acidity should be expected.

One reason may be that the groundwater quality is also influenced by processes taking place below the root zone. By passing through different layers in the ground, denitrifying processes can take place which both decrease (biological) and increase (oxidation of ferro and sulphide compounds) the acidity of the water.

Another reason is perhaps that sampling of groundwater for pH analyses which, though it seems very simple, is rather complicated because of evaporation of carbon dioxide and low accuracy in pH measurements. Groundwaters with pH values below 7 normally have an excessive amount of carbon dioxide in relation to equilibrium with the atmosphere, so that carbon dioxide tends to evaporate. This will result in measurements of higher pH than in the original sample. Further, transporting the sample to the laboratory, pouring the sample out into a beaker to adjust the temperature of the sample before measuring the pH, leads to further evaporation of carbon dioxide. In the Danish interlaboratory calibration, the pH appears to be the worst of all determinands examined, both for samples sent to

Table 17.7 — Percentage occurrence of pH values in soils in Western Jutland, 1975 to 1984 (Landsudvalget for Planteavl)

	pH in soils (0–20-cm depth)			
	<5.5	5.5–5.9	6.0–6.4	6.5–6.9
1975	27	43	23	6
1976	28	42	22	7
1977	26	40	25	8
1978	24	40	25	9
1980	20	40	29	9
1982	19	40	29	10
1984	17	37	32	12

Fig. 17.6—Coherent pH and hydrogen carbonate concentration at 93 new boreholes established in 1965 to 1979 in area 75, Ringkøbing county

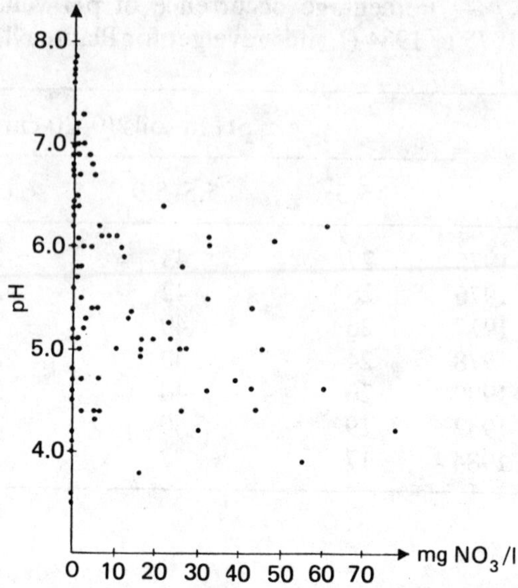

Fig. 17.7 — Coherent pH and nitrate concentrations at 112 new boreholes established in 1965 to 1979 in area 75, Ringkøbing county

laboratories within a few hours, and for samples measured on the spot by laboratory staff at The Ministry of the Environment's reference laboratory.

A third problem in relation to pH values in the water is the time of year when samples are taken. Boreholes will normally be established in April to December and the water samples will be taken in the same period. However, it is known that the nitrate content will be highest during January to March, and to find the most acid water the sample must be taken during the same period.

In the near future a monitoring programme of groundwater will be started in different areas of the country. At the same time many procedures of sampling and analyses will be modified. The practice of taking water samples in the summer period will be changed, and instead several samples will be taken over the year. The winter samples are most interesting, because a high nitrate will be found and presumably also low pH values in the shallow groundwaters.

Today it can only be said that it is probable that an acid excess may exist, but it is hoped to be proven in the near future.

17.6 CONCLUSIONS

- The use of nitrogen models for Danish agriculture to calculate the acid production in the fields and the use of lime has shown an excess production of acid. Precipitation will only increase the acid load by about 1% of the total amount of acid produced in the field.

- It cannot be excluded that ammonia evaporation contributes to the nitric-acid formation in the atmosphere when exposed to solar radiation. The ammonium content in rain will further increase the acidity of rain about four times, when nitrifying bacteria and oxygen transform the ammonium into acid in the soil, or in watercourses, or lakes.
- The existing pH data have so far not shown any significant signs of increased acid content in groundwaters in the most vulnerable parts of western Denmark, but data from new boreholes for water show that nitrate-containing waters in sandy areas are acid, with pH values below 6.5, while nitrate-free waters all show pH values between 7.8 and 3.5.

REFERENCES

Agricultural Statistics (Landbrugsstatistik. Copenhagen: Danmarks Statistik. Yearly editions 1960–1982 (in Danish with English translation of all tables).

Aslyng, H. C. (1976) *Climate, Soil and Plants.* DSR forlag KVL Copenhagen (in Danish).

Aslyng, H. C. & Hansen, S. (1984) *Nitrogen balance in crop production — simulation model NITCROS*, Royal Vet. and Agric. Univ., Copenhagen.

Bådsgård, J. (1983) Investigation of 442 single water wells in Ringkøbing, not published.

Christensen, W. (1962) Views on the geochemical evolution in upper soil layers in Denmark, Medd. fra *Dansk Geologisk Fore.* Copenhagen **15**, p. 112 (in Danish).

Dam Kofoed, A. (1983) Manure and its use, *Agrologisk Tidsskrift Marken*, **10**, p. 14 (in Danish).

Havas, M., & Hutchinson, T. C. & Likens, G. E. (1984) Red Herrings in acid rain research. *Environ. Sci. Technol.* p. 18, 176 A.

Heidam, N. Z. (1983) European Monitoring and Evaluation Program Bulletin 1981 National Agency of Environmental Protection Air Pollution Laboratory (in Danish).

Heidam, N. Z. (1985) European Monitoring and Evaluation Program Bulletin 1982. Roskilde, National Agency of Environmental Protection Air Pollution Laboratory (in Danish).

Jensen, J. (1962) *Investigation of Content of Nutrients in Precipitation.* Tidsskrift for planteavl, pp 894–906 (in Danish).

Jørgensen, V. (1974) *Content of nutrients in precipitation.* Report No. 1168 from Statens Planteavlsforsøg (in Danish).

Jørgensen, V. (1979) *Chemical composition of air and precipitation in Danish agriculture land* Report No. 1434 from Statens Planteavlsforsøg (Danish with English summary and English text in tables and figures).

Jørgensen, S. S. & Lassen, R. (1984) Use of nitrogen in Danish agriculture, *Ugeskrift for jordbrug* **25**, p. 687 (in Danish).

Landsudvalget for planteavl *Survey of land investigations, yearly reports of nitrogen from agriculture,* 2, p. 6 (in Danish).

National Agency of Environmental Protection (1984b) *Acidification project: Emission, transport and deposition* (in Danish).
National Agency of Environmental Protection (1984c) *The NPO Report.*
Schrøder, H. & Simonsen, J. (1985) *Assesment of measures against leaking of nitrogen from agriculture,* 2, p. 67 (in Danish).
Tjell, J. C. (1983) *Mass balance of nitrogen in Danish agriculture,* Report from Lab. tech. Hyg. Technical Univ. of Copenhagen.

18

Pollution from agriculture in the area of the Welsh Water Authority

W. R. Howells and **R. Merriman**

18.1 INTRODUCTION

Agriculture is one of the major industries in Wales. In 1983 the total revenue from agriculture was calculated as £693m (ADAS pers. comm). The dairy and beef sectors together accounted for 63.1%, sheep and wool, 19.1%, and pigs, poultry and eggs, 8.8% of this total. Agriculture in Wales has undergone considerable changes in the last 20 years, there being a change in emphasis from the non-intensive mixed farming to specialisation in certain sectors. These changes have resulted in dairy herds typically doubling, being concentrated in the areas of faster growth of grass.

This intensification of agriculture has coincided with an increase in the number and severity of pollution incidents caused by discharges from farms, particularly dairy, and to some extent, intensive beef units.

The Welsh Water area does not correspond exactly with that of Wales (Fig. 18.1) but it does include approximately 93% of the Welsh dairy farms. The principal dairy areas in the Welsh Water area include Dyfed, where approximately 55% of Welsh milk is produced, Clwyd (18%), Gwent (8%) and Glamorgan (5%). A further 6.8% is produced in Montgomery within the Severn Trent Water area — Fig. 18.1 (ADAS pers. comm). Parts of Hereford and Worcester and the Cheshire plain also fall within the Welsh Water area, the latter being an important dairy-farming area.

The discharge of farm wastes is now one of the most serious pollution problems facing Welsh Water, as demonstrated by the increasing number of prosecutions instituted by the Authority (Table 18.1).

18.2 NATURE OF POLLUTION

18.2.1 Slurry

The main discharges of concern to Welsh Water are silage liquor and cow slurry, which together account for over 90% of serious pollution incidents.

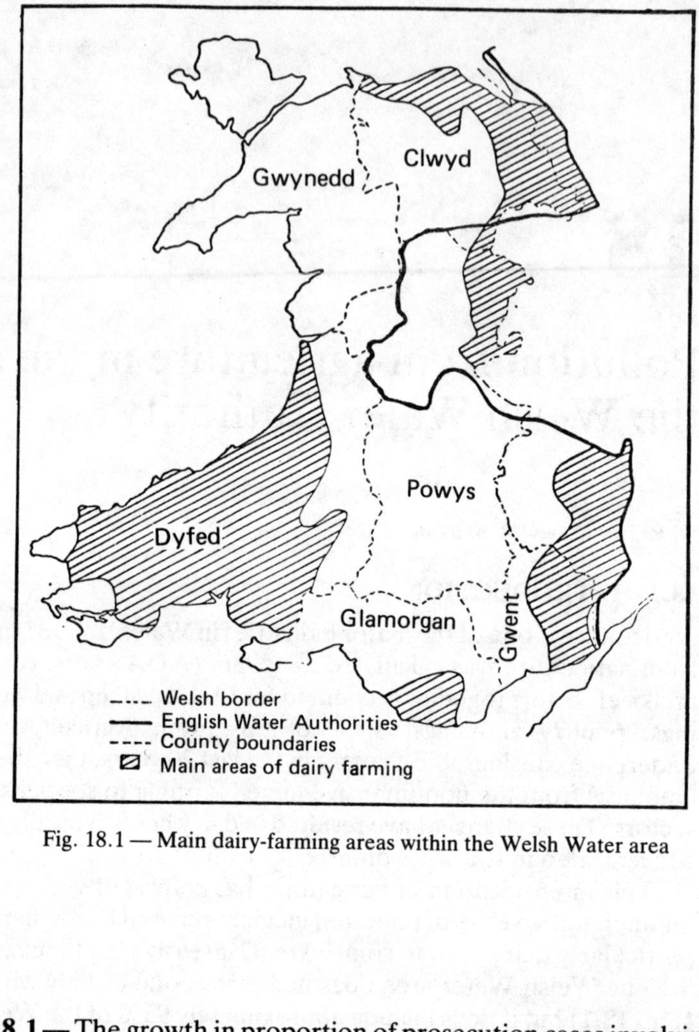

Fig. 18.1 — Main dairy-farming areas within the Welsh Water area

Table 18.1 — The growth in proportion of prosecution cases involving farms

Year	Total no. of prosecution cases	No. of prosecution cases involving farm wastes	
1978	20	6	30% of total
1982	30	17	57% of total
1983	34	22	65% of total

The increase in severity is associated with increased herd sizes, with the cows predominantly housed in cubicles over winter rather than bedded on straw or left outside on pasture land. This practice, coupled with the increased use of silage rather than hay as winter fodder, produces large quantities of very liquid slurry, far more likely than the drier dung produced in the past to drain into watercourses unless strictly controlled.

Many farmers apparently still do not appreciate the polluting nature of slurry, particularly that of the liquid fraction. Undiluted cow slurry is, in terms of biochemical oxygen demand, some 40 to 80 times stronger than untreated domestic sewage and the liquid fraction, although often referred to as 'only coloured water', has about half the polluting strength of homogeneous slurry. Sufficient care is not taken by farmers in the storage or disposal of liquid slurry, often as a result of inadequate advice or publicity in relation to this problem.

The slurry is usually stored in open lagoons, which may also receive very large quantities of rain-water, spring-water, or water running off land or concrete surfaces due to inadequate provision for separation of this water at the time when farm buildings, yard areas, and so on, were constructed. Sufficient care is not given to the separation of this clean water from sources of slurry. The over-loading of storage systems by excess water is a significant cause of the great majority of slurry-pollution incidents in Welsh Water and is exacerbated in areas of high rainfall. Rainfall in Wales is typically above the national average and it is not unusual for slurry volumes to be increased fivefold due to the inadequate separation described above.

As a result of this overloading, some farmers have provided holes or breaches in the walls of slurry lagoons to allow the seepage of liquid slurry, thus greatly increasing the chances of serious pollution, as most farm holdings are situated near watercourses or springs intended originally to provide drinking water. A study in Dyfed has shown that during 1983, of the 24 incidents of slurry discharges considered as serious, 16 involved major discharges via man-made holes, breaches, sluices or valves, in lagoon walls, whilst none involved secure lagoons which may have overflowed due to heavy rainfall experienced earlier in the year. Similarly in 1984, seven of the 13 serious slurry discharges in the same area involved discharges via man-made holes, breaches, and so on.

18.2.2 Disposal of Slurry

So far as is known, there is no economic way at present of treating farm slurry. Anaerobic digestors are expensive to set up, require regular attention, and appear generally unsuitable for those problem farms where storage systems are overloaded by excess water. Barrier ditches/soakaway systems are unsuitable for the treatment of strong organic wastes such as slurry, whilst the topography of the region is such as to restrict the areas of very shallow gradients required for such systems to treat even weaker wastes, such as milking-parlour washings.

The only known practical method of disposal available in virtually all cases is by spreading slurry over land if ground conditions permit. Disposal over land in winter is becoming less common as stocking densities increase, as this necessitates keeping land in good condition. Such disposal is not without its problems unless carefully controlled, and large areas of land are required. If the land is waterlogged or frozen at the time of application, or becomes so due to excessive slurry application or rainfall, the slurry is not incorporated into the soil and may run off into watercourses.

The specialisation and intensification of farming restricts the periods for spreading slurry even during favourable dry periods. It is now common practice for farms to aim for two, three, or even four cuts of grass for silage each year, which necessitates intensive use of the land, and farmers are reluctant to spread slurry on the land used for silage production until after the last silage is harvested. This may well be relatively late in the year when rainfall can then further restrict effective spreading over land.

18.2.3 Silage Liquor

Silage production is less vulnerable to unpredictable weather than haymaking, and earlier cutting enables fast regrowth of grass. These advantages, coupled with changes in animal housing, have resulted in a fourfold increase in silage production in Wales in the past ten years (ADAS pers. comm). Production may well increase further following the introduction of the milk quotas and as farmers are encouraged to produce additional silage rather than feed 'concentrates' to the cows.

Silage liquor itself is extremely polluting, in terms of oxygen-demand it is up to 200 times more polluting than untreated domestic sewage. It is well known that efficient wilting of grass prior to ensiling can virtually eliminate silage liquor and is often favoured by farmers. This is not always possible, however, due to adverse weather conditions. Pollution from silage liquor usually arises due to one or both of two causes:

- inadequate storage of liquor, particularly at those farms where it is anticipated there may be little or no liquor production,
  and
- failures in the base or walls of the silo itself or of the liquor collection system.

The latter problem has become more prevalent in recent years, presumably as silos have aged and become damaged by the aggressive nature of the liquor, but particularly in the last two years where there has been a shift to the production of unwilted silage, the increased liquor causing increased corrosion of silos and drains. The recommendations contained in the farming press to make unwilted silage and to rely upon straw to mop up liquor may have led to very serious problems arising at some farms.

18.3 EFFECTS OF FARM WASTE DISCHARGES

A vareity of interests, and in some cases, very large numbers of people, are affected adversely by the improper disposal of farm wastes.

18.3.1 Tourism

Wales is an important area for tourism, comparable in many ways with that of agriculture. The Wales Tourist Board (pers. comm.) estimate that 12 million visitors entered the country in 1983, and the total revenue for tourism in the same year was estimated to be £510m.

The most important areas for tourism are given in Table 18.2 and coincide, to some extent, with the main areas for dairy farming in Wales. Polluting discharges from farms can threaten all forms of aquatic life, can cause aesthetic nuisance in watercourses, including occasionally those which run over beaches, and may thus reduce an area's appeal from a tourism viewpoint.

Table 18.2 — Importance of tourism and dairy farming in Welsh counties

County	Tourism revenue for 1983 in £m	Percentage of Welsh milk production 1982 to 1984
Gwynedd	175	6
Dyfed	150	55
Clwyd	75	18
Powys	70	8
Gwent	10	8
West Glamorgan	10	
Mid Glamorgan	10	5
South Glamorgan	10	

18.3.2 Fisheries

There are very important and valuable game, and to a lesser extent coarse, fisheries in the Welsh Water area. It was estimated that in 1975, the 137 000 anglers licensed to fish in the area spent approximatley £30m in pursuit of their sport (European Parliament Committee on Agriculture, 1983). Consideration of the numbers of jobs which exist within the management of fisheries in the public and private sectors, in the manufacture and sale of fishing tackle, and indirectly in the hotel or other trades, suggests that angling generates significant full-time and seasonal employment, mostly in the rural areas. In addition, there are commercial fisheries inshore, and in estuaries, which are dependent upon adult salmon and sea trout (*Salmo trutta trutta*) successfully reproducing and their progeny surviving in Welsh river systems.

The fisheries in the Welsh Water area are under threat for a number of reasons — disease, poaching, acidification of upland waters, over-exploitation by anglers and commercial fishermen, and land drainage. Discharges of farm wastes appear to be having a significant and increasing effect on fisheries. In Dyfed alone in 1983, at least 110 000 fish are thought to have been killed by farm-waste discharges. Cases have been recorded where angling associations have needed to suspend angling following serious pollution.

Slurry and silage liquor are potentially harmful to fish, due to their very high oxygen-demand which may depress oxygen levels in watercourses, and their high ammoniacal nitrogen content which may be directly toxic. In addition, solid material contained in the wastes may, when gross pollution occurs, severely damage or kill eggs and juvenile fish, affect fish food, and blanket the stream bed. The nutrient content of the waste may also increase primary production leading to chemical and biological changes.

Nursery streams in particular may be rendered unusable or severely damaged by one of more of these factors, the juvenile fish usually being more susceptible to the effects of farm wastes than adults (Alabaster & Lloyd, 1980) and their loss not readily apparent.

18.3.3 Farming Itself

Although not always appreciated, many farmers themselves suffer as a result of farm-waste discharges. Water used for crop irrigation and stock watering in particular, can be rendered useless for these purposes. It is estimated that over half the complaints received by Welsh Water relating to farm pollution originate from the farmers themselves. Farmers often express serious concern about the possible spread of waterborne infections to their stock from polluted streams.

18.3.4 Potable Water

Most potable water in the Welsh Water area is supplied by the Authority, though two water companies operate in north-east Wales. Many important sources of potable water have been severely affected by farm wastes, the frequency and severity of incidents increasing significantly during the last two years. The main problems have arisen in Clwyd, Dyfed, and in the south east of the area and have brought about serious problems to Welsh Water and consumers alike.

From the consumers' viewpoint, although the risk to health is minimised by the treatment afforded to the water, the disinfection processes may increase the unpleasant taste which may already exist in the water due to farm pollution. In addition to the general public, particular problems may be caused to hospitals, kidney-dialysis patients, and food-processing industries, which may require water of a very high quality. Increased costs may be imposed due to requirements for additional treatment.

From Welsh Water's viewpoint, operational problems have been considerable and have involved the abandonment of a number of sources, the temporary shutting down of important abstractions, switching of supplies, and development of alternative sources to replace those temporarily or permanently lost. Increased capital and running costs have been incurred at a time when cash savings are urgently required. For example, in 1983 it was estimated that in Dyfed alone nearly £500 000 had to be spent to combat the effects of farm pollution, mainly due to a stream diversionary scheme at one underground source, costing £380 000. The problems are often compounded because the majority of interruptions to supply due to farm pollution occur during the holiday season when demand for water is higher. For example in Dyfed, of 13 'shut downs' due to farm pollution in 1983 involving direct abstraction from rivers, nine occurred between July and September.

The serious impact of farm discharges and farming operations on potable supplies can best be underlined by considering two specific sources in Dyfed.

18.3.4.1 Abstractions from the River Eastern Cleddau at Canaston Bridge near Haverfordwest, Dyfed

This abstraction (Grid Reference SN 065 149) has a maximum output of approximately 83 × 10^3 m^3/d (18.3 million gallons per day (mgd)) serving a resident population of 65 000 (plus a substantial tourist population), with water also being available for use by industries situated around Milford Haven. Improved abstraction and treatment facilities were commissioned in October 1980 at a cost of £2.2m, largely to accommodate increased industrial demand. The scheme includes an existing regulating reservoir at Llysyfran to supplement river flows in the summer period. During the design of the scheme, serious pollution of the river by farm wastes was extremely rare but the source was nevertheless protected by a bankside storage reservoir of approximately 25 × 10^3 m^3 (5.5 million gallons).

During 1980 and 1981, serious cases of farm pollution were recorded in the river, necessitating shutting down the source on three occasions. Chronic farm pollution also has had a measurable effect on water quality. Whereas in 1978, the river was classified as 1A according to the National Water Council (NWC) classification as representing water of the highest quality and suitable for all uses, by 1980 this had deteriorated to 1B and a major tributary, the Afon Syfynwy, at the same time deteriorated from Class 1A to Class 2. A major effort was made between 1981 and 1983 to visit as many farms as possible within the Syfynwy catchment which were thought to have been responsible for the deterioration in water quality (Fig. 18.2). The catchment is an intensive dairy-farming area and the annual rainfall in the study area is above average: 1300 to 1500 mm. Analysis of the more serious problems revealed that inadequate separation of clean spring surface water or roof water was the most prevalent cause of the pollution problems.

To provide early warning of organic pollution, raw-water quality is now automatically monitored at the abstraction point for pH, conductivity, turbidity, and ammonia. High ammonia levels cause an alarm at the continuously-manned water treatment works served by the abstraction. Pollution incidents continued to rise, and in 1983 abstraction had to be suspended on 11 occasions, of which ten were attributable to farm pollution.

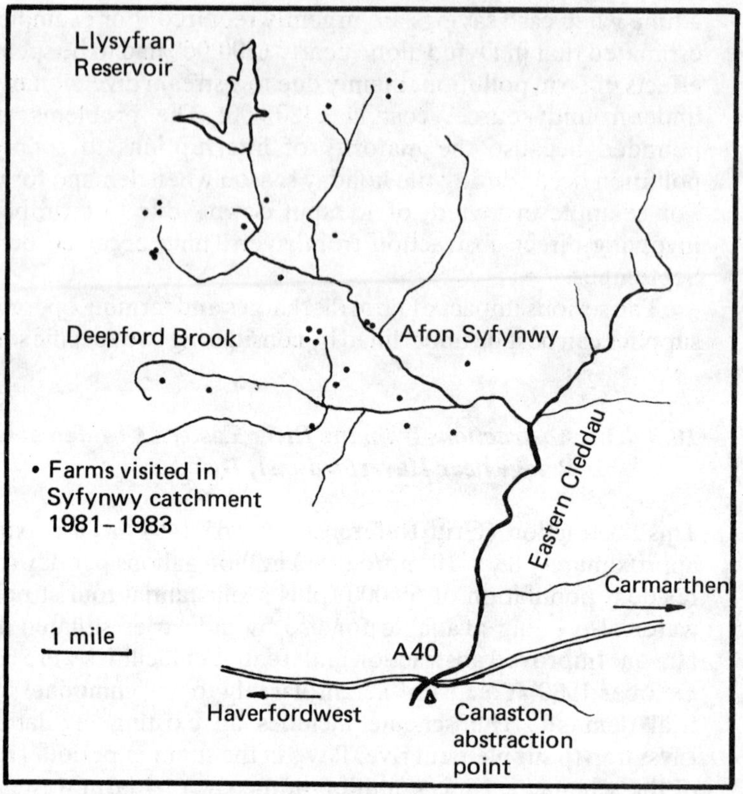

Fig. 18.2 — Llysyfran river regulating scheme

On six of these occasions the period of closure was at least six hours. Various operational modifications have been undertaken, including the provision of a chlorine dioxide disinfection unit which reduces the production of taste-forming compounds. This, due to increased operating costs, is only used at times of poor raw-water quality. Large-scale publicity campaigns have been undertaken and visits to farms have been made in an attempt to reduce the risk of pollution.

During 1984, abstraction had to be suspended on only four occasions. One, however, lasted 18 hours at a time of peak demand and, despite some alternative supplies being available, stored water almost ran out. Alternative sources of water are not readily available. A possible solution would be to increase the capacity of the Llysyfran reservoir and lay a new main directly from it to treatment. This would, however, cost in excess of £20m and have serious implications for the river below. Even this would not necessarily solve the problems altogether and discharges from several farms draining into Llysyfran Reservoir are now posing a threat to the quality of its water.

18.3.4.2 Borehole abstraction at Morfa Bychan, Pendine, Dyfed

The average output from this source (Grid Reference SN 224 075) is approximately 3.6×10^3 m^3/d (0.8 mgd) and the water is obtained from a highly-fissured limestone aquifer. Direct connections exist between three surface streams contaminated by farm waste and the aquifer. No problems were recorded when the source was developed in 1971. During the intervening period, considerable intensification of some of the farms in the catchment has led to silage liquor and slurry discharges contaminating the streams, particularly between 1981 and 1983.

Although rainfall in this coastal area is not high (1100 to 1200 mm/yr), episodes of heavy rain have caused major problems due to slurry spread over land being washed into the streams. Serious problems have also been caused by direct discharges of farm wastes and soil being washed off fields, problems no doubt increased by the disturbance of the land due to agricultural activities. Despite uprating the plant, effective treatment of the raw water is difficult due to its high content of suspended matter which blinds filters and may reduce the efficiency of the disinfection process.

In order to minimise the effects of farm pollutions, a major scheme was undertaken in 1983 to divert two of the streams away from the aquifer they were known to connect with, at a cost of £380 000 (Fig. 18.3). This work has now been completed and the quality of the abstracted water has improved substantially. Purchase of land around the third stream and diversionary works is planned, as is raw-water storage at the treatment plant, at a cost of £220 000. This work is necessary because further changes in land use in the area cannot be ruled out, and it is not possible to be sure that diversionary works on their own will be totally effective.

18.4 DIFFICULTIES IN OVERCOMING PROBLEMS

Whilst the increased problems have arisen largely through the intensification of the dairy and beef sectors, it is considered that many of the problems could have been avoided, or at least reduced in severity, if attention had been given to the following subjects.

18.4.1 Publicity

Sufficient publicity is still not given to the polluting nature of some discharges, notably the liquid fraction of slurry, or the necessity to keep clean water separate from sources of slurry or silage liquor. The former ADAS Wales Farm Waste Committee was a forum which served to highlight relevant problems, and its demise is regrettable.

18.4.2 Consultation

There is a serious lack of consultation over farm developments at the planning stage. The majority of farm developments, which may directly or indirectly give rise to pollution, do not require planning permission and the recent changes in the General Development Order do little to improve the situation.

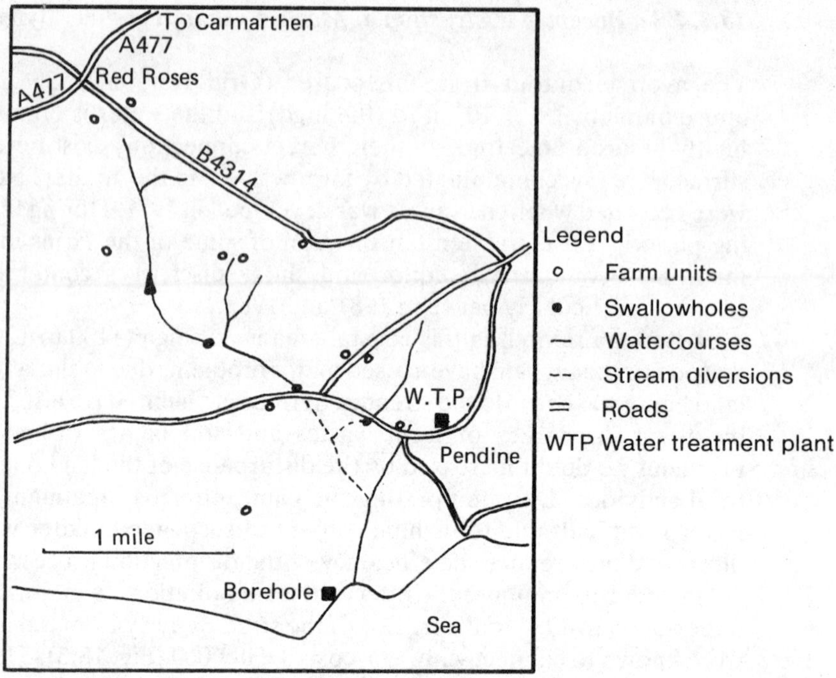

Fig. 18.3 — Morfa Bychan Diversionary Works

Following a Departmental review, the prior approval procedure for grants for farm development was revised in 1980. This means that in most cases grant-aided schemes are now proceeded with, before the farmer can apply for the grant. The farmer includes a declaration in his application to the effect that he has 'taken into account the conservation and amenity of the countryside'. It is only if a farmer chooses to consult in advance of the work being carried out that ADAS has an opportunity for seeking to assess any potential adverse environmental impacts. Prior notification now exists only in respect of farmland in National Parks or in Sites of Special Scientific Interest.

The abolition of 'prior approval' for grant-aided schemes was instituted against the advice of Welsh Water and has had unfortunate consequences in some areas. The opportunity for inspection of plans and discussions is lost and farmers then usually carry out the work without the benefit of ADAS advice, and water authorities very rarely have an opportunity to comment or advise on potentially-sensitive schemes.

The lack of consultation and requirement for planning control for most farm developments was recognised by the Royal Commission on Environmental Pollution (1979). Many of their subsequent recommendations have not been supported by Government (Department of the Environment, 1983). The benefits of such consultation can be demonstrated by the liaison which presently exists between Welsh Water and the Pembrokeshire Coast

National Park Authority in West Wales, where it is the policy of the National Park Authority that a farmer seeking to develop within the area of the Park should obtain Welsh Water approval before being considered by the National Park Authority for grants for farm developments. The area of the Park is small and potato production is the main agricultural practice and, to date, only about 20 such consultations have been received. Welsh Water have, however, agreed to all but one of the schemes virtually as proposed, to the satisfaction of the National Park Authority and the farmers concerned.

The Select Committee on the European Communities (1984), in examining those issues arising from consideration of the European Commission's Draft regulations on improving the efficiency of agricultural structures, noted the implications of the abolition of 'prior approval', but also noted that in numerous instances farmers do not apply for a grant when undertaking development work. Therefore even if prior approval was reinstated, some farm development projects would not be monitored unless a notification procedure existed for all developments, regardless of whether a grant was being sought or not. The Select Committee also noted that the recent reduction in the level of grants may well render the availability of grant aid less important to farmers.

18.5 LEGISLATION

The earlier legislation (namely Rivers (Prevention of Pollution) Acts 1951–1961) was generally adequate to control most farm-waste discharges, but not seepages to underground strata or the spreading of slurry on land. Although in many ways the implementation of several more sections of Part II of the Control of Pollution Act, 1974, has strengthened the legislation, some sections of this Act give cause for concern.

Under the provisions of Section 31 of the Act, the Minister of Agriculture, Fisheries and Food and the Secretary of State for Wales have published a 'Code of Good Agricultural Practice' (1985). The water industry in England and Wales had for some time been concerned that this Code might be used by some farmers as a defence against prosecution if they could demonstrate that the act concerned constituted 'good agricultural practice'. Water authorities have, however, been encouraged by the degree and extent of consultation afforded them in the preparation of the Code. The Water Authorities Association's comments have been presented in a series of meetings with the Ministry of Agriculture, Fisheries and Food, the Farming unions, and the County Landownwers Association (CLA), and substantial progress has been made towards allaying many of the water industry's fears. It is clear that all parties now hope that the Code will be used in a constructive way to encourage good practices and to avoid pollution. If this turns out not to be the case, then the Water Authorities Association would look to MAFF and Welsh Office Agricultural Department (WOAD) to take the necessary action in amending the Code appropriately.

The spreading of farm slurry is, for various legal reasons, not generally controllable by consent under the Control of Pollution Act 1974. Section 51 of this Act is intended to be the most appropriate means of preventing or controlling those adverse effects which might otherwise result from the unrestricted spreading of farm slurry onto land. Under this section, water authorities may ask the Secretary of State to serve a notice requesting an occupier to prevent specific agricultural practices. This section of the Act is now in force but it is not likely to be extensively used, as it could have serious implications in respect of land use generally and there could, in addition, be claims for compensation against the Authority by the farmers affected, or even the need for compulsory purchase by water authorities of large areas of land. It therefore seems most unlikely that there will be widespread use of this Section by water authorities or by the Secretary of State.

The control of the use and storage of polluting matters under the provisions of the Rivers (Prevention of Pollution) Acts, 1951–1961, was extremely limited. Much more extensive powers are available under the provisions of the Control of Pollution Act (COPA) 1974, without recourse to court orders, in particular as a result of future regulations to be made under Section 31 (4) and (5). The purpose of these regulations will be to designate areas within which prescribed activities shall not be carried out without consent and to lay down appropriate precautions in respect of such activities. The power to make these regulations is now available but it is understood that they are unlikely to appear, even in draft form, until later in 1985. Their potential effectiveness is therefore a matter for speculation at this time.

Section 46 (4) — (7) of the Control of Pollution Act (COPA), now in full operation, confers upon water authorities wide powers to take direct action to forestall and remedy pollution. The necessary cost of such operations would, generally speaking, be recoverable from those responsible for the actual or likely pollution. However no guidance has yet been received on the scope and operation of the powers available under this section.

Some of the difficulties which water authorities are faced with in controlling activities in important water-supply catchments may, therefore, remain even though most of COPA II is implemented.

18.6 MILK QUOTAS

The recent imposition of the European Community (EC) milk quotas will probably place a considerable burden upon many dairy farmers, particularly those who have significantly increased their herd sizes in recent years. Pollution cases in the past have shown that during a farm expansion, it is common practice for the herd size to be increased without a corresponding increase in waste handling and storage facilities. Many dairy farmers may, in the future, be in a position of having reduced incomes, and although herd sizes may also be reduced, it is still likely that waste handling/storage facilities will be inadequate. In such circumstances, expenditure to improve these facilities may not be forthcoming.

Although silage production seems certain to rise, the increase may be accelerated in the short term as a result of the imposition of milk quotas. The MAFF advice has been to feed grass and silage rather than 'concentrates' in an attempt to reduce farmers' costs. The risk of pollution may be significant as expenditure to deal with increased silage liquor production is likely to be minimal.

The changes in the capital grant scheme announced on 11 December 1984 may well exacerbate the situation. Despite the 'Government's declaration of achieving a closer integration of conservation and agricultural policies through high rates of grant for certain agricultural works of particular benefit to the environment' (MAFF, 1984), the standard rate of grant for waste-disposal systems has been cut, further reducing the incentive or capability of many farmers to improve deficiencies or overloaded systems.

Mention has previously been made of the report of the Select Committee on the European Communities (1984) into the environmental aspects of the European Commission's draft regulations on improving the efficiency of agricultural structures. The proposed regulation is intended to replace, *inter alia,* Directive 72/159 of the EC which deals with the modernisation of farms. In this Directive there are no specific references to environmental considerations and, during the last decade, the European Community has continued to pursue a policy favouring high levels of production. The House of Lords Select Committee considers that the draft regulation is still too closely orientated to production.

The MAFF are now giving consideration to the introduction of new capital-grant schemes to implement the new EC Regulation which will replace Directive 72/159. It is likely that the Regulation will be passed by the Council of Ministers in the Spring of 1985 and, though no special provision is likely to be made for conservation aspects, it is hoped that the UK will take conservation into account when 'development plans' are produced under the provisions of the Directive.

In so far as this particular problem is concerned, there should clearly be means for grant-aiding waste-disposal and waste-handling facilities, and information regarding waste-disposal problems should be available to the water authorities at an early stage and before any problem arises. Any scheme should ensure that a farmer seeking grant aid is obliged to consult with the water authority before he starts work.

18.7 REMEDIES

Water authorities are charged under statute with the protection of a variety of interests, including fisheries, water resources, and with the duty of controlling water pollution. So far as agriculture is concerned, Welsh Water considers that there a number of essential objectives which should be achieved if this problem is to be reduced in severity. These include the following.

- Full consultation at the planning stage should occur for all farm developments involving directly or indirectly the production, handling, or storage of waste products.
- Farmers should be fully informed of the polluting nature of materials likely to be generated on their farms and the best ways to contain, dispose of, or treat those wastes so as to minimise the risk of pollution.
- Measures to store and dispose of waste materials on farms should be eligible for grant aid.

If these objectives were achieved, farmers would, it is hoped, generally appreciate the need for proper handling, care, and storage of farm wastes and, given the correct advice, be likely to install suitable systems at the outset of any improvement scheme. Welsh Water and other water users would then benefit from the reduced risk of pollution, as would the farmers themselves.

18.8 EFFORTS OF WELSH WATER

Welsh Water recognises that the achievement of many of the objectives mentioned above is in the hands of Government and MAFF. The Authority will therefore continue to press for improvement in legislation and in the procedures concerned, and will further develop its already close working relations with WOAD and MAFF. At the same time it is recognised that a close relationship and understanding with the farming community is absolutely essential, and the Authority is therefore currently concentrating on measures to further these relationships.

(i) During recent years Welsh Water has been working closely with ADAS, the farming unions, and the CLA, to make farmers aware of the problems caused by farm waste discharges and to publicise their effects, together with ways of overcoming them. The publicity has included the insertion of advisory articles in ADAS or farming-union journals and the production of advisory leaflets. A 'cartoon calendar' for 1984 was produced and distributed to farmers in areas where most pollution problems are caused. The calendar was designed to provide topical advice in an easily-readable form. Radio and TV interviews have also been given, together with the production of articles for local newspapers and, in addition, farming groups have been addressed on the subject.
(ii) The Authority was one of the main organisers of 'Welsh Muck' 1984 at Haverfordwest on 22 March, 1984, the essential theme being the safe handling, storage and disposal of farm wastes, in particular slurry, so as to minimise the risks of water pollution. A major ADAS demonstration featuring the role of concrete and the hazards of the use of chemicals on farms is planned for 1985.
(iii) An Agricultural Liaison Officer has been appointed in south-west Wales

whose role is to liaise closely with, and advise, the farming community on ways to prevent pollution as a result of agriculture-related activities.

18.9 SUMMARY

Pollution by farm wastes, principally silage liquor and slurry-based discharges, appears to be increasing in both frequency and magnitude in the Welsh Water area. The problem seems to be related to the intensification of farming practices. Effluent storage and handling systems have become overloaded as farms have expanded. Problems are often compounded because insufficient attention is paid to the separation of clean surface, spring and rain-water.

The polluting nature of the resulting large volumes of liquid slurry does not appear to be fully appreciated by farmers and, consequently, sufficient care is not always given to its disposal. At the same time, the safe disposal of slurry is often hindered by the high stocking densities which exist on some farms and the reservation of grassland for silage production at a time when the disposal of slurry over land can be most effective.

Several important interests are adversely affected by farm pollution in Wales, including tourism, fisheries, farming itself, and the supply of potable water to the public. Consequently, Welsh Water has faced increasing problems and incurred substanial costs in recent years due to the effects of farm pollution. Minor water sources have had to be abandoned and major sources have been temporarily shut down.

The implementation of further sections of the Control of Pollution Act 1974 is an important development which, it is hoped, will lead to some improvement. At the same time, it is considered that a major factor is the lack of consultation and control over most farm developments. It appears that many farmers do not have the advice of MAFF or the water authority in incorporating satisfactory water separation, effluent handling, and storage facilities. The water authorities and MAFF are now looking at ways in which this situation can be improved.

Welsh Water has made a major effort in the last three years to make farmers and farming bodies aware of the problems and of ways in which they may be overcome.

REFERENCES

Alabaster J. S. & Lloyd R. (1980) *Water Quality Criteria for Freshwater Fish,* Butterworths.

Department of the Environment, Central Directorate of Environmental Protection, Agriculture and Pollution (1983) Paper No. 21 The Government Response to the Seventh Report of the Royal Commission on Environmental Pollution.

European Parliament, Committee on Agriculture, Working Group on

Fisheries — Working Document (1983) *The Social and Economic Importance of Salmon and Sea Trout Fishing in England and Wales.*

Ministry of Agriculture, Fisheries and Food (1984) Press Release on Capital Grant Schemes, 11 December.

Ministry of Agriculture, Fisheries and Food, Welsh Office Agriculture Department (1985) *Code of Good Agriculture Practice.* MAFF (Publications).

Royal Commission on Environmental Pollution (1979), Seventh Report, *Agriculture and Pollution*.

Select Committee on the European Communities (1984). *Agriculture and the Environment*, London, HMSO.

19

Non-point source planning and implementation in Wisconsin

J. G. Konrad, J. S. Baumann and J. A. Ott

19.1 INTRODUCTION

Sources of water pollution may be classified as either point or non-point sources. In general, point sources have discrete, identifiable locations and have discharges which can usually be measured directly, or otherwise quantified. Non-point sources are land areas where activities on the land, including the management of the land, result in the transport of pollutants, contaminants, or constituents, generally by runoff water, to lakes, streams or groundwater. Non-point sources are more difficult to identify and cause chronic degradation of water quality.

Point and non-point sources differ by the means and extent to which they deliver pollutants to waterbodies. Pollutants from point sources usually are discharged dirctly to waterbodies with fairly constant concentrations and amounts, while pollutants from non-point sources may follow transport paths which result in partial deposition of the pollutants before delivery to receiving waters, and the concentrations and volumes vary greatly by season and year. Non-point source pollution problems vary greatly between geographic regions of the United States and between individual states; however, agriculturally-generated pollutants are the most pervasive.

Sediment, of all non-point source pollutants, comprises the greatest volume of materials transported. The largest portion of the sediment which enters surface waters in the United States arises from cropland erosion. Other significant sources of sediment are stream banks and urban areas undergoing construction. The regional areas most affected by sedimentation are associated with the production of row crops such as corn (upper mid-west) and soybeans (mid-west and lower Mississippi River). Wheat producing areas (great plains and north-west), which are prone to both wind and water erosion, are also major problem areas.

In areas of the United States where livestock-based agriculture is prevalent, water-quality problems associated with organic and nutrient loads also exist. In the upper mid-west, non-point source pollutants from both croplands and livestock operations have degraded many of the surface-water resources. Other areas where livestock-related, non-point source problems exist include the states of Washington, New York, Pennsylvania, Vermont, and California, the area along the south-east coast, and the area draining to Chesapeake Bay. Figure 19.1 summarizes the locations of these

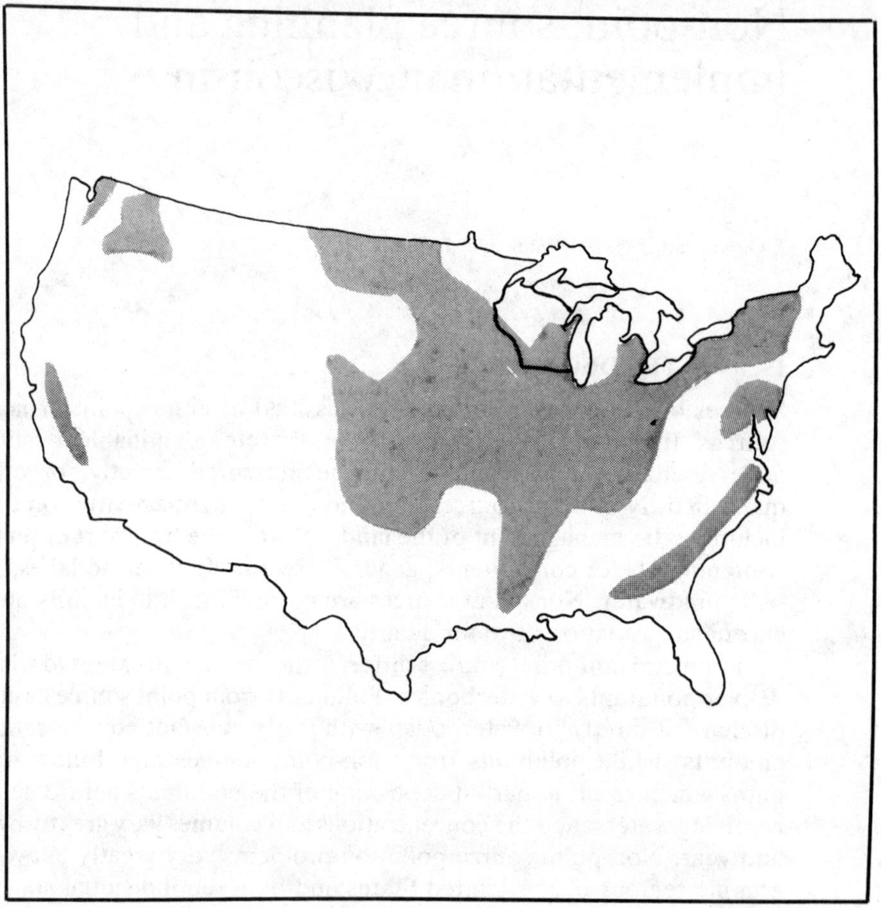

Fig. 19.1 — Areas of greatest potential rural non-point source pollution in the contiguous United States. The location of the State of Wisconsin is also indicated.

major non-point source areas in the United States.

Since major portions of Wisconsin are in this critical area, officials recognised years ago that satisfactory water quality for fishing and swimming will not be reached in many lakes and streams unless an aggressive programme for controlling urban and rural non-point sources is pursued. The degradation of smallmouth bass (*Micropterus dolomieui*) and trout fisheries, accelerated eutrophication of inland lakes, and impaired water quality of the near-shore waters of Lake Michigan, are examples of the water-quality problems that require the control of both point and non-point sources. Water resources such as these are vital to Wisconsin's economy because of their important recreational use.

19.2 WISCONSIN NONPOINT SOURCE WATER POLLUTION ABATEMENT PROGRAM

The Wisconsin Legislature recognised this need and responded in 1978 with the creation and funding of the Wisconsin Nonpoint Source Water Pollution Abatement Program. The programme was tailored to the non-point source needs in urban and rural areas of Wisconsin by incorporating aspects of various existing programmes, as well as devising new approaches.

Three premises were established as the basis for the Wisconsin Program:

- First, because pollutants originate from many different types of non-point sources, effective non-point source pollution control must be comprehensive. This means that all critical non-point sources must be controlled to accomplish the desired water-quality objectives.
- Second, because of the comprehensive nature of Wisconsin's non-point source problems, the components of soil conservation programmes do not meet the goals of the Wisconsin Program. Therefore, an effective non-point source management programme cannot rely solely on existing soil-conservation programmes.
- Third, a non-point source control programme is, by definition, a water-quality programme, and therefore requires strong technical involvement and leadership by the state water-quality agency. This leadership is essential, particularly in identifying water-quality objectives, and using assessment techniques such as catchment-area models.

Overall responsibility for the Wisconsin Program is assigned to the Wisconsin Department of Natural Resources, which administers both resource management and environmental protection (including water quality) programmes. Cities, villages, and counties are assigned the responsibility for local implementation in project areas. In rural areas, this framework is designed to maximise local agency contact with individual landowners and is based as much as possible on existing agencies and institutions. In urban areas, this framework is designed to maximise city and village involvement.

Six years of programme experience has shown that the successful blend of many agencies and units of government results in a comprehensive programme that assures that the objectives of non-point source pollution control are achieved.

19.2.1 Key programme components

The Wisconsin Program includes seven major components:

(i) programme purpose and objectives;
(ii) criteria for project selection;
(iii) project objectives and critical source identification;
(iv) development of implementation strategies;
(v) state budget support;
(vi) management practices package approach; and
(vii) accomplishment tracking.

19.2.2 Programme purpose and objectives

The basic purpose of the programme is to control systematically non-point source pollution so surface-water and groundwater quality goals can be met within a reasonable timescale. The programme is designed to deal with the varying nature of non-point sources throughout the state, including sediments from croplands, construction sites, stream banks and grazed woodlots, and nutrient loads from barnyard runoff, cropland erosion, manure spread on croplands, and runoff from city lawns and streets.

The three major programme objectives are:

(i) to identify the most effective approach for achieving specific water-quality objectives, and to provide adequate financial and technical assistance to landowners and operators to assist in the installation of approved non-point source-control practices;
(ii) to provide co-ordination between non-point source pollution control and other elements of the state's water quality programme; and
(iii) to focus limited technical and financial resources in critical geographic areas.

The third objective warrants specific attention. Unlike many erosion-control programmes, the Wisconsin Program: (a) concentrates on entire hydrological units rather than on random or political boundaries; (b) deals with *all* urban and rural categories of non-point sources rather than selected categories; and (c) relies on systematic processes to identify, rank, and select critical catchment areas† (watersheds) and portions of catchment areas to receive comprehensive attention.

Single-source management programmes will achieve many on-site land management objectives and may achieve some pollution control. However, these programmes, with scattered installations, often are of limited value in solving pollution problems arising in larger hydrological units. The Wisconsin Program concentrates available funds for technical and educational support in to projects on selected catchment areas where maximum comprehensive improvements in water quality can be achieved.

All categories of non-point sources — animal wastes, croplands, forest and woodlands, construction sites and urban lawns and streets — are

† 'Watershed' in the USA is used in the same way as 'catchment area' in the UK.

addressed. Other resource problems are identified and solutions to these problems are co-ordinated to make the programme as comprehensive as possible.

19.3 PRIORITY WATERSHED APPROACH

Early in the Wisconsin Program, data from the entire State were analysed to assess comprehensive non-point source control needs and to aid in project selection. Major land-use areas and associated types of land management were compared with data on water resources. The resulting geographical distribution supported the need to focus the programme further on the most critical watersheds.

This critical area in Wisconsin contains about 40% of the State's Class I and II trout streams‡ (about 7200 km), about 50% of the State's smallmouth bass streams (about 5600 km) and about 120 deep-water recreational lakes. The area consists of about 40% of the surface area of the State (8700 km^2) and almost all the major urban centres. Approximately 130 individual watershed projects, averaging 40 000 ha in size, will need to be implemented within this area. This area is illustrated in Fig. 19.2.

This hydrological unit approach, called the *Priority Watershed Approach*, allows all categories of urban and rural non-point sources within specific critical areas of a catchment area to be identified and controlled through the installation of management practices. Specific areas within a catchment area that contribute pollutants to lakes and streams are collectively called *Priority Management Areas*.

In addition to problem and source identification, the priority watershed approach has proved to be an effective framework for project implementation. This approach, implemented through *Priority Watershed Projects*, concentrates available educational, financial, and technical resources in those critical catchment areas where maximum water-quality benefits will result from the investment of money and personnel. To date, there are 26 Priority Watershed Projects in varying stages from project development to final implementation. Each project requires one year for identification of critical problem areas, three years for landowner sign-up, and five more years for installation of control practices.

19.3.1 Criteria for project selection

Six criteria are used for Priority Watershed Project selection:

- the severity of water quality problems;
- the magnitude of the pollutant load likely, causing the pollution and the potential to significantly reduce that load;
- the landowners' willingness to participate in the programme;
- the willingness and capability of local agencies to carry out their role in the programme;

‡ Class I = excellent; Class II = good; Class III = fair.

- the willingness and capability of local agencies and units of government to control other sources of pollution, such as by enacting construction site erosion control ordinances; and
- the potential public use and benefits that will result from the proposed watershed project.

The process emphasises projects where water-quality problems are critical, and land management practices are the most feasible means of control. Although areas of high cropland erosion are considered in the selection process, the emphasis is placed on areas where water-quality problems are the result of non-point sources, not solely on the severity of cropland erosion.

19.3.2 Priority watershed project objectives and critical source identification

Selection of a Priority Watershed Project is followed by an eight to nine year planning and implementation process. An implementation plan is prepared based on a detailed inventory and assessment of critical source areas in the catchment and the project's water-quality objectives. Generally, about one year is required to complete the assessment and prepare the plan. The Priority Watershed Plan is a consolidation of water quality and land use information so the specific causes and critical areas contributing to the water-quality problem can be identified and the most practical means of controlling the pollution can be developed. The plan guides the Priority Watershed Project and details procedures and responsibilities to aid local staff in working more effectively. It can also serve an important educational function by showing the cause-and-effect relationship between land management and water quality.

Central to each Priority Watershed Project are the water-quality objectives identified for its lakes and streams. The determination of critical pollutants, significant sources, the level of desired non-point source pollutant-load reduction, and the measurement of accomplishments, are all based on these specific water-quality objectives. In addition, the severity of water-quality problems and the attainability of water-quality objectives are primary factors in the selection of projects.

An understanding of pollutant impacts on water resources is essential in determining water-quality objectives. The objectives must be based on potential use. However, with objectives related to non-point sources, the type of impairment rather than the numerical criteria commonly used for in-stream standards, is more important. Impairments such as degraded fish habitat due to sedimentation of the bottom substrate, which commonly occurs in many of Wisconsin's trout streams, do not relate well to numerical standards. Identification of water-quality problems and objectives in Wisconsin depends, to a large degree, on biological and physical techniques which relate to the type of impairment and use. Reliance on chemical determinands alone could easily result in many impaired uses being overlooked. Biological indicators often integrate fluctuations in chemical characteristics and retain an overall measure of water-quality impacts for a long period of time. Identification of water-quality objectives in this manner requires the efforts of aquatic biologists and fish managers.

Three techniques are especially important in identifying objectives on the basis of impaired stream fisheries. One technique is the Hilsenhoff Biotic Index (1982). This technique evaluates tolerant and intolerant aquatic insects (macroinvertebrates) and relates most closely to the in-stream effects of oxygen-demanding materials such as manure. A second technique is a stream classification system based on potential use (Ball, 1982). In-stream fish habitat conditions and impairments are the main components of this technique. The third technique is fish population and distribution surveys. These fish surveys most closely relate to the public perceptions and recreational uses of the stream. Chemical monitoring and subjective professional judgement are used to identify and quantify problems in lakes.

In Wisconsin, some of the water-quality objectives identified for Priority Watershed Projects are:

- protection of the near-shore waters of Lake Michigan;
- rehabilitation of a warm-water fishery;
- rehabilitation of a cold-water fishery such as the upgrading of a trout stream through habitat improvement;
- protection of a desired warm-water fishery;
- protection of a desired cold-water fishery;
- rehabilitation of an inland lake; and
- protection of an inland lake.

With the variety of dairy and cash-crop farming and urban land uses present in Wisconsin, water-quality problems are seldom caused by a single type of non-point source. Thus, a categorical approach, one that deals just with one category of sources, such as eroding croplands, will not be effective in controlling non-point source pollution. An approach involving all landowners and all sources, regardless of significance, is inefficient and not cost-effective. The key is to identify the critical land areas and operations producing most of the pollutant load. In many areas of Wisconsin about 25% of the land area produces about 75% of the non-point source pollutant load. For example, in one of the Priority Watershed Projects, an inventory and assessment of barnyards showed that a 59% reduction in barnyard phosphorus could be achieved through the control of only 27% of the 327 operations (Pfender, 1984).

19.3.3 Inventories

A comprehensive assessment of all non-point sources is conducted prior to implementing a Priority Watershed Project. Barnyards, fields where manure is spread, eroding streambanks, eroding croplands, construction sites, and existing urban areas, are all inventoried. These inventories enable more efficient use of time and money during implementation. For example, 25 to 50 barnyards can be inventoried in the time required to design and install barnyard runoff controls on one or two barnyards. Thus, substantial time and money are saved by not designing and installing practices for barnyards that might have been considered significant using less-detailed or more-subjective inventories.

A number of tools are used to analyse the information collected during inventories. All barnyards in a Priority Watershed Project are now assessed using an animal-lot runoff model (Young *et al.*, 1982) and are ranked in order of pollutant load. To estimate sediment loads from cropland areas, the Department is developing an operational version of the ANSWERS model (Beasley & Huggins, 1982). Information and models from the International Joint Commission (IJC) — Pollution from Land Use Activities Reference Group (PLUARG) (Bannerman *et al.*, 1979; Novotny *et al.*, 1979) and Nationwide Urban Runoff Program (NURP) studies are being used in assessing urban areas (Bannerman *et al.*, 1983).

19.4 DEVELOPMENT OF IMPLEMENTATION STRATEGIES

Just as important as the identification of critical sources and areas where non-point source controls are necessary to meet water-quality objectives is the often overlooked aspect of developing detailed strategies for project implementation. Currently, implementation strategies include detailed landowner contact lists based on the results of the catchment area inventories. These lists are accompanied by a preliminary assessment of the severity and extent of non-point sources for each operation on the lists. As a result, county staff can concentrate their efforts on the 20 to 30% of the operations that are critical, rather than individually contacting all landowners in the catchment. Project implementation strategies also identify and schedule educational activities, outline fiscal management procedures, discuss preliminary project budgets, and estimate staff needs.

19.4.1 State budget support

Wisconsin provides financial support in three major categories:

- cost-share funds for landowners and municipalities to install management practices;
- aids for local governments to fund additional technical assistance, education and information and financial and project management; and
- administrative and planning funds for state administration and preparation of Priority Watershed Plans.

Individual management practices are cost-shared, the State paying 50 to

70% of the installation cost. Higher cost-share rates are used for practices where the capital costs for installation are high and the off-site water-quality benefits exceed the landowner's on-site benefits.

19.4.2 Management practice package approach

Since effective management practices must operate as systems, the Wisconsin cost-share agreements must contain *all* management practices necessary to control non-point sources on each participating farm or municipality. The landowner or land manager may not limit participation to the practices most directly useful to him or her. Many installed practices and non-structural controls would not be applied without the 'systems package' requirement.

19.5 ACCOMPLISHMENT TRACKING

Wisconsin's Program also includes progress or accomplishment tracking. Accomplishment indicators have been used to some degree in all projects and are being used to a greater degree in new projects. The accomplishment indicators used: (a) relate directly to the water-quality objectives and the pollutant causing the problems; (b) relate to the type and significance of the sources to be controlled, so that reductions in pollutant load can be calculated; (c) provide feedback to the implementing governmental unit so progress can be determined on a frequent basis, and (d) provide sufficient detail on the location and level of control to guide and interpret monitoring results.

One project aimed at improving a trout fishery exemplifies the usefulness of these tracking techniques. The stream was degraded by low levels of dissolved oxygen. Barnyard-runoff control served as the indicator of accomplishment. Because the pollutant load for each barnyard was calculated prior to the implementation phase of the project, each time a new cost-share agreement was signed involving barnyard-runoff control, the cumulative pollutant load reduction can be determined. Because project managers always know the accomplishment status of the project, they can better direct available staff and other resources to the remaining non-participating critical landowners.

19.6 APPLYING PROGRAMME PRINCIPLES

Following the establishment of the Wisconsin Nonpoint Source Control Program in 1978, the initial months were used to refine and further develop the principles and policies previously discussed, and to establish the procedures to carry out the programme. The programme elements were put in to operation in 1979 with the initial group of five Priority Watershed Projects. In 1984, planning was initiated on a group of nine new Priority Watershed Projects. These projects, along with 17 other ongoing projects, are all utilising the same programme elements. These programme elements have been continually reviewed, have proved to be sound, and have been refined and expanded as necessary to meet new situations or to incorporate expanded understanding and technical expertise.

In 1983, two of the first group of five Priority Watershed Projects selected in 1979 concluded the period for landowners to sign cost-share agreements. Although the installation of management practices will continue for an additional five-year period in both projects, water-quality improvements projected to be achieved during this period can be estimated. The following summary of the Elk Creek Priority Watershed Project illustrates the way water quality objectives and non-point sources identified during the planning phase of a project are later used in the evaluation process. Because a less detailed inventory was used for this project than those currently used, the indicators are similarly less detailed than those now used.

The primary land use in the Elk Creek Priority Watershed Project is dairy farming, with very little cash cropping. The watershed is steeply sloped and has an area of 29 000 ha. Figure 19.2 shows its location in west central Wisconsin.

Water-quality objectives for the Elk Creek watershed project are:

(i) the reduction of very high faecal coliform levels;
(ii) the improvement of degraded trout habitat, and
(iii) the reduction in sediment to a downstream lake.

The trout fishery in Elk Creek and its tributaries is marginal. Most of the streams are Class III (fair) trout fisheries. They are stocked with adult trout and support very few carry-over fish. The impoundment at the lower end of the catchment area has required dredging several times during the past 25 years.

There are 150 signed cost-share agreements with landowners in the Elk Creek project. These cost-share agreements contain a number of manure, stream bank, and cropland management practices. Since only about half of the management practices specified in these agreements are installed at this time, it is not yet possible to measure extensive improvements in water-quality in the lakes and streams within these catchments. However, several indicators point to the improvements that can be expected. These indicators are based on the reduction in the amount of pollutant generated by a specific non-point source and relate directly to the major water-quality problems identified.

The 'per cent of the animal units under signed cost-share agreements' is one of two indicators of accomplishment used. This indicator represents the per cent reduction in the amount of manure carried in runoff from barnyards. An animal unit is a means to equate different types of farm animals. About 70% (9356 out of 13369) of the animal units are under signed cost-share agreements. The percentage is fairly uniform in the six sub-catchments in the Elk Creek Watershed.

The 'per cent of eroding stream bank under signed cost-share agreement' is the other indicator of accomplishment used. This indicator relates to fish

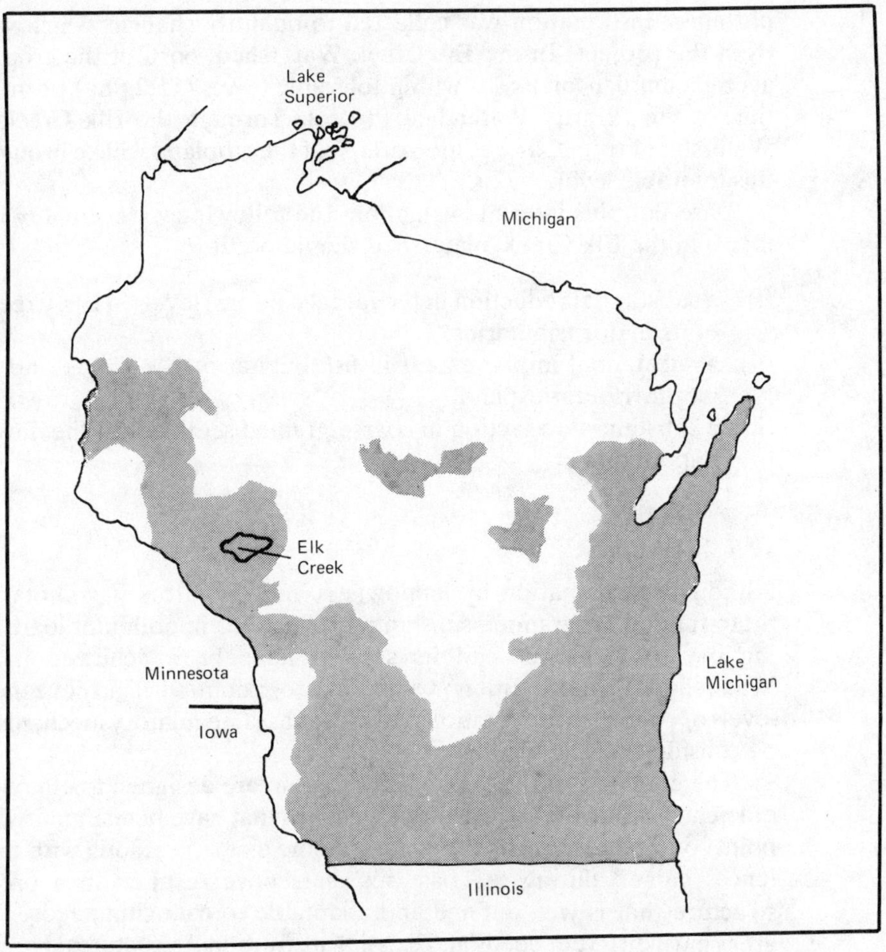

Fig. 19,2 — Areas of greatest potential rural and urban non-point source pollution in Wisconsin. The location of the Elk Creek Priority Watershed Project is also indicated.

habitat impairment. For Elk Creek, 82% (50 214 m out of 61 572 m) of the eroding stream bank is under signed cost-share agreements. The percentages are similar for the various sub-catchments.

These accomplishment indicators project substantial reductions in the amount of pollutants from non-point sources in this Priority Watershed Project; however, the evaluation also identified a number of critical operations whose operators chose not to participate. For example, ten sizeable livestock operations (97 to 232 animal units) will not have barnyard runoff controlled. If these ten had signed cost-share agreements, the percentage of animal units under signed cost-share agreements would be at least 72% for each sub-catchment.

Although cropland erosion was considered a source of secondary importance, information was collected to quantify changes which will result from the project. In the Elk Creek Watershed, 66% of the cropland had average annual soil losses within tolerable levels (11.2 t/ha) prior to initiation of the Priority Watershed Project. Through the Elk Creek Priority Watershed Project, an additional 11% of the cropland will be brought below this tolerable level.

Based on this interim evaluation, the following water-quality improvements in the Elk Creek Watershed should occur:

(i) a substantial reduction in the faecal coliform levels in Elk Creek and all of its major tributaries;
(ii) a substantial improvement in fish habitat in Elk Creek and all of its major tributaries; and
(iii) a substantial reduction in coarse-grained sediment to the downstream impoundment.

19.7 SUMMARY

Although participation by landowners and operators is voluntary in this State-funded programme substantial reductions in pollutant load, depending on the pollutant and its source, have been achieved in Priority Watersheds Projects. But no voluntary programmes will achieve the desired levels of control in all situations. In those cases, regulatory mechanisms must be considered.

The elements of the Wisconsin Program are designed to effectively and efficiently achieve water quality objectives that have been impaired by non-point source pollutants. These programme elements, along with the experiences gained during the past six years have resulted in a programme structure that is well defined and adaptable to changing needs. Different areas have different needs and existing institutional structures. However the principles used as the foundation for the Wisconsin Nonpoint Source Control Program can be applied to the development of effective programmes to control a variety of non-point source problems.

REFERENCES

Ball, J. (1982) Stream Classification Guidelines for Wisconsin. In: *Water Quality Standards Handbook*, US Environmental Protection Agency. Washington, DC (A2-A30).

Bannerman, R., Baun, K., Bohn, M., Hughes, P. E. & Graczyk, D. A. (1983) *Evaluation of Urban Nonpoint Source Pollution Management in Milwaukee County, Wisconsin — Vol. I Urban Stormwater Characteristics, Pollutant Sources and Management by Street Sweeping*. PB 84-114164, Wisconsin Department of Natural Resources, Madison, 206 pp.

Bannerman, R., Konrad, J. G., Becker, D., Simsiman, G. V., Chesters, G., Goodrich-Mahoney, J. & Abrams, B. (1979) *The IJC Menomonee River Watershed Study — Vol. III Surface Water Monitoring Data.* EPA-905/4-790-029C. US Environmental Protection Agency, Region V, Chicago, Il. 322 pp.

Beasley, D. B. & Huggins, L. F. (1982) *ANSWERS — Users Manual*. EPA-905/9-82-001, US Environmental Protection Agency — Region V, Chicago, Il. 54 pp.

Hilsenhoff, W. L. (1982) *Using a Biotic Index to Evaluate Water Quality in Streams*. Tech. Bull., No. 132. Wisconsin Department of Natural Resources, Madison. 22 pp.

Novotny, V., Chin, M. A. & Tran, H. (1979) *The IJC Menomonee River Watershed Study — Vol. IV Description and Calibration of a Pollutant Loading Model — LANDRUN*, G. Chesters & J. Diamantis (eds). EPA-905/4-79-029D. US Environmental Protection Agency, Region V, Chicago, Il. 144 pp.

Pfender, J. A. (1984) *The Kewaunee River Priority Watershed Management Plan.* Wisconsin Department of Natural Resources, Madison. 122 pp.

Young, R. A., Otterby, M. A. & Roos, A. (1982) *An Evaluation System to Rate Feed Lot Pollution Potential.* ARM-NC-17, Ag. Res. Ser., US Department of Agriculture, North Central Region, Peoria, Il. 78 pp.

20

Tackling agricultural pollution — The difficulties facing conservationists in Great Britain

C. Newbold, G. Fry and A. Cooke

20.1 INTRODUCTION

The conservation of freshwater ecosystems is defined as the management of water bodies with the intention of maintaining their scientific and wildlife interest or rehabilitating their physical, chemical, or biological quality. A pollutant is defined as any substance or activity which disrupts, alters or destroys the physical, chemical, or biological quality of a site.

This chapter will outline those conservation problems in fresh water, which are caused by agricultural pollution, including afforestation. Where possible, appropriate ameliorative and remedial management will be considered.

20.2 SELECTION OF SITES OF NATURE-CONSERVATION INTEREST

The Nature Conservancy Council (NCC) is the goverment body which promotes nature conservation in Great Britain. It gives advice on nature conservation to government and all those whose activities affect the wildlife and wild places of Great Britain. It also selects and manages a series of National Nature Reserves (NNRs). The NCC has also identified other important places known as Sites of Special Scientific Interest (SSSI). In these areas the Wildlife and Countryside Act, 1981, (Department of the Environment 1981) requires owners and occupiers to tell the NCC in advance if they intend to carry out a damaging operation, so that the NCC can discuss an alternative, less-harmful approach, or offer financial help within a nature-conservation management agreement.

Although it is very important to conserve special sites it must also be ensured that the remainder of the countryside is managed sympathetically for wildlife. Few species can survive only in isolated sites, most rely on the countryside around them. The NCC's role in the wider countryside is one of advice and persuasion.

The Wildlife and Countryside Act, 1981, also specifically protects a number of wild plants and animals, so the NCC, in the wider countryside, attempts to prevent the killing, injury, taking, or selling of specially protected animals such as the otter, (*Lutra lutra*), badger (*Meles meles*), and red squirrel (*Sciurus vulgaris*).

It is also an offence intentionally to disturb these animals in their places of shelter. All wild birds, their nests and eggs are also protected with the exception of pest and sporting species. Specially-protected plants must not be uprooted, picked or sold, and uprooting any wild plant is illegal.

It is possible, however, to claim exemption if anyone can show that the damage to any species (with the exception of Bats Order: Chiroptera) was the incidental result of a lawful operation and could not reasonably have been avoided.

20.3 AGRICULTURAL POLLUTION

Using the broad definition considered in this chapter, agricultural pollution comes in many guises. The main problems are now considered.

20.3.1 Phosphorus

The effect of enrichment (eutrophication) on freshwater systems is well documented (Lund, 1972; Moss, 1983). Attention is often focused on the removal of phosphorus, as a means of reversing the process of enrichment. This, and other rehabilitation techniques, have been reviewed by Jørgensen (1980). Phosphorus-stripping at sewage works in Norfolk's Broadland, an area of shallow lakes, fens and wooded islands, is being carried out for environmental and nature-conservation interests (Phillips, 1984). A recent nutrient budget on Esthwaite Water and North Head Fen National Nature Reserve, Cumbria, England, also pointed to the need for phosphorus stripping from Hawkshead Sewage-Treatment Works. North West Water are committed to a removal programme. Sewage by-pass techniques are being used to avoid enrichment at Bosherston Lake National Nature Reserve in Pembrokeshire, Wales. A nutrient budget is being researched for Loch Leven National Nature Reserve in Kinlross, Scotland. Previous studies during the International Biological Programme (IBP) (Morgan, 1970) suggest that pollution from sewage-works effluent is the main source of enrichment with phosphorus.

In all of the above studies approximately 80% of the phosphorus loading has arisen from sewage sources. Phillips (1984) suggests that the 20% loading arising from agricultural sources in Norfolk's Broadland is insufficient to cause enrichment. This seems to be true also for Esthwaite Water, Bosherston Lake, and Loch Leven.

In less-populated areas there are suggestions that phosphorus enrichment from agricultural sources is sufficient to cause gross changes to the nature-conservation interest of a site, with a loss of macrophytes and increases in pollution-tolerant species. Certain limnological conditions favour the enrichment process from agricultural sources. Loch Eye in Ross and Cromarty, Scotland, is naturally eutrophic, lying geologically on phosphorus-rich Old Red Sandstone. The fertile black soils in the catchment have been converted to arable in the last decade. The loadings of phosphorus from such a change in land use may have been sufficient to 'tip the balance' in favour of an algal-dominated system. Limnologically the loch has a very low rate of flushing. Similarly Lochs Rescobie and Balgavies in Angus, Scotland, seem to have a long period of retention. In the 1940s records show that they contained 10 and fifteen *Potamogeton* species respectively. The total *Potamogeton* flora in Britain is 21 true species. Many of those species found in Lochs Rescobie and Balgavies are sensitive to enrichment processes. By the mid 1960s their number had been reduced to six, and by 1969 to one, the pollution tolerant *Potamogeton pectinatus*. The appearance of *Zanichellia palustris*, and the dominance of algae in the 1970s, confirmed their gradual decline in conservation interest. Lochs Rescobie and Balgavies are two in a series of three isolated kettle-hole lochs. The catchment área is composed of phosphorus-rich Old Red Sandstone. The agriculture is a mix of conifer plantations and fertile grazing land at lower altitudes.

The enrichment of Lochs Davan and Kinord in Deeside, Grampian Region, Scotland, is at a more formative stage. The macrophyte *Elodea canadensis* covers much of the shallow Loch Davan to the exclusion of other species. The indications are that the main form of enrichment comes from the effluent of an intensive piggery unit, but an expensive nutrient budget may have to be undertaken to confirm this view.

20.3.1.1 Ameliorative and remedial management

Nutrient budgets would confirm the relative importance of geology and land use for Lochs Eye, Rescobie, and Balgavies. If agricultural interests were found to be important, the only reliable means of rehabilitating such lakes would be to 'put the clock back' by reversing the present farming trends. This means paying large amounts of compensation to take land out of agricultural and forestry production. Compensation payments can be made available under the provisions of the Wildlife and Countryside Act 1981.

One tentative means of reducing inputs of particulate phosphorus would be to plant a wetland buffer-strip around a loch shore. This acts as a crude biological filter. The New Zealand government, for this same reason, insist on a fringing wetland edge along rivers important as trout fisheries, (McColl, *et al.*, 1975). The use of straw placed in cages in inflow steams as a means of removing phosphorus requires more thorough investigation (Barrett, P. R.

F. pers comm, 1985), but this could be a very useful technique where there are small, relatively isolated, lochs and lakes.

Piggery and cattle effluent from intensive units, slurry from all too-frequently broken slurry lagoons are point sources of agricultural pollution, and can be controlled under the Control of Pollution Act, 1974. In all cases pollution has to be proven.

The recent re-introduction by NCC of the otter into an East Anglian river in co-operation with Anglian Water and a local farmer was almost thwarted by an upstream land owner. A very large release of slurry from the upstream farm caused heavy fish mortalities along the stretch chosen for the re-introduction (Jefferies, D. pers comm, 1984). The polluter was successfully prosecuted by the Water Authortity.

Persistent pollution from an intensive piggery unit was also causing local damage to the ditch flora and fauna within Holme Fen NNR Cambridgeshire. The farmer was threatened with prosecution by the Water Authority and the pollution ceased (Massey, M. pers comm, 1984).

20.3.2 Nitrogen

Cuttle (1982) has indicated that nitrogen inputs from agricultural land are as high as 80% of the total loading entering Loch Leven NNR, Kinross, Scotland. Similar figures have been reported for Hickling Broad NNR, and the River Bure both in Norfolk, England (Moss, 1978). All their catchments are, in the main, composed of basic soils. Levels of nitrate nitrogen occasionally reach 10 mg/l in Loch Leven, and peak loadings give figures generally in excess of these levels in Hickling Broad and the River Bure. The phenomenon of die-back of reeds (*Typha angustifolia*) was attributed to similar loadings in the Berlin Le Havel lakes (Sukopp *et al.*, 1975; Schindler, 1978). Such loadings disturbed the sclerenchyma/parenchyma balance in the stem of *Typha angustifolia* causing it to break more readily. This work seems to have gone largely unnoticed, perhaps because attention has been mainly focused on phosphorus (see section 20.4). Recently, Crook *et al.*, (1983) have found that the rhizomes of *Phragmites australis* in Norfolk's Broadland are similarly weakened. Breaks most often occur where the edge of the reedswamp forms a floating 'raft' of vegetation. This, in part, is the reason for the large-scale decline of reedswamp in the Broadlands of Norfolk, (Boorman *et al.*, 1977).

It is probable that other emergent and submerged aquatic plants can be stressed by high levels of nitrogen. Equally, it is known that elevated levels of nitrate-nitrogen can cause methaemoglobinaemia in human babies (albeit rarely in the United Kingdom), but it is not known whether the young of aquatic mammals are similarly affected.

20.3.2.1 Ameliorative and remedial management

Pollution by nitrate nitrogen in fresh waters arises largely from applications of fertilizer running off the land. It is a diffuse source and its control, whilst desirable, is difficult to achieve. Rigorous timing of the application of inorganic fertilizers and careful use of manures and slurries from intensive cattle and pig units will go a considerable way to reducing the nitrate losses. These measures are all too dependent on the individual farmer. The apllication of slurries to hard, frosted ground is still a common cause of excessive runoff.

Waldendorp *et al.*, (1966) found scarcely any losses of nitrogen fertilizer when ammonium nitrate was applied to grassland in spring and summer, except when heavy rainfall followed application. However, 30 to 60% of the nitrogen applied during the autumn and winter was leached.

Harris & Colbourn (1985) found similar losses leached from winter wheat during autumn and winter. Slow-release fertilizers have been suggested as a means of preventing this loss and this may be a practical means of reducing losses during the autumn/winter. During the spring period of growth, however, the crop requires large quantities of readily-available nitrate and this will not be provided by slow-release fertilizers.

With more-intensive agricultural practices, both in amounts of stock kept and increased yields of crops, the losses from the land are likely to increase rather than decrease.

Denitrification processes along the muddy edges of wetland margins are known to be high (McColl & Gibson 1979). The use of natural biological filters, so reducing inputs to lake or river, is under-exploited in Britain. Indeed, there has been an active programme of wetland removal (see sections 20.10 and 20.11).

Another solution might be to make farmers pay for the enormous cost of diluting or treating nitrogen-rich water for potable supply. The cost to the Anglian Water area, serving the highly-agricultural counties of Norfolk, Suffolk, Essex, Cambridgeshire, Bedfordshire, Northamptonshire, and Lincolnshire, was approximately £3m in 1984 (Tester, D. pers comm, 1985).

20.3.3 Silage liquor pollution

Ammonia levels in silage liquor are sufficiently high that leakage from silage clamps and towers into streams causes fish deaths. A recent survey undertaken by the Department of Agriculture and Fisheries for Scotland (DAFS) (Taylor & Brownlie, 1982) showed that all 100 silos/clamps in the survey had some fault which caused leakage.

20.3.3.1 Ameliorative and remedial management

The following recommendations were made by DAFS.

- Training courses for building contractors in aspects of silo construction conducive to effluent control, should be considered.
- The agricultural colleges should be asked to consider the preparation of an advisory leaflet on recommended remedial measures for the various types of structural defect identified in the survey.
- Advisers should place greater emphasis on the need to investigate drainage systems in the vicinity of sites for new silos.
- Further research work into methods and materials for sealing cracks and joints in silo floors and walls is required.

20.3.4 Fish farming

Freshwater fish farming in 1984 produced approximately 8700 tonnes of trout, 78% of which was for the table (Lewis, 1984). This production is classed as agriculture and is supported by grants from the Ministry of Agriculture, Fisheries and Food.

The polluting impacts of fish farming are well-documented (Alabaster, 1982; Solbé, 1982; Hambrey, *et al.*, 1985). Deleterious increases is suspended solids, elevated levels of BOD, ammonia and phosphorus, reduction in concentrations of dissolved oxygen, the abstraction of large volumes of river water (occasionally the total flow of the river), and the escape of alien fish, has produced a strong lobby from the water authorities and angling interests against these excesses.

20.3.4.1 Ameliorative and remedial management

A recent 'code of practice', produced by the water authorities and the National Farmers Union (1984), will, if followed, reduce some of the pollution. However, fish farmers remain strongly independent and they often flout the effluent conditions set by a water authority. The Nature Conservancy Council, in conjunction with the local county council, have, for a small stretch of the River Itchen, Hampshire, and River Hull, Humberside, successfully applied for an Article 4 directive from the Secretary of State for England and Wales under the Town and Country Planning General Development Order No 289 (1977). This directive can extend planning controls to those operations normally outside the Town and Country Planning Act (1971). Normally all agricultural operations are exempt. These extended planning controls can only be successful in preventing the building of a new fish farm or its subsequent expansion. They cannot control the over-abstraction of river water or its pollution. This is dependent on a water authority, which cannot hope to monitor or 'police' a fish farm effectively 24 hours a day.

20.3.5 Crayfish farming

The crayfish is one of the largest freshwater invertebrates in Great Britain, measuring, when adult, some 12 cm. Its distribution is confined to the calcium-rich rivers of England and Wales (Jay & Holdich, 1981).

In 1977 the Nature Conservancy Council asked MAFF to control the import of the larger, and alien, American crayfish, *Pacifastacus leniusculus*, as it was known to be a carrier of the fungus *Aphanomyces astaci.* This causes Crayfish plague to which the native Crayfish, *Austropotamobius pallipes*, is highly susceptible. Imports of the American crayfish were unfortunately then, outside any existing legislation. The Disease of Fish Act, 1981, and the Import of Live Fish Act, 1983, (MAFF, 1983) came too late.

An expanding crayfish industry is now established in England and Wales. Grant-aid supporting the industry is available from MAFF.

20.3.5.1 Ameliorative and remedial management

In 1982 the NCC tried to control the spread of the disease by making it an offence under the Wildlife and Countryside Act, 1981, knowingly to introduce American crayfish into the wild. The fungal spores could not be controlled and this may be the reason for the increased incidence of the disease during the last eighteen months. The disease is now present in the Bristol Avon, the Hampshire Avon, the Lee and the Whitewater, the River Wey, Dorset, and the Kennet, Wiltshire. It has, as far as can be ascertained, exterminated the native crayfish in these rivers. Certain refuges containing the native crayfish could survive unaffected by the disease well into the future. Article 4 directives will have to be sought to prevent the development of an American crayfish farm in such catchments.

20.3.6 Irrigation

Land drainage, the movement towards arable crops, and the hot dry summers of 1976 and 1984 have increased demands for irrigation water. The number of abstraction licences consented by Anglian Water for irrigation purposes rose from 2897 in 1969, to 2950 and 3382 in 1976 and 1983, with an increase in abstraction volume from 155.9 to 172.9 to $224.3 \times 10^3 \, m^3/d$ respectively (Anglian Water Authority, 1983). In the grazing marsh districts of Romney, Kent, Broadland, Norfolk, and the Gwent Levels, South Wales, changes to arable have resulted in the use of ditch water for irrigation. These ditches once provided wet fences for cattle by delineating field boundaries. The stable water levels provided a rich habitat for aquatic wildlife. The fluctuating spring water levels found in ditches draining arable crops and intensive grass leys, and the drying out of such ditches in summer have destroyed aquatic wildlife (Mountford & Sheail, 1982).

20.3.7 Land drainage

Land drainage may not be the usually accepted form of agricultural pollution affecting the aquatic environment, but the digging of deeper drains has resulted in five forms of aquatic pollution.

(i) County Naturalists Trust Reserves outside a drainage scheme have partially dried out. This is due to the permeability of the soils (George, 1976).
(ii) Perched water tables have been destroyed, draining areas outside the scheme (Gosling & Baker, 1980).
(iii) Saline water tables have also been breached (Dirscoll, 1983), affecting freshwater fauna and flora and yielding unsuitable water for irrigation.
(iv) The deepening of drains has exposed pyritic peats causing ferrous-rich 'tides' of water (Bloomfield, 1972). These affect invertebrates, smother aquatic plant-growth, and reduce the penetration of light (NCC, File 1982). They also cause pollution and increasing sedimentation in the main river.
(v) Sulphide-rich peats have been exposed which through oxidation/reduction processes release sulphuric acid. Oscillations in pH between 7.4

and 3.2 have had disastrous effects on the flora and fauna of Calthorpe Broad, Norfolk (Gosling, 1971; Gosling & Baker, 1980). Small areas of West Sedgemoor, Somerset, have suffered similar problems (Williams, R. J., pers comm 1982).

20.3.7.1 Ameliorative and remedial management

Certain ameliorative measures can be undertaken; but sometimes the cost prohibits action. Perched water tables can be protected by 'clay walling', or sealing in the water table. This has been achieved on a 200-ha reserve at Woodwalton Fen NNR, Cambridgeshire, at an approximate cost of £0.4m. This was paid for by the Middle Level Commissioners (a local drainage board) with a grant from the European Economic Community (EEC). Pumps feeding iron-rich water into Martham Broad, part of the Hickling Broad NNR, Norfolk, have been moved so that the water now bypasses the reserve. This cost to the Nature Conservancy Council was £7000 in 1983–84 (NCC Tenth Annual Report, 1984). Presumed intrusion of saline water into Hickling Broad NNR has had to be accepted. Water, rich in sulphuric acid, exposed through deep drainage on a small area of West Sedgemoor SSSI, prior to the establishment of the SSSI, has to be accepted. Fluctuations in acidity at Calthorpe Broad NNR, Norfolk, are crudely treated by liming. The Ministry of Agriculture, Fisheries and Food and the farmer cannot be made to pay for actions outside the reserve which affect the perched water table in the reserve.

One way of avoiding such effects would be to survey the area prior to any drainage work for its soil types, hydrology, and wildlife interest and produce drainage schemes which are sensitive to these environmental conditions (Newbold, in press). By optimising water tables, irrigation requirements would be reduced.

20.3.8 River engineering

One of the most damaging forms of agricultural 'pollution' is that of river engineering to serve the needs of land drainage. The river engineer's aim is generally to minimise both the quantity of water which flows onto the land, and the extent and period of inundation. Channel capacity is increased, obstacles are removed which contribute to the roughness coefficient of the river. Projecting bushes and trees, islands, sand bars and lush aquatic growth, all have been removed in an effort to create the most economical shape to maximise water flow — the trapezoidal channel.

The destruction of habitats in rivers to serve the needs of land drainage increased during the 1960s and 1970s. The lowland rivers of England and Wales suffered most (Table 20.1). The Great Ouse, the Rivers Welland and Nene flow through flat, fertile, river valleys before they help drain the fens of Cambridgeshire; 44% and 47% of their respective lengths have been channelised (Brookes, 1981, and pers. comm 1984). In contrast only 11% of all rivers in the mountainous county of Cumberland (Cumbria) have been channelised.

There are no figures available for Scotland, but channelisation is organised, generally, on a need basis by the setting-up of a consortium of farmers

Table 20.1 — Channelised river lengths in each local land drainage district (1930 to 1980) (Brookes. pers. comm. 1984)

Water authority area/ local land drainage district	Main river length (km)	Channelised length (km)	Percentage of main river channelised (%)	Area of district (ha)	Channelised length per 100 000 ha (km)
ANGLIAN					
Essex	1553	295.7	19	364 788	81.0
Great Ouse	1238	549.9	44	859 343	63.9
Lincolnshire	1210	326.9	27	528 239	61.8
Norfolk and Suffolk	1371	444.7	32	529 540	83.9
Welland and Nene	1131	532.9	47	398 427	133.7
	6503	2150.1	33	2 680 247	
NORTHUMBRIAN	1422	350.1	25	921 089	37.9
NORTH WEST					
Cumberland	1447	162.2	11	459 034	35.3
Lancashire	1791	232.7	13	525 968	44.3
Mersey and Weaver	2265	282.2	12	457 372	61.7
	5503	677.1	12	1 442 374	
SEVERN TRENT					
Severn	1663	401.6	24	1 142 173	36.0
Trent	1737	744.7	43	1 043 346	71.4
	3400	1146.3	34	2 185 519	
SOUTHERN					
Hampshire	679	23.8	4	273 544	8.7
Isle of Wight	120	55.4	46	38 101	145.4
Kent	1051	420.9	40	476 853	88.3
Sussex	870	244.3	28	308 555	79.2
	2720	744.4	27	1 097 053	

SOUTH WEST					
Cornwall	537	177.0	33	462 816	38.2
Devon	931	168.6	18	613 594	27.5
	1468	345.6	24	1 076 410	
THAMES					
Conservancy	3669	550.3	15	994 899	55.3
Lee	1062	30.2	3	196 287	15.4
	4731	580.5	12	1 191 186	
WELSH					
Dee and Clywd	1021	114.6	11	294 839	38.9
Glamorgan	667	176.9	26	195 426	90.5
Gwynedd	1316	309.6	24	404 692	76.5
South West Wales	1474	329.5	22	626 362	52.6
Usk	422	86.8	21	174 082	49.8
Whye	601	161.5	27	418 393	38.6
	5501	1178.9	21	2 113 794	
WESSEX					
Avon and Dorset	696	170.5	24	427 640	39.8
Bristol Avon	707	167.3	24	222 018	75.4
Somerset	825	256.8	31	308 602	83.2
	2228	594.6	27	958 260	
YORKSHIRE	1774	598.6	34	1 349 164	44.4
LONDON EXCLUDED AREA	336	138.0	41	99 462	38.7
Totals for England and Wales	35 586	8504.2	24	15 114 558	

wanting to 'improve' their local stretch. The water authorities of Scotland are distinct in function and law (Water (Scotland) Act 1967) from the English and Welsh water authorities. The water authorities in England and Wales exercise a general supervision over all matters relating to land drainage. They are responsible for engineering works on 'main' rivers. In general, it is safe to state that the rivers in Scotland have not been modified to the degree they have in England and Wales.

20.3.8.1 Ameliorative and remedial management

The conservationist, with his knowledge of the habitat requirements of plants and animals, has in the last ten years been assisting the water authorities of England, Wales and Scotland in more effectively furthering the nature-conservation interest in rivers. Equally the river engineers have shown an enthusiasm and ingenuity in the way they have built in conservation considerations to their land-drainage brief.

There are numerous guidelines bringing together these once mutually-exclusive activities. The most significant are those published by the Water Space Amenity Commission (1981), the Nature Conservancy Council (Newbold *et al.*, 1983) and the Royal Society for the Protection of Birds in conjunction with the Royal Society for Nature Conservation (Lewis & Williams, 1984).

In many water-authority areas such sympathetic management of rivers is very much the issue of, 'too little, too late'. However, some water authorities are beginning to recreate river habitats where they once destroyed the 'naturalness' of the river to create their trapezoidal channel. If this commitment to conservation is sustained, the lengths of channelised rivers considered in Table 20.1 will have to be reduced.

20.3.9 Afforestation

20.3.9.1 Acidification

The uplands of Britain receive the largest total annual rainfall and often are subject to mists or low cloud. These factors added together mean that atmospheric pollutants are deposited on the uplands more so than elsewhere. Acidifying agents, mainly the oxides of sulphur and nitrogen and their derivatives, are deposited in rain and directly from the atmosphere. Although the concentrations of these pollutants are greatest in the east (so that the rain in eastern Britain is more acid than in the west), the high rainfall in the west produces a higher level of total deposition of these substances (Fowler *et al.*, 1982).

Upland areas overlying slow-weathering rocks are vulnerable to acidification, and it is now clear that blanket afforestation with conifers in these areas is an additional stress factor increasing the acidity problem (Harriman & Wells, 1984). The main mechanisms by which plantations increase the acidity of runoff to streams and lakes are in doubt, although the following are likely.

- More efficient removal of acidifying pollutants from the atmosphere by the large surface area presented by forests compared with moorland, is followed by runoff into water bodies.

- Uptake of nutrients by trees removes some of the soil's ability to neutralise acid inputs via rain.
- As forests mature, the increased rates of evapotranspiration concentrate acidity in the remaining groundwater.
- Ploughing and other changes to the hydrology of catchments result in less opportunity for rain/soil interaction by increasing the rate of surface runoff and therefore decreasing neutralisation.

Whatever the main mechanism, it is now generally accepted that afforestation exacerbates the acid-rain problem. There are many examples from Wales, the Lake District, and Scotland, where wildlife interests have suffered after afforestation of catchments (Harriman & Morrison, 1982; Stoner *et al.*, 1984).

Acidity and/or consequential high concentrations of dissolved metals have caused losses of aquatic invertebrate and fish populations; these are linked with reduced populations of dipper (*Cinclus cinclus*), and are implicated in wildlife-conservation problems in a number of other habitats (Fry & Cooke, 1984; Ormerod *et al.*, 1985).

Although acidification is one of the major threats to upland aquatic environments posed by afforestation, it is one of the least amenable to management. If the water, draining from plantations in sensitive areas, passes unmodified into important freshwater habitats, it will cause damage.

20.3.9.2 Alteration of hydrology

This occurs mainly through ploughing and restructuring drainage patterns. The immediate effect is to change the spate cycle of rivers and streams to one of more rapid flood followed by a quick return to low-water conditions (Robinson, 1980). In the long term the trees, through evapotranspiration, deprive the catchment of water. This effect may result in a net water loss of 20 to 30% in heavily afforested catchments and may seriously threaten streams with desiccation in the dry summer months. Alteration of the drainage system with increased rapid runoff will also increase erosion and, hence, alter the physical characteristics of upland streams (Newson, 1980). Small streams can be destroyed completely when hills are ploughed.

20.3.9.3 Chemical runoff

During the normal forest cycle, trees may be treated with a number of chemicals, particularly fertilizers and pesticides. These can, and do, get into watercourses especially after heavy rainfall (Harriman, 1978). The long-term effects of the addition of such chemicals may be difficult to quantify but are undesirable on an SSSI. For example, the fertilization of forests by rock phosphate has contaminated oligotrophic waters, leading to algal blooms, or increased growths of the rush *Juncus bulbosus* var *fluitans*.

20.3.9.4 Siltation

The silt loading of recently afforested streams can be extremely high even on areas of hard rock such as granite. This increased silt loading of streams is a product both of the initial ploughing prior to planting, and catchment erosion resulting from a greatly accelerated runoff. Silt deposits can be observed in afforested catchments, deeply blanketing slow stretches of rivers and lake beds. The fine silt is an unsuitable habitat for many of the invertebrate organisms characteristic of stony streams; it has been blamed for loss of spawning grounds for fish and can drastically alter the species-composition of macrophyte communities (Mills, 1980; Smith, 1980).

20.3.9.5 Shading

As the forest grows, small streams and rivers can become shaded by the trees resulting in loss of light and heat from the sun. In addition, streams may become littered with pine needles coating the bed. The combined effects of these physical changes is to reduce general productivity and the diversity of stream-dwelling plants and animals (Mills, 1980; Smith, 1980).

20.3.9.6 Ameliorative and remedial management

Many of the effects of upland afforestation on freshwater habitats are a result of ploughing operations. Dialogue with the Forestry Commission has disclosed that on well-drained slopes ploughing is less important for drainage, but is regarded as of benefit for establishing a tree crop. Weed suppression and planting-up operations in large forests are greatly aided by ploughing. Alternatives to ploughing are few and more costly. Nevertheless, one promising technique uses a machine that, instead of producing a continuous drainage channel, turns over individual sods of earth providing an ideal planting site, without having a major impact on the drainage pattern. This technique also has the advantage of rendering the forest less prone to windthrow.

Where ploughing takes place, great care should be exercised in protecting small streams and the natural drainage pattern. Ditches and drains should not be allowed to enter streams or lakes directly, but they should be diverted to soakaways or containment ditches so that water is forced to percolate through soil before entering the water body. The design of these needs to be carefully considered to avoid erosion and to cope with flood conditions. Careful planning of ploughing and drainage work will combat the worse effects of siltation and erosion and, to some extent, chemical runoff.

To prevent afforestation from increasing the acidity of watercourses, it is essential that ditches and drains are not allowed to run directly into streams and lakes. Several lochs in Scotland have suffered from increased acidity after drainage ditches from plantations were channelled into them, and some of these have been selected for rehabilitation experiments including the use of lime to reduce acidity. For nature conservation this is not generally considered a suitable option as liming itself will alter the chemical composition of the water to provide conditions different from those which gave rise to the scientific interest. Liming is also costly in remote areas and a perpetual management commitment.

To prevent shading and reduce acidification, buffer strips should be planted along all streams, regardless of size. These buffer strips may be planted with trees native to the area or be allowed to develop naturally. The minimum width of the buffer strip differs from situation to situation, but should be wide enough to avoid shading of the stream when the plantation matures. Buffer strips do not need to be parallel to streams, but can make use of natural features so that they are wider in some places than others. The provision of buffer strips has the added advantages of readily identifying the presence of watercourses thus furthering their protection, providing a rich source of food in the form of detritus for invertebrates and acting as a protecting strip during thinning and logging operations. However, it must be emphasised that regardless of the width of the buffer strip, its effectiveness as a means of ameliorating the effects of acidification and siltation is severely limited if drainage ditches are fed directly into streams.

20.3.10 Pesticides

20.3.10.1 Insecticides

Following the work on Clear Lake in the late 1950s (Hunt & Bischoff, 1960) a large amount of literature has accumulated on the effects of organochlorine insecticides on aquatic organisms. Build-up of insecticides in the food chain can lead to the elimination of species at the higher trophic levels, and to those sensitive species at lower trophic levels. This serious conservation problem is now well recognised and in western countries the more persistent insecticides are only allowed limited use. The voluntary ban in Britain on the use of dieldrin and DDT in recent decades has in the main worked. However, otters and other wildlife were directly poisoned by dieldrin pollution in the River Tone in Somerset during 1972. Dieldrin residues were found in the spraints of otters living in the River Taw in Devon (Jefferies, D., pers comm, 1985). Haas & Cooke (1983) also found increasing dieldrin residues in 1981 compared with the previous five years in the kingfisher (*Alcedo atthis*). Dieldrin is still used as a sheep dip by some hill farmers (Ministry of Agriculture, Fisheries & Food, 1982) and by some means or another finds its way into watercourses.

The short-term persistence, high toxicity to vertebrates and aquatic invertebrates, of the alternative organo-phosphorus and carbamate compounds is for the same reason a regular pollution problem in upland streams (Rangeley, J., pers comm, 1984). The effects of spray drift onto fresh waters of these and other pesticides, particularly the synthetic pyrethroids with their high toxicity of fish, also gives cause for concern (Perring, 1985).

20.3.10.2 Herbicides

The widespread use of herbicides in agriculture and forestry can cause a local decline in aquatic vegetation when the herbicide accidentally drifts onto waterways. Emergent vegetation is more likely to be affected than submerged vegetation. The conservation hazards are, however, far greater when herbicides are used for the control of submerged and emergent plants. A large amount of literature has accumulated on the use, mis-use, toxocity, (Alabaster, 1969) and the potential environmental hazard of such chemicals (Brooker & Edwards, 1974; Newbold, 1975; Newbold, 1976).

20.3.10.3 Ameliorative and remedial management

Careless throwing of containers with pesticides in them into ponds and streams can cause severe local poisoning. There are various 'codes of good agricultural practice' produced on the correct disposal of unwanted pesticide containers, the correct use of sprays on cropped land, and on the correct use of herbicides in or near watercourses. (A full list is available from MAFF.) Contravention of such 'codes' could lead to a prosecution under the Control of Pollution Act, 1974 (Department of the Environment/Welsh Office, 1974). The development of more efficient sprays, such as the Controlled Droplet Application, would go to a long way in minimising the wastage of pesticides and consequent damage to the countryside. The removal of some of the more volatile compounds from the proposed registration of cleared pesticides would also remove an increasing area of concern amongst conservationists (Perring, 1985).

20.4 CONCLUSIONS

Although some of the more important agricultural pollution problems, and their effects on the conservation of fresh waters have been touched upon, it is not possible to be comprehensive. There are probably many other less obvious problems which are not yet known about and will only be detected by careful monitoring of the environment. It is only through such biological surveillance that new threats to fauna and flora may be detected early enough and remedial measures sought. The freshwater environment is certainly a delicate sensor of environmental contamination. For this monitoring role, constrained as it is by strict financial budgets, gratitude is expressed to the water industry of England, Wales, and Scotland.

A difficult issue facing the conservation fresh water is that a waterbody can only be conserved with a proper understanding of, and some form of dialogue with, those managing the catchment area. As has been shown some ameliorative measures are possible without grossly affecting catchment area use, in others a decision has to be taken whether conservation or some other use of the land or water in the catchment has priority.

REFERENCES

Alabaster, J. S. (1969) Survival of fish in 164 herbicides, insecticides, fungicides and wetting agents, and miscellaneous substances. *International Pest Control*, **11**, 29–35.

Alabaster, J. S. (1982) Report on the EIFAC workshop on fish farm effluents. Silkeborg, Denmark 18–26 May 1981, EIFAC Technical Report (41) 166pp.

Anglian Water Authority (1983) Spray irrigation. In: Draft Report on Demand Forecasts.

Bloomfield, C. (1972) The oxidation of iron sulphides in soils in relation to the formation of acid sulphate soils and of ochre deposits in field drains. *Journal of Soil Science,* **23,** pp 1–16.

Boorman, L. A., Sheail, J. & Fuller, R. M. (1977) The *Phragmites* Die-Back problem. Progress Report. Institute of Terrestial Ecology Project No 539.

Brooker, M. P. & Edwards, R. W. (1974) Review paper: Aquatic herbicides and the control of water weeds. *Water Research,* **9,** pp 1–15.

Brooks, A. (1981) *Channelization in England and Wales.* Discussion Paper No 11. Department of Geology, The University of Southampton.

Crook, C. C., Boar, R. R. & Moss, B. (1983) The decline of reedswamp in the Norfolk Broadland, causes, consequences and solutions. Norwich: Broads Authority (BARS 6). Thomas Harvey House, 18 Colgate, Norwich, Norfolk, NR3 1BQ, England.

Cuttle, S. P. (1982) Sources of nitrogen in surface waters within the catchment of Loch Leven, Kinross-shire. Work done by Soil Science Department, Edinburgh School of Agriculture. Nature Conservancy Council Report Edinburgh, 12 Hope Terrace, Edinburgh, EH9 2AS, Scotland.

Department of the Environment/Welsh Office. (1974) *Control of Pollution Act 1974* Her Majesty's Stationery Office.

Department of the Environment. (1981) *Wildlife and Countryside Act 1981* Her Majesty's Stationery Office.

Driscoll, R. (1983) Improvements in land management: their effects on aquatic plants in Broadland. *Watsonia*, **14,** (3), pp 276–277.

Fowler, D., Cape, J. N., Leith, I. D., Patterson, I. S., Kinnaird, J. W. & Nicholson, I. A. (1982) Rainfall acidity in Northern Britain. *Nature*, London, **297**, 383–386.

Fry, G. L. A. & Cooke, A. S. (1984) Acid Deposition and its implications for Nature Conservation in Britain. *Focus on Nature Conservation No 7.* Nature Conservancy Council, Northminster House, Peterborough, PE1 1UA, England.

George, M. (1976) *Is wildlife being drained away?* Conservation Review (12) Spring. Published by Society for the Promotion of Nature Reserves, Lincoln, England.

Gosling, L. M. (1971) *Calthorpe Water acidity — background, effects, causes and suggested treatment.* Unpublished report Nature Conservancy Council, East Anglian Regional Office, 60 Bracondale, Norwich, NR1 2BE, England.

Gosling, L. M. & Baker, S. J. (1980) Acidity fluctuations at a Broadland Site in Norfolk. *Journal Applied Ecology*, **17,** (2), pp 479–490.

Haas, M. B. & Cooke, A. S. (1983) Pollutants in Kingfishers. In: *Institute of Terrestial Ecology Annual Report 1982*. Merlewood, Grange over Sands, Cumbria, England.

Hambrey, J. B., Newbold, C. & Smith, I. R. (1985) *Nature Conservationand Freshwater Fish Farming*. Chief Scientists Directorate Note. Nature Conservancy Council, Northminster House, Peterborough, PE1 1UA, England.

Harriman, R. (1978) Nutrient leaching from fertilized forest watersheds in Scotland. *Journal of Applied Ecology*, **15**, 933–942.

Harriman, R. & Morrison, B. R. S. (1982) Ecology of streams draining forested and non-forested catchments in an area of central Scotland subject to acid precipitation. *Hydrobiologia*, **88**, 251–263.

Harriman, R. & Wells, D. E. (1984) Causes and effects of surface water acidification in Scotland. Proceedings of the Institute of Water Pollution Control Conference Paper No 7, 1984. IWPC, Leadson House, 53 London Road, Maidstone, Kent, ME16 8JH, England.

Harris, G. L. & Colbourn, P. (1985) *Brimstone Farm — Surface Layer Nitrate Leaching*. Agricultural and Food Research Council, Unit 12, Pioneer Road, Faringdon, Oxfordshire, England.

Her Majesty's Stationery Office (1971) *Town and Country Planning Act 1971*.

Her Majesty's Stationery Office (1977) *Town and Country Planning General Development Order 1977*.

Her Majesty's Stationery Office (1967) *Water (Scotland) Act 1967*.

Hunt, E. G. & Bischoff, A. I. (1960) Inimical effects on wildlife of periodic DDD applications to Clear Lake. *California Fish and Game, 46,* 91–106.

Jay, D. & Holdich, D. M. (1981) The distribution of the crayfish *Austropotamobius pallipes* in British waters. *Freshwater Biology,* **11,** 121–129.

Jørgensen, S. E. (1980) Lake Management. *Water Development Supply and Management* (14)167 pp. Pergamon Press, Oxford.

Lewis, G. & Williams, G. (1984) *Rivers and Wildlife on rivers*. Royal Society Protection of Birds, Royal Society Nature Conservation. Sandy, Bedfordshire, 295 pp.

Lewis, M. R. (1984) *Fish farming in the United Kingdom*. University of Reading, Department of Agricultural Economics and Management. Miscellaneous Study No 71.

Lund, J. W. G. (1972) Eutrophication. *Proceedings Royal Society London*, B **180**, 317–382.

McColl, R. H. S. & Gibson, A. R. (1979) Downslope movement of nutrients in hill pasture, Taita, New Zealand. *New Zealand Journal of Agricultural Research,* **22,** 279–286.

McColl, R. H. S., White, E. & Waugh, J. R. (1975) Chemical run-off on catchment converted to agricultural use. *New Zealand Journal of Science*, **18** pp 67–84.

Mills, D. H. (1980) *The Management of Forest Streams*. Forestry Commission Leaflet No 78, HMSO.

Ministry of Agriculture Fisheries and Food (1981) *Diseases of Fish Act 1981* Her Majesty's Stationery Office.

Ministry of Agriculture Fisheries and Food (1982) *Survey of Dieldrin residues in food.* The seventh report of the steering group on food surveillance. The Working Party on Pesticide Residues. Food Surveillance Paper No 7.

Ministry of Agriculture Fisheries and Food. (1983) *Import of Live Fish Act 1983* Her Majesty's Stationery Office.

Morgan, N. C. (1970) Changes in the fauna and flora of a nutrient enriched lake. *Hydrobiologia*, **35,** 545–553.

Moss, B. (1978) The Ecological History of Medieval Man-made Lake Hickling Broad, Norfolk, United Kingdom. *Hydrobiologia*, **60**(1), 23–32.

Moss, B. (1983) The Norfolk Broadland: Experiments. *Biological Reviews* **58**, 521–561.

Mountford, J. O. & Sheail, J. (1982) *The Impact of Land Drainage on Wildlife in the Romney Marsh.* The availability of baseline data. Institute of Terrestial Ecology. Project No 718. Interim report to the Nature Conservancy Council, Monks Wood Experimental Station, Abbots Ripon, Huntingdon, Cambridgeshire, PE17 2LS, England.

National Farmers Union (1984) *Code of Practice — Fish Farming and Good Husbandry Practices.* NFU London.

Nature Conservancy Council File (1982) Internal Report; *The Effects of iron ochre on freshwater plants and invertebrates within grazing marsh systems in Norfolk.* East Anglia Regional Office, 60 Bracondale, Norwich, NR1 2BE, England.

Nature Conservancy Council (1984) *Tenth Annual Report.* Her Majesty's Stationery Office.

Newbold, C. (1975) Herbicides in Aquatic Systems. *Biological Conservation*, **7,** (2), 97–118.

Newbold, C. (1976) Environmental Effects of Aquatic Herbicides. Proceedings of Symposium on Aquatic Weeds. British Crop Protection Council Monograph 16, pp 78–90.

Newbold, C. (In Press) Nature Conservation, River Engineering and the Drainage Area of a river scheme. *Biologist*. Institute of Biology.

Newbold, C., Purseglove, J. & Holmes, N. T. H. (1983) *Nature Conservation and River Engineering*. Nature Conservancy Council, Peterborough, 36 pp.

Newson, M. D. (1980) The erosion of drainage ditches and its effect on bed load yields in mid-Wales: reconnaissance case studies. *Earth surface Processes*, **5,** 275.

Omerod, S. J., Tyler, S. J. & Lewis, J. M. S. (1985) Is the breeding distribution of Dippers influenced by stream acidity? *Bird Study*, **32**, 33–40.

Perring, F. (1985) Comment — Pesticides Safety Legislation. *Natural World*, p 6. Winter/Sping 1985.

Phillips, G. L. (1984) A Large-Scale Field Experiment in the Control of Eutrophication in the Norfolk Broads. *Water Pollution Control*, **83**, (3), 400–408.

Robinson, A. (1980) *The effect of pre-afforestation drainage on the streamflow and water quality of a small upland catchment.* Report 73; Institute of Hydrology, Wallingford, Oxfordshire, OX10 8BB, England.

Schindler, H. (1978) Verbreiting von *Typha angustifolia* an der Berliner Havel unter besonderer Berucksichtigung der Bestandsveranderung. Dipl. arb. am Fackber Biologie der FU Berlin.

Smith, B. D. (1980) The effects of afforestation on the trout of a small stream in Scotland. *Fisheries Management*, **11**, 39–58.

Solbé, J. F. (1982) Fish farm effluents. In: Alabaster, J. S., *EIFAC Technical Report*, **41**, pp. 29–55.

Stoner, J. H., Gee, A. S. & Wade, K. R. (1984) The effects of acidification on the ecology of streams in the upper Tywi catchment in West Wales. *Environmental Pollution*, (Series A), **35**, 125–157.

Sukopp, H., Markstein, B. & Trepl, L. (1975) Rohrichte unter intenswem Gross stadteinfluss. Beitr naturle, Forsch Sudw DH, (34) pp. 371–385.

Taylor, N. & Brownlie, T. G. (1982) *Report on Survey of Fodder Silos.* Department of Agriculture and Fisheries for Scotland, Edinburgh.

Water Space Amenity Commission (1980) *Conservation and Land Drainage Guidelines. Drainage Flood Protection and Sea Defence Works, England and Wales.* WSAC, London.

Woldendorp, J. W., Dilz, K. & Kolenbrander, G. J. (1966) The fate of fertliser nitrogen on permanent grassland soils. Proceedings *General Meeting European Grassland Federation*, **1**, 53–68, Wageningen, Netherlands.

21

Remedies to problems caused by agriculture — the engineering solution

V. R. Phillips

21.1 INTRODUCTION

The title the author was invited to speak on includes the phrase 'The Engineering Solution', but the word 'engineering' has been interpreted broadly (as engineers often need to!) and, as will be seen, 'The Technological Solution' might be a more accurate title. By the distinction it is meant to imply that in the author's judgement it is unlikely that there will be monolithic hardware-based solutions to the various problems of fresh waters caused by agriculture. On the contrary, a range of technological approaches will need to be brought to bear, and even then these technologies will need to be backed up by careful agricultural management practice.

As will be seen, in many cases appropriate technology is still only at the development stage, and so until this technology can be applied routinely on a farm scale, control of pollution risks via careful management is all the more important. The Ministry of Agriculture, Fisheries and Food (MAFF) have in fact recently published a Code of Good Agricultural Practice (MAFF, 1985), which sets out the precautions recommended for minimising water pollution from agriculture. When a pollutant accidentally enters a watercourse, the onus is on the farmer to prove that he was following Good Agricultural Practice, either as set out in the MAFF Code or in some other Code as defined by Part II of the Control of Pollution Act, 1974. Failure to provide this evidence will lay him open to prosecution by the Water Authority under Part II of the Act.

21.2 THE PROBLEM

What then *are* the problems which agricultural use of land can cause with fresh waters (watercourses and/or groundwater)? The author has taken them to be excessive concentrations of any of the following: nitrate ion, phosphate ion, potassium ion, organic matter — both insoluble (in suspension) and soluble, soil solids (in suspension), pathogens (for example certain viruses and bacteria), heavy metals, and pesticides.

The definition of 'excessive' depends on the use to which the water is subsequently to be put (Barnden, 1984).

21.3 THE TECHNOLOGICAL SOLUTIONS

21.3.1 Nitrate

21.3.1.1 Nitrate from fertilizers

In 1982 about two million tonnes of fertilizer nitrogen (mainly in the form of ammonium nitrate) were used in the UK, mostly on cereals and grassland. The fraction of this nitrogen which ultimately ends up biochemically bound within the crop depends on many interacting factors, for example, soil type, crop type, weather, time of year of application of the fertilizer and rate of application (kg N/ha), but is typically 60%. Detailed discussions of such factors can be found (MAFF, 1981; 1982a; Royal Society, 1983), but certainly even in favourable circumstances a proportion of this fertilizer can, by leaching down through the soil, end up as nitrate in freshwater streams and/or in deep strata from which drinking water is abstracted via boreholes. With both cereals and grazed grassland, up to one third of applied nitrate can be lost by leaching (Royal Society, 1983; Ryden & Garwood, 1984).

How can these losses of nitrate to watercourses be reduced? Certainly, reducing the amount of nitrate *entering* the watercourses appears in general a more promising approach than trying to remove nitrate which has already been allowed to enter, that is, once again, prevention is preferable to cure.

Although nitrogen exists in soils in a variety of chemical forms (for example, ammonium ion, organic nitrogen compounds, nitrate ion), the uptake of nitrogen by plants is mostly as nitrate (and to a lesser extent as ammonium ion), and in general this uptake occurs only when the plant is undergoing vegetative growth. The key to minimising leaching of nitrates to watercourses, then, is to ensure that as far as possible the nitrate form of nitrogen exists in the soil only at times when the plant can utilise it, and only in amounts that the plant can utilise fully. This objective applies equally to applied nitrate and to nitrate formed in the soil by bacterial action.

Thus if fertilizer is already in nitrate form (for example as ammonium nitrate) when applied to the land, the timing and rate of application are all-important in minimising leaching. Detailed ADAS recommendations (MAFF, 1985a) are available for fertilizer timing and dose rate for different crops on different soils, although these doses are optimised for crop yield rather than for minimum leaching.

Methods for predicting the optimum dose rate and time of application are still being improved via mathematical modelling (Rothamsted Experimental Station, 1984; Royal Agricultural Society of England, 1985) and more account is now being taken of leaching criteria as well as crop yield. Two other approaches have been proposed to help farmers optimise the *timing* of spring fertilizer application. The 'T°-sum' is a simple calculation for predicting the optimum time to fertilize grass, but its value under UK conditions seems limited (Archer & Unwin, 1984). More fundamentally, but still needing further development, microscopic monitoring of the development of the embryo ear in cereals may give guidance to timing the application of fertilizer (Agricultural and Food Research Council, 1983; Dodgson, 1985).

Meanwhile, England (1984a; 1984b) has modelled the effect of imposing a tax on fertilizer as a means of reducing the amounts used. However, he finds that, for cereals considered alone (England, 1984a), even if the price of fertilizer were doubled, it would not be worthwhile to the farmer to reduce his fertilizer use by more than 10%. Even when a range of typical arable crops are considered together (England, 1984b), the corresponding figure is not more than 25%.

In most forms of fertilizer other than ammonium nitrate (for example, animal slurries, sewage sludge, urea, anhydrous ammonia), little of the nitrogen is in the form of nitrate at the time of application, but the other forms of nitrogen are converted slowly to nitrate, after application, by nitrifying bacteria in the soil. Thus with these forms of fertilizer a slower, steadier supply of nitrate for use by the plant is available. The soil's own pool of organic nitrogen compounds similarly contributes a slow, steady supply of nitrate via bacterial nitrification.

Competing always with the plant for nitrate are the soil's natural *denitrifying* bacteria, which reduce nitrate, via nitrite, to nitrous oxide and nitrogen gas, which can be wasted through being lost to the atmosphere (Royal Society, 1983; Ryden & Garwood, 1984).

21.3.1.2 Technological control of the availability of nitrate in the soil

A number of techniques exist for *controlling* the rate, after land application, at which nitrate, either from ammonium nitrate or from other forms of fertilizer, is made available for plant uptake.

For ammonium nitrate, the simplest technique is of course to apply it in fractional doses at two or more intervals during the plant growth period — rather than applying the whole dose at the beginning of the period (Royal Agricultural Society of England, 1985; Kanwar *et al.*, 1984; Morrison, 1984). Kanwar *et al.* (1984), using a computer model calibrated against field data, found that fractional dosing of maize should reduce the losses of nitrate through leaching by up to 23%, depending on rainfall pattern.

Resin-coated granular ammonium nitrate is also available — the resin coating reduces the rate of solution of the nitrate in water. However, this form is as yet much more expensive than normal ammonium nitrate and therefore is not yet widely used.

In cases where nitrification is producing nitrate faster than the plant can assimilate it, synthetic nitrification inhibitors may be used, which reduce the rate of action of nitrifying bacteria and hence prevent the formation of excess nitrate which may be lost by leaching. However, at present it is more expensive to the farmer to use an inhibitor than it is to make up with more fertilizer any nitrate lost by leaching.

In late winter and early spring, nitrate leaching may be reduced by taking advantage of the fact that, as long as root temperatures stay below 12 °C or so, roots absorb ammonium ion more readily than nitrate ion. Therefore at this time of year ammonium fertilizer (for example ammonium sulphate) may be used instead of nitrate.

Nitrate levels in rivers are normally highest during late autumn and early winter (Society of the Chemical Industry, 1984). Reasons for this include:

(i) the higher rainfall in winter, which causes a greater flow of leaching-water down through the soil;
(ii) nitrate formed by nitrification accumulated in the soil through the summer, especially under grassland and especially in dry summers; and
(iii) in the case of arable land, autumn cultivations of the soil, which encourage nitrification.

In this third respect, a less vigorous cultivation means less nitrate loss by leaching, over the winter period. Thus workers at Letcombe Laboratory (Royal Agricultural Society of England, 1985) found that, on clay soil, the nitrate leaching from direct-drilled winter wheat was 40% less than that from winter wheat sown after ploughing.

Another way partially to reduce winter leaching of nitrate from arable land is by incorporating the straw into the soil after the summer harvest. Bacteria and fungi decompose the straw and, in doing so, immobilise nitrogen from nitrate produced by nitrification, which would otherwise be lost by leaching. Typically this straw can immobilise 20 kg N/ha. This nitrogen then gradually becomes available to plants once more over several years (Rothamsted Experimental Station, 1984; Audsley, 1984.

21.3.1.3 Nitrogen-fixing plants

Some plants, such as peas, beans, and clover, fix their own nitrogen, that is their roots are nearly always associated with certain bacteria (for example *Rhizobium*) which convert atmospheric nitrogen gas directly to ammonia for use, after nitrification, by the plant. Such plants need little or no added fertilizer, and so if their use can be maximised, direct leaching of fertilizer nitrate can be reduced (although there will still be some leaching losses, such as from the roots of the plants after they die).

An example of the use of nitrogen fixers is in forage for livestock, where a carefully managed mixed white clover/grass sward can, without any added fertilizer, produce the same forage yield as a pure grass sward which received 200 kg nitrate-N/ha yr (Royal Society, 1983; Morrison, 1984). The clover fixes enough nitrogen not only for itself but also for the grass growing with it.

Can this concept of growing a nitrogen-fixing plant together with a non-fixing plant be extended? Research is already under way at Rothamsted on tropical crops (Rothamsted Experimental Station, 1984). Perhaps in the long term in the UK, cereals grown mixed with a nitrogen fixer, with the two resulting types of grain being separated within the combine harvester, will be seen.

Inocula containing Rhizobia bacteria are now commercially available for adding to soils which are initially low in Rhizobia. Meanwhile, breeding and genetic research continues, to improve both the Rhizobia themselves and the plants which will grow with them (Royal Society, 1983). In the long term such work may lead to true nitrogen-fixing cereals.

21.3.2 Phosphorus and potassium

Phosphate and potassium ions are applied to agricultural land either as simple fertilizers (MAFF, 1976; 1978), or as constituents of animal slurries (MAFF, 1982a) (phosphorus and potassium) or sewage sludge (MAFF, 1982b) (phosphorus only). In all but the lightest soils, phosphates and potassium are only very slowly lost by leaching down into the soil, and any of either not taken up by the crop is bound strongly in the top soil (Hucker, 1981), but both can still reach watercourses by runoff across the soil surface. However, if, when spreading animal slurries or sewage sludge on land, the hydraulic limit described in section 21.3.3.3 of 50 m^3/ha (per 3 weeks) is observed, this should also serve to prevent either phosphate or potassium from entering watercourses by runoff.

21.3.3 Organic matter

21.3.3.1 Background

Cases where agriculture causes excessive soluble or insoluble organic matter to enter watercourses usually arise from careless storage or spreading of animal slurries (that is mixtures of faeces and urine without bedding) or of silage effluent (ADAS, 1984). The Biochemical Oxygen Demand (BOD) of undiluted animal slurry is typically 20 000 mg/l, while that of silage effluent can be as high as 80 000 mg/l.

Traditionally, animals were bedded on straw, and farmyard manure was produced. This presented no particular BOD hazard as it was stackable rather than fluid, and also its BOD was usually reduced during composting before land application. However, with the increasing intensification of livestock farming and the drive to reduce labour requirements, slurry has become much more common, and arises in greater amounts, albeit on fewer sites. This fluid form of waste requires more careful storage and spreading to reduce the risk of runoff or leaching into watercourses.

(Straw bedding of animals is likely to make a comeback in the next few years, at least in some parts of the UK, to improve animal welfare and because of a possible ban on straw-burning. If this does happen, the risks of water pollution from livestock wastes will be lessened.)

In the early days of slurry systems (1960s, 1970s), research and development was carried out on treatment techniques (particularly aeration and anaerobic digestion) (Baines *et al.*, 1973; Hobson *et al.*, 1981) for reducing the BOD of slurries sufficiently to allow direct discharge to watercourses. However, reducing the BOD to this extent proved to be too expensive and so land disposal of slurry, either untreated or partially treated with respect to

BOD removal, is now the norm. The soil's naturally-occurring bacteria then complete the job of BOD removal.

Because of economies of scale, more complete treatment of slurries before disposal to land or water is economically more feasible for extremely large livestock units, for example, over 10 000 pigs. Such an approach has been taken in Eastern Europe, but is judged not to be appropriate for the UK.

In the UK, a few livestock farmers, by arrangement with a water authority, do dispose of untreated slurry direct to a sewer.

21.3.3.2 Slurry storage

Various forms of store exist for storing animal slurries awaiting spreading on land (MAFF, 1980). Earth-bank lagoons and above-ground tanks of bolted steel panels on concrete bases are common types.

In the winter, weather and soil conditions frequently prevent land spreading, and besides, as discussed in section 21.3.1.1, the fertilizer value of slurry is best realised, and nitrate leaching is minimised, if the slurries are spread on the land in spring. Therefore stores may need to be sized to hold slurry arising over as much as six to nine months. Overflowing of undersized stores causes many of the water-pollution incidents attributable to farm wastes (ADAS, 1984) and reflects poor planning and/or management of storage requirements. For example some farmers make no effort to prevent clean rain-water from roofs, open concrete areas, and so on, ending up in the slurry store, thus unnecessarily adding to the slurry itself.

A British Standard on retaining structures for agricultural products, including reception pits and above-ground tanks for slurry (steel or concrete) is currently under preparation. Once the Standard is available, it should further reduce the risk of water-pollution incidents from poorly-designed slurry stores.

21.3.3.3 Slurry spreading

Land spreading of slurries is usually carried out using some form of tanker (MAFF, 1983). Weather conditions, soil type, topography and dose rate will all influence the risk of runoff of spread slurry, and hence of water pollution. In general, as for inorganic fertilizers, spring is the best time of year for slurry spreading.

The ADAS recommends (MAFF, 1983a) that the maximum dose rate on any one occasion be 50 m^3 slurry/ha (that is a depth of 5 mm) and that there should be a period of at least three weeks between successive applications.

There is scope for the development of devices to help the driver of the spreader achieve a rate of 50 m^3/ha, or any other given rate. Presently tractors and/or tankers have little such instrumentation or control, and the driver must rely on his own judgement about spreading rate.

Accurate information on the spreading rate achieved is also useful when estimating the amounts of nutrients added to the land via spread slurry, and when subsequently estimating the reduction in rates of *inorganic* fertilizer which can be made. A new commercially-available meter† may also help here, as it is claimed that it allows farmers quickly and easily to estimate the inorganic nitrogen content (principally ammonium) of a slurry. The reliability and accuracy of the meter is now being assessed within ADAS.

One quarter of all UK pigs are in intensive livestock units which do not themselves have sufficient land for slurry spreading without exceeding the ADAS recommendations on maximum rates of spreading. However, provided such enterprises make arrangements with neighbouring farmers to spread on *their* land, then these recommendations can be met everywhere in the UK except for one or two small areas of very intensive pig production.

In the Netherlands the problem is much more acute and in certain areas government-operated 'manure banks' have been set up. Farmers with excess manure either inform the 'bank manager' or deposit it themselves in the bank. Other farmers in the neighbourhood who need manure can then obtain it from the bank. Costs are shared between government, feed companies and farmers. As manure volumes have increased, disposal areas have had to be extended. Often slurry is now transported 70 or 80 km.

Similar precautions about spreading time and rate apply to sewage sludge as to animal slurries (MAFF, 1982b). However, the spreading of sewage sludge on agricultural land is a far more localised practice than is the spreading of animal slurries. Transport costs normally prevent the spreading of sewage sludge more than a few miles from a sewage works.

21.3.3.4 Composting

For the livestock farmer whose wastes arise as slurry but who has straw available to him, composting is a possibility. Straw and slurry must be intimately mixed and then formed in to windrows. The composting process reduces BOD and pathogen concentrations, as well as reducing overall bulk and converting the waste to a stackable form (Biddlestone & Gray, 1984). A machine is commercially available which mixes 500 l/min of slurry with 80 kg/min of straw and throws out the mixture for forming windrows.‡ Audsley (1983) has estimated the costs of on-farm composting to be typically £15/tonne of compost.

21.3.3.5 Slurry separation

The popularity of slurry systems, as opposed to straw-based waste systems, stems from their reduced labour requirements, and these in turn stem from the ability to handle slurry as a liquid, which can, for example, be pumped along pipes, or drained along channels. Unfortunately, raw slurries do not always behave as well as clear liquids might (for example if the slurries have a high fibre content or if some bedding material does enter them) and pumps and pipes can sometimes become blocked. Separating machines have therefore been developed, at the National Institute for Agricultural Engineering (NIAE) and elsewhere (Pain *et al.*, 1978), for removing fibre and

† Agros Ltd., Ovagen 1, S-533 03, Kallby, Sweden.
‡ Armix. Alfred Lyons & Sons Ltd., Rugeley, Staffordshire, England.

any bedding material from slurries. Typically all particles above 1 mm are removed. The separated solids are dewatered sufficiently to be stackable, while the separated liquid can be treated with more confidence as a liquid: thus it can be spread on land through sprinkler irrigation equipment, and at application rates which are more easily measured and controlled than for raw slurries. With fibre now removed, the slurry is less prone to runoff, being absorbed more rapidly by the land, and also its nutrient content is now less variable than for raw slurry.

The separated solids are low in BOD (Baines *et al.*, 1973). They can be spread on land using a muck spreader, or, better, composted or vermicomposted (Edwards, 1983) and sold for horticulture.

21.3.3.6 Dirty water

This name covers such effluents as drainage from cattle yards and piggery yards. Its BOD (typically 2000 mg/l) is lower than that of neat slurries, but it is a very common form of effluent and, perhaps because some farmers feel that it does not merit particular care, it is the cause of a significant proportion of water-pollution incidents (ADAS, 1984). Methods of disposal of dirty water recommended by ADAS (MAFF, 1985b) include barrier ditches and, after settlement of solids, low-rate irrigation to land at no more than 5 mm depth per hour.

21.3.37 Silage effluent

Except for waste milk, silage effluent has, at 80 000 mg/l, the highest BOD of any farm effluent. Prevention of water pollution (MAFF, 1984) therefrom centres on:

(a) making the silage in such a way as to minimise the creation of effluents, for example, by wilting the forage in the field before ensiling, and/or by incorporating with the silage edible absorbents such as straw, and
(b) constructing proper leak-proof silos and equipping them with adequately-sized and sealed sumps to collect and store any effluent which does drain from the silo.

Final disposal is usually by spreading on land, at not more than 25 m^3/ha, at not less than three-weekly intervals, although some trials on feeding to livestock have been conducted.

Silage effluent is acidic and therefore very corrosive to concrete and steel, and searching of cracks in silos. Research is under way at the West of Scotland Agricultural College into economic sealing methods which are proof against silage effluent, for use in the construction of new silos or in the repair of existing ones.

21.3.3.8 Aeration and anarobic digestion of animal slurries

Research and development is under way at the NIAE and elsewhere (Evans *et al.*, 1975; Williams *et al.*, 1984; Pain *et al.*, 1984) on the application of both aeration and anaerobic digestion of animal slurries. Currently the major motivations are odour control in the case of aeration and energy production (via biogas) in the case of anaerobic digestion. However, both techniques do also necessarily reduce the BOD of slurries (Hobson *et al.*, 1981; Evans *et al.*, 1975; Williams *et al.*, 1984) and so could be used if it was necessary to do so to an extent before spreading on land. Examples of typical annual costs with current technology are £3 per pig place per year for aeration, and £30 per cow per year for anaerobic digestion.

21.3.4 Soil solids

Soil solids may enter watercourses as a result of soil erosion caused by heavy rain. Morgan (1985) has reviewed this subject, but he stresses that there has been little basic research on soil erosion under UK conditions. He concludes that it is not yet possible to make soundly based and economic recommendations to UK farmers on how to prevent soil erosion, although his work on controlled clod formation via carefully-planned tillage is promising.

21.3.5 Pathogens

As with other organic matter, the major source of these potential water pollutants (such as *Salmonella, Leptospira*) is animal slurries. Jones (1983) reports a survey showing that, of herds producing waste as slurry, 10% of cattle herds and 23% of pig herds produce slurries containing salmonellae. There is little research on the extent to which such pathogens in animal waste might actually enter watercourses after such wastes have been land spread. In one of the few studies, Evans & Owens (1972) found that, for pig slurry on a sandy clay loam soil, even when the hydraulic limit for spreading on land was observed, concentrations of faecal bacteria in the drainage water rose 20- to 60-fold within two hours of spreading, but returned to normal levels within 36 hours.

Jones (1983) has outlined measures to reduce pathogen levels *before* land spreading. For example, *Salmonella* numbers are reduced by 90% by storing slurry for four weeks prior to spreading it on land. Since survival rate increases with slurry total solids, separation of the slurry before storage (see section 21.3.3.5) further reduces the survival rate. Aeration and anaerobic digestion of slurries can also kill salmonellae (Jones, 1983; Demuynck *et al.*, 1984). Numbers of *Salmonella* in waste continue to fall after land spreading: Jones (1983) reports that *Salmonella* numbers are dramatically reduced during the first four weeks after spreading of slurry on grassland.

21.3.6 Heavy metals

Pigs for fattening commonly have copper or zinc salts added to their diet as growth promoters and most of these are subsequently excreted. Therefore if pig slurry is spread on land, copper and zinc compounds are added to the soil. Similarly, if sewage sludge is land spread, other heavy metals (such as cadmium, nickel, lead) may be added to the soil. The full range of possible metals is wide (MAFF, 1982b) and depends on the origin of the sludge.

There is considerable evidence (MAFF, 1982b; L'Hermite & Dehandtschutter, 1981) that heavy metals are strongly bound by topsoils. Although this may lead to problems of toxicity to crops, it does mean that leaching of heavy metals is not normally a potential cause of water pollution. Heavy metals might, however, enter watercourses via runoff of pig slurry or of sewage sludge which has been land spread at too high a dose rate. Therefore the hydraulic limit described in section 21.3.3.3 should be observed for this reason as well.

21.3.7 Pesticides

The term 'pest' can include weeds, insects, fungi, nematodes, mites etc. and the range of chemical agents used as pesticides is very wide (MAFF, 1985c). Chemical pesticides are available in a variety of different formulations (Mathews, 1979), for example, dusts, granules, aqueous suspensions, emulsions, and microcapsules. The farmer's choice of a formulation depends on convenience and on the mode of action of the particular chemical pesticide. For example, some herbicides act through the roots of the plant and are therefore applied to the ground, while for winged insects a spray of insecticide droplets (Mathews, 1979) is usually used.

Water pollution by chemical pesticides can thus occur by a variety of routes. Examples are

- all types — by direct contact between the pesticide and the watercourse;
- sprays or dusts — by windborne drift;
- sheep-dip — by runoff across the soil surface.

There is a corresponding variety of techniques for prevénting pollution of water by pesticides. A basic requirement is, of course, to follow strictly all the makers' and ADAS's recommendations for storage and use (MAFF, 1985c; 1984a; 1984b; 1984c). Wherever possible, a form of pesticide and/or method of application which avoids the risk of drift should be considered (Gilbert, 1984). Application via the spraying of droplets is nevertheless likely to remain common, but even here recent developments are reducing the risk of drift. For example, new types of spraying device (Mathews, 1979) produce a narrower and more controllable range of droplet sizes, so that droplets smaller than 150 μm, which are prone to drifting, can be avoided. Also, recent work at NIAE (Marchant & Green, 1982) and elsewhere has shown that electrostatic charging of pesticide droplets can double the proportion of them which reach their target.

Pesticides vary greatly in their persistence in the environment after application. In many cases, great persistence is unnecessary and less-persistent pesticides are increasingly being developed. An example is sheep dips, where the UK sale of dips containing the persistent organochlorine chemical, Lindane, recently ceased. Organophosphorus compounds are replacing it: these degrade in the soil over a few weeks, after the used dip has been disposed of by land spreading, or in a soakaway, well away from watercourses (MAFF, 1984c). Such less-persistent pesticides reduce the risk of water pollution via runoff or leaching.

Concern over the non-specificity of chemical pesticides, over the tendency of pests to become resistant to them, and over their persistence (particularly earlier types) is leading to the development of *biological* pesticides. For example, a virus (Entwistle, 1984) fatal to a fly which is a serious pest of pine trees, and a bacterium fatal to a wide range of caterpillar pests of many crops are already commercially available (Environmental Data Services Ltd, 1985) in the UK. Development is also under way on various other agents, such as nematodes and fungi, as biological pesticides (Rothamsted Experimental Station, 1984).

Although this technology is still young, the development and use of such biological pesticides is likely to accelerate, because for example they are highly specific to a given pest, their toxicological testing is relatively cheap and, so far at least, they appear to present no environmental hazards (Environmental Data Services Ltd, 1985). The virus and bacterium mentioned above are both naturally-occurring organisms, but genetic engineering offers the chance further to widen the scope for effective biological pesticides.

21.4 CONCLUSIONS

For the next year or so at least, prevention of water pollution from agriculture will still depend greatly on careful agricultural management. Even with the existing levels of technology, there is scope in many instances (for example slurry storage and spreading) for reducing the risks of water pollution by improved operating practices. The implementation of the Control of Pollution Act 1974 Part II will press farmers to become more aware of such risks and to review and upgrade their practices.

Further research is needed, to *quantify* better the reductions in pollution which may be achieved by improved practices such as *by how much* can improved control of timing and dose rate of fertilizer applications reduce nitrate levels in rivers?

In five years' time, new technology will have 'designed out' the need for much of the management care presently necessary. For example, biological pesticides, with their promise of much smaller environmental hazards, should by then be in much more common use, as should nitrogen-fixing plants.

ACKNOWLEDGEMENTS

The author thanks Dr J. C. Ryden, Animal and Grassland Research Institute, and Mr V. C. Nielsen, ADAS Farm Waste Unit, for reading drafts of this chapter and offering helpful criticisms of them.

REFERENCES

Agricultural Development and Advisory Service Farm Waste Unit (1984) *A report on the 1983 survey of water pollution incidents caused by Farm Wastes in England and Wales*. MAFF, Reading.

Agricultural and Food Research Council (1983) *Better British Wheat.* Booklet, June 1983, London.

Archer, F. C. & Unwin, R. J. (1984) The timing of the first nitrogen application to maximize the yield of early grass. In: (Riley, H. & Skjelvag, A. O., eds) *Impact of climate on grass production and quality.* Proceedings 10th general meeting of the European Grassland Federation, As, Norway.

Audsley, E. (1983) *Straw/sewage-sludge compost.* Feasibility Study for Water Research Centre. WRC Report No. 595-M.

Audsley, E. (1984) *The economic consequences on the farm of a ban on straw burning: farm profit and practices.* Div. Note DN 1234, Natl. Inst. Agric. Engng, Silsoe, June.

Baines, S., Evans, M. R., Hissett, R. & Hepherd, R. Q. (1973) Principles of treatment of animal slurries. *The Agric. Engr.,* Summer, 72–76.

Barnden, A. D. (1984) Environmental Quality Objective theory in practice. *J. Effl. and Wat. Treat.,* November, pp. 409–416.

Biddlestone, A. J. & Gray, K. R. (1984) Composting: Practical experience with farm scale systems†. Paper presented at the EEC Symposium on *'Composting of Agricultural and other Wastes'*, Oxford, March 1984.

Demuynck, M., Nyns, E. J. & Naveau, H. P. (1984) A review of the effects of anaerobic digestion on odour and on disease survival†. Paper presented at the EEC Symposium on *'Composting of agricultural and other wastes'*, Oxford, March 1984.

Dodgson, G. (1985) *Farmer's Weekly*, 25 Jan., p. 19.

Edwards, C. A. (1983) Earthworms, organic waste and food. *Span,* **26** (3), 106–108.

England, R. A. (1984a) *Reducing the nitrogen input to cereals: I. Effect of nitrogen price.* Div. Note DN 1219, Natl. Inst. Agric. Engng., Silsoe, January.

England, R. A. (1984b) *Reducing the nitrogen input to cereals: II. Price effects at farm level.* Div. Note DN 1254 Natl. Inst. Agric. Engng., Silsoe, November.

Entwistle, P. F. (1984) Viruses for insect pest control. *Span.* **26** (2), 59–62.

Environmental Data Services Ltd. (1985) ENDS Report No. 120 (January 1985) pp. 9–11.

Evans, M. R., Hissett, R., Ellam, D. F. & Baines, S. (1975) Aerobic treatment of piggery waste prior to land treatment: a case study. In: *Managing Livestock Wastes.* St. Joseph, Michigan, ASAE, pp. 556–559.

Evans, M. R. & Owens, J. D. (1972) Factors affecting the concentration of faecal bacteria in land-drainage water. *J. Gen. Microbiol.* **71**, 477–485.

Gilbert, A. J. (1984) Applying pesticides as non-liquids. *Span.* **27** (1), 16–18.

Hobson, P. N., Bousfield, S. & Summers, R. (1981) Methane production from agricultural and domestic wastes. *Applied Science,* London.

† Since published in *Composting of agricultural wastes.* Gasser, J. K. R. (ed.), Elsevier, London, 1985.

Hucker, T. W. G. & Catroux, G., eds. (1981) Phosphorus in sewage sludge and animal waste slurries. *Proceedings of an EEC Seminar*, Groningen, June, 1980. D. Reidel, Publ., London.

Jones, P. W. (1983) Reducing the risk of spreading infectious diseases with slurry. In: *Getting more out of muck*. Agricultural and Food Research Council, London.

Kanwar, R. S., Baker, J. L. & Johnson, H. P. (1984) *Simulated effects of fertilizer management on nitrate loss with tile drainage water for continuous corn*. Trans. ASAE pp. 1396–1404.

L'Hermite, P. & Dehandtschutter, J., eds. (1981) Copper in animal wastes and sewage sludge. *Proceedings of EEC workshop,* Bordeaux, October 1980, D. Reidel, Publ., London.

Marchant, J. A. & Green, R. (1982) An electrostatic charging system for hydraulic spray nozzles. *J. Agric. Engng. Res.*, **27**, 309–319.

Matthews, G. A. (1979) *Pesticide application methods*. Longmans, London.

Morgan, R. P. C. (1985) Soil erosion measurement and soil conservation research in cultivated areas of the UK. *The Geographical Journal,* March 1985.

Morrison, J. (1984) Improving the efficiency of grassland production. In: *Grassland Research Today*. Agricultural and Food Research Council, London.

Ministry of Agriculture, Fisheries & Food (1976) *Phosphatic fertilizers.* ADAS Leaflet No. 442.

Ibid (1978) *Potassium and sodium fertilizers*. ADAS Leaflet No. 443.

Ibid (1980) *The storage of farm manures and slurries*. ADAS Booklet No. 2273.

Ibid (1981) *Nitrogen fertilizers*. ADAS Leaflet No. 441.

Ibid (1982) *Soil Nitrogen Index*. ADAS Leaflet No. 803.

Ibid (1982a) *Profitable utilization of livestock manures*. ADAS Booklet No. 2081.

Ibid (1982b) *The use of sewage sludge on agricultural land.* ADAS Booklet No. 2409.

Ibid (1983) *Equipment for handling farmyard manure and slurry*. ADAS Booklet No. 2126.

Ibid (1983a) *Advice on avoiding pollution from manures and other slurry wastes*. ADAS Booklet No. 2200.

Ibid (1984) *Silage — Avoiding pollution*. ADAS Booklet No. 2429.

Ibid (1984a) *Guidelines for the use of herbicides on weeds in or near watercourses and lakes*. ADAS Booklet No. 2078.

Ibid (1984b) *Farm Chemical stores*. ADAS Leaflet No. 767.

Ibid (1984c) *Guidelines for the disposal of unwanted pesticides and containers on farms and holdings*. ADAS Booklet No. 2198.

Ibid (1985) *Code of Good Agricultural Practice*. January 1985.

Ibid (1985a) *Fertilizer recommendations 1985–86*. ADAS Reference Book No. 209.

Ibid (1985b) *Dirty water disposal on the farm.* ADAS Booklet No. 2390.

Ibid (1985c) *Agricultural Chemicals Approval Scheme: Approved Products for Farmers and Growers 1985.* Reference Book No. 380.

Pain, B. F., Hepherd, R. Q. & Pittman, R. J. (1978) Factors affecting the performance of four slurry separating machines. *J. Agric. Engng. Res.,* **23**, 231–242.

Pain, B. F., West, R., Oliver, B. & Hawkes, D. L. (1984) Mesophilic anaerobic digestion of dairy cow slurry on a farm scale: first comparisons between digestion before and after solids separation. *J. Agric. Engng. Res.,* **29**, 249–256.

Rothamsted Experimental Station (1984) *Report for 1983.* Harpenden.

Royal Agricultural Society of England (1985) *Better use of nitrogen — the prospects for winter cereals.* Proceedings of a conference held at the National Agricultural Centre, Stoneleigh, on 23 January 1985.

Royal Society (1983) *The Nitrogen Cycle of the United Kingdom.* London.

Ryden, J. C. & Garwood, E. (1984) Evaluating the nitrogen balance of grassland. In: *Grassland Research Today.* Agricultural and Food Research Council, London.

Society of the Chemical Industry (1984) *Agriculture and Water Quality.* Abstracts of papers presented at meetings held 21 February 1984 and 20 March 1984, at the SCI, London.

Williams, A. G., Shaw, M. & Adams, S. J. (1984) The biological stability of aerobically-treated piggery slurry during storage. *J. Agric. Engng. Res.,* **29**, 231–239.

22

Agricultural pollution — the farmers' view

M. Payne

22.1 IS AGRICULTURE A SPECIAL CASE?

It may well be asked whether agriculture differs from other industries in the approach which should be adopted to the pollution which stems from it. There are a number of aspects which are peculiar to agriculture and which must be recognised in a full analysis of farming pollution.

Agriculture is an extensive industry accounting for over 80% of the land surface of England and Wales. It has continued in its fundamental form of large numbers of small units for many centuries, although in recent years a number of far-reaching changes in management practices have been introduced, particularly in animal husbandry. Nonetheless the industry continues to be organised on a fragmented basis with relatively large numbers of small businesses providing an intimate contact with the land on which production takes place. Water undertakers use agricultural land as a catchment for the commodity they have been established to supply to the ever-increasing urban populations. In the absence of conurbations typical of an industrial society water undertakings would hardly be required, and local provision for water of adequate quality would still be possible. This dual use of land as both the surface from which to derive water supplies and a production medium, plays a large part in bringing the practice of agriculture into conflict with those responsible for maintaining the quality of water supplies. The two uses are regulated by different legislation, different government departments, and different national policies which demonstrate only minimal areas of co-ordination. While water supply for human use may be regarded as the key use for water, it is only one aspect of environmental quality relating to water. Society in Great Britain is placing value on maintaining and restoring the quality of its environment, a trend to which agriculture must not be blind.

Agriculture can be identified as forming a continuum over much of the land's surface as compared with the relatively small and often isolated sites occupied by other industries. Over and above this, and more fundamental, is agriculture's integration with the natural environment. A typical industrial enterprise is isolated from the outside environment by walls, floors and roofs, and operates in a regulated environment. It may use natural raw materials, or it may discharge its waste to the outside world, but it functions largely independently of nature. In contrast, a farm depends on soil, rain, sunlight, and atmospheric gases for its productivity. While any or all of these can be artificially supplemented or replaced in certain forms of agricultural practice, agriculture consists fundamentally of the exploitation of these features of the natural environment to produce food and clothing for human beings. In countries where agriculture is well developed, it has largely become the natural environment of the land surface, a fact which is both well recognised and much valued in the visible effects of agriculture as manifested in the landscape of the UK. The effect continues into the less visible aspects of the environment, and as a consequence, water draining from agricultural catchments will always differ from catchments where other forms of land use prevail.

The integration of agriculture and the natural environment provides an interface with water-resource management almost as extensive as the area of agricultural land itself. The practice of agriculture cannot be isolated from that of water-resource management. The situation of each farm is different, varying with farm enterprise and management, soil type and underlying geology. The bulk of water draining from agricultural land does so naturally in a totally uncontrolled and uncontrollable manner, in contrast to typical industrial circumstances. Comprehensive management of the water resources as it leaves a particular production unit may be a realisable objective in the industrial context. In the agricultural context, it could only be achieved by full control of agricultural practices. The *relatively* small number of point sources from which discharges from agricultural buildings occur are susceptible to conventional control, but an entirely different concept is required in the context of agricultural land.

It is therefore the author's submission that agriculture is a special case and that it must be recognised as such in considering pollution in the diffuse sense. This thinking is embodied in Part 2 of the Control of Pollution Act 1974 (which has entered force only recently) by the provision of a special defence of good agricultural practice linked to the publication of a statutory code by the Minister of Agriculture. Thus, a new course is already taken in considering the control of pollution from agriculture. Although the approach itself was conceived and incorporated in to legislation more than eleven years ago, it is hard to take a view on whether it will provide a satisfactory system for the future until there is experience of the operation of this approach in practice.

22.2 THE COMPETING CLAIMS OF WATER RESOURCES AND AGRICULTURE

Where the unrestricted practice of agriculture presents difficulties in terms of quality for water resources, a policy must be evolved to resolve the conflict. The options available fall in to four categories:

(a) no environmental change tolerated;
(b) no environmental pollution (change with unwanted effects) tolerated;
(c) limited environmental pollution tolerated;
(d) virtually unlimited environmental pollution tolerated.

There is an important distinction to be recognised between categories (a) and (b). This is set out in the Tenth Report of the Royal Commission on Environmental Pollution (RCEP) (1984) which finds the following a useful definition of pollution:

> The introduction by man into the environment of substances or energy liable to cause hazards to human health, harm to living resources and ecological systems, damage to structures or amenity, or interference with legitimate uses of the environment.

The Commission then goes on to add:

> It is implicit in the above definition that pollution is not simply the presence in the environment of an alien substance or other unnatural disturbance; there must also be an unwanted effect. Substances introduced into the environment become pollutants only when their distribution, concentration or physical behaviour are such as to have undesirable or deleterious consequences. For many substances whether a particular discharge or emission is considered to be pollution depends not only on the nature of the substance, but also on the circumstances in which it occurs, and often on the attitude of people affected and on value judgements.

An important question to consider here is who acts as proxy for society as a whole in reaching decisions on what constitutes pollution, and the answer will usually lie with the elected government of the day, or in practice with officials in government or other departments.

In the early days of the industrial revolution, little effective control on water pollution existed and the policy operated therefore fell mainly in to category (d). Since that time, the control of pollution has received more attention. However, pollution still exists and is tolerated or 'controlled' via trade-effluent consents for industrial or commercial discharges, and for releases from public sewage-treatment works. Therefore, the present policy falls within category (c). Category (a) is not an available option in the UK since some change in the environment in general and water quality in particular inevitably accompanies significant agricultural and other human activity. The scope of debate is therefore restricted to categories (b) and (c). While present policy falls in to category (c), the questions raised under the broad subject of agricultural pollution may range rather wider. In certain respects, agriculture is failing to comply with the present law. The records of serious pollution incidents arising from point sources within agriculture, such as silage liquor or slurry, testify to this. However, even if all these problems were to be resolved overnight and agriculture were to comply with the present water pollution law as it is currently perceived, important questions would remain, such as the problem of nitrates. As is indicated later in this chapter, it is possible that nitrates have no significantly damaging effects. Assuming for the moment that this was to be the case, a logical policy would not require them to be controlled under either categories (b) or (c). Yet there is pressure for their release to be restricted regardless of whether they have harmful effects. Such pressure cannot be justified if pollution-control policy is to remain in category (c). It would be a large step indeed from the present overall situation to aim for a target set by a policy in category (b).

However, taking now the opposite assumption in the case of nitrates, namely that significant damage does accrue from the enhanced level which they now attain in agricultural drainage water, the question which would require to be answered would be whether the damage experienced exceeded that deemed acceptable by the current UK policy (category (c) policy). If the damage experienced related to human health, it would be likely to be deemed unacceptable and steps would be taken to control either the pollution or its effects. When the RCEP considered this matter in its Seventh Report (1979), it attempted to balance the respective costs to the agricultural and water industries of attempting to reduce nitrate content of water prior to it leaving agricultural land, and reducing the nitrate content of water at the point where it enters public supply. Whilst the calculations made by the Royal Commission may now be out of date, the principle of clearly identifying the costs attributing to any chosen policy before a decision is taken is to be heartily commended.

22.3 AGRICULTURE — THE INDUSTRY

In developing appraisals of the impact of policy options on agriculture, it may be helpful to have some feeling for the position of the industry in the economy and life of the nation. For many years, agriculture was the nation's largest industry, although it has recently been eclipsed by the North Sea oil industry. However, as oil production declines, agriculture may regain its former position.

At the behest of successive governments since World War II, the industry has increased its production from less than 50% to 80% of national requirements of temperate foods, and even now it is only in certain commodities that surpluses exist. The effect of this increase in production (from the nation's finite land resources) has benefited Great Britain's balance of payments to the extent of many billions of dollars. At the same time, real prices of foodstuffs to the consumer have fallen steadily and the diminishing real returns to farmers have forced amalgamations of farms on a massive scale, together with substantial investment in plant and machinery to reduce labour requirements and unit costs. Even today, the annual increase in food prices to the consumer is still below the rate of inflation, and increases to farmers are smaller still. The farmer has, therefore, had to run quite fast for many years in order to retain his relative position, and to sprint in order to improve himself. It is important to appreciate that the move to larger farms, increased mechanisation, and the loss of jobs from the countryside has been achieved at the behest of the Government and under the pressure of diminishing real prices from farm profits over the past three decades or more. It is only with the most recent change of UK Minister of Agriculture that government exhortations to increase production have been suddenly muted. The industry is aware that change is upon it, but has yet to be told what society now requires of it. Farmers are understandably anxious about their future livelihoods, and such investment as is taking place is generally orientated towards cutting production costs for the future. Against this background, any pressure for additional restrictions on agricultural practices is unlikely to receive a warm welcome.

Despite its economic uncertainty, the industry has to its credit the creation and maintenance of the British countryside. It is vital to the rural economy of many areas of the country. In the present interregnum while agricultural policy is reshaped, it would be helpful if the industry could be reassured that essential environmental objectives will be encompassed by the new policy, and farming will not therefore be receiving conflicting messages as to the demands society wishes to make of it, nor placed at a disadvantage with respect to direct competitors in Europe. The industry has responded to society's needs in the past and has the ability to do so again if it is given clear guidance as to the demands made of it, and policy instruments used to direct it are shaped accordingly. Chief amongst these policy instruments is the price-support mechanism by which virtually the whole of agriculture is regulated in common with most developed countries. Historically, this has been geared to increasing output. However, in the future, society may wish to direct agriculture in directions other than towards output and may therefore wish to provide support other than through product prices. Water quality may well be a legitimate consideration in determining policy in this area.

22.4 CURRENT PROBLEMS

The most apparent agricultural pollution problems experienced by the water industry are those recorded in the annual statistics collected by the Ministry of Agriculture, Fisheries & Food from the ten regional water authorities in

England and Wales. Figures for 1974 to 1984 are set out in Table 22.1. From these annual totals, it appears that the scale of farm pollution has increased from 61 incidents in 1974 to 2961 in 1984, a rise of such proportions that it is hard to accept the figures as an objective assessment. Changes within the agricultural industry cannot explain such a rapid increase, and it appears that other factors are tending to distort the number of incidents reported. When it is remembered that it was in 1974 that the water authorities were first established, and that in the early years they were largely preoccupied with industrial as opposed to agricultural pollution, it is possible to understand how the figures shown might have arisen. Further examination of the figures for 1984 indicates that these relate to all incidents of farm pollution, and not just serious ones as in previous years. The total for serious incidents in 1984 is probably of the order of 1387, with the breakdown figures correspondingly reduced. However, even in the absence of any reliable element of comparability between one year and another, the figures are probably adequate to indicate certain trends, and it does seem probable that agricultural pollution has increased over the period, but is now tending to stabilise or possibly decline.

Table 22.1 — Total number of serious farm pollutions in England and Wales

	1974	1975	1976	1977	1978	1979	1980	1981	1982	1983	1984
Silage	15	50	69	162	41	189	378	600	322	473	663
Cows	3	94	114	339	72	532	690	790	657	874	1640
Pigs	0	6	6	54	65	27	184	267	205	186	186
Other	43	121	84	397	135	412	413	579	668	150	472
Total	61	271	273	952	313	1160	1665	2236	1852	1683	2961

Source: Regional Water Authorities

The breakdown of the figures between different sectors of the industry is also of interest. When it is considered that silage is largely made for the benefit of cattle, and these figures grouped with those attributable to cows, it can be seen that the dairy industry is implicated to a large extent. A considerable amount of expansion has taken place on dairy units within the period covered by Table 22.1, and there are numerous examples of farms where increases in cow numbers have not been accompanied by increases in waste storage and disposal facilities. However, the introduction of dairy quotas in April 1984 will have had the effect of arresting the tendency to further expansion. It would therefore be expected that since the majority of the increase in incidents in the past five years appears to be attributable on the basis of the water authority figures to the dairy industry, it should be possible to begin to reverse the trend in pollution incidents over the coming years. Silage itself may pose a special problem. The economic pressure on the profitability of many dairy units may tend to push farmers towards feeding greater quantities of home-grown forage, and particularly silage, rather than foodstuffs imported on to the farm. If increased quantities of silage are to be made, clearly the potential to increase pollution will exist. Additional to this is the possibility that some producers may attempt to handle more silage with their existing facilities, again tending to increase the risk of pollution. As might be expected from the distribution of the dairy industry, most of the problems experienced are in the west of the country, in the areas of the South-West, Wessex, Welsh and North West Water Authorities.

In contrast to the increasing figures for incidents arising from cows and silage shown in Table 22.1, it can be seen that incidents from pigs and other causes have remained relatively constant since about 1980. This tends to suggest that recording of incidents by water authorities may now be more reliable and that the position in the dairy industry has deteriorated over the past five years. While this is only one possible interpretation of figures whose accuracy must be suspect, it is believed to be a realistic interpretation giving some hope for the future.

If the pollution figures for 1984 are broken down into more detailed categories (Table 22.2), it can be clearly seen that the newer management practices of handling animal manures as liquid slurry (rather than solid straw-based farmyard manure) and feeding silage instead of hay are giving rise to a high proportion of the incidents. It would therefore appear that while the farming community has been quick to appreciate the economic benefits of switching to more modern management practices, it has tended not to appreciate the increased potential for pollution which these practices possess. This manifests itself partly through the provision of inadequate handling facilities for silage and slurry, and partly in poor management even where facilities are adequate. Traditional practices have tended to give rise to fewer problems, reflecting both the relatively innocuous nature of the products handled and the thorough understanding farmers have of the correct treatment of the materials. Only when an entire generation has grown up with the modern methods is such a thorough understanding of the new problems likely to be achieved. Even then, greater demands on management ability may still leave increased scope for error.

Solutions to the continuing problem of chronic pollution from agriculture are not easy to come by. The fragmented nature of the industry, together with the scale of the problem poses great difficulties. The key to the situation lies with education of the individual farmer. With such a high proportion of young farmers passing through formal agricultural training, the potential for educating new farmers entering the industry is present and can be used to provide a thorough understanding of the facilities required on a given site and the correct methods of management. Just as the problem of education of a changing 'stock' of farmers can most easily be tackled through the new intake, so the problem of inadequately designed facilities must also be understood. Once the need for properly-designed facilities is realised, it is not particularly expensive to design adequate pollution-control features in to a new installation. However, modification of existing facilities is a costly exercise and may stretch the resources of some farms to breaking point. It may therefore be necessary to provide substantial grant aid in order to bring

Table 22.2 — Summary of farm pollution incidents for 1984 in England and Wales

Cause of pollution	Total incidents		Serious incidents	
Slurry stores				
Inadequate storage	270		166	
Burst/leaking store	189	656	75	320
Poor management	197		79	
Runoff from solid manure stores				
Inadequate storage	58	122	8	26
Poor management	64		18	
Silage effluent				
Inadequate storage	223		165	
Leaking silo	126	573	89	359
Leaking storage tank	72		35	
Poor management	152		70	
Runoff from land	149		77	
Yard water				
Cattle	630		328	
Pigs	156	817	100	442
Poultry	31		14	
Failure of treatment system	70		36	
Sheep dip	16		13	
Mineral fertilizers	6		1	
Pesticides	50		27	
Vegetable washings	19		7	
Oil spillages	89		15	
Miscellaneous	261		64	
Total	2828	1387		

Source: MAFF Farm Waste Unit

existing units up to a satisfactory standard. A major defect with the current agricultural grant aid scheme is that prior approval of proposed works is no longer required. This results in the Agricultural Development and Advisory Service (ADAS) having a reduced impact on the design of schemes, leading to the subsequent refusal of grant aid in some cases. In the author's view, this represents a poor use of resources. It would be possible to ensure both that grant was not refused, and that adequate facilities were installed if part of the resources themselves were devoted to a suitably modest form of prior approval of farmers' plans for new installations.

A further avenue which might usefully be explored is the provision of better co-ordinated advice to farmers by ADAS and water authorities. At present, there is a major division of responsibility between the ADAS officer, whose prime duty is to the farmer and whose organisation handles grant-aid matters, and the water authority officer whose chief responsibility is to the water authority and who is charged with controlling any pollution which may arise from farm works. There is a need to bridge the gap between these two groups of officers with individuals able to command the respect and trust of farmers, who possess an understanding and knowledge of agriculture, and with a full knowledge of grant-aid matters. It is pleasing to note that both Welsh Water and Wessex Water have already appointed farm liaison officers in an attempt to fill this need.

22.5 NITRATES

In recent years, concern has been felt about rising levels of nitrates in both rivers and groundwater. It appears that about 75% of this nitrate originates from agricultural land, and that concentrations tend to be higher in the south and east of England than in the north and west. Concern has been expressed, particularly in respect of possible human health hazards, namely methaemoglobinaemia (blue baby disease) and stomach cancer. Nitrates in water have been positively implicated in blue baby disease, but recognition of the problem has lead to its avoidance in the UK, where there have been no cases whatsoever since 1972 in spite of the high levels of nitrate reached following the 1976 drought. Where it has occurred, it has been mainly in bottle-fed infants and in association with other factors such as low stomach acidity and high bacterial populations. So far as the link with stomach cancer is concerned, this is at the present time hypothetical. It is interesting to note that stomach-cancer rates are falling in all developed countries, and that within the UK, stomach-cancer rates are lowest in East Anglia where nitrate levels in drinking water are highest. Thus, at worst, the impact of nitrates is subsidiary to other factors which are leading to an overall decline. However, recent research results suggest that the probability of a significant effect from nitrates is small (Forman *et al.*, 1985).

There are also some indications that problems may be caused for nature conservation in certain situations by nitrate release, although the Royal Commission on Environmental Pollution (1979) considered that this was not a major difficulty wthin the UK.

Nitrate release from agricultural land is sometimes attributed to direct runoff or leaching of fertilizer nitrogen. The evidence indicates that this is not the case, and work carried out at Letcombe Laboratory (Dowdell & Webster, 1980) using ^{15}N as a tracer has demonstrated that the actual loss of fertilizer nitrogen within four years of application is quite small. This is supported by other research findings. The main release of nitrates takes place in the autumn and winter, and it is now understood (Barraclough *et al.*, 1984) that this consists largely of nitrogen mineralised from the soil's organic content (which can typically contain 7000 kg N/ha). This has important implications for approaches to the problem, and demands a fundamental examination of soil processes under agricultural management.

One of the main effects of agriculture in managing the land for the production of food lies in an increased nutrient flux. This derives from both the addition of nutrients by man to replace those removed in products taken from the land, and in the liberation of nutrients from the organic content of the soil through oxidation following tillage operations. As a consequence of its solubility, nitrogen in the form of nitrate tends to be leached from the topsoil in water, while in the form of ammonia it is lost to the atmosphere as gas. Virtually any form of agriculture is likely to increase the quantities of nitrogen lost from the soil, although it is only in the last decade or so that concentrations of nitrate both in rivers and groundwater have started to cause concern.

The problem appears to be fundamentally an interaction between rainfall pattern and the quantities of mineralised nitrogen present in the soil. Under UK conditions, there is normally a soil-moisture deficit in agricultural land during the summer, and as a consequence little rainfall leaches through the soil and is lost as drainage water. However, with the onset of autumn and winter, soil-moisture deficits disappear, and excess rainfall is carried into rivers and aquifers. Where excess winter rainfall is high, nitrate tends to be highly diluted, and the concentration in water supplies does not approach limits currently in operation. However, in areas where excess winter rainfall is less, the same quantity of nitrate may be less diluted and thus achieving concentrations closer to tolerable limits. It appears that the total quantity of nitrate carried by rivers is relatively similar when considered in relation to the total agricultural area of the catchment in both the east and west of England. The other factor governing the amount of nitrate leached is the quantity of nitrogen mineralised to nitrate which is available in the soil in autumn and winter. Under arable farming systems producing crops such as cereals, there is very little uptake of nitrate by crops after June. However, mineralisation of nitrogen from the soil's organic reserve remains high in the late summer under the influence of soil temperature, and a substantial accumulation may be present when excess rainfall begins to drain from the land.

Measures intended to limit the application of nitrogenous fertilizers to agricultural land have been suggested as a means of tackling the increasing quantities of nitrate being leached. It appears highly unlikely that such measures could be effective due to the role played by the reserve of nitrogen contained within the soil's organic matter. This provides a buffer of immense proportions, the depletion of which could not be countenanced as it would imply destruction of soil fertility. An interesting experiment has been conducted at Rothamsted Research Station where a field has remained unfertilized for over 100 years. During this time, about 2500 kg N/ha have been lost through leaching, and this loss continues at a rate of about 15 kg/ha yr. A further reason to question the effect of limitation on inputs of nitrogen fertilizer is the likely response of farmers to such a situation. It is not unlikely that they would turn to other sources of nitrogen, such as leguminous crops, to optimise their position. The release of nitrogen from a previous legume crop might well pose a greater problem for water resources than conventional nitrogenous fertilizers. Such fertilizers provide readily-available nitrogen in a large flush which can be timed to coincide with the main growth period of a crop. The release of nitrogen from the breakdown of legumes (or most other organic matter) would inevitably be more prolonged, and therefore less efficient and likely to lead to a greater leakage from the system.

It must be recognised that agriculture is now operating at a higher intensity of both inputs and outputs than has been the case in the past. This 'high pressure' system inevitably imposes greater strains on a situation which is inherently leaky. As knowledge of soil processes grows, it is likely that forms of agricultural practice which tend to retain nitrogen in the soil will be better identified, and increasingly introduced for the benefit of both the agricultural industry and water resources. Such practices may include the following.

- *Direct drilling/reduced cultivation*
  Letcombe Laboratory and British Geological Survey research have shown that nitrate leaching from direct-drilled crops may be only half of that from conventional deep ploughing.
- *Winter cereals*
  The sowing of winter cereals provides a growing crop uptake for some of the residual nitrate in the soil profile in the autumn and can reduce leaching losses.
- *Early planting of winter cereals*
  Sowing winter cereals as early as possible encourages yet more of the residual nitrate in the soil to overwinter in the crop, where it will not be vulnerable to leaching loss.
- *Avoiding unnecessary autumn seed bed nitrogen*
  Applications of nitrogen in the autumn may be agronomically desirable in some situations, but care must be taken to ensure that they are only used in appropriate circumstances.

All these steps relate to positive soil and crop management in order to reduce the availability of nitrogen for leaching. Such steps have the potential to make a significant contribution to reducing nitrate losses from agricultural land, and they are likely to be more acceptable and perhaps more effective than crude attempts to limit nitrogen inputs.

It must be remembered that much of the current popular interest in nitrates stems from the imminent implementation in the UK of EEC legislation regulating the nitrate content of drinking water. The level to be introduced, 50 mg/l, is half that presently operated in the UK. Attempts to identify the basis on which the EEC level was determined have revealed no scientific basis for it. Medical advice to British Ministers is that 100 mg/l is safe, and in this context the farming industry finds it hard to understand why its livelihood should be threatened by this suggestion of harsh measures for which there is no established need. Until a rational case can be demonstrated, farmers may tend to feel victimised as a result of pressure by ill-informed protagonists of draconian measures. In the meantime, it is recognised that prudent contingency plans should be drawn up.

22.6 CONCLUSIONS

As one of the country's largest industries, it is not surprising that some pollution problems should stem from agriculture. It appears that there has been little progress in combatting acute pollution problems in recent years, and that greater efforts will be needed in the future. The example of nitrates has been considered in the realm of chronic pollution, and the close relationship of the industry with the natural environment in the form of soil processes identified.

In tackling these agricultural problems, two essential points stand out. Firstly, agriculture is an industry which is effectively managed by state institutions, and policies need to be designed for the direction of the industry which better integrate society's diverse demands than has been the case in the past. Secondly, there are nearly 250 000 agricultural holdings in England and Wales alone, representing a very wide diversity of situations from agri-business farms covering thousands of hectares, to subsistence small-holdings. Such an industry cannot be effectively regulated in the absence of a reasonable need to do so which can achieve a degree of consensus between the regulators and the regulated.

REFERENCES

Barraclough, D., Geers, E. L. & Maggs, J. M. (1984) Fate of fertiliser nitrogen applied to grassland. 2. Nitrogen 15 leaching results. *J. Soil Sci.*, **35,** 191–199.

Dowdell, R. J. & Webster, C. P. (1980) *Lysimeter Studies of the Fate of Fertiliser Nitrogen in a Shallow Arable Soil Overyling Chalk.* ARC Letcombe Laboratory Report, pp. 50–51.

Forman, D., Al-Dabbagh, S. & Doll, R. (1985) Nitrates, Nitrites and Gastric Cancer in Great Britain. *Nature,* **313,** 620–625.

Royal Commission on Environmental Pollution (1979) Seventh Report. *Agriculture and Pollution,* HMSO.

Royal Commission on Environmental Pollution (1984) Tenth Report. *Tackling Pollution — Experience and Prospects.* HMSO.

23

Regulatory remedies for agricultural pollution

P. Worthington

23.1 INTRODUCTION

The resources of the environment form the basis of and constitute the limits to further economic and social develoment. There is a need to integrate protection of the environment in to the planning and development of certain economic activities.

Farming is an activity which is perhaps the most basic to human needs and one where man's accord with his environment has best been preserved. It is the urban authorities which look to the countryside for sources of clean water, at some distance away from their own polluted runoff. However, by virtue of the vastness of the catchment of agricultural land the relatively-small amounts of chemicals used in its husbandry add up to considerable amounts, drawing attention to the significance of the use of land as an important aspect of environmental protection.

The interrelations between agricultural activities and the environment are complex and involve various disciplines, nevertheless there are cases in which environmental needs will involve the slowing down or reduction of agricultural growth or the rapid development of cleaner practices.

23.2 THE PROBLEMS

The use of chemicals in agriculture does not always result in water-pollution problems. Hence the obvious first step when considering pollution-control possibilities is to identify the type, magnitude, frequency, and source of the problem. In a river basin many diffuse or non-point sources contribute nutrients and other potential contaminants to surface waters. The relative contribution from various sources is important in determining the types of controls that are needed and the locations to which those controls should be applied.

Potential pollutants from agriculture result from leaching, surface runoff, erosion and discharges of animal wastes. The relative importance of these contributions varies considerably and depends upon the type of agricultural operation and management, and site-specific geographical characteristics.

Many of the more apparent risks to the aqueous environment from farming activites are caused by occurrences such as accidental spillages, seepages, overflows, most of which could be removed by a tighter control of the siting and construction of chemical storage areas and vessels, silage units, manure bays, and so on, and by persuading or encouraging farmers to deal responsibly with manures and wastes. There is generally no need to reinforce prevention of pollution legislation which would deem the accidental discharges unlawful. Indeed most direct discharges of agricultural effluents are covered by the same legislation which exists for industrial and urban discharges.

Therefore the more pressing of environmental problems in land use arise from the dispersed use of chemicals, especially nutrients which percolate, leach or are washed out into aquifers or watercourses. Levels of nitrate in aquifers are reported to be causing problems in specific areas in almost every Member State in the Community. The increased usage of fertilizers and over-concentration in the intensive-livestock sector have been identified as contributory causes. It is generally true that the greater the nitrogen applied, surplus to crop uptake, the higher the risk of contamination of water resources.

The use of pesticides in agriculture creates problems which have been recognised for many years by the intervention of the public authorities to control the sale and use of such products. Community measures already exist in this field; the use of certain products known to be dangerous is now prohibited by Council Directive 79/117, and conditions are prescribed for classification, packaging and labelling (Council Directive 78/631).

23.3 TYPE OF ACTION REQUIRED

Farming covers large areas in the Community — about 60% of its surface area with some five million farms. Community farmers use about 25% of all the pesticides and about 17% of the fertilizers consumed on a world basis. Farming in Europe has become increasingly a science- and technology-intensive industry. In October 1984 the Commission, together with the Irish Authorities, held a symposium in Dublin where scientific and agricultural experts discussed environmental problems caused by the use of agrochemicals. This led to useful conclusions which can, we hope, lay the foundations for Community initiatives and for co-operation on a European level in tackling the key problems on which action needs to be taken.

As land is a very limited natural resource, its use is subject to careful planning. To a large extent land-use planning decisions determine the future quality of the environment, often for a considerable number of years. Not only, therefore, is action required at the level of individual farming activities, but also on the level of large-scale agricultural planning.

A strategy in keeping with diffuse sources must therefore be formulated and implemented. Action is required concerning both individual farms and large-scale agricultural planning procedures.

23.4 POSSIBLE REMEDIES

The use of pesticides is thought to be necessary to maintain crop yields and thus to keep down costs of agricultural production and the price of food. There is no doubt these chemicals have played a big role in the increase in agricultural productivity in recent years. On the other hand, pesticides are often highly-toxic chemicals, released directly into the environment which may pose serious risks especially with regard to their long-term effects.

Correct timing and dosage of pesticides will ensure maximum uptake by plants and minimum losses to the soil. This is primarily a question of sound environmentally-sensitive advice to the farmer.

The role of the agricultural advisory and extension services in providing advice to farmers on the safe handling, rational use and, where appropriate, proper disposal of pesticides cannot be overemphasised. Together with all those involved in the marketing and distribution of pesticides to farmers, they have a duty to ensure that farmers are conscious of their responsibilities, have access to the necessary advice, and equip themselves to apply pesticides correctly without unnecessary adverse effects on the environment.

Prevention is better than cure; a progressive reduction of pesticide usage to a minimum consistent with efficient agricultural production is desirable. This could be achieved by a strategy designed to favour, where possible, the use of environmentally less-harmful products, the elimination of excessive usage, as well as the development and application of other methods of control of pests involving less reliance on chemicals.

There is also a clear need to reduce the total use of nutrient-supplying agrochemicals and to ensure that when and where they are used they are applied in quantities and concentrations just sufficient for the crop requirements and at the correct times when the desired results may be achieved, and not when they cannot.

Thus measures are required which will:

- advise farmers on product choice and methods and equipment for applying fertilizers. The responsibility for information rests, on the one hand with manufacturers and suppliers of fertilizers (obligations to provide instructions for use), and on the other hand with the agricultural advisory services provided by the public authorities in Member States. A

special effort in the dissemination of information and in education is desirable in the case of fertilizers;
- monitor more closely the methods and equipment used to apply agrochemicals;
- subject intensive-stock rearing to environmental impact assessment procedures and, where necessary, reclassification as an industrial activity with associated responsibilities towards waste and effluent disposal;
- ensure that environmental protection forms part of agricultural development programmes for specific regions;
- pay special attention to sensitive areas of the environment by ensuring that activities are compatible with the environmental objectives for these zones, and by encouraging the development of farming techniques which take account of these objectives;
- in relation to forests, establish better information on the possible effects on the environment of intensive production and the latest methods in forestry management.

Legal action to control and reduce water pollution is provided at Community level by directives (and a decision) which are administered within each Member State by individual national legislation. At the present time most of the obvious sectors of water pollution at point discharges are covered by Community legislation.

Point sources are sources which can be related to specific points of discharge, such as industrial effluents piped to watercourses and effluents from sewage treatment works.

Non-point sources are diffuse in nature and spread over a large area. Examples are dry and wet atmospheric fallout, land runoff or leachate, erosion, and corrosion. Domestic sewage, because of the diffuse origin of its constituents, is often regarded as a diffuse source although the sewerage-collection system inevitably leads to a point discharge.

Non-point sources are difficult to control if control is defined as the ability to establish and enforce effluent-discharge standards. However, many non-point sources are controllable if appropriate management practices are utilised. More emphasis for non-point source control should perhaps be on 'best management practices' rather than on the collection, treatment and effluent standards approach used for control of point sources, although the latter may sometimes offer the most suitable control, for example in intensive stock-rearing. Enforcement of good practice in special zones might be the subject of possible future legislation but would have to encompass other important fields in addition to agriculture such as industry, transport, regional policy, and so on.

23.5 EXISTING CONTROL

Up to now 12 European Community (EC) directives on the protection of the aquatic environment have been adopted. Four directives contain standards for surface waters according to use (drinking water, bathing water, freshwater fish, shellfish (Council Directives, 1975a, 1975b, 1978 & 1979 respectively). These directives lay down quality objectives for particular uses of water.

The 'dangerous substances directive' (Council Directive, 1976) is an example of control applied to specific substances by aiming for the elimination of pollution caused by List I substances and for the reduction of pollution caused by List II substances. The directive is a framework directive; four subsidiary directives regulate List I substances, two for mercury (Council Directives 1982 & 1984), one for cadmium (Council Directive, 1983) and one for hexachlorocyclohexane — (Council Directive, 1984a).

Apart from mercury and cadmium, List I refers to families of groups of substances, rather than to individual substances identified on the basis of toxicity, persistence, and bioaccumulation.

Neither List I nor List II contain nitrate although List II refers to ammonia in the context of its oxygen-demanding properties. However, the 'groundwater directive' (Council Directive, 1981) includes ammonia in List II.

Where materials containing List II substances are spread on land for disposal or tipping for the purposes of disposal, the 'groundwater directive' requires that there should be prior investigation and, provided all technical precautions for preventing pollution by these substances are observed, may grant authorisation. Hence a partial means of control of over-application of organic manures already exists. Control of other activities is left to the discretion of Member States.

At the moment there is some effort to limit by regulation inputs of nitrogen to soils in areas of high levels of nitrate in underground waters in Germany, The Netherlands, and Denmark.

In lower Saxony, Germany, inputs of manure per hectare of farmland are already restricted by local regulations (called 'Gülleerlass') (Runderlass, 1983). In November 1984 the Netherlands government prohibited new construction or enlargement of buildings in the southern provinces where the density of production is high; this prohibition is temporary (until January 1986) until measures to counter the problem have been found. In December 1984 the Danish government presented to Parliament a plan under which the storage and disposal of animal waste would be controlled.

In England and Wales new powers under Part 2 of the Control of Pollution Act (1974) (COPA2), commenced in mid-1984, which gave water authorities, central government, the courts and others, additional powers to control effluents discharged and activities carried out in the course of agriculture.

Under the new provisions it is an offence to allow polluting material to enter water. As long as good agricultural practice, as recommended in a Code (Code of Good Agricultural Practice, 1985), was followed, no offence would arise.

This Code, published and approved by the Minister of Agriculture, Fisheries and Food, and the Secretary of State for Wales, recommends that organic manures are applied at rates which match the capacity of the crop to utilise the nutrient as far as practicable and at times which are optimal for plant usage of the nutrient content.

Recommendations for inorganic fertilizer allow rates which produce optimum plant yields, guidance for each crop being given in a separate publication — RB209 (1985). However, under COPA2 water authorities and the government have powers to prohibit certain agricultural practices in particular places where they would be harmful, for example application of slurry on land near to a borehole or overlying a shallow aquifer.

23.6 OTHER POSSIBLE DEVELOPMENTS

In the European Community the Council of Ministers (Agriculture) in the course of its discussions in 1984 on the new agricultural structures policy examined a number of questions relating to the protection of the environment, in particular the provision for Member States to pay national aids in environmentally-sensitive zones, to maintain incomes of farmers who undertake to conserve or improve the environment in certain ways. This measure was finally adopted by the Council in March 1985 as Article 19 of Regulation 797/85 (Council Regulations, 1985); in this context, the Commission declared that it considered the measure to be a provisional one and undertook to present by the end of 1985 proposals for a common system.

In announcing in January 1985 its intention to examine the prospects for the common agricultural policy, the Commission stated that one of the aims should be 'the creation of a modern and efficient agriculture which continues to exploit its potential to improve productivity in the interests both of farmers and consumers but which, at the same time, respects the environment and conserves the heritage and landscape and species of Europe'.

Some commentators have criticised the levels of price support accorded to certain agricultural products, on the grounds that the expansion and intensification of production has been detrimental to habitats and species. Others have argued that price support has helped to maintain the traditional pattern of farming in Europe, including the family farm.

It is worth pointing out that the level of support for the intensive animal production sector, (that is pigs, poultry, eggs, veal) has been much lower than for other sectors. The bias away from these sectors is repeated in the structured part of the policy. Here investment aid for the production of poultry, eggs, and veal, has been banned since 1972 (Directive 72/159/EEC). Aid for pig production has been limited to smaller units and conditioned on the availability of sufficient land for the safe absorption of the pig slurry.

Aid is now also restricted to smaller herds in the dairy sector and, of course, dairying itself is being contained by the milk quota system where Community farmers are being encouraged to reduce milk supplies. In the recent EEC Regulation 797/85 of 12 March 1985 on improving the efficiency of agricultural structures, aids for investments required to protect the environment by Member States are authorised (Article 19 refers). As an example a condition for receiving investment aid for cattle housing would be that adequate storage and facilities are provided for dealing with the resulting cattle slurry. These waste arrangements would be considered part of the investment and aided accordingly.

At the same time the Commission is continuing with its 'prudent' price policy where price increases for farm produce would continue to be well below the level of inflation. All of these limitations will tend to reduce the level and intensity of farm production.

When a deeper involvement of environmental factors in land-use planning has been achieved a more direct action to protect the aqueous environment may then be considered.

23.7 CONCLUSIONS

Agriculture has begun to be recognised as an important potential source of water pollution. There is sufficient information to identify regulatory measures which can be used to mitigate the effect that agriculture might have. Principally they are:

- the establishment of a code of agricultural practice, which would incorporate environmental protection measures, to form the basis of advice to farmers;
- a more rigid enforcement of a code of agricultural practice in zones of special risk to environmental resources;
- guidelines on fertilizer usage which minimise loss to the environment whilst ensuring economic crop yield;
- establishment of a means of communicating and issuing advice to agriculturalists on the proper control of the application of agrochemicals and on certain practices which minimise environmental disturbance;
- reclassification as industrial wastes, wastes produced by intensive farming activities;
- incorporation of environmental considerations into long-term agricultural planning procedures.

Despite the title and content of this chapter, it would be wrong to conclude that problems caused by land use may only be solved by agricultural policy and regulatory legislation. Much remains to be done in water management by river basin authorities and agricultural advisers themselves.

REFERENCES

Code of Good Agricultural Practice (1985) Ministry of Agriculture, Fisheries and Food, Alnwick, Northumberland.

Control of Pollution Act 1974. Her Majesty's Stationery Office, London.

Council Directive of 16 June 1975a concerning the quality required of surface water intended for the abstraction of drinking water in Member

States. *Official Journal of the European Communities* No. L 194, 25. 7.75, p. 26.

Council Directive of 8 December 1975b concerning the quality of bathing water *Ibid*, No. L 31, 5.2.76, p. 1.

Council Directive of 4 May 1976 on pollution caused by certain dangerous substances discharged into the aquatic environment of the Community, *Ibid,* No. L 129, 18.5.76, p.23.

Council Directive of 18 July 1978 on the quality of fresh waters needing protection or improvement in order to support fish life. *Ibid*, No. L 222, 14.8.78, p. 1.

Council Directive of 30 October 1979 on the quality required of shellfish waters, *Ibid*, No. L 281, 10.11.79, p. 47.

Council Directive of 19 December 1981 on the protection of groundwater against pollution caused by certain dangerous substances. *Ibid*, No. L. 20, 26.1.80, p. 43.

Council Directive of 22 March 1982 on limit values and quality objectives for mercury discharges by the chlor-alkali electrolysis industry. *Ibid*, No. L 81, 27.3.82, p. 29.

Council Directive of 26 September 1983 on limit values and quality objectives for cadmium discharges. *Ibid*, No. L 291, 24.10.83, p. 1.

Council Directive of 8 March 1984 on limit values and quality objectives for mercury discharges other than the chlor-alkali electrolysis industry. *Ibid*, No. L 74, 17.3.84, p. 49.

Council Directive of 9 October 1984a on limit values and quality objectives for discharges of hexachlorocyclohexane. *Ibid*, No. L 274, 17.10.84, p. 11.

Council Regulations of 12 March 1985 on improving the efficiency of agricultural structures. *Ibid*, No. L 93 of 30.3.1985, p. 1.

RB209 *Fertilizer Recommendations 1985–86* (ISBN 01 1242709 X) HMSO London.

Runderlass Vollzug des §15, Abs. des Abfallbeseitigungsgesetzes; hier: Massnahmen gegen die Überdüngung mit Gülle und Geflügelkot RdErl. d. ML. of 13.4.1983 — 318.2 — 03 14 55, in: Niedersächsisches Ministerialblatt No. 22/1983 pp. 437–439.

Discussion

Chairmen: **Professor R. W. EDWARDS**, Deputy Chairman, Welsh Water
S. C. AGNEW, Chief Engineer, Scottish Development Department

During the four discussion sessions on agriculture, which have been summarised together here, Irrigation, Fish Farming, Land Drainage, Nitrogen, Pesticides, Organic Wastes, and the Control of Agricultural Activities were discussed, but the last of these, and Nitrogen, featured most prominently.

IRRIGATION

A number of questions referred to the changing patterns of irrigation, its effects and control.

Trends; effects of low flows

Mr D. Hammerton (Clyde River Purification Board) contested a suggestion by Dr Hardwick that irrigation was unlikely to increase pollution. Abstraction for irrigation must decrease the river flow to some extent and thus the dilution of any pollutants received downstream.

Dr Hardwick agreed that this could be so and stressed the need for further study of both sides of the equation — the quantity of pollutants and the quantity of water. He hoped that Welsh Water and the English Water Authorities would be able to make use of the Control of Pollution Act to tackle the problems of abstraction as well as those of pollution.

Mr Hammerton continued that he had been made aware of a new trend for farmers to sign contracts with supermarkets for delivery of specific vegetables on pre-arranged dates. In order to meet such deadlines increasing irrigation must surely occur. **Dr Hardwick** confirmed that there were economic forces at work which could promote this trend but he judged that any resultant increase in irrigation would happen slowly. The cost-benefit relationship was so fine that the management of the markets must be equally subtle.

Mr R. Furniss (South West Water) reported that in south west England a number of rivers, primarily in agricultural catchments, were failing their river-quality objectives from time to time. This could partly be due to decreased dilution. Summer base flows in some of these rivers seemed now to be only half what they used to be. Possible causes included increased spray irrigation, land-drainage activity and climatic change.

Legislation

The Scottish River Boards Association, **Mr Hammerton** stated, was very concerned about the lack of control of abstractions in Scotland. The Spray Irrigation Act was so cumbersome, difficult and expensive to apply that it was almost useless. Scotland needed the same powers to control abstraction as those in England and Wales, in order to safeguard the river flows and to make sure that (in the interests of farmers) the available resources were fairly shared. **Mr W. Halcrow** (Forth River Purification Board) endorsed much of the above but said that the Spray Irrigation Act (Scotland), applied by Order, was not entirely useless. It had the virtue of allowing the Board to prohibit abstraction when the minimum acceptable flow was reached. However, to achieve regulation effective legislation was needed and the farmer must pay for his use of the natural resource.

ABSTRACTION FOR FISH FARMING

This subject was raised by **Dr W. R. Howells** (Welsh Water). Dr Hardwick had briefly referred to the degree of control of discharges from fish farms but had not mentioned the complete lack of control of water authorities in England and Wales over the abstraction of water to fish farms where fish were produced for the table. Some of these abstractions could be very significant, affecting major salmon rivers and the authorities' own abstractions. For years water authorities, without help from MAFF, had sought to bring such abstractions under the controls applied elsewhere under the Water Resources Act 1963.

Dr Hardwick appreciated the legal difficulties and agreed that there was a problem. There was a lot of concern over the balance of responsibility, particularly for rivers, between the water authorities and MAFF who have a proper concern also for fish in the rivers and for flood control. The Chief Executives of the water authorities and MAFF were discussing such problems.

LAND DRAINAGE

In answer to a question from **Mr M. Cominetti** (South West Water), **Dr Newbold** said that the river-engineering schemes he had described were all on main rivers and had been initiated by water authorities.

Schemes for reducing need for irrigation

In Dr Newbold's presentation he had put forward a new design for a drainage channel which would require some loss of agricultural land along the edge of the field. **Dr J. Hellawell** (Severn Trent Water) suggested that this loss of land would hardly win the support of farmers, unless there was evidence of a net gain, in the form of increased productivity, and the reduced requirement for irrigation. Was the Nature Conservancy Council to carry out an experimental field trial for example? **Dr Newbold** had no hard evidence for a benefit from the proposed system but made the point that in East Anglia the present shortage of water meant that crops were stressed and yields would fall without irrigation. He suggested that his scheme, with its maintencance of fairly high water tables, was an alternative means for maintaining yields. In the Somerset Levels the idea of bordering river channels with emergent berms was being tried out, but this was in an area of grassland, not arable farming.

NITROGEN

Discussion of the effects of nitrogen in fresh waters ranged widely, from the patterns of usage of fertilizers, the mechanisms of release of nitrogen from soils into surface or ground waters, and the effects of the released material, to the value of present legislation concerning nitrogen.

Dr K. Robinson (North of Scotland School of Agriculture) reminded **Dr Phillips** that the use of ammonium nitrate permitted a two-stage release of nitrogen, remembering that stimulation of the nitrifying micro-flora would be required.

Secondly, in his understanding, organo-nitrogen compounds, not nitrate, were the end-product of nitrogen fixation.

Leaching to surface waters

Acidification

Mr Cominetti asked **Dr Forslund** whether she had found that acid discharges to surface waters were of sufficiently low pH to cause fish kills. In south west England serious fish mortalities had been observed following severe droughts. The cause had been assumed to be the result of sulphuric acid in water draining from fields, but Dr Forslund's findings suggested that this should be re-examined. **Professor Edwards** added that episodes of acid pollution in Wales tended to occur in the autumn too, but these were linked with the presence of aluminium. **Dr Forslund** replied that she did not know whether such fish kills occurred in Denmark. The peak leaching of nitrate (and the lowest pH) occurred during the winter, when the dilution available was at a maximum.

Professor Henin expressed surprise at the amount of acid said to have originated in the decomposition of organic matter, considering the buffering capacity under well-aerated conditions. In waterlogged soil some acidity might be expected. **Dr Forslund** had not, herself, calculated the hydrogen-ion balance relating to carbon dioxide production. Reference to the source of the data was given in Chapter 17.

Denitrification

Following on from this **Professor Edwards** suggested that the extension of land drainage in Britain might have significantly reduced the possibilities for denitrification below the root zone. It was not possible to examine this question in the data from Denmark, **Dr Forsland** explained, because the area suffering acidification and high concentrations of nitrate was a sandy area not requiring land drainage. Where drainage was required, in the more chalky soils, any effect would be chemically buffered. **Mr Young** added that there was little doubt that drainage water contained higher concentrations of nitrate, presumably because there was less opportunity for denitrification.

Professor Henin asked whether **Mr Young** had studied the leaching of nitrogen from sloping land where sub-surface runoff could occur. In such cases the leaching of nitrogen was thought to be very different from the pattern seen in downward movement. **Mr Young** pointed out that although specific studies on this subject had not been carried out, experience with other observations could be utilised. In Triassic Sandstone ammonia (from sewage effluent distributed onto the soil surface) was converted to nitrate almost stoichiometrically during vertical movement. In Chalk, however, (where the organically-rich effluent saturated the Chalk), nitrification occurred at first and denitrification took place later when anaerobic conditions arose. The chance alternation of wet and dry conditions favoured these reactions.

Professor Henin approved the suggestion of Dr Newbold that rivers used extensively for land-drainage should have berms, because these areas would permit further denitrification.

Mobilisation by cultivation

Mr M. B. Green (Severn Trent Water) wondered whether cultivation using processes such as direct drilling which disturbed the soil less then ploughing, would lessen the release of nitrate. **Mr Young** confirmed that in the short term this was probably so, but direct drilling could not be used on a field indefinitely, because conditions for seedling growth would gradually deteriorate in the resulting mat of old roots. Ploughing at that stage would be similar in effect to the ploughing-up of old grassland.

Origins of leached nitrogen

Mr Young expanded on a point made by Mr Payne in presenting Chapter 22, that only a small fraction (6%) of ^{15}N-labelled fertilizer appeared in leachate in Chalk, while 52% entered the crop and about 30% was bound up in the unharvested biomass. The lysimeter studies referred to indicate that the plots also produced quite large amounts of unlabelled nitrogen throughout the experiment. Taking Mr Payne's figure of a reserve of 7 t N/ha in undisturbed soil, it seemed that the turnover was equivalent to about 3% of the reserve.

Mr Payne emphasised the importance of the point. It was essential to know the origins of the nitrogen in the leachate. If the bulk of the nitrogen released from agricultural land came from recently-applied fertilizer then steps directed towards controlling the rate of application of fertilizer could have a major impact. If, however, the nitrate was largely from the soil's organic reserve, then by stopping the addition of fertilizer this reserve would eventually (perhaps over more than a century) be depleted; concentrations in the leachate would fall, but a long time prior to this point productive agriculture would have ceased.

Leaching to ground waters
Prediction of future levels of nitrate
Dr A. Geake (British Geological Survey), while accepting that in the Chalk of southern England solutes tended to move down through the unsaturated zone in a simple non-dispersive manner, said that the situation in eastern England was more complex. In the unsaturated zone of Chalk in Norfolk, below long-term arable land, there was downward movement of course, but the main peaks were held back. This cast doubt on the use of simple models of nitrate leaching losses (to predict future nitrate levels in groundwater). In the top four metres the 1984 nitrate peak was at a lower concentration than the 1976 levels, despite similar applications of fertilizer and the fact that the investigations were carried out at the same hydrological time of year. This difference might be accounted for by complex solute movement, by the predominantly earlier sown winter cereals in recent years, or by improved strains of crops and cultivation techniques. **Mr Young** accepted her comment but pointed out that the model used by WRc did incorporate an optional dispersion coefficient. Although he may have over-simplified the situation the results of water authorities, WRc, and the British Geological Survey all supported the notion of a mass downward movement of nitrate. There were three problems:

- how to estimate the rate of release of nitrate from soils under given agricultural conditions;
- how to model or predict the way in which the solution moves through an aquifer system;
- how to connect the above two problems.

Following on from this point **Mr D. R. H. Price** (Anglian Water) questioned Dr Harwick about predictions of changes in land use, in relation to the quantities of nitrate leached. Were there, for example, any forecasts available of the likely change from permanent grassland to arable? **Dr Hardwick** replied that MAFF had considered the consequences of the milk-quota situation and grain mountains. It was concluded that it will be very difficult to find alternative *farming* enterprises that could be guaranteed to have any long-term viability. Predictions were difficult but he saw no great or rapid change in farming's use of land on a national scale, nor any obvious case for ploughing up grassland, at present. Some farmers might pull out of agriculture altogether and concentrate on tourism or secondary industries.

Possibilities of legislation

In view of the increasing level of nitrate or other undesirable chemicals ground water and the probability that EC standards for potable water would be exceeded, **Mr Cominetti** asked whether a rush to invoke Section 51 of the Control of Pollution Act 1974 was anticipated. (Section 51 enables the Secretary of State to serve a notice requesting abstention from certain agricultural practices.) This might have particular relevance in areas having major aquifers. In **Mr Young's** opinion, invoking Section 51 would normally not be appropriate because of the long lag period between control and observable benefit, but there might be cases where this legislation could usefully protect groundwater quality.

Legislation/guidelines

WHO guidelines

Mr J. I. Waddington (World Health Organization) reported on a WHO Working Group meeting held in 1984 to look again at the draft guidelines on nitrate. The working Group concluded that the nitrate limit of 10 mg/l as N should be retained. For the methaemoglobinaemia problem, the 'blue-baby syndrome', they said that where waters exceeded the limit, alternative supplies of water for infants should be found. With regard to gastric cancer, the Working Group felt that the limit was a wise one, although there was no linkage suspected between nitrate and cancer at concentrations below 20 mg/l as N.

EC Directive on Drinking Water

Mr T. D. George (Fertiliser Manufacturers' Association) supported the acquisition of more data on the effects of nitrate, without which any requirement to reduce concentrations of nitrate to 50 mg/l (as NO_3) could not be challenged in a proper manner. **Dr Hardwick** thought that it was a question of balancing risks and benefits. As concentrations of nitrate rose, obviously there would be an increasing risk, however slight, of methaemoglobinaemia in infants. The question of cancers was far more difficult but presumably, if concentrations rose high enough, there would be a risk. It would therefore be stupid to let concentrations rise unchecked for ever. Possibly the costs of improving UK safety standards were going to be very considerable. He felt that the EEC had been too cautious and subsequently caused a degree of expenditure that they had not foreseen. The MAFF and the Department of the Environment (DoE) were looking at the situation again and trying to evaluate what the costs to agriculture and the water industry might be of the options before them. This was an area where new knowledge would be most helpful.

Costs of control

Mr L. A. Greene (Strathclyde Regional Council) referred to the Royal Commission's 1978 Report on nitrate. The Anglian region had been studied intensively and the Commission had compared the costs of reducing nitrate release by restricting agriculture with the costs of installing nitrate-removal plant at water-treatment works in affected areas. Had there been a more recent analysis taking these options (or the provision of alternative supplies) into consideration? **Dr Hardwick** reported that DoE was funding a study by an agricultural consultant on this very point. There was indeed a pressing need for these figures.

PESTICIDES

The discussion on pesticides and their effects on fresh waters moved quickly from questions about their occurrence and prediction of future levels, to the EEC legislation. The validity of the figures for pesticides in the EC Drinking Water Directive was in serious doubt.

Occurrence; consumption; problems

Dr R. F. Packham (Water Research Centre) referred to the survey carried out by Dr Croll (Chapter 13) and suggested that unlike surface waters, where a wide range of herbicides could generally be expected, this might not be so in groundwaters. Negative findings, or very low concentrations, would also be useful in analysing the behaviour of pesticides. **Dr Croll** confirmed the pattern for surface waters, although pesticides could be rare in rivers in less arable areas. Herbicides were only found in about 12 of 300 groundwaters sampled, occurring where direct infiltration to the aquifer could take place, but not in non-arable areas, or in older waters.

Fashions changed in everything, including pesticides, **Dr Packham** continued. Twenty years ago organo-chlorine pesticides were the concern. In the UK we still seemed to act retrospectively, discovering chemicals in water first, and then trying to find out their uses and rates of consumption. What chemicals should we now be looking for and concerned about? (**Dr Harwick** explained that MAFF could still not supply detailed information on pesticide usage but its Regional Offices had been asked to provide as much help as possible to the water authorities in this regard.) **Dr Croll** hoped that Anglian Water had taken a first step in the right direction in looking for, particularly the MCPA group of herbicides which had been little investigated.

Professor Henin thought that the problems caused by pesticides were common to most countries. In France the average rate of usage of pesticides wa 1 kg/ha, or 2.8 kg/ha if one only considered the 17 million hectares of cultivated land. To comply with the EC Directive only 1% of the pesticides applied could be permitted to enter the groundwater. In fact, the amount found was generally below the EC limit, except for certain areas such as those with highly fissured rocks. The problems were caused by point-source pollution: overflows when filling tanks, discarded containers and so on.

Prediction of future levels of pesticides

Dr Croll believed that leaching into groundwater was not wholly insignificant. If pesticides behaved like nitrate one could calculate the possible future concentrations. For nitrate, applied at 200 kg/ha, 100 mg/l could be appearing in underground waters now, and this might rise to 200 mg/l. Phenoxyacetic acids were applied at 2 kg/ha; therefore one might expect 1 mg/l to leach from the soil. One per cent of this gave 10 μg/l, well over the EC Directive limit and getting towards the kind of concentration (50 μg/l) that could cause taint and odour in drinking water and impair the use of the water for spray irrigation.

Mr M. Payne questioned the assumed parallel drawn by Dr Croll between the behaviour of nitrate and pesticides. There was not a 7 t/ha reservoir of pesticides in soil as there was of nitrogen.

Legislation/quidelines

WHO guidelines

Mr Waddington outlined the WHO Guidelines on pesticides and compared these with the EC Directive on Drinking Water. In preparing the Organics Section of the WHO Guidelines, only for eight substances were toxicological data sufficient for limits to be set. Because it was necessary to consider levels far below what can be examined using experiments on animals and anticipating very long-term exposure, the technique of linear multi-stage extrapolation modelling was used. The Guidelines stressed that this gives a level of uncertainty of one or two orders of magnitude. The sum of the limits of the eight pesticides was 134 μg/l.

EC Directive on Drinking Water

Comparing the sum of the WHO limits of 134 μg/l for eight chemicals with the EC's 0.5 μg/l for all pesticides including insecticides and herbicides, and PCBs and PCTs, one could be excused for thinking that there was something wrong somewhere. (**Mr Waddington** also felt that the EC should explain how these substances are to be measured to give practical effect to this limit.)

Mr Payne was particularly concerned about the values contained in the EC Directive, which **Dr Packham**, supported by **Dr Croll**, had earlier described as a nonsense. It was worrying for agriculture and for everyone else that values had been incorporated in legislation (to come into force in mid-1985) on which scorn was clearly poured. How had such a situation arisen? On what criteria were the levels based? **Dr Croll** said that, apart from pesticides, most of the numbers in the Directive were sensible. The UK water industry had never been told the criteria on which the levels were set. Levels set by the US Environmental Protection Agency and WHO, where the principles were known, seemed perfectly reasonable. The EC level was about right for organo-chlorine insecticides but the blanket level ignored the widely varying toxicities seen in other pesticides. The water industry had asked that pressure be put on the European Commission to examine this particular value.

Bromide/THMs

Small amounts of bromide can significantly affect the amount of tri-halomethanes formed during water treatment. **Mr V. W. Howells** (Severn Trent Water) asked whether this sort of problem had been observed with bromide in the Netherlands. **Mr de Heer** agreed that bromide ion was a big problem when chlorine or ozone was used to disinfect supplies of potable water. The natural concentrations of bromide in some Dutch waters (in areas with upward seepage of salt water) are so high that, even without the addition of (methyl) bromide from horticulture, ozonisation and chlorination cannot be used in case bromoform and other toxic substances are formed.

ORGANIC WASTES

It was generally agreed that, in terms of pollution of surface waters, the discharge of organic matter (farmyard wastes, silage liquor) presented the most serious problem. A discussion followed of the need for research in the treatment of these materials, and who should fund such work.

The principal problem

Referring to the major increase in serious farm pollution incidents reported between 1974 and 1984 and discussion by Mr Payne (Table 22.1), **Mr Hammerton** wondered how much of this increase had been due to the development of intensive animal husbandry with its production, from a small area, of wastes perhaps 100 times stronger than sewage. **Mr Payne** suspected that the increase in intensive farming was substantial but not as great as in the decade befroe 1974. It was the dairy industry which contributed the bulk of the problem and there improvements were needed, rather than pigs or poultry of other, intensive, livestock units.

A comment by **Mr T. G. Brownlie** (Department of Agriculture and Fisheries for Scotland) confirmed the above. In surveys carried out by DAFS and MAFF it was clear that most of the manure generated by intensive livestock units was transported to the land of other farmers where it was needed as fertlizer, thus making good use of the manure and preventing pollution.

Research

Dr S. C. Warren (Water Research Centre, Environment) began a brief discussion of research requirements by stating that Welsh Water and Wessex Water had pressed WRc for some time to work on the development of low-cost treatment of farm wastes. The Centre had always resisted because (a) methods would not be cheap — due to the sheer bulk of the material to be treated — and (b) WRc would have no means for persuading the farmers to use the techniques developed. Should WRc get involved? Were there other, more appropriate, bodies who should do the work? **Dr Howells** certainly thought that it was within WRc's remit because it was a problem costing Welsh Water some millions of pounds and both WRc and its predecessor, the Water Pollution Research Laboratory, had undertaken research on a wide variety of treatment problems. **Dr Hardwick** thought that publicity was needed, not research, and pressure to apply known remedies. **Mr H. Fish** (Natural Environmental Research Council) also thought that WRc should not become involved in agricultural research. The polluter should pay, not those who paid water rates. If the farming industry was to be subsidised in its research and development it was the responsibility of the Minister of Agriculture, Fisheries and Food. If MAFF could not afford it, the farming industry must afford it, just as any other industry had to meet its own research and development costs, and the costs of their pollution control. If pressure came from extra environmental expectations from the public, perhaps they should pay, but by a general tax bill, not a special water bill. **Dr Howells** acknowledged the common sense of Mr Fish's argument but Welsh Water felt that this area was just as important as some of the other areas currently within the Centre's programme. We must get out of the straight-jacket idea that all we could do with farm slurry was spread it about!

Mr Brownlie, following on from the points made by Mr Fish, said that agriculture was not in a position to pass on extra costs to the consumer.

CONTROL OF AGRICULTURAL ACTIVITIES

Throughout the discussions on the effects of agriculture on fresh waters there were questions on the approaches which could be adopted by the regulatory authorities in attempting to control this form of pollution. Advice and aid to farmers was discussed, the latter being linked to planning approval, good agricultural practice, and the prevention of pollution by establishing protection zones or areas in which land use was tightly controlled.

Approaches

Mr Furness introduced the two attitudes to dealing with pollution prevention and control — the conciliatory approach and the heavy hand. South West Water was committed to the first, especially when dealing with farmers. In dealing with fisheries offences however the authority prosecuted even minor offenders. Could the cost-effectiveness of these two approaches be tested on a pair of catchments? **Dr Howells** thought not. One had to be reasonable. All prosecution decisions were judged on a series of criteria. Welsh Water did not automatically prosecute when fish were killed but would do so (whether or not a fish-kill occurred) if gross negligence was indicated. **Mr Merrimen** added that the even-handed approach was essential. They could never present any other attitude to the National Farmers' Union. They were working hard with the NFU and ADAS to develop good working relationships and prevent problems.

Mr Brownlie agreed that co-operation was essential. We should identify those practices that cause the most damage and seek solutions at the least cost to agriculture.

Mr Green suggested that the level of obvious agricultural pollution incidents was so high that it threw doubt on the effectiveness of the friendly approach. **Mr Payne** responded that in all environmental matters the NFU strongly supported the spirit of the Wildlife and Countryside Act 1981 that co-operation, preferably voluntary co-operation was the approach to be adopted. This should equally be applied to matters concerning freshwater pollution, in the great majority of cases. There would be exceptions, but voluntary programmes (as Dr Konrad showed in Chapter 19) could be extremely successful.

Dr Newbold, however, put 'voluntary co-operation' into a different light when he reported that NCC in re-scheduling 1800 Sites of Special Scientific Interest in England and Wales have been asked by farmers for compensatory payments in approximately 600 cases! **Mr Payne** agreed that compensation was involved, but why should farmers pay for the conservation inheritance of the nation?

Mr Hammerton believed that in Scotland relationships between the water industry and farming were good. This situation was fostered by such things as the Scottish Farm Wastes Liaison Group, by close contact with the Agricultural Colleges, by articles in the local farming press and on local radio. By these means the River Purification Board hoped to advise every farmer of, for instance, the hazards of discharges of silage liquor. The latest returns showed that serious pollution incidents in Scotland have been substantially reduced.

Advice

Sound advice **Mr Furness** maintained, was the key to the South West Water's farm pollution control programme in co-operation with ADAS, and several speakers had referred to the importance of sound advice in various aspects of the protection of fresh waters. He asked for comments on the proposal that ADAS should charge for their services. As **Mr Payne** understood it the possibility of charging in various ways was being explored, but the NFU does not welcome the proposal. Information from water authorities could effectively be disseminated by ADAS. **Dr Hardwick** explained that the aim was to cover about £25m of the £125m cost of ADAS and to reach such a point in two to three years. It was not clear yet how this would be done and there were areas, such as pollution control, where a careful assessment must be made of the possible consequences of charging.

Good agricultural practice

Dr Howells knew that many other sectors of British Industry would also have welcomed a Code of Good Practice in relation to their own activities. He suggested that its introduction in the case of agricultural was a reflection of the power of the agricultural lobby, but **Mr Payne** did not accept this. The NFU's experience as a political lobby indicated that it was essential to produce fair and reasoned arguments. Good cases brought about useful influence.

Mr Worthington, responding to a question from **Dr Newbold**, confirmed that most countries merely had codes for fertilizer application for different crops. There was very little advice on intensive agriculture or the disposal of organic manure.

Grant aid/planning

Mr Halcrow suggested that all farming activities in terms of changing land use and intensification should be brought under planning controls, with proper participation by water authorities, and this was something that should be brought to the attention of the Government. **Dr Howells** supported Mr Halcrow and hoped that some form of 'prior approval' could be incorporated in the new capital grants scheme so that water authorities can become aware of developments.

Dr Hardwick made a plea that the term 'prior approval' should not be used in this context. 'Prior approval', an internal MAFF procedure in which ADAS decided whether a proposal conformed to their criteria, was now limited to a small sample of proposals. What the water authorities needed, in his opinion, was a mechanism directing the farmer to have prior *consultation* with bodies like the water authorities to ensure that when grant applications were made to the Ministry, they complied with the authorities, requirements.

'Planning', **Mr Brownlie** said, was recognised as a solution to an immediate problem, but he thought that it was a long-term measure. Since very few new agricultural enterprises were coming into operation, effort should be applied to the present problems.

Mr Hammerton suggested that modern farms were becoming more like industrial premises and should perhaps be treated as such. **Mr Payne** admitted that some intensive livestock units were remarkably like industrial enterprises and some were remote from actual farms. Where they sought to discharge effluents in the manner of an industrial enterprise they were controlled as one. In most cases, however, their wastes would be returned to land in the normal recycling process.

Grants in the USA

Dr Konrad explained the reasons for State financial assistance in the Wisconsin Program (Chapter 19). Firstly there was a 30 to 50-year history of Federal support of agricultural activities, including conservation practices. So the farmer expected financial support if he was asked to do something. Secondly, to help justify the public expenditure, the types of financial assistance and the practices supported were limited to those practices having benefits which were primarily off-site, to the public resource, not to the landowner. This was why there was an obligation on the public sector to support what the landowner was being asked to do.

EEC grants

Moving to the European scene, **Mr Cominetti** asked Mr Worthington whether EEC improvement grants could be made to individual farmers. **Mr Worthington** said that these were available in certain areas such as hill farming.

Techniques for preventing pollution

Dr Phillips had emphasised that 'prevention was better than cure' in the context of agricultural pollution, a sentiment that met with the support of other contributors to the discussion. However, **Mr Cominetti**, while approving, was concerned that this was not going to be easy to apply to the existing 250 000 UK farming units. Apart from silage, one of the biggest problems seemed to be that there was no provision on most farms to keep the surface water and foul systems separate. **Dr Phillips** suggested that, with financial assistance, modifications could be made in existing farms. He would apply his maxim by, for example, preventing wastes with a high BOD from entering watercourses.

Aquifer protection zones

Mr Hammerton had noted that the Danube countries were experiencing problems of high levels of nitrate in the groundwaters of areas given over to intensive agriculture. He complemented Dr Homonnay's recommendations for action. In Czechoslovakia he had observed three zones for protection around each relevant borehole. In the central circle no fertilizers or pesticides could be applied; around this were two areas of restriction, with limitations on land use lessening with distance from the borehole. He understood that farmers would be compensated for the loss of crops resulting from the protective restrictions. Was there the possibility of applying this technique in the UK? **Dr Homonnay** added that such systems were indeed applied in Hungary and should effectively protect underground resources. The systems could, however, be uneconomically expensive. For the UK situation, **Mr Young** said that the expense was the limitation. Water undertakings might have to purchase the land to enable them to restrict farming activities. Some water companies did this. **Mr Hammerton** suggested that there was a case for trial areas in the UK. **Dr Croll** added that aquifer protection had been examined in the Anglian Water region, but to be effective perhaps many tens of thousands of hectares of good agricultural land above the major aquifers would have to be returned to low or medium productivity grassland, at great cost.

Advanced agro-techniques

In response to a question from **Dr Phillips** concerning the promotion of advanced agro-techniques for limiting pollution, **Dr Homonnay** said that in Hungary there was an Agricultural Engineering Consulting Organisation responsible for such optimal techniques. For some decades too much fertilizer had been applied. The Organisation took into account the special characteristics of each area before recommending the best ploughing and fertilizer techniques. For both economic and environmental reasons fertilizer dosage rates were being reduced.

Reduced agricultural demand

Mr Fish suggested that the situation was in the process of balancing itself: the EEC was over-producing grain and milk; subsidies were too expensive and would be cut back, and farming must become more economic, requiring less input to achieve the appropriate reduction in output.

Part 4
Forestry

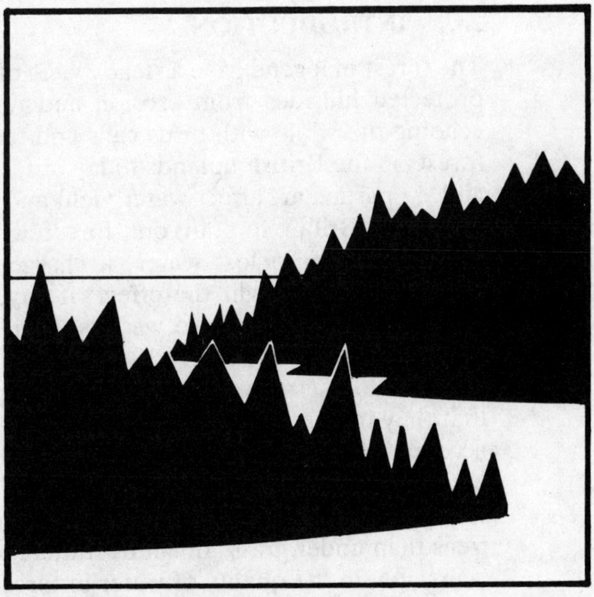

24

Forestry and fresh waters: Problems and remedies

W. O. Binns

24.1 INTRODUCTION

The forest of legend gave a steady yield of gin-clear water the year round; it protected hillsides from erosion and reduced flooding, its sunlit glades echoing the while with birds calls and teeming with animal life. Managed forests in the British uplands today differ somewhat from that idyllic place and, in particular, affect water yield and quality strongly. Law (1956) was the first in Britain to point out, forcefully, that forests were costing a great deal of money in lost water, a charge greeted with disbelief by many foresters who thought the effect, if any, would be small (Laurie, 1956). Surprisingly, the evidence was also challenged by some physicists. Early forestry text books (Brown, 1882; Nisbet, 1905) make little reference to any effects of forests on water quality and are quite categorical in their assertions that they increase the soil moisture and water supply. It is, however, now accepted by all scientists (and most foresters), following the intensive work by the Institute of Hydrology in recent years (Calder, 1979; Calder & Newson, 1979), that catchments in upland Britain yield less water under trees than under grass. In contrast there was little interest, up to five or so years ago, in the quality of water in forested streams and the rivers fed by them. Foresters were of course aware of the dangers of erosion following cultivation and drainage, and of operations such as road-making, but thought that the full forest canopy would, over the life of a crop, have a stabilising and beneficial effect.

Forestry can affect water yield and quality in many ways. The initial operations can disturb the existing balance: drainage and cultivation alter the response of a catchment to rainfall and can introduce solids into the runoff; the removal of vegetation (usually moorland or rough grazing) and its replacement by a crop initiates long-term changes; herbicides, insecticides, and fertilizers may be used and can get into the water system. Other effects appear later on: thinning during the life of the crop will temporarily affect water yield and may reduce quality by incidental damage to drains and watercourses; the forest may, by its nature, affect the chemistry of throughfall and stem-flow and, by its physical presence, lower stream-water temperature and prevent the growth of bank-side vegetation; and finally, the clear felling of the crop at the end of the normal rotation abruptly removes the intercepting canopy and returns the yield to somewhere near its pre-planting value.

This chapter examines the impact of forest operations and the influence of the forest itself on water yield and quality, and discusses ways in which the forester can mitigate or prevent undesirable effects. Foresters have sometimes claimed that they are restoring the original cover of woody vegetation which occupied most of Britain in prehistoric times, but that is not relevant to the question being addressed: the concern is with the changes which follow the re-afforestation of ground which has long been treeless, in particular major water-gathering catchments in the uplands.

24.2 WATER YIELD

Blackie and Newson (Chapter 26 of this book) have covered the effects of forests on water yield in detail. These are summarised in the model proposed by Calder & Newson (1979) to account for the total evaporation (TE) of a partially-forested catchment:

$$\begin{aligned} TE &= E_t\,(1-f)+f\,[aP+(1-w)E_t] \text{ mm/year} \\ &= E_t+f\,(aP-wE_t) \text{ mm/year} \end{aligned}$$

Where E_t = evaporation from the catchment if grassed;
f = the proportion of the catchment with a forest canopy;
P = annual rainfall averaged over the catchment;
a = the rainfall fraction lost by interception and re-evaporation;
w = the fraction of the year for which the canopy is wet.

For general purposes f is assumed to be two-thirds of the gross forested proportion.

Pyatt (1984) provided a table giving values of a, w, and E_t for different values of P and a graph of annual runoff for a range of rainfalls and proportions of afforestation.

The consequences of afforestation for water yield can thus be deduced quite easily.

24.2.1 Management options

The forest manager has little room for manoeuvre, given that a catchment is to be afforested. Trees in the uplands are almost all planted for commercial, timber-production purposes, though considerable attention is paid these days to landscape, recreation, and conservation (Crowe, 1978; Forestry Commission 1982, 1983, 1984a and b). Most of the trees will therefore be conifers, since there are no broadleaves which can be grown profitably on any scale in the uplands. A deciduous conifer, such as larch (*Larix spp.*), will intercept less rain than an evergreen one, but not all that much less: leafless twigs seem to intercept rain quite efficiently, though there is little direct evidence from the British uplands.

The only change in operations by which the manager can influence water yield is to increase the frequency or intensity of thinning (removing a proportion of the trees periodically to give intermediate returns and concentrate subsequent growth on a smaller number of trees) or to shorten the length of time the crop is on the ground (which increases the proportion of the crop's life in which the trees are small and have no discernible effect on interception).

There are three factors which militate against thinning and can shorten the ideal rotation: high winds and frequent gales that cause windthrow, particularly of thinned crops; long hauls to markets which reduce the profit on small roundwood which is the main product of early thinnings; and high road costs, which discourage construction for as long as possible.

Optimum felling ages for the three most important upland conifers, Sitka spruce (*Picea sitchensis*), Lodgepole pine (*Pinus contorta*), and Japanese larch (*Larix kaempferi*), are given in Table 24.1. These show that, for thinned crops, about 28% of the Sitka spruce will be felled by 50 years, and about 82% by 55 years. For unthinned Sitka, however, the ages fall by between five and ten years and the same is true for the other two species. In fact the figures are not truly comparable because the sites with severe exposure and waterlogged soils, on which crops are no longer thinned, are generally higher up the hills and therefore have lower yields, with the exception of some small forests in places such as Devon and Cornwall where yields are high but crops are unstable.

Since thinning has already become less frequent than it used to be, the increased water yield from recently thinned stands (Holstener-Jørgensen, 1967) is unlikely to help much; and in any event recent research suggest that the increased turbulence and roughness offsets this in the windy British uplands (Calder, pers.comm.). However, the relatively small number of trees per hectare in British forests may tend to keep interception amounts down for the early part of the rotation: the highest values have been recorded from closely planted, unthinned Norway spruce (*Picea abies*)

Table 24.1 — Optimum felling ages for crops thinned by management tables

Yield† class	Sitka spruce Age (yr)	Sitka spruce % in yield class‡	Lodgepole pine Age (yr)	Lodgepole pine % in yield class	Japanese larch Age (yr)	Japanese larch % in yield class
2&4	—	—	75	13.6	60	7.8
6	60	3.6	65	37.4	55	20.5
8	60	8.6	60	32.7	50	27.8
10	55	20.7	55	10.6	50	22.2
12	55	33.1	55	2.6	45	13.6
14	50	16.9	50	0.5	45	6.7
16	50	7.4				
18	50	3.0				
20	45	0.8				
22 & over	45	0.3				

†Yield class is the maximum value of the quotient $\frac{\text{Total yield of timber}}{\text{age}}$ in m^3/ha

‡Forestry Commission crops only; there are some crops of each species in check, i.e. with no yield class.

(Delfs, 1967). As already mentioned the small amounts of broadleaves in uplands are likely to have little impact on the total water yield from forested catchments.

Thus decisions on where new forests should be and what proportion of any catchment should be planted will influence water yield more than manipulations by the forester. It is, however, worth remembering that rainfall increases markedly with altitude. It follows that if the effects of afforesting different proportions of a catchment are being considered, the appropriate mean rainfall for the area to be planted should be used in each calculation.

24.2.2 Ground preparation effects

The main impact of draining and cultivation will be not so much on the total yield of water but on its distribution: the shape of the hydrograph is altered and the response becomes markedly more rapid following draining (Robinson, 1980). Water yield may be increased first by release of water stored naturally in the peats and then because of the lower storage capacity of drained ground; the second effect will persist until offset by the increasing interception of the growing trees. In contrast, deep ploughing of mineral soils on hard heathlands should increase the storage capacity, and lower flood peaks. This phase may last for up to a third of the life of the crop and, in a forest where there was a roughly equal proportion of stands of different ages, will go some way to offset the loss of yield which has probably been worst in forests where most of the planting was done over a short number of years.

24.3 WATER QUALITY

24.3.1 Ploughing

It has been mentioned already that the disturbance following forest draining and cultivation affects water quality directly, because mineral matter —

sediment — and organic compounds get into streams via drainage ditches. This is likely to be worst on peaty gley soils where both mineral and peat layers are disturbed and where the underlying rocks are subject to erosion, for example weathered schists and boulder clays. The systems used can tend to increase both flashiness and erosion. Figure 24.1 shows peatland plough-

Fig. 24.1 — Shallow double-throw ploughing to provide planting ridges on peaty ground. Good weed suppression and easy planting, but rapid runoff of rain falling into furrows

ing and Fig. 24.2 downhill tine ploughing; in steep country it may be unsafe to plough across slopes. There is clear evidence (Painter *et al.*, 1974; Robinson, 1980) that considerable erosion takes place following ploughing and may persist for many years after draining has been completed. Battarbee *et al.* (1985) also report an increase in the rate of accumulation of loch sediments following planting. To reduce this problem it is now Forestry Commission policy not to allow drains to discharge directly into streams.

24.3.2 Chemicals

Most forest soils in the uplands are fertilized with phosphate, or phosphate and potash, at least once during the crop's life. Table 24.2 shows statistics for some recent years. The notable features are the relatively small areas

Fig. 24.2 — Single-throw downhill tine ploughing on mineral soil, with cross drains nearer the contour. Good local weed suppression, easy planting and improved vertical percolation, but likelihood of increased erosion for some years

Table 24.2 — Fertilizer use in Forestry Commission forests in recent years (hectares†)

Year	N	P	PK	Total‡
1971–2	200	13 600	9200	23 050
1973–4	650	32 050	18 300	51 000
1976–7	1000	17 000	7050	25 050
1980–1	1650	16 450	23 050	41 150
1982–3	1350	10 350	24 150	35 850

†To nearest 50 ha
‡Use of other nutrient combinations, e.g. NP, is trivial.

treated with nitrogen, the peak in total area treated in the mid-1970s, and the increasing importance of PK as opposed to straight P fertilizer, which reflects the increasing amounts of peat and peaty soils being planted. In 1982 to 1983 three-quarters of the fertilizer used was on established crops, rather than on newly-planted ones. Phosphate is applied as coarsely-ground rock phosphate at 375 kg or, more recently, 500 kg/ha to supply 50 or 70 kg P, and potassium as 200 kg crude potassium chloride per hectare, to supply 100 kg K.

Almost all this fertilizer is applied by helicopter and it follows that an appreciable proportion — between 10 and 20 per cent — will fall into drains or streams and, though the coarsely-ground rock phosphate used in all aerial work is only slowly soluble, a rapid increase in the existing and usually low phosphate concentrations is inevitable. The peak falls fairly rapidly, but the new steady concentration of phosphate may be ten times the original for some months (Robinson, 1980) and could be sufficiently high to encourage algal blooms in reservoirs, lakes, or freshwater lochs. Some other studies have shown that the concentrations return to somewhere near their original level after a year or so (Truesdale & Taylor, 1978). When fertilizers are applied to established crops less will be lost in this way and, in some instances, it may be applied before ploughing which will also reduce losses in water.

Since there are important economies of scale in aerial fertilizing there is a conflict of interests. Good liaison between foresters and Water Authorities or River Purification Boards over fertilizer programmes is a first essential, and the likelihood of algal blooms could be reduced by restricting application of phosphate, in sensitive catchments, to the autumn. Increased potassium concentrations in forest areas are unlikely to cause any problems, and as little nitrogen fertilizer is applied in such catchments its use is unlikely to affect water quality on any scale.

Herbicides are widely used in forestry to control competing weeds, especially grasses. There have been marked changes in recent years in the range of compounds used and glyphosate ('Round-up') now accounts for at least half the area treated annually by the Forestry Commission, propyzamide and 2,4-D accounting for a further third. Glyphosate and 2,4-D are both harmful to fish, but the former is fixed so effectively in the soil that risks come from spillage or misuse, and 2,4-D taints water in such small concentrations that great care is always taken in its use. Propyzamide is only slightly toxic to fish. Asulam is used on bracken and is therefore important in some catchments: again it is only slightly toxic to fish. The Forestry Commission publishes a guide to herbicide use (Sale *et al.*, 1983) and the 1985 revision will contain a section on precautions to be taken in water-supply catchment areas, though it is concerned more with taint of drinking water than with stream biology.

Insecticides are used for two main purposes: to protect newly-planted trees on felled ground against beetles, and to deal with potentially-devastating outbreaks of individual insects. Insecticides have been used from the air to kill Pine looper (*Bupalus piniarius*) on Scots pine (*Pinus sylvestris*) on a number of occasions, and recent outbreaks of Pine beauty (*Panolis flammea*) on Lodgepole pine in north Scotland have been controlled by aerial spraying (Holden & Bevan, 1981). Foresters, however, prefer not to prop up a species which is subject to regular attacks by a pest or diesease, but to change over to one which is less susceptible. An example is the gradual replacement of Scots pine by Corsican pine (*Pinus nigra* var. *maritima*) in places where the former is badly damaged by the Pine looper. Furthermore, naturally-occurring viruses may be encouraged or spread deliberately, for example to control sawflies on pines. As with herbicides the main problems arise from spillage or misuse, such as ill-advised washing out of containers.

There are few chemical fungicides used in British forestry. Fomes root and butt rot (*Heterobasidion annosum*) is controlled in pines by treating newly-cut stumps with a suspension of a competing fungus, *Peniophora gigantea,* and in other species in the same way but with a solution of urea.

All chemicals have the potential to contaminate water and upset stream biology, but the most likely cause is accident or carelessness. Co-operation between foresters and the appropriate authorities is essential when any large-scale applications are planned.

24.3.3 Thinning and clear felling

Where trees are planted close to streams, or where ditches discharge directly into them, forest operations will almost inevitably affect water quality. Following the guidelines prepared by Mills (1980) it is now Forestry Commission policy not to plant or replant close to streams and new forest drains do not discharge directly into them. It will, however, be a long time before the trees along all forest streams have been cut down and the banks left clear when replanting: the act of cutting trees and extracting timber near forest streams may thus continue to have damaging effects on water quality for some time to come. Foresters have to control contractors when they are doing the felling and, if necessary, impose conditions in the contract on where operations such as the removal of branches and the cutting off of the tops should take place, and should also forbid the dumping of such residues in streams.

24.3.4 Forest influences

The widespread evidence of acidification of fresh waters in Scandinavia (Overrein *et al.,* 1980) is believed by most people to be due to emissions of acidic pollutants (Overrein, 1983). Acidification of fresh waters is, however, a long-term, natural process (Pennington, 1984) which may be hastened or increased by industrial pollution (Catt, 1985). Recent research in Galloway by Battarbee *et al.* (1985) using diatom frequency in loch sediments to infer pH at the time of deposition and dating them by means of a lead isotope, has strongly suggested that acidification rates increased between 80 and 150 years ago. He examined different hypotheses to account for this change, such as afforestation, change of vegetation from grass to heather, podsolisation and acid depositions, and rejected all but the last.

There is, however, strong circumstantial evidence for increased acidity and poorer biology in forest streams compared with the equivalent moorland streams (Harriman & Morrison, 1981; Stoner & Gee, 1985). This seems to occur only over hard acid rocks, and even then not in all places. Over richer rocks there may be some increased acidity but the fish are still present. The processes causing this change are not properly understood, partly because most of the work has been observational rather than experimental; research now in hand, which examines various remedial treatments or changes in practice, should throw some light on them. Hornung (1985) has reviewed the processes which make for soil acidification, and Miller (1985) has looked at the possible causes of stream-water acidification. The most likely causes of the latter are interception of deposition, acidification of throughfall, soil acidification, and changes in soil hydrology.

Miller (*loc. cit.*) points out that while it is easy to demonstrate an increase in the capture of many elements on an inert intercepting surface, compared with an open raingauge, there are as many examples of neutralisation as of acidification by forest canopies and, moreover, there is evidence of an age (or perhaps size) effect on the chemistry of throughfall under forest stands. Older trees of many species tend to acidify throughfall while younger ones neutralise it, suggesting a physiological basis to acidity exchanges by the forest canopy. Hornung (in press) found acidification to occur under 35- and 45-year-old Sitka spruce in Wales, confirming the effect of large trees under British conditions. Soil acidification occurs under many trees, broadleaves as well as conifers, and part of this may be due to cation-anion imbalance in nutrient uptake, which will be perpetuated where trees are harvested (Nilsson *et al.*, 1982). However, without the addition of a mobile anion, such as sulphate, acidity generated by this process will remain in the soil. Changes in hydrology will increase with the growth of a forest, due both to an efficient imposed drainage system and greater soil drying, which together not only allow rainfall to travel more rapidly through drains to streams, but also open up cracks which reduce residence time of water in the soil. Changes in hydrology are self-evident within the forest, soils are dry and peats show cracking. Bache (1983) points out that provided acid water penetrates into the mineral-rich C horizon, neutralisation will take place and this phenomenon has been shown to occur under Sitka spruce ground in North Wales (Hornung, in press).

Miller (1985) makes the important point that it does not matter much whether the acidification of stream water is a consequence of intercepted deposition, acidification of throughfall, acidification of soil, or changes in hydrology: if the drainage water can be held in contact with the C horizon long enough to ensure neutralisation, the problem is solvable in principle. Current Forestry Commission recommended practice should help, but developments in cultivation and drainage patterns near streams are needed.

24.4 FOREST AREAS

The current area of productive woodland in Britain is about 2 000 000 ha as shown in Table 24.3.

With a land area of 22.7m ha, Britain has now over 9.5% of its land surface with some form of woodland cover and the figures should reach 10% before the end of the century. The cover is of course not at all even: Fig. 24.3 shows forests in Britain, and the large forests are in the high rainfall, hilly parts of the country, with the exception of the old Royal Forests in the New

Table 24.3 — Forest and woodland in Great Britain 1984, thousand ha (Forestry Commission 1984c)

	Coniferous high forest		Broadleaved high forest			Total		
	FC‡	Private	FC	Private	Coppice†	productive woodland	Unproductive woodland	All woodland
England	206	191	40	381	37	855	90	945
Wales	130	43	6	53	2	234	10	244
Scotland	515	338	4	72	—	929	71	1000
GB	851	572	50	506	39	2018	171	2189

† All except 1000 ha in private ownership

‡ FC = Forestry Commission

Fig. 24.3 — Forestry in Great Britain 1976. Forestry Commission, all land; private forests, dedicated and approved woodland only. Islands with little forest have been omitted and the smallest woods do not show up on this scale.

Forest and the Forest of Dean, and the Breckland of East Anglia. Some counties are very heavily forested, for example by 1982 forests covered 22% of Dumfries and Galloway, while in 1947 the figure was only 4% (Davies, 1982).

Figure 24.3 is, however, misleading as it shows all Forestry Commission land, planted or not, but only those private forests which are managed under some form of agreement. Thus in England, Wales and Scotland many private forests of long standing are not shown and furthermore many small forests do not show up at this scale. The figure nevertheless shows areas where large forests could have major effects on water yield and quality.

24.5 CONCLUSIONS

Foresters are stuck with the unpalatable fact that their activities cause a marked reduction in water yield, though Calder and Newson (1979) show that only 12 out of 32 large (>10 million m^3) reservoirs in England and Wales had more than 10% of the catchment afforested and only three out of 40 in Scotland. It nevertheless follows that careful consideration should be given to new afforestation schemes where catchments are either reservoired or where rivers are used directly for major supply schemes (though there are periods in winter when water spills over the dams of many reservoirs). It is probably even more important where water is used for hydro-electric schemes, since virtually every cubic metre can be used for power generation.

Problems with acidification of water are likely only in forests overlying hard acid rocks: but landscape requirements and the physical effects of forests on water temperature and stream-bank vegetation mean that all streams will be treated differently in the future, for the benefit of naturalist, rambler, conservationist, fish - and fisherman. Though it is rash to make confident predictions, the evidence available suggests more and more that the increased water acidification associated with new forests could be dealt with by changes in drainage and cultivation practices, and by leaving unplanted margins along streams, the size of the margin depending on topography and stream size.

REFERENCES

Bache, B. W. (1983) The implication of rock weathering for acid neutralization. In: *Ecological effects of acid deposition,* 175–187. National Swedish Environment Protection Board, PM 1636.

Battarbee, R. W., Flower R. J., Stevenson, A. C. & Rippey, B. (1985). Lake acidity in Galloway: a palaeoecological test of competing hypotheses, *Nature,* **314** (6009), 350–352.

Brown, J. (1882) *The Forester.* 5th edition, Blackwood.

Calder, I. R. (1979) Do trees use more water than grass? *Water Services,* **38**, 11–14.

Calder, I. R. & Newson, M. D. (1979) Land-use and upland water resources in Britain — a strategic look. *American Water Resources Association Bulletin* **15**, 1628–1639.

Catt, J. A. (1985) Natural soil acidity. *Soil Use and Management,* **1**, 8–10.

Crowe, Sylvia (1978) *The Landscape of forests and woods.* Forestry Commission Booklet 44.

Davies, E. J. M. (1982) Forestry in Dumfries and Galloway. *Forestry,* **55**, 79–89.

Delfs, J. (1967) Interception and stemflow in stands of Norway spruce and Beech in West Germany. In: Sopper, W. E. and Lull, W. H. (Editors) *International symposium on forest hydrology,* Pergamon, Oxford, 179–183.

Forestry Commission (1982) *The Forestry Commission and Recreation.* Policy and Procedure Paper No 2.

Forestry Commission (1983) *The Forestry Commission and Landscape design.* Policy and Procedure Paper No 3.

Forestry Commission (1984a) *The Forestry Commission's Objectives.* Policy and Procedure Paper No 1.

Forestry Commission (1984b) *The Forestry Commission and Conservation.* Policy and Procedure Paper No 4.

Forestry Commission (1984c) *Forestry Facts and Figures 1983–84.* Forestry Commission, Edinburgh.

Harriman, R. & Morrison, B. R. S. (1981) Forestry, fisheries, and acid rain in Scotland. *Scottish Forestry,* **35**, 89–95.

Holden, A. V. & Bevan, D. (1981) *Aerial application of insecticide against Pine beauty moth.* Occasional Paper 11, Forestry Commission (reprinted 1982).

Holstener-Jørgensen, H. (1967) Influences of forest management and drainage on ground water fluctuations. In: Sopper, W. E. and Lull, H. W. (Editors), *International symposium on forest hydrology,* Pergamon, Oxford, 325–332.

Hornung, M. (1985) Acidification of soils by trees and forests. *Soil Use and Management,* **1**, 24–27.

Hornung, M. (in press). Precipitation-Canopy and soil-water interactions. In: *Proceedings of joint meeting on acid rain.* Institute of Water Pollution Control and Institution of Water Engineers and Scientists. Birmingham, March 1984.

Laurie, M. V. (1956) The effects of forests in water catchment areas on the water losses by evaporation and transpiration. In: *Report 12th IUFRO Congress, Oxford,* **1,** (11), 82–85.

Law, F. (1956) The effect of afforestation upon the yield of water catchment areas. *Journal of the British Water Works Association,* November, 489–494.

Miller, H. G. (1985) The possible role of forests in streamwater acidification. *Soil Use and Management,* **1**, 28–29.

Mills, D. H. (1980) *The management of forest streams.* Forestry Commission Leaflet 78.

Nilsson, S. I., Miller, H. G. & Miller, J. D. (1982) Forest growth as a possible cause of soil and water acidification — an examination of the concept. *Oikos,* **39**, 40–49.

Nisbet, J. (1905) *The Forester,* Blackwood.

Overrein, L. N. (1983) The European acid rain situation. In: *Ecological effects of acid deposition,* 43–52. National Swedish Environment Protection Board, PM 1636.

Overrein, L. N., Seip, H. M. & Tollan, A. (1980) Acid precipitation — effects on forests and fish. Final report of the SNSF-Project 1972–1980. *SNSF-Project FR 19/80,* 1–175.

Painter, R. D., Blyth, K., Mosedale, J. C. & Kelly, M. (1974) The effect of afforestation on erosion processes and sediment yield. In: *Effects of man on the interface of the hydrological cycle with the physical environment.* Ass. int. Hydrol. scient. Publication No. 113, 62–68.

Pennington, W. (1984) Long-term natural acidification of upland sites in Cumbria: Evidence from post-glacial lake sediments. *52nd Annual Report Freshwater Biological Association,* Ambleside.

Pyatt, D. G. (1984) *The effect of afforestation on the quantity of water run-off.* Forestry Commission Research Information Note 83/84 SSN.

Robinson, M. (1980) *The effect of pre-afforestation drainage on the streamflow and water quality of a small upland catchment.* Institute of Hydrology Report 73, 69 pp.

Sale, J. S. P., Tabbush, P. M. & Lane, P. B. (1983) *The use of herbicides in the forest — 1983.* Forestry Commission Booklet 51.

Stoner, J. H. & Gee, A. S. (1985) Effects of forestry on water quality and fish in Welsh rivers and lakes. *Journal of the Institution of Water Engineers and Scientists,* **39**, 27–45.

Truesdale, G. & Taylor, G. (1978) Quality implications in reservoirs filled from surface water sources. *Prog. Wat. Tech.* **10**(3/4), 289–300.

25

Aspects of current forestry practice and silvicultural trends in West Germany affecting fresh waters

E. F. Bruenig

25.1 FORESTS AND WATER CONSUMPTION IN GERMANY

Forests cover almost one third of the surface area of the Federal Republic of Germany. Their distribution is very uneven and concentrated in areas away from industrial conurbations. Favourite forest sites are super-humid mountain and hill areas, often with more shallow soils, and humid to sub-humid flat lowlands where agriculture has pushed the forests on to less favourable, mostly sandy and sometimes stiffly clayey, oligotrophic and often degraded soils (Fig. 25.1). Both categories contain important catchment areas in which water is stored, regenerated, and collected to supply big cities, such as Hamburg or Munich, and industrial conurbations, such as the Ruhr and Rhein-Main regions. Both lowland and hill sites have hydrological problems. The hilly areas were largely denuded during the 12th to 18th centuries, then reforested (often with spruce of unsuitable provenance, which now suffer from stress due to the climate and man-made pollution) on fragile soils. These soils, on exposure, are susceptible to erosion. The largely Pleistocene lowlands, similarly denuded through misuse since prehistoric times, were reforested during the 18th and 19th centuries, essentially and necessarily with pine and spruce species suited to the area. These now also suffer from pollution in addition to the consequences of natural acidification and natural climatic stresses. There are problems associated with the recreation of favourable soil conditions, prevention of leaching, and restoration of the previously-lost organic matter in the form of mild humus, which requires adequate biological activity in the soil. Most forest ecosystems on the less-favourable soils in the lowlands are subject to processes of acidification.

In West Germany, about $28\times10^9\,m^3$ water, or 14% of the annual precipitation, were used by man in the early seventies (Table 25.1). Of this nearly $4\times10^9\,m^3$ were used in households and were required to be of drinking quality. It is expected that the total water use will reach $70\times10^9\,m^3$ at the turn of the century (Olschowy, 1978). About half the water used in private households is supplied from forested catchments. Forests therefore play an important role as a major part of the area in which water is stored and is regenerated (Fig. 25.2). Correspondingly, hydrological research in woodlands has been conducted in Germany for more than a century and Germany may be considered the birthplace of woodland hydrology (Brechtel, 1982).

25.2 SOCIAL VALUE OF WATER SUPPLY FUNCTION OF FORESTS

With the present rate of water utilisation of approximately one third of the precipitation minus evapotranspiration, and 50% of drinking water coming from forested catchment areas, forests fulfil an important social function. In Germany the absolute amount of annual water yield is yet of minor importance. The major benefits from forest vegetation are that the quality of water is better and the water flow more even from forested water catchment. As long as the loss of water by evapotranspiration is valued less than these benefits, the social balance of forestry with respect to the hydrological functions is positive. This is particularly the case on sites with fragile and easily eroded soils, uneven rainfall distribution, and high risks of landslides and floods (Brechtel, undated).

The 'Arbeitsgemeinschaft Deutscher Waldbesitzerverbände' initiated a project group which prepared the first social cost/benefit balance of forestry in the Federal Republic of Germany (AGDW, 1976). The assessment included the costs and benefits of the influences of the forests on water quality, water balance, and water supply. The calculation assumed as usual that the forest ecosystem affects the water cycle by two major processes:

- purification by filtering and biological processes in humus and mineral soil;
- regulating runoff by evapotranspiration, storage and gradual release.

If the saving in cost is very conservatively estimated at only 10% of total water-production costs for purified drinking water and maintenance of

Fig. 25.1 — Distribution of forested land in the Federal Republic of Germany concentrated on mountains, hills and poor and degraded soils in lowland plains (e.g. Lueneburg Heath)

water bodies, the total annual social benefit of these two functions is calculated as at least DM 250m (1975 levels). Future levels of savings in the cost of water and of other social benefits are impossible to predict. Much will depend on the developments of the technology of water use. The future

Table 25.1 — Approximate estimate of water balance and water use in the Federal Republic of Germany (for 248 577 km^2 surface area) in the early seventies

Category	mm	10^9 m^3	%
Annual precipitation	803	200	100
Evapotranspiration	415	103	52
— from plants	315	78	
— from non-vegetated surfaces	100	25	
Runoff	388	97	48
— surface	276	67	
— groundwater	112	28	
Human use	112	28	14
— industry		11	
— household		4	
— hydroelectric		13	
— agriculture and others		1	

The present rate of 28×10^9 m^3 or 28% of the runoff used by man may be doubled by the beginning of the next century unless the efficiency of use of water is improved by better technology and greater care.

condition of the German forest will strongly affect water quality and discharge patterns. A major uncertainty of prediction is the unpredictable course of the presently-observed forest decline ('neuartige Waldschäden'). Currently noticeable tendencies of change of silvicultural practices towards more diverse and high-performance forests will also affect the hydrological functionality of future forests.

25.3 FOREST REGULATIONS CONCERNING FOREST WATER FUNCTIONS

Water management is generally regulated by the State Water Law in each of the ten German Länder, and Berlin. Forest management plans contain prescriptions accordingly if a gazetted water-catchment protection area falls within the management unit area. A water protection forest is generally defined as serving the maintenance of the purity of groundwater, lakes, and streams ('stehende and fließende Gewässer'). The principal water-protection functions of forests are as follows:

- forests are relatively natural, clean areas, emitting less water-polluting

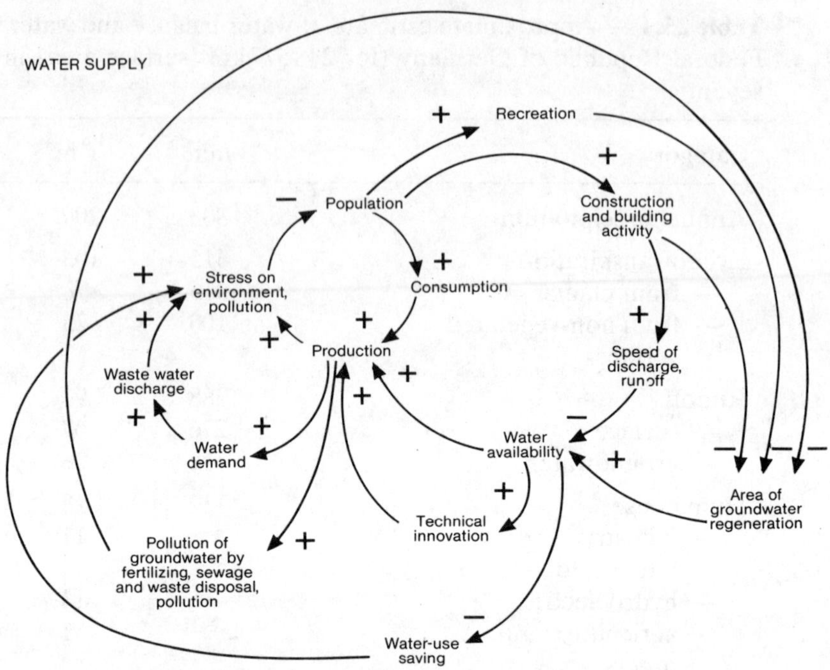

Fig. 25.2 — The interlinkings of the national water supply sector (as illustrated in chapter 'Water conservation, treatment and regeneration', (translated, amended and redrawn by the author) BMI, 1983)
+ connected elements change in the same direction
– conversely, when one increase, the other decreases, and *vice versa.*

substances than any other land cover (except perhaps unmanaged scrubland);

- forest stands and soil purify water by filtering, absorption, and biological decomposition;
- forests lessen the extreme patterns of runoff, reducing peaks during periods of extremely high rainfall and snow-melt and maintaining, under certain conditions at least, water flow in streams during periods of drought;
- forests readily accept and rapidly store water, so adding substantial storage capacity to reservoirs;
- forests effectively cover and protect soils against erosion by water and wind;
- riparian forests are mechanical barriers to fast-flowing floods in flood plains, and improve the biological self-cleaning capacity of water bodies.

There are several categories and zones of water-protection. The most severely controlled water-protection zone is around waterworks, where even forests are not permitted. In some legally disputed cases, water authorities have even demanded replacement of forests by scrub. In the less-severely protected catchment area, forestry is permitted, but certain activities are prohibited (logging with heavy machinery, construction of asphalt roads, and intensive applications of fertilizers and biocides, are examples). Otherwise, in this zone, a major concern is the continued maintenance of a functionally-effective forest structure. Management plans usually prescribe: no formally fixed rotation period; individual tree harvesting; avoidance of clear-felling; natural regeneration; complex mixed stand structure and avoidance of mono-layer, uniform canopy. Recently concern has arisen that the efficient and very substantial removal of pollutants from the air, especially by conifers and aerodynamically-rough canopies, channels more pollutants into the stand, the soil, and finally into the runoff and ground-water, and thereby damages vegetation, soil and water supply.

25.4 FOREST CONDITION AND WATER CYCLE

Forests affect the water cycle in two ways: quantitatively and qualitatively. The quantities of water interception, throughfall, stemflow, evapotranspiration and runoff are influenced by the floristic and architectural structure of the forest stand, and the age of the top-canopy trees. The aerodynamic features of the canopy have an important effect. Aerodynamically-rough canopies with protruding elements intercept and absorb more direct and diffuse solar radiation. There is more scatter of radiation within the canopy and reflected and re-radiated energy is again intercepted within the canopy. A greater proportion of incoming solar radiation is absorbed by the emergent tree crowns and by the canopy as a whole. Within the rougher canopy, there is greater heterogeneity of surface temperatures and infrared transfers. Advected sensible heat, water vapour and fog are more effectively intercepted and finally absorbed. Similarly, large forest tracts can be aerodynamically rough if they consist of structurally heterogeneous forest stands of varying height. This pattern produces strong clothes-line, turbulence and oasis effects. Again the net result is an overall increase of exchange intensity of water vapour and thermal and radiative energy.

The quantitative features and effects on the hydrological cycle of tree species and genotypes vary greatly between sites and between periods with different climatic conditions. For example, the relative performance of two species, such as Norway spruce (*Picea abies*) and Douglas fir (*Pseudotsuga menziesii*), or Norway spruce and beech (*Fagus sylvatica*), may be quite different and sometimes even reversed in prolonged dry or wet periods. As a result, data sets vary, interpretation and generalisations are difficult, and predictions of performance of species in mixture very dangerous. Table 25.2 gives an example of the ranges of rates of evapotranspiration of species and Table 25.3 of the range of variation of interception rates due to stand density and amount of rainfall in a mature Scots pine (*Pinus sylvestris*) stand.

A further source of variation between species and sites affecting the water cycle is rooting intensity and depth, and related to this, soil and parent material texture and structure. Lehnhardt and Brechtel (1980), summarising their own and other quantitative data on rooting depth of beech, oak (*Quercus* spp), red oak, alder (*Alnus* spp), Norway spruce, Scots pine, larch (*Larix decidua*), and Douglas fir, demonstrated the wide range of rooting depth in relation to age of the trees and type of soil, which is as wide as that in tropical forests (Fig. 25.3). For example, Norway spruce, which is quite wrongly often considered to be generally shallow-rooted, is so on gley and gley-type soils and on mineral-poor, summer-dry sandy podzols (10 to 50 cm depth). However, the species is distinctively deep-rooted on well-drained, deep and fertile brown forest soils where rooting depths are between 1.3 m and more than 3.2 m. Such misconceptions as this, misunderstandings and false beliefs are common in the practice and science of temperate and tropical forestry and ecology alike, and are among the major obstacles to rational silvicultural and environmental management.

Another source of variation is the condition of the soil surface. Hydrophobic litter, humus, carbon and surface conditions after drought, and perhaps burning, may change infiltration rates drastically. Undergrowth and ground vegetation may compete with the main canopy for water supply and add substantially to the total transpiration (Table 25.2). A stand of Scots pine with dense grassy ground-cover on moderately dry, sandy soil may evapotranspire annually more than annual precipitation and, consequently, groundwater is depleted rather than replenished under such stands (Brechtel and Scheele, 1982). The influences of patterns of forest-stand distribution, within-stand structures and climatic conditions on snow accumulation and snow-melt is another important source of variation in evaporation and runoff, which has been reviewed more recently for conditions in the uplands of the state of Hesse by Brechtel (1984).

In Germany, even more important from a water-use point of view is the effect of forests on water quality. Forests filter and partly retain nutrients and other substances from the air either as dry or wet desposition. They also retain minerals and organic matter from inflowing surface and sub-surface water and from groundwater. The substances are absorbed, incorporated, recycled, stored, and in part eventually discharged again. Fog and rain drip and stemflow carry to the ground the (often toxic and noxious) substances deposited on the plant; these, with leachates from the plant itself, may eventually enter overland and subsurface flow. Forests add chemical substances to the runoff and thereby affect water quality. Well-known examples are the extremely acid, tea-coloured waters, rich in poly-phenols and low-molecular humic compounds draining from tropical and temperate podzols and peats. Flushes of apparently antibiotic, or at least unpleasant, solutes from tropical forests can be so intense that river shrimps panic and temporarily but hurriedly abandon their usual aquatic habitat for safer streambank shelter (McDowell, pers. comm.). Evidence of flushes of humic acids has also been reported from tropical Kerangas rainforest in Borneo, and Caatinga in the Amazon basin which are thought to be associated with

Table 25.2 — 'Guesstimate' of approximate range of evapotranspiration from mature forest stands under conditions of ample water supply
(The variation is due to conditions of soil, site and canopy structure.)

	Range of evapotranspiration (mm)
During vegetation period	
Beech and Norway spruce	200– 500
Scots pine	50–>350
Douglas fir	250– 600
Selection forest (mixed fir, spruce, beech)	350– 500
Ground vegetation	50–>100
Outside vegetation period (Winter)	
Beech, oak	50– 100
Scots pine	100– 150
Norway spruce, Douglas fir	150– 200

Table 25.3 — Relation between stand density, regulated by thinning regime, and interception for two cases of rainfall intensity in a 120-year-old Scots pine stand (from Brechtel, 1971)

Stand basal area	Relative density of basal area stocking in % full stocking	Interception at two rainfall intensities in mm precipitation per day, in % precip.	
m²/ha	%	3.6 mm	7.1 mm
5	15	1– 2	<1
8–11	30	10–17	1– 3
14–17	50	24–27	5– 7
19–22	70	32–39	9–10

climatic variation and natural forest-stand dynamics (Bruenig, 1966; Bruenig & Schmidt-Lorenz, 1985).

The potential threat of acidification and direct discharge of largely low-molecular acid humic substances from soils into the drainage has been recognised by foresters for a hundred years, but has recently received increasing attention in the course of the 'acid rain' syndrome and has become the subject of public discussion in national and international media. Major concerns are the expansion of conifer plantations and the air pollution. The latter has been reduced in the EC but increased in the Comecon in recent years. On balance, pollution in Europe as a whole has

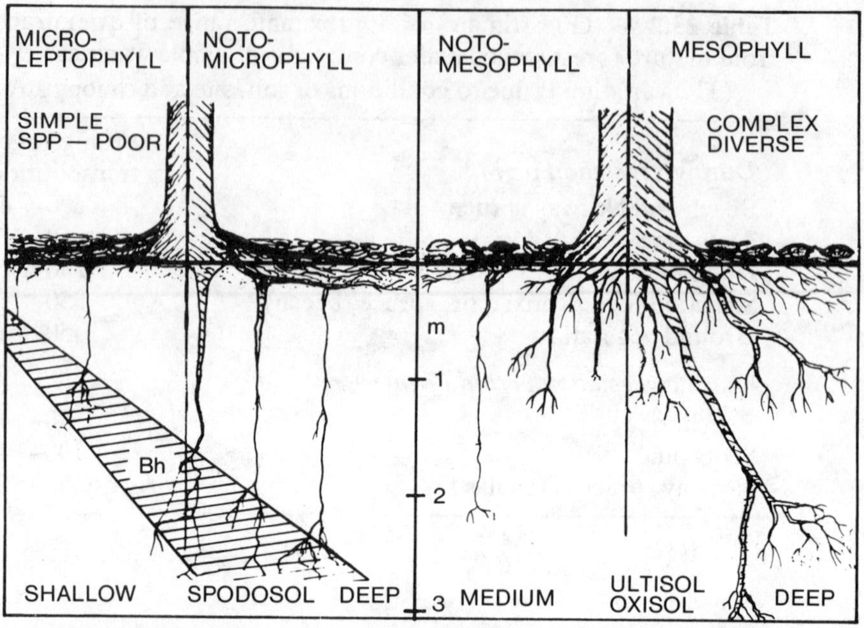

Fig. 25.3 — Dependence of rooting depth of a tree species or of forest stands on physical soil conditions (structure, texture and profile pattern of horizons) (The example is from the author's field work in tropical rainforests in Sarawak and Brunei, but is equally indicative of conditions in temperate forests and forest soils (compare, for example, Fig. 1 in Lehnardt & Brechtel, 1980). The increasing degrees of sclerophylly and xeromorphic canopy physiognomy from deep, mesic, loamy-clayey soils on the right to shallow xeric and more oligotrophic, sandy (or stiff gleytype clay) soils on the left indicates the increasing probability of deficiency of plant-available water and the greater likelihood of occurrence of drought stress during bright periods of low rainfall.)

remained evenly high. It is well known and documented that the effect of forest-tree species on soil acidity depends on the natural edaphic and climatic soil conditions, the state of the soil due to human interference, and floristic composition of the stand, the silvicultural management, and the utilisation regime. Degradation of soils by repeated burning, litter raking, full tree harvesting in short rotations, or unsuitable choice of species may cause increased sensitivity to natural or man-induced acidification. Examples in Germany are the sandy soils in the Lueneburg Heath, on which the Auermuehler Production Programmes were developed as a stabilising remedy. Predicting the development of the soil and of the water yield under certain forestry practices requires a comprehensive dynamic ecosystem appraisal. A useful summarising review of the subject in English has recently been given by Hornung (1985). The effects of acidification on stream water are reviewed by Miller (1985), corroborating the findings of Hornung.

Ulrich (1981) and Matzner (1983) point out that gradual acidification of forest soils in the course of organic matter accumulation, modification and decomposition of litter, amplified by the input of acid-forming substances, not only releases toxic substances into the soil water, but also makes soils less elastic with regard to episodic flushes of acids which occur particularly after dry, hot summers, such as those of 1967 and 1969. Such flushes lead to peaks of transport of mobilised iron, aluminium, heavy metals and nitrate into the drainage and groundwater. Evers (1985) reports results of measurements of inputs of NO_3, SO_4, Cl, NH_4, K, Ca, Mg and Mn with precipitation in ten coniferous forest locations in south-west Germany. He concludes tentatively that in areas with high proton inputs (more than 750 mol/ha in coniferous stands) on mineral-poor soils with low base saturation acidification, increased leaching of bases and release of ionised aluminium and heavy metals are potential threats to the health of the stand and the quality of groundwater.

25.5 CURRENT SILVICULTURAL RECOMMENDATIONS AND TRENDS

Bonnemann and Röhrig (1971) state that, other matters being equal, fully-stocked forest stands (total vegetation) consume (evapotranspire) water at almost equal annual rates, irrespective of species composition and species phenology. The authors conclude a brief review on the matter by stating that effective increases of utilisable water discharge are only possible at the expense of the objective of maximum timber yield. They suggest the following successively more drastic silvicultural measures, whose effectiveness depends on site conditions:

- removal of undergrowth and intermediate storeys in multistoreyed stand;
- heavy thinning to reduce canopy cover to the point at which evapotranspiration from the ground is still subdued and erosion still prevented;
- strip-wise management to provide maximum shelter against direct sunlight and increase snow accumulation;
- as a last resort, replace forest by scrub and herbaceous vegetation, and apply anti-transpirants.

With respect to 'forest decline' the report of the Sachverständigengruppe (1972) recommended the application of fertilizer and lime to sites on which forest growth seems to be retarded by pollutants. However, while the normally small-scale and low intensity of fertilizing in forestry does no environmental harm, the large-scale and heavy fertilizing and liming against pollution effects may cause undesirable and serious environmental side-effects. It may put the nutrient state of the soil even more out of balance, cause undesirable releases, and besides is no lasting remedy against forest decline.

A comprehensive approach to improvement by a bio-cybernetic design of forest stands began in the early seventies with the Auermuehle Production Programmes (APP) (Bruenig, 1975; 1980). These aim at improving simultaneously ecological and economic functions, the dynamic stability and self-regulating capacity of the ecosystem, and the resistance (resilience) of trees and stands against damage. Hydrological improvements result from:

- suitable and target-oriented stand structure, density and species richness (less unproductive evaporation and transpiration);
- less need for chemical treatments (less need for herbicides, pesticides, lime, and fertilizers);
- higher biological activity in the soil (richer supply of leaf and root litter, more rapid recycling, less acidification and leaching).

This approach considers the network of interdependencies within the stand ecosystem (Fig. 25.4). Its general feature is to produce more diverse and dynamic stands, the canopy more open in youth (Fig. 25.5) and more closed, aerodynamically less rough, toward and during economic maturity. The aim is not only to improve the hydrological situation for the stand itself, but also the quality of the water discharge by:

- avoidance of acid humus accumulation;
- diversity of litter and soil life;
- continuity and balance of accumulation and dissipation (growth and decomposition) processes.

The principles are illustrated by the general features of the example of conventional yield table silvicultural in Scots pine and the new, successional-rotational and more complex and diverse silviculture of the Auermuehle approach on the same site (Fig. 25.6).

25.6 SILVICULTURAL OPTIONS: LIMITATIONS AND POSSIBILITIES

With respect to securing and improving quantity and quality of water supplies from forests, silvicultural measures which influence hydrological condition effectively are:

- choice of species;
- method of soil preparation and stand establishment;
- growing stock density and structure.

Generally, choice of species and silvicultural techniques affect the quantity of water discharge substantially only if large areas are involved. Forestry in Germany is an established going concern with a great diversity and complexity of forest conditions. Any changes of silvicultural practice may immediately affect water cycle, performance and stability of an individual stand or a series of stands, but will become effective for large tracts of forests only over long spans of time. Short-term effects on the water cycle are (at least at the beginning) more or less negligible and much less than in areas of large-scale afforestation or deforestation.

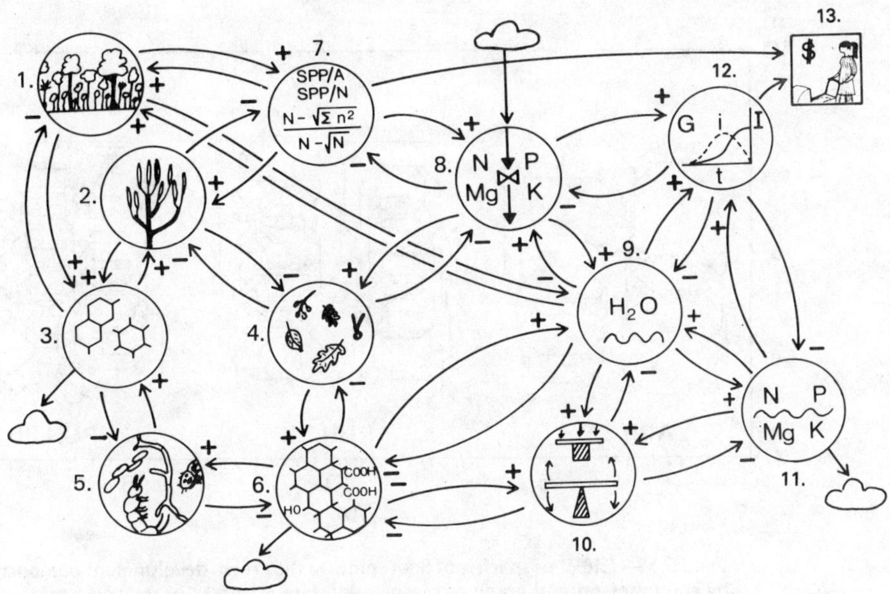

Fig. 25.4 — Some of the interdependent features and interacting feed-back properties in natural forest and plantation ecosystems which affect the water cycle

1. *Stand structure* (complexity of architecture and organisation of the ecosystem);
2. *Physiognomy* of the canopy trees;
3. *Biochemical* composition of the plant tissues;
4. *Litter* composition and amount;
5. *Organisms* as consumers and decomposers on and in the soil (including pests, diseases, N-fixers and humus formers);
6. *Soil-humus* as stabilising and absorbing complex;
7. *Species-richness* (spp/A and spp/N) and *floristic diversity*;
8. Biomass as *nutrient filter* of inputs from the environment;
9. *Water* availability and cycling (nutrient 'pump', drought);
10. *Stability* features (static and dynamic); referring to dynamic equilibrium states with elastic or persistent stability of survival, resilience and resistance;
11. *Nutrient* cycling, balance and outflow (loss);
12. *Growth, increment and yield* as resultant of the interacting system process;
13. System structure (incl. amenity) and processes effect *human welfare* and *happiness*.

The Federal Ministry of Food, Agriculture and Forestry had a working group of experts established in 1971 to 1972 to report on the relationship between forestry and the environment in the Federal Republic of Germany. The report (Sachverständigengruppe, 1972) stated, with respect to the volume of water supply from forests, the following:

- Clear-felling affects the water cycle and yield on the site and in small catchments or around water bodies. It has very limited consequences over the large areas because only around 1% of any large forest tract will be clear-felled in any one year, and the vegetation cover will quickly be restored naturally and by reforestation.

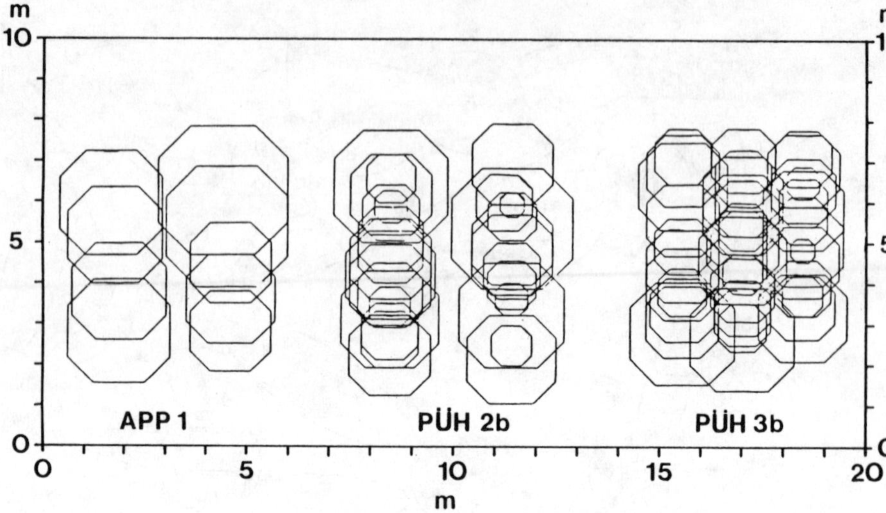

Fig. 25.5 — Effect of spacing of Scots pine on the crown development demonstrated by computer-printed graph of sample plot data at age 12 years, using mean crown projection diameter.
APP 1: Auermuhle Production Program, planted 3×1 m
PUH 2b: conventional planting at 1.2 × 0.3–0.4 m, every second row removed at the age of 5 years
PÜH 3b: as before, but left at initial spacing (from Bruenig *et al.*, 1985a)
The wide spacing in APP 1 stimulates biological soil activity and ground vegetation. The dense canopy in the sample plot PÜH 3b not only increases the interception of light and water, but also retards stem and crown diameter growth and it initiates the phase of massive litter and humus accumulation, which peaks at pole-stand age.

- The water discharge (surface and subsurface runoff) of most catchments varies only marginally in relation to tree species and the structure of stands and forest tracts in the order of ±5%.

Clear-felling, wide initial spacing of trees and early heavy thinning and pruning, reduce interception rates for light, moisture, and pollutants. They increase penetration of radiation, rain, and air. The amounts vary with the patterns of branching and wind turbulence in the canopy. Increased light and eddy penetration has a strong effect on the vigour, vitality and stability of individual trees and stands. Wide spacing and thinning improve the biological activity in the soil, the quality of the humic matter in the soil, and the soil moisture conditions and the quality of the water discharge. However, the effect on the volume of water discharge by comparison is rather small and ephemeral. The hydrologically-important aspects of maintaining a more open canopy, a complex structure, and a rich ecologically favourable ground vegetation and a consequently diverse litter supply to the soil are:

- improvement of water quality by avoiding or reducing litter and raw-

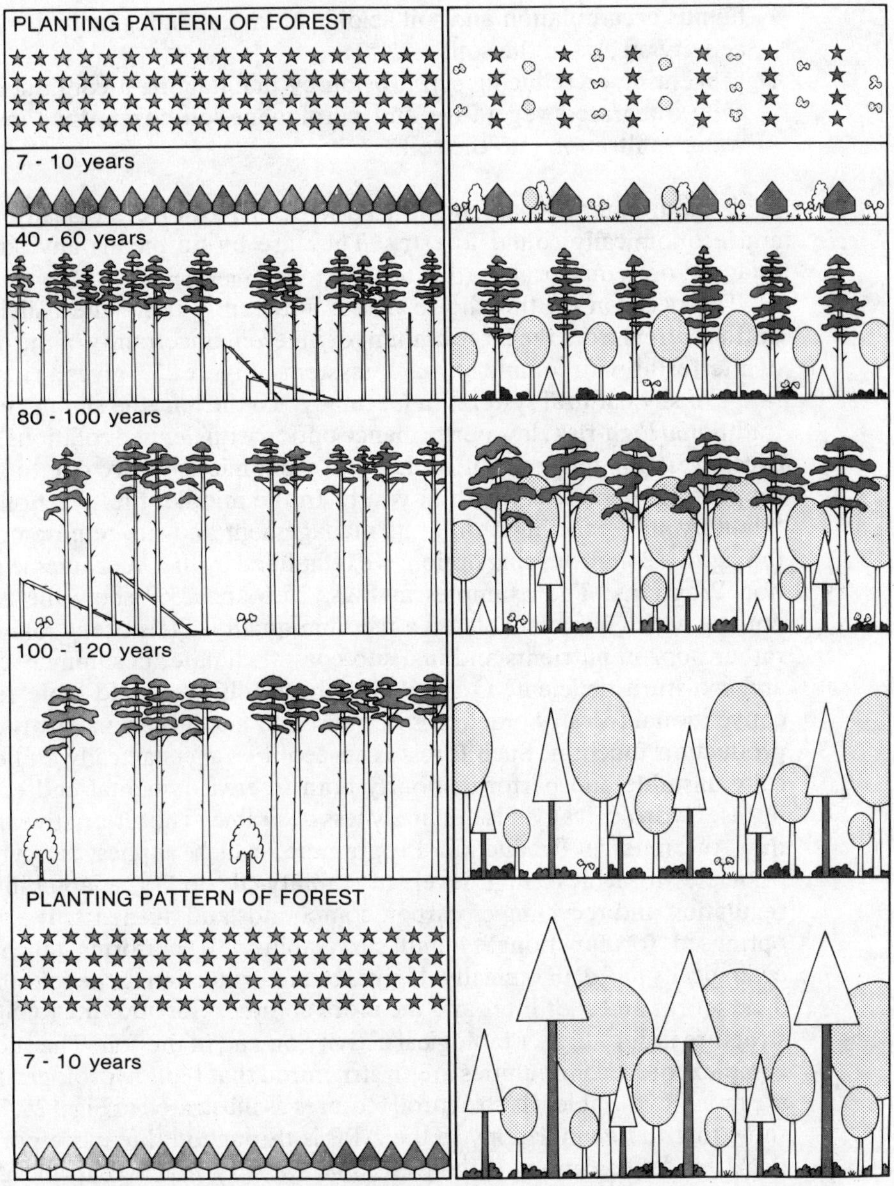

Fig. 25.6 — Structure and rotational features of conventional yield table normal forest (left), and of the ecologically and economically integrated, high-performance management system Auermuehle Production Programme (APP) on a moderately nutrient supplied and summer-dry sandy soil on a natural oak-birch (beech) forest site in the Pleistocene lowland of Lower Saxony.

The main distinctive features with respect to hydrology are:

Left	*Right*
Mono-layered	Multi-layered
Uniform monoculture	Diverse, successional
Raw humus accumulation	Rapid decomposition
High density of stems	Low density of stems
Low self-regulation	High self-regulation
Clear-felling	Integrated regeneration
Successive crops	Overlapping rotations
Aerodynamically smooth	Aerod. mod. rough

humus accumulation and soil acidification, and inducing high biological activity on and in the soil;

- preventing or reducing soil erosion by maintaining a continuous vertically integrated vegetative and permanent soil cover, thus improving water infiltration into the soil.

Both requirements conform with basic principles of ecologically adapted and economically sound forestry. They are by no means any additional financial or managerial burden to forest management.

The decision on the choice of the apparently best-suited silvicultural option suffers from the large amount of inherent uncertainty of the outcome of interfering with complex dynamic systems. Figure 25.7 gives an idea of the range of silvicultural systems in Germany. To the left, the simple, uniform, traditional high-risk, low performance option with defined rotations, usually excessively long as a result of delayed economic maturity due to severely suppressed diameter growth in youth. In the middle, the selection system 'Plenterwald', exacting both in its management and site requirements. To the right, a compromising option as exemplified by the Auermuehle Production Programs. The examples in Figs. 25.6 and 25.7 show the range of options which are available on a medium quality, Pleistocene loamy sand, rather poor in nutrients and in a sub-coastal climate, in summer often dry and moisture deficient. On the left, the traditional yield-table forest of conventional forestry practice is optimised to maximise a narrowly-defined production function. Such forest is susceptible, mechanically and ecologically unstable and performs poorly both in environmental and economic terms. Hydrologically it is definitely less desirable. The alternative option is the Auermuehle Production Programme, which applies bio-cybernetic principles to achieve high levels of stability, flexibility, adaptability, self-regulation and recycling of carbon compounds and nutrients. It is broadly optimised for functionality and survivability. It performs much better ecologically and economically. It gives better protection to the soil; the more diverse leaf and root litter and the more complex floristic and architectural structure induce higher biological activity on and in the soil. The successive complex forest communities are so structured that their hydrological effects are more favourable. Its structural features as illustrated in Figs. 25.5 to 25.7 show that the main canopy in the APP is throughout life more permeable than the dense canopy of conventional forestry practice at all ages, but particularly during the ecologically and economically critical thicket and pole stages, that the main economic crop grows fast and the shortened rotations overlap. Edaphic conditions are maintained by continuous but species-rich, diverse and well-structured cover.

The complex forest ecosystem of the APP is richer than the yield table forest in animal and plant species which stabilises both forest stand and soil. The total evapotranspiration per area unit is probably not less than in the yield table forest. Its productive transpiration may possibly even be somewhat higher. Its capacity to filter toxic and noxious emissions, and aerosol and solute nutrients, such as sulphuric and nitrogenous compounds, into the

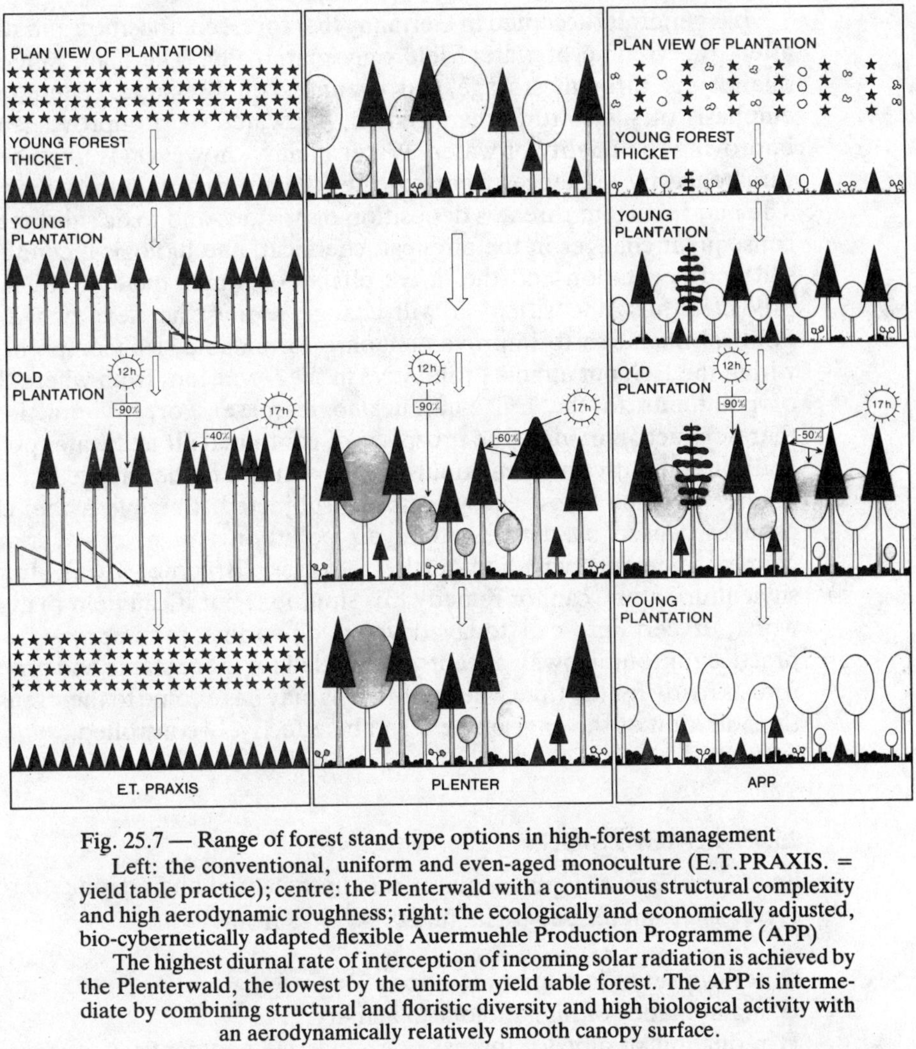

Fig. 25.7 — Range of forest stand type options in high-forest management

Left: the conventional, uniform and even-aged monoculture (E.T.PRAXIS. = yield table practice); centre: the Plenterwald with a continuous structural complexity and high aerodynamic roughness; right: the ecologically and economically adjusted, bio-cybernetically adapted flexible Auermuehle Production Programme (APP)

The highest diurnal rate of interception of incoming solar radiation is achieved by the Plenterwald, the lowest by the uniform yield table forest. The APP is intermediate by combining structural and floristic diversity and high biological activity with an aerodynamically relatively smooth canopy surface.

system's cycle is most probably higher than in a repetitive uniform, monoculture, but also higher is its capacity to buffer and adapt. The APP forest, as a result of continuously high biological activity in and on the soil, is less prone to produce and discharge mobile, low-molecular humic acid flushes into the drainage. The effect of the APP on the volume yield of utilisable water may be negligible, but its effect on water quality may be more substantial. It may help to alleviate the currently topical problem of acidification of fresh waters, and to revitalise those waters which have already been biologically damaged. In a state of uncertainty and unpredictability of environmental and economic developments, the APP, and similar systems, also help to reduce risk of failure to achieve ecological and environmental objectives by being so broadly optimised that they possess a high level of elasticity, adaptability, self-regulation, and buffering capacity.

It is generally accepted in Germany that forest composition and structure affect the volume of water yield only marginally. The main influence on quantity is through storage and evenness of discharge. Therefore, the emphasis of silvicultural management for water yield improvement is on improving the quality of water. Water quality, however, is more seriously threatened by influences from outside forestry than by forestry practices. The major current threat is deposition of noxious and toxic substances, the consequent changes in the physical, chemical, and biological conditions of soil and vegetation and the direct effects on water quality. It cannot be predicted how the situation will change during the next two decades. Political obstacles to improvements are formidable, if incomprehensible, within the EC, but almost prohibitive in the Comecon, from where the flow of pollutants to the EC continues to increase. Forestry practice must therefore accommodate the prospects of continued air and water pollution, and consequently chemical inputs, which will add to the natural stresses and strains to which forest ecosystems are subjected. Possible global climatic changes caused by deforestation, air pollution and natural variation of climatic elements will add further stresses (Bruenig, 1975). Improved silviculture alone cannot remedy this situation, but it can help to avoid the worst. It certainly can today develop blueprints for improved forestry practice for tomorrow. Concepts are needed for more stable and productive forests ready for the time when politicians may have come to their senses and the pollution of the environment can be effectively controlled.

25.7 CONCLUSIONS

Forests in Germany play a major role in the aerial and edaphic freshwater cycle. Their major beneficial functions arise from:

- soil conservation and protection;
- maintenance of high biological activity;
- reduction of albedo, increase of advective and convective (forced and free) turbulent airmass exchange.

Consequently, atmospheric and edaphic water flows are smoothed and water quality for human consumption increased. Both are important social services by the forest which, in Germany and elsewhere, is threatened by:

- natural tendencies to acidification;
- inadequate silvicultural systems and practices;
- expected trends of climate in Europe toward greater variability with more frequent extremes (drought, flood, cold, heat, gale) as a result of global air pollution and removal of tropical and boreal forests (Bruenig *et al.*, 1985b);
- stress and strain on the forest ecosystems from noxious and toxic substances.

The situation is serious and urgently requires effective action to:

- improve scientific research (quality rather than quantity);
- introduce better-adapted silvicultural practices and systems;
- stop emission of pollutants from industry, traffic, and households;
- stop deforestation and general environmental degradation throughout the world.

However, the urgently-needed action is obstructed globally by:

- political opportunism and preoccupation with power;
- scientific prostitution and preoccupation with prestige;
- professional dogmatism and traditional reliance on pre-scientific experience and empiricism;
- bureaucratic indolence and ignorance;
- public misinformation, misapprehension, and misunderstanding of the relevance of their own activities and the complexity of interdependencies and feedback mechanisms, which affect the environmental quality in general and the freshwater quality in particular.

REFERENCES

AGDW-Arbeitsgruppe (1976) Entwurf für eine erste Sozialbilanz des Waldes. Hamburg-Bonn, unpublished report, mimeogr., pp. 40.

BMI (1983) Abschlußbericht der Projektgruppe *Aktionsprogramm Ökologie.* Umweltbrief Nr. 29. Bonn, Bundesministerium des Inneren, pp. 127.

Bonnemann, A. & Röhrig, E. (1971) Waldbau auf ökologischer Grundlage, von A. Dengler, Vol. I. *Der Wald als Vegetations-typ und seine Bedeutung für den Menschen.* Hamburg-Berlin, Paul Parey, pp 229.

Brechtel, H. M. (1971) *Zur Bedeutung der gebietshydrologischen Forschung für die Landschaftsplanung, Landschaft und Stadt, 3. Jahrgang,* Verlag Eugen Ulmer, Stuttgart.

Brechtel, H. M. (1982) *Influence of forest cover on the hydrologic regime of mountain streams-watershed conservation and restoration, water yield management.* FAO/EFC Working Party on the *Management of Mountain Watersheds*, 13th Session. Hann. Muenden, 182, 31 pp.

Brechtel, H. M. (1984) Möglichkeiten und Grenzen einer Steuerung der Schneeansammlung und Schneeschmelze durch forstliche Maßnahmen. *Deutscher Verband für Wasserwirtschaft und Kulturbau* Mitt. 7: 331–354.

Brechtel, H. M. (undated). Sozialfunctionen des Waldes. Hess. Min. Landesentwicklung, Umwelt, Landwirtschaft und Forsten, Hess. *Forstl. Vers Anst.*, Hann. Muenden, p. 12.

Brechtel, H. M. & Scheele, G. (1982) Bewirtschaftung von Grundwasser durch land- und forstwirtschaftliche Maßnahmen. *Deutscher Verband für Wasserwirtschaft und Kulturbau*, Darmstadt, p. 48.

Bruenig (Brunig), E. F. (1966) *Der Heidewald von Sarawak und Brunei — Eine Studie seiner Vegetation und Oekologie.* Habilitation Thesis, University of Hamburg, Hamburg, pp. IV+117.

Bruenig, E. F. (1971) On the ecological significance of drought in the equatorial wet evergreen (rain) forest of Sarawak (Borneo). *Transact. First Aberdeen-Hull Symposium on Malesian Ecology.* Univ. Hull, Dep. Geogr., Misc. Ser. No. 11: 66–97.

Bruenig, E. F. (1975) Grundsätze zum umweltgerechten Wiederaufbau des privaten Wirtschaftswaldes im Sturmschadensgebiet in Nordwestdeutschland. Landwirtschaft — Angewandte Wissenschaft. Landwirtschaftsverl. Hiltrup, p.p. 311.

Bruenig, E. F. (1980) The means to excellence through control of growing stock. *Proceedings, 1st Weyerhaeuser Science Symposium*, Tacoma, 30.4.–3.5.1979. Weyerhaeuser, Tacoma, 1980, IX + pp. 224, 201–224.

Bruenig (Brunig), E. F. & Schmidt-Lorenz, R. (1985) Some observations on the humic matter in kerangas and caatinga soils with respect to their role as sink and source of carbon in the face of sporadic episodic events. Hamburg University, SCOPE/UNEP Special Volume **58**:107–122.

Bruenig, E. F., Stenmanns, F., Kreysa, J. & Schneider, T. W. (1985a) Auermühler Produktionsprogram für Kiefer. Eine erste Analyse der Sturmschadenaufforstung nach 1972. *Allg. Forst-zeitschr.* **40**, 9/10: 185–189.

Bruenig, E. F., Degens, E. T., Kempe, S., Spitzy, A., Flohn, H. & Bach, W. (1985b). Ist eine Klimaänderung unausweichlich (Is a change of climate inescapable?). *Die Umschau*, **85,** (3), 152–159.

Evers, F. H. (1985) Ergebnisse niederschlagsanalytischer Untersuchungen in südwestdeutschen Nadelwaldbeständen. *Mitt. Verein Forstl. Atandortk. und Forstpflanzenzüchtung,* Stuttgart, Nr. 31: 31–36.

Hornung, M., (1985) Acidification of soils by trees and forests. *Soil Use and Management,* **1,** (1), 24–28.

Lehnhardt, F. & Brechtel, H. M. (1980) Durchwurzelungs- und Schöpftiefen von Waldbeständen verschiedener Baumarten und Altersklassen bei unterschiedlichen Standortverhältnissen. *Allg. Forst-Jagdztg.* **151**, 6/7: 120–127.

Matzner, E. (1983) Die Auswirkungen von Immissionsschäden auf den Waldboden. In: *Die Erhaltung des Waldes — eine nationale und internationale Aufgabe.* Wille, G. & Staa, H. van (eds.). Innsbruck, Wagner'sche Universitätsbuchhandlung, Kommissionsverlag, p. 77–96.

Miller, H. G. (1985) The possible role of forests in streamwater acidification. *Soil Use and Management,* **1,** (1), 28–29.

Olschowy, G. (1978) *Natur- und Umweltschutz in der Bundesrepublik Deutschland.* Hamburg-Berlin, Paul Parey Verlag, 926 pp.

Sachverständigengruppe Forstwirtschaft (1972) *Forstwirtschaft und Umwelt in der Bundesrepublik Deutschland. Ausschuß Entwicklung und Zusammenarbeit in der Forst- und Holzwirtschaft.* Bonn, Ministerium für Ernährung, Landwirtschaft und Forsten, Dec. 1973, revised 9.4.1973, p. 63.

Ulrich, B. (1981) Die Rolle des Waldes für die Wassergüte. *Allg. Forstzeitschr.* **36,** (42), 1107–1109.

26

The effects of forestry on the quantity and quality of runoff in upland Britain

J. R. Blackie and M. D. Newson

26.1 INTRODUCTION

Climate, precipitation, geology, land form, soils and vegetation are the major factors controlling the quantity, chemistry, and time distribution of runoff from any given area. The considerable interdependence among these factors precludes the development of universally-applicable relationships between runoff and any one of them. Simple rainfall–runoff relationships which are apparently valid in one area become ludicrous in others where the controlling factors differ radically. Equally it is not possible to quote simple relationships between vegetation type, precipitation, and runoff without consideration of the other factors. Failure to grasp this basic point resulted in much controversy in early studies of the effects of vegetation on runoff when attempts were made to extrapolate perfectly valid local observations to other environments.

To address the problem expressed in the title of this chapter, therefore, it is necessary first to consider the variability of the British upland environment and the types of land use practised there.

26.2 UPLAND BRITAIN

The uplands of Britain, variously defined, comprise approximately one third of the land surface (Fig. 26.1), lying mainly in the north and west of the country. The underlying geology is complex and highly variable, ranging from shales and mudstones through carboniferous limestones and schists, to granites and hard volcanic rocks. Land form varies from the relatively gently-contoured plateau uplands of south-west England, mid-Wales, the Pennines and southern Scotland, to the steep, rocky, glaciated mountains of

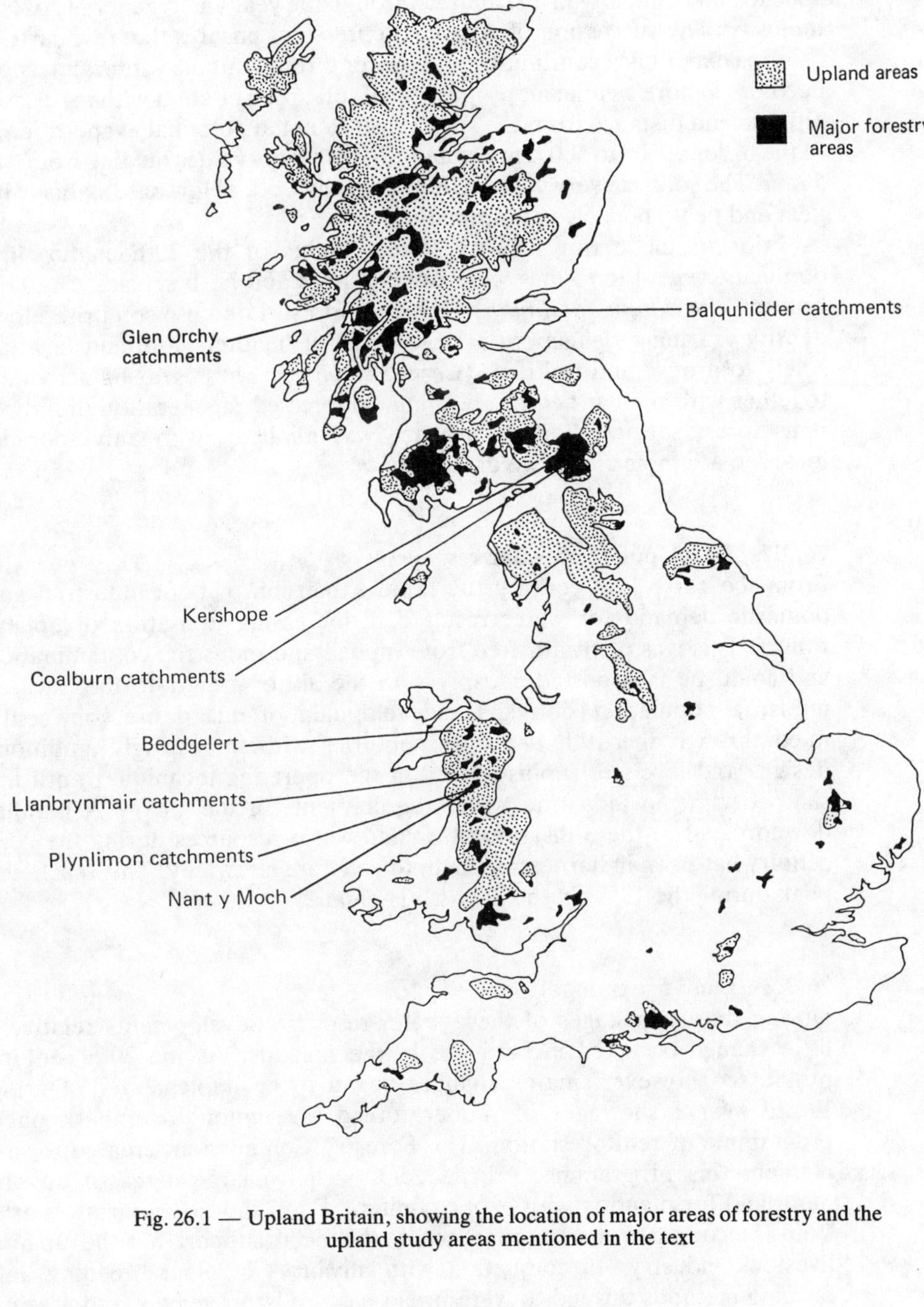

Fig. 26.1 — Upland Britain, showing the location of major areas of forestry and the upland study areas mentioned in the text

North Wales, Cumbria, and the Scottish Highlands. A cool, wet, windy, oceanic climate prevails throughout the area. Precipitation, in the range 1000 to 3000 mm, is well distributed through the year and is generally of low intensity. The succession of Atlantic depressions ensures that few parts of the area experience continuous snow cover throughout the winter, but snow becomes a more significant proportion of total precipitation with increasing altitude and distance from the west coast. Penman potential evaporation is of the order of 300 to 500 mm per annum, with daily values ranging from 0 to 5 mm. The soils are very variable but generally acid, of low base status, with peat and peaty podsols predominating.

Prior to the major forestry developments of the 20th century the dominant vegetation types were upland grass and herb species, heather, bracken (*Pteridium aquilinum*) and scrub forest. Land use comprised low-density grazing by sheep with large areas of heather moorland devoted solely to grouse and red deer (*Cervus elaphus*). These grazing activities, together with regular heather burning, suppressed regeneration of indigenous forest species. Soil disturbance was minimal, with only sporadic localised attempts at surface drainage.

26.2.1 Development of water resources

From the early 19th century the rapid expansion of both industrial and domestic demand for water resulted in increasing utilisation of upland runoff. This was plentiful, free from human and industrial contamination and could be transported cheaply with the aid of gravity to the lowland industrial population centres. The development of upland reservoirs continued through the 19th and 20th centuries with increasingly ambitious designs and levels of sophistication in the operating techniques until the sharp decline in industrial water requirements in the 1970s. A parallel development in the utilisation of upland water resources during the 20th century has been in harnessing them to generate electricity. This reached a peak during the 1950s in the Scottish Highlands.

26.2.2 Land use changes

During the early stages of these water-resource developments relatively little change occurred in land use in the uplands. As the 20th century progressed however, major changes began to be implemented. During World War I shortages of timber forced the nation to embark on a programme of reafforestation. The Forestry Commission, created to implement this programme, proceeded to acquire large areas of upland moorland for planting with exotic conifers. Food shortages during World War II ecouraged a new approach to financial support for the upland livestock industry. In conjuction with advances in plant breeding and farming methods this led to widespread pasture improvement and grazing intensification. Recreational use of the uplands also increased rapidly.

It was the development of plantation forestry that was of greatest interest to the water industry. Initially it was regarded with favour as a means of minimising water pollution by humans and farm livestock, a view endorsed by the post-war government (Minsitry of Health, 1948). As the rate of forestry development increased, however, the debate turned more towards the effects on the quantity of runoff.

26.3 EFFECTS ON RUNOFF QUANTITY

As reservoir design and operating rules became more sophisticated better estimates of the catchment runoff were required. Initially these were based on empirical relationships with rainfall. These were reasonably accurate in the area from which they were derived as long as the land use remained stable. The realisation that the land use changes described above could affect the runoff volume, and hence the operational efficiency of the reservoirs, caused increasing concern.

26.3.1 Early studies

Prior to the 1950s no accurate first-hand information on the effects of afforestation had been obtained in upland Britain. Studies in USA, Switzerland, Africa, and elsewhere had been carried out in very different conditions and could not be extrapolated readily. Most of the American studies, in any case, had been concerned with the inverse problem, namely, the effect on streamflow of forest removal. Hibbert (1967) had summarised these early studies and concluded that forest clearing did result in an initial increase in runoff. It remained an article of faith with the international forestry fraternity, however, that forests were the best cover for water catchment areas.

When Law (1956) published the results of his study at Stocks showing a 290 mm reduction in runoff from his 450 m^2 lysimeter under spruce, as compared with grassland runoff in the area, and proceeded to translate this reduction into economic terms the debate took on a much greater urgency. Considerable effort and resources were then allocated to studies of forest hydrology and hydrometeorology in British upland areas.

26.3.2 Catchment studies

A paired catchment study at Plynlimon (Fig. 26.1) confirmed that runoff from a forested catchment in this wet Welsh upland area was significantly lower than that from grassland. Mean annual streamflow from the former, under a 70% forest cover, was found to be some 300 mm or 15% less than that from the grassland catchment, although both received virtually the same annual precipitation of 2400 mm (Institute of Hydrology, 1976). 'Natural' lysimeter studies within the forested catchment (Calder, 1976) produced results comparable with Law's and compatible with the catchment figures. The continuing accumulation of data from the Plynlimon catchments (Fig. 26.2) verify the basic differences in runoff.

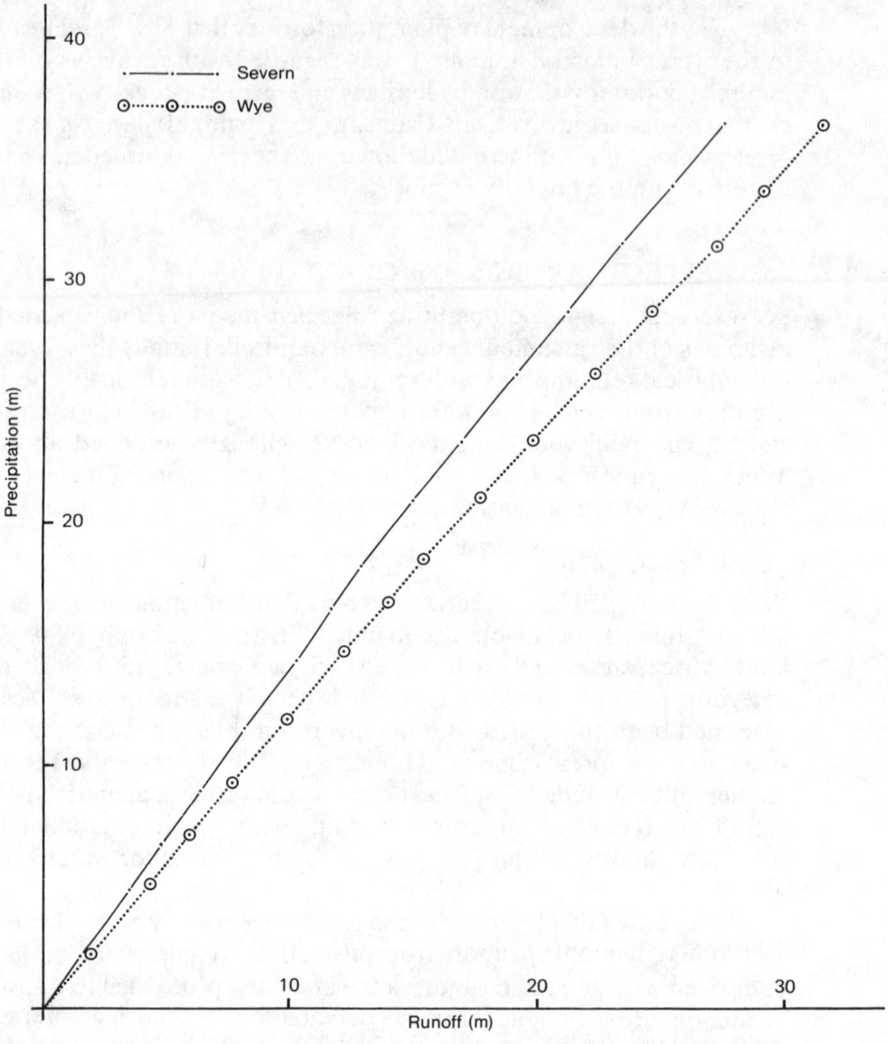

Fig. 26.2 — Cumulative annual totals of precipitation and runoff for the Severn (70% forest) and Wye (grassland) catchments at Plynlimon

26.3.3 Process studies

Concurrently, detailed studies of the physical processes occurring within forest canopies (Rutter, 1966; Stewart, 1976; Shuttleworth, 1977) began to elucidate the processes responsible for this observed streamflow reduction. As had been suggested by Law (1957), the key process was the interception of rainfall by the forest canopy. The qualitative understanding of the effects of this which emerged runs as follows.

- Under dry conditions the transpiration rate from coniferous forest is marginally lower than that from grassland under the same conditions.

- Forest canopies intercept more of the incoming rainfall than grassland.
- The forest canopy is aerodynamically much rougher than grassland. This results in more turbulence, that is more efficient mixing of the air above the canopy, and consequently more opportunity for evaporation.
- The rate of this evaporation from a wet canopy is considerably greater than the transpiration rate from a dry canopy under the same conditions.

26.3.4 Prediction methods

The implications of this model of canopy behaviour are far-reaching. They make it possible to reconcile many apparently contradictory empirical observations, and to develop methods of predicting the water use of forest in wet areas using precipitation and meteorological data only. The efficiency of what is, in effect, a 'clothes line' evaporation process is such that it can occur during as well as after rainfall, winter and summer, night and day. Roberts (1978) studied the complementary process of transpiration in Norway spruce close to Calder's lysimeter on Plynlimon, and concluded that interception and transpiration were mutually exclusive processes. Thus Calder and Newson (1979) were able to derive a simple expression to predict the increase in catchment losses due to afforestation by assuming that a single interception ratio was applicable throughout the uplands during wet conditions, and combining it with the Penman prediction of transpiration for dry periods. Applying the expression required only rainfall data and an agricultural handbook (MAFF, 1967) as input information. The method was used to produce a map of predicted loss increases for the catchments of all of Britain's major reservoirs, assuming the levels of afforestation which were being predicted at the time (Forestry Commission, 1977). Despite the acknowledged limitations of the method the predictions resulted in the water industry taking a guarded position on forestry expansion on upland reservoir catchments (Centre for Agricultural Strategy, 1980).

26.3.5 Additional research requirements

It is now generally accepted that there are enhanced evaporation losses from forest relative to grassland in the wet, windy upland areas of Britain, and hence lower quantities of water available for runoff. Predictions of the loss differences relative to grassland can be made from rainfall and meteorological data.

It is recognised, however, that more information is required on the snow-interception characteristics of forest canopies and on the hydrometeorological characteristics of heather and bracken before accurate predictions can be made for the more extreme areas. This refers particularly to the Scottish Highlands where current projections suggest most of the new afforestation in the next 50 years will take place, and where heather rather than grassland is the dominant existing cover. To meet this requirement a new programme of research was initiated by the Institute of Hydrology in 1981, with funding from a Consortium of environmental, water, and forestry interests. The programme comprises a paired catchment study at Balquhidder (Fig. 26.1), together with studies of snow interception, and of the interception and transpiration characteristics of heather.

26.3.6 Other aspects

Whilst evapotranspiration controls the volume of water available for streamflow, there are many other factors controlling the movement of the residual rainfall through the catchment system to the river or reservoir. These affect its distribution in time and also its quality. The increasing importance of these aspects of runoff to the water industry and other downstream interests is mirrored in the increasing emphasis now being placed on studying the effects of afforestation on them.

26.4 EFFECTS ON RUNOFF TIME DISTRIBUTION

The time distribution of runoff is determined principally by the distribution and intensity of effective rainfall. The runoff hydrograph is modified by the catchment topography, by soil infiltration rates and storage capacities, and by the storage-discharge relationships of shallow aquifers which control the baseflow component.

26.4.1 Seasonal effects

The extent to which different forms of land use modify both the seasonal and the storm hydrographs is not easy to determine. Paired catchment studies present problems of interpretation in this respect since soils, topography, and aquifer storage are rarely identical. For example, the long-term means from Plynlimon (Fig. 26.3) suggest a fairly uniform reduction in flow throughout the year from the forested catchment. With no information on the pre-afforestation seasonal flow pattern, however, this is not conclusive evidence that no change has occurred. Ideally this problem should be studied on a before-and-after basis.

A study of this type in relation to afforestation was initiated on the Coalburn tributary of the Irthing, in the northern Pennines, in 1966. After a five-year calibration period this catchment was ditched and planted. Obviously it will be some time before it will yield any firm evidence on the long-term seasonal effects, but it has already provided valuable information on the effects on the storm hydrograph.

26.4.2 Pre-planting treatment

Afforestation in upland Britain has ventured boldly on to the most difficult soils, suffering mainly from poor drainage and nutrient deficiency. To achieve reasonable growth during the establishment phase it has been necessary to dig furrows and drainage ditches across the plateaus and down steep hillsides, creating in many areas soil disturbance for the first time in 5000 years. An early investigation of the effects of drainage ditches on deep moorland peat soils (Conway & Millar, 1960) established that gravitational drainage of such soils is almost impossible. Restricted drawdown of soil

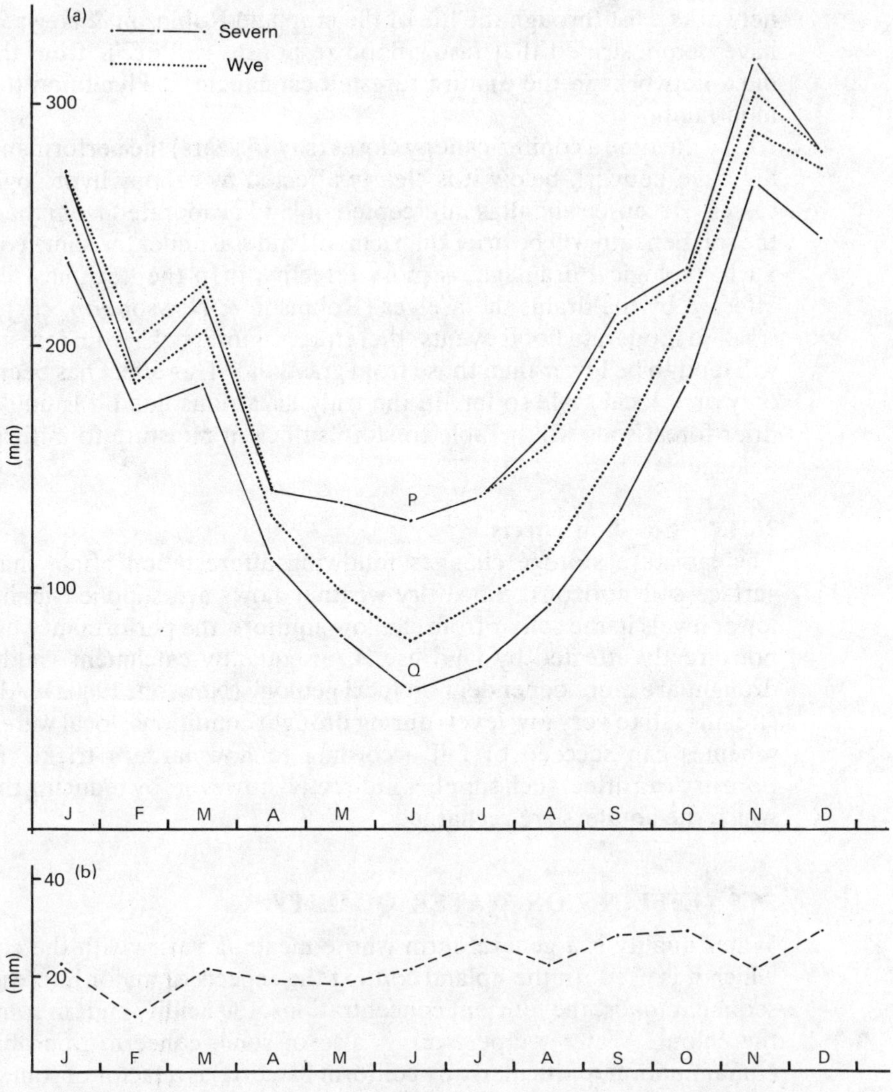

Fig. 26.3 — Mean monthly values, 1969–83, from the Severn (70% forest) and Wye (grassland) catchments at Plynlimon (a) Precipitation, P, and runoff, Q (b) Runoff differences between Wye and Severn

water occurs only immediately around the ditches. Consequently during heavy or prolonged rainfall the ditches act merely as collectors for surface, or near-surface, runoff.

26.4.3 Flood hydrograph modification

The changes observed in the flood hydrographs on the Coalburn catchment (Robinson, 1980) confirmed Conway & Millar's findings: the ditching prior to planting leads to reduced time-to-peak and higher peak flows. The ditch networks exist through the life of the crop and Robinson & Newson (1985) have demonstrated that faster flood responses still occur from the dense ditch networks in the mature forested catchment at Plynlimon than from moorland.

By the time a conifer-canopy closes (say 15 years) the performance of the drainage network below it is clearly affected by canopy hydrological processes. Because rainfall is intercepted on and evaporated from the canopy, the soil beneath will be drier than a moorland soil under the same conditons. Such 'biological drainage' is more effective than the 'technical drainage' effected by the drains themselves (Robinson & Newson, *op. cit.*). During small to moderate flood events, therefore, runoff peaks from forested areas will tend to be lower than those from grassland. This effect has been proved only on a local scale so far. In the truly hazardous flood it is unlikely that drier forest soils will be able to store sufficient moisture to mitigate flood damage.

26.4.4 Low-flow effects

The moisture storage changes following afforestation affect mainly the surface soil horizons. Since dry-weather flows are supplied mainly from lower levels in the soil or from shallow aquifers, the performance of these is not directly affected by land use. Consequently catchment yields during drought are more dependent on local geology (Newson, 1980a). All upland streams fall to very low levels during drought conditions; local water-supply schemes can succeed or fail according to how large a trickle remains. Forestry can affect such supplies indirectly, however, by reducing the rate at which the aquifers are recharged.

26.5 EFFECTS ON WATER QUALITY

Water quality is a general term whose meaning varies with the context in which it is used. In the upland context the aspects of major interest are the sediment loads, the nutrient concentrations, the acidity and, in some areas, the colour. Water temperature is also of some concern. Microbiological contamination, particularly by coliform bacteria, is a factor of some importance in grassland catchments, but not in forested ones (McDonald & Kay, 1981) unless they are used also for recreational activities.

26.5.1 Sediment yields

Steep mountain streams possess considerable energy and it is inevitable that any available loose material will be transported to the reservoir or the lowland reaches. In their stable, 'natural' state most upland streams are supply limited. Following the land preparation for afforestation described above, catchment yields for both fine and coarse sediments increase partly because of the changing flood regime, but mainly because the soil disturbance makes available a large new supply of material for transport.

Studies on the Coalburn catchment (Robinson & Blyth, 1982) suggest that the yield of fine sediments peaks rapidly after site preparation, then stabilises at a higher level than that observed before the treatment. Under the 30-year-old forest on Plynlimon, intensively ditched at planting time, suspended-sediment yields are double those from the grassland catchment, and yields of bedload gravel are five times as high (Newson, 1980b). Preliminary results from the recently initiated study at Balquhidder in Scotland suggest similar values for the differences there between mature forest and moorland.

Obviously the magnitude of the increases in sediment load associated with forestry will vary with the conditions and management practices on each site. Nevertheless the point remains that the soil disturbance during site preparation results in a long-term increase in the supply of sediment available. The extent to which sediment yields are further modified during clear-felling and subsequent re-planting, forms an important part of the next phase of the ongoing studies by the Institute of Hydrology at Plynlimon and Balquhidder.

26.5.2 Stream chemistry

The chemical changes occurring in water between its arrival in the catchment as precipitation and its departure as streamflow are extremely complex. Geochemistry and soil chemistry are basic site-specific factors. The hydrological pathways followed by the water are equally important however, since these determine the materials with which the water can interact, and the duration of contact. Thus rapid storm runoff is likely to emerge with a different chemical imprint from that found in baseflow. Vegetation also plays an important part. It is the first point of contact for some of the precipitation and it is a major factor in determining the nutrient status of the soils. As has already been indicated, it is also a major factor in determining the volume of water passing through the system and the pathways it follows.

Land use change, therefore, can be expected to cause changes in the chemistry of the runoff through the direct effects on vegetation and soils, and the indirect effects on hydrology.

26.5.3 Nutrient concentrations

The changes of most concern to the water industry, and indeed to the upland land user, are changes in nutrient concentrations. Increases in these represent potential increases in treatment costs to the former and potential productivity losses to the latter. Nitrate concentrations in upland forested areas are unlikely to exceed the statutory limits for potable water, but any significant increase would cause some concern since it would reduce the dilution effect on downstream inputs. Phosphate concentrations are of concern because of their established link with algal growth (Holden, 1976), in lowland impoundments at least, and the fear that significant increases could result in eutrophication in upland reservoirs which would radically increase treatment costs (Youngman & Lack, 1981).

Afforestation in its mature, stable phase has been shown to have little effect on the nutrient concentrations in the runoff from the Plynlimon catchments. Monitoring of the nitrogen, phosphorus, and potassium concentrations indicated very low mean values of 0.35, 0.01, and 0.2 mg/l respectively from both the forested and the grassland catchments, the only significant difference being that the nitrate-N concentrations from the grassland were marginallly lower than those from the forest (Roberts *et al.*, 1984).

Despite these low concentrations in the stable state, very considerable quantities of nutrients, particularly nitrogen in its organic forms, are potentially available from upland soils (Batey, 1982). These can only be released, however, if much more rapid mineralisation is induced by, for example, major soil disturbance. Afforestation, with its extensive soil disturbance at both the planting and the felling stages, could result in increasing losses. A literature review in 1979 revealed a wealth of information on the effects of forest management on water quality in North America and other areas, but only sparse data from the British uplands or any comparable environments (Blackie *et al.*, 1980). Consequently a programme of research was initiated to determine the magnitude and duration of the effects on water quality of the critical phases of forestry and of upland grassland improvement.

With funding and support from the Natural Environment Research Council, the Department of the Environment, the Water Research Centre, the Welsh Office and the Forestry Commission, this co-ordinated programme is being undertaken by the Institute of Hydrology, the Institute of Terrestrial Ecology, the Freshwater Biological Association, and the Water Research Centre. It comprises paired catchment studies, process studies of the soil/water effects, and studies of the impact of the changes identified on aquatic biota and on water-treatment requirements.

Efforts are being concentrated initially on the 'major disturbance' phases of forestry, namely, land preparation and initial planting on Llanbrynmair Moor (see Fig. 26.1), clear felling at Beddgelert, Kershope, and Plynlimon, and remedial fertilizer application in Glen Orchy. Also included in the programme are the effects of grassland improvement at Plynlimon and Nant y Moch.

Results from these studies, initiated in 1981 to 1982, will take some time to emerge but it is already evident that the soil and vegetation disturbance associated with the initial planting, clear felling and grassland improvement operations results in some increases in the concentrations of NO_3-N, PO_4-P and K.

Baseline studies of the nutrient concentrations from the moorland and mature forested catchments at Balquhidder are being undertaken by the Macaulay Institute prior to initial planting and clear-felling operations there. In addition the Department of Agriculture & Fisheries for Scotland (DAFS) Freshwater Fisheries Laboratory and the Forth River Purification Board are monitoring the aquatic biota in the Balquhidder streams.

When these studies all come to fruition the present dearth of hard information on the effects of upland management practices on water quality and their consequences for the water and fishery industries should be rectified in part at least.

26.5.4 Acidity

Perhaps the most contentious parameter in connection with upland water quality is its acid status. Newson (1981) has pointed out that an acid stream environment is typical for the British uplands, dating from the post-glacial leaching of bases from soil profiles and the growth of peat. The limited buffering potential of upland soils makes the streams particularly sensitive to changes in the acidity of incoming precipitation. Such changes, their regional extent and the factors giving rise to them are currently the subject of intensive study and are outside the scope of this chapter. Land-use change, however, can result in changes in the acid status of runoff and it is this aspect that is dealt with here.

Effects of runoff acidity on fish stocks were first recorded in Scotland (Harriman & Morrison, 1982). Subsequently there have been reports linking decline of fish numbers and water quality in upland Wales (for example Crisp *et al.*, 1980). In every case forested streams have been shown to exhibit lower mean pH values and a tendency to record acid flushes when rain storms follow prolonged dry periods. Given the low base status of the soils, the factors contributing to an increase in acidity following afforestation are closely linked with those which cause changes in the hydrological response. The large surface area of the forest canopy and the way it promotes efficient mixing of the air above it makes it a very efficient scrubber of dry pollutants from the airstream, These are flushed from the canopy by subsequent storm events. This effect, in conjunction with the concentration effect of canopy evaporation losses, results in greater concentrations in the net rainfall reaching the soil. Thereafter the more efficient drainage by the forest ditch system ensures that the rapid response component of runoff will be in contact with the soils for less time, thus reducing the neutralising effect of any remaining bases. In some areas the presence of forest has been shown to accelerate drainage even further by drying and cracking the soil (Binns, 1979).

Obviously the magnitude of the change in pH resulting from the above effects will be site-specific, both in terms of the soils and the degree of atmospheric pollution. The point remains, however, that runoff acidity will increase when areas with base-deficient soils are afforested.

26.5.5 Temperature and colour

These aspects of water quality have received less attention than the others. The effects of afforestation on stream temperature are essentially a reduction in the diurnal range as a result of shading. In the Plynlimon catchments, for example, afternoon temperatures in the forested catchment streams may be 6 °C lower than those in the grassland. This effect does not persist far beyond the forest boundary and is mitigated within the forest by clearings. Interest in temperature stems mainly from its possible effects on fish and their food chain.

Colour is a localised effect not specifically related to forestry. Where it is a problem it proves to be a costly one in treatment terms. The causes are obviously to be found in the chemistry of the soils, rock or even the vegetation of the catchment and may be dependent on the acid status of the stream.

26.6 CONCLUSIONS

From a situation of comparative ignorance 30 years ago advances have been made to a stage where it can be stated categorically that afforestation of former grassland in the wet, windy areas of upland Britain results in increased evapotranspiration loss. The processes giving rise to this increase are understood and, given adequate rainfall and meteorological data, the magnitude of the increase can be predicted. For those areas where the existing vegetation is heather rather than grass, and where snow is a significant proportion of the precipitation, knowledge is less complete. Studies are currently in hand, on both the catchment and the process scale, to rectify this situation.

Some changes in the time distribution of the reduced total runoff have been identified. Intensive ditching prior to planting results in a more rapid rise to higher flood peaks in storm events. Whilst the magnitude of runoff decreases after closure of the forest canopy, this more rapid flood response persists throughout the growth cycle. No firm evidence is yet available on the effects on seasonal runoff distribution, but consideration of the processes leading to enhanced evaporative loss suggests that post-drought recharge rates are reduced.

Any intensification of land use involving disturbance of soil and vegetation leads to changes in the quality of runoff. In the case of upland afforestation, the intensive surface ditching practised in first crop planting, results in considerable increases in sediment loads. At Coalburn suspended-sediment concentrations stabilised at double those observed prior to planting, after an initial peak. Bedloads from the mature forested catchment at Plynlimon are some five times those recorded from grassland.

Under mature forest, stream nutrient concentrations are very low and comparable to those from unimproved grassland. These concentrations increase during the major disturbance phases of planting and clear felling. The magnitude and duration of these increases, their effects on the stream biota, and their implications for water treatment are being investigated in a co-ordinated series of studies.

In what is generally a fairly acid environment, afforestation can result in a further decrease in stream pH values. Thus when increases in atmospheric pollution occur, the effects will appear first in streams emanating from forested catchments.

The time is approaching when it will be possible to predict quantitatively the hydrological and hydrochemical effects of land use change in upland areas. Such predictions will facilitate the identification of undesirable consequences before any change is implemented. Extending the concept further, such predictions could be used to determine for each catchment the combination of usage which best meets the requirements of all interested parties.

REFERENCES

Batey, T. (1982) *Nitrogen cycling in upland pastures of the U.K.* Phil. Trans. Roy. Soc. Lond., B 296, 551–556.

Binns, W. O. (1979) The hydrological impact of afforestation in Britain. In: *Man's impact on the hydrological cycle in the United Kingdom*. Ed. Hollis, G., Geo. Abstracts.

Blackie, J. R., Ford, E. D., Horne, J. E. M., Kinsman, D. J. J., Last, F. T. & Moorhouse, P. (1980) *Environmental effects of deforestation, an annotated bibliography*. Occasional Paper No. 10. Freshwater Biological Association.

Calder, I. R. (1976) The measurement of water losses from a forested area using a 'natural' lysimeter. *J. Hydrol.,* **30**, 311–325.

Calder, I. R. & Newson, M. D. (1979) Land use and upland water resouces in Britain — a strategic look. *Water Resources Bulletin,* **15,** (6), 1628–1639.

Centre for Agricultural Strategy (1980) *Strategy for the U.K. forest industry*. CAS Report 6, Univ. of Reading.

Conway, V. & Millar, A. (1960) The hydrology of some small peat-covered catchments in the northern Pennines. *J. Instn. Wat. Engrs.,* **14**, 415–424.

Crisp, D. T., Cubby, P. R. & Robson, S. (1980) *A survey of fish populations in the streams of the Plynlimon experimental catchments*. Freshwater Biological Association, Teesdale Unit.

Forestry Commission (1977) *The wood production outlook in Britain*. Edinburgh.

Harriman, R. & Morrison, B. R. S. (1982) Ecology of streams draining forested and non-forested catchments in an area of central Scotland subject to acid precipitation. *Hydrobiologia,* **88**, 251–263.

Hibbert, A. R. (1967) Forest treatment effects on water yield. In *Forest Hydrology,* (ed. W. E. Sopper & H. W. Lull), 527–543. Pergamon Press.

Holden, A. V. (1976) The relative importance of agricultural fertilisers as a source of nitrogen and phosphorus in Loch Leven. In: MAFF Tech. Bull. 32, 306–314.

Institute of Hydrology (1976) *Water balance of the headwater catchments of the Wye and Severn, 1970–75*. Inst. Hydrol. Report 33.

Institute of Hydrology (1984) *Research Report 1981–84*.

Law, F. (1956) The effect of afforestation upon the water yield of catchment areas. *Jrnl. Brit. Waterworks Assn.,* **38**, 489–494.

Law, F. (1957) Measurement of rainfall, interception and evaporation losses in a plantation of Sitka spruce trees. IUGG/IASH: Gen. Assembly, Toronto, II, 397–411.

McDonald, A. T. & Kay, D. (1981) Enteric bacterial concentrations in reservoir feeder streams: baseflow characteristics and response to hydrograph events. *Wat. Res.,* **15**, 951–968.

Ministry of Agriculture, Fisheries and Food (1967) *Potential Transpiration.* MAFF Tech. Bull., 16, HMSO, London.

Ministry of Health (1948) *Gathering grounds. Public access to gathering grounds, afforestation and agriculture on gathering grounds.* HMSO, London.

Newson, M. D. (1980a) Water balance at selected sites. In: *Drought Atlas*, Eds. J. C. Doornkamp, K. J. Gregory & A. S. Burn. Inst. Brit. Geogrs., London. 37–38.

Newson, M. D. (1980b) The erosion of drainage ditches and its effect on bedload yields in mid-Wales: reconnaissance case studies. *Earth Surface Processes,* **5**, 275–290.

Newson, M. D. (1984) *Slope and channel runoff processes in upland catchments: interfaces between precipitation and streamflow acidity.* Cambria (in press).

Roberts, J. M. (1978) The use of the "tree cutting" techniques in the study of the water relations of Norway spruce, Picae abies. *Jnl. Exptl. Botany,* **29**, 465–471.

Roberts, G., Hudson, J. A. & Blackie, J. R. (1984) Nutrient inputs and outputs in forested and grassland catchments at Plynlimon, mid-Wales. *Agric. Water Management,* **9**, 177–191.

Robinson, M. (1980) *The effect of pre-afforestation drainage on the streamflow and water quality of a small upland catchment.* Institute of Hydrology, Report 73.

Robinson, M. & Blyth, K. (1982) The effect of forestry drainage operations on upland sediment yields: a case study. *Earth Surface Processes and Landforms,* **7**, 85–90.

Robinson, M. & Newson, M. D. (1985) Hydrological aspects of forest and moorland in an area of heterogeneous peat soils. *Jnl. Int. Peat. Soc.* (in press).

Rutter, A. J. (1966) Studies on the water relations of *Pinus sylvestris* in plantation conditions. *J. Appl. Ecol.,* **3**, 393–405.

Shuttleworth, W. J. (1977) A simplified one dimensional theoretical description of the vegetation–atmosphere interaction. *Bound. Layer Meteorol.,* **14**, 3–27.

Stewart, J. B. (1976) Evaporation from the wet canopy of a pine forest. *Wat. Resour. Res.,* **13**, 915–921.

Youngman, R. E. & Lack, T. (1981) New problems with upland waters. *Water Services,* **85**, 13–14.

27

Pollution of surface waters by the wood processing industry

P. Heinonen

27.1 INTRODUCTION

Finland's economic development is based to a large extent on its considerable reserves of timber. Forest industries still comprise the foundation of the major economic activities of the country. In 1980 the proportion of overall exports accounted for by this section of industry was 42%. At over 100 locations throughout the country, different municipalities have developed and prospered as a result of the activities of the wood-processing industry. However, this development has also introduced some problems and disadvantages. Environmental effects, in particular pollution of the atmosphere and of watercourses, have become ever more problematical with the increasing size of the production plants. Acquisition of the necessary raw material has also resulted, in some areas, in major alterations to the natural environment.

During the early years of industrial development, the effects of loading on the environment were not considered a major problem. Watercourses and the atmosphere were regarded simply as agents of dilution and dispersion of wastes generated in the production process. The first indications of the deleterious effects of effluents were noticed by professional fishermen, whose income was reduced due to changes in fish-population structure. Nowadays the level of knowledge of pulping industry wastes and their loading effects in recipient waters is very much higher. The effects of these wastes on watercourse status can be examined objectively and compared, for example, with other sources of loading, such as municipal wastes and non-point loading. At the same time the planning of water-protection programmes can be carried out on a more realistic and reliable basis.

27.2 THE PULP AND PAPER INDUSTRY IN FINLAND

A total of almost 70 mechanical wood-processing plants were operational in Finland in 1984. At the same time the number of pulp and paper mills was somewhat over 50, which from the point of view of the environment, and in particular of watercourse protection, present considerably greater problems. The majority of the mills are situated inland, near to the forest reserves, which further increases the problems for watercourse protection. The production capacity of chemical wood processing has increased considerably during the past decade. Production of bleached kraft pulp in Finland has increased to 2 900 000 tonnes annually and that of paper to 5 100 000 t. By contrast the production of sulphite pulp has decreased, being only 965 000 tonnes in 1980.

The watercourse loading from pulp and paper mills has traditionally been estimated with the aid of effluent volume figures and analyses of suspended solids and the seven-day biochemical oxygen demand (BOD). With increasing watercourse-eutrophication problems in many countries, measurements of phosphorus, and nitrogen loading have also been developed and applied to the estimation of pollution effects. The loading of wood-processing industries on the recipient watercourse has always been considerable. Loading increased proportionally with increased production levels until 1965, when it halted and turned in the early 1970s to a lower level. The present loading rate (Fig. 27.1), measured as BOD_7, is about the same as that in 1950 (Junna, 1984).

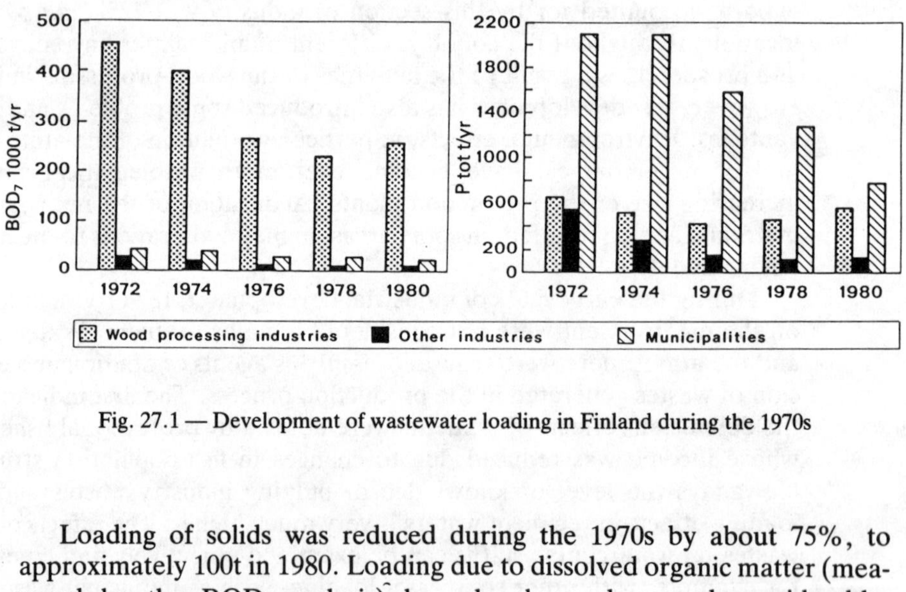

Fig. 27.1 — Development of wastewater loading in Finland during the 1970s

Loading of solids was reduced during the 1970s by about 75%, to approximately 100t in 1980. Loading due to dissolved organic matter (measured by the BOD_7-analysis) has also been decreased considerably, although the level of nutrients remained almost constant throughout the 1970s.

The volume of effluents has also decreased. These decreases have been achieved either by internal technical operations in mills, or by external means, such as wastewater handling systems. The nature of this development can be seen in Table 27.1, in which the development of specific loading figures during the 1970s has been calculated for different lines of production.

Table 27.1— The amount (Q m^3/t) and loading content (BOD_7 kg/t) of pulp and paper mill effluents in relation to production during the period 1972–1980 (National Board of Waters, 1983)

	1972		1975		1978		1980	
	Q	BOD_7	Q	BOD_7	Q	BOD_7	Q	BOD_7
Kraft pulp, bleached	344	71.7	288	49.5	194	36.0	170	32.6
Sulphite pulp, bleached	488	138	425	114	401	97.0	395	94.6
Paper, wood-free	109	9.4	98	8.0	75	5.4	73	5.6
Paper, wood-containing	63	14.4	55	14.0	50	11.7	41	10.8

The major part of this positive development has been achieved by improvements in process techniques. Pulp and paper mills are usually equipped only with mechanical wastewater treatment. Of the 34 pulp and paper mills considered as major polluters of the watercourse network at the beginning of 1985, less than ten were equipped with biological wastewater-treatment plants.

27.3 EFFECTS OF EFFLUENTS ON WATERCOURSE QUALITY

Effluents of the pulp and paper industry have a readily-observable effect on watercourse quality in Finland. The utilisability of watercourses is significantly reduced in the vicinity of all the larger factories. Waters allocated to grade V (poor water quality) of the five-grade general water quality classification scale, amounted at the beginning of the 1980s, to about 130 km^2. The majority of these badly polluted areas were situated near to pulping mills. In the following, the watercourse effects of wood-refining industries are examined with the aid of examples from different branches of the industry, although it must be borne in mind that the actual processes taking place in the water are of a very complex nature and depend on several different phenomena simultaneously.

27.3.1 Solids and BOD_7

The high concentration of organic matter in pulp-mill effluents causes considerable problems for the oxygen economy of watercourses. This is particularly evident in Finland, where the recipients of effluents are usually lakes. In addition it must be taken into consideration that, even in southern Finland, the lakes are covered for about five months of the year with ice, which prevents oxygen replenishment.

The oxygen-depleting effects of solid materials are usually long-lived. Furthermore, sedimentation of this material on the lake bottom destroys the bottom structure of the lake and prevents the normal development of flora and fauna. The major single factor with regard to the oxygen economy of the water mass as a whole is dissolved organic material, of which a part is degraded very rapidly. Figure 27.2 presents an example of how pulping wastes follow the bottom contours of the lake during the season of ice cover, and also demonstrates the effects of the oxygen-depleting properties of the effluent. The observation site was in southern Lake Saimaa, the largest

Finnish lake. The nearest pulping mill, from which these effluents originate, is situated about 40 km (25 miles) from the observation site.

The rate of oxygen consumption is initially very high, decreasing rather rapidly as a function of time, but causing major disturbances in oxygen economy over a wide area even after this attenuation. The normal chains in the aquatic environment are disturbed, for example as a result of the radical change in benthic populations. Both the quality and quantity of fish species undergo changes to the extent that waters in the vicinity of pulp and paper mills are, without exception, totally unsuitable for professional fishing (Heinonen & Falck, 1971). The effects of pulping effluents in river waters are of a similar nature. However, the continuous dissolution and mixing of oxygen from the air effectively prevents the risk of total oxygen depletion. In summer, during the period of minimum flow and maximum water temperature, totally oxygen-free areas have been recorded in small river watercourses in the immediate vicinity of pulping mills. The daily variation in oxygen concentration may also disturb aquatic life.

The deleterious effects of solids in river waters can be seen in stretches with low flow rates, where the natural sediments may be destroyed very rapidly. The sedimentation rate of suspended solids may, in exceptionally bad cases, reach as much as several centimetres annually. Methane production in anaerobic fibre deposits during warm periods may cause large rafts of floating fibrous material on the water surface, and thus generate a new source of internal loading which travels down the watercourse with consequent reduction in its utilisation potential.

27.3.2 Nutrients

The disadvantages of the effluents of chemical wood processing were for a long time equated solely with their capacity to deplete the oxygen content of recipient waters. When eutrophication was later identified as deleterious to watercoruses, reducing their utilisation potential, attention was also directed to nutrient loadings. Of the two major nutrients required for primary production, namely phosphorus and nitrogen, phosphorus concentrations in particular are high in pulping effluents. The nutrients are, however, at least partly adsorbed on solid particles and therefore cannot be utilised by phytoplankton. Many other factors in effluents have a toxic effect, which actually inhibits primary production in the vicinity of the discharge site. The natural phytoplankton population of the lake water cannot accommodate either the effluent concentration as such, or a particular factor such as pH or the concentration of one component, for example organochlorine compounds produced in bleaching. This phenomenon, which has been demonstrated by experiments with algae in many laboratories, can also be observed in the recipients of most pulp and paper mills (Fig. 27.3).

As in the case of municipal effluents, discharge of pulping wastes to the recipient causes increased nutrient concentrations immediately downstream from the point of discharge. Despite this, the level of primary production decreases as a result of pulping effluents. In many cases primary production is totally prevented. Only when the effluents have been sufficiently aged and

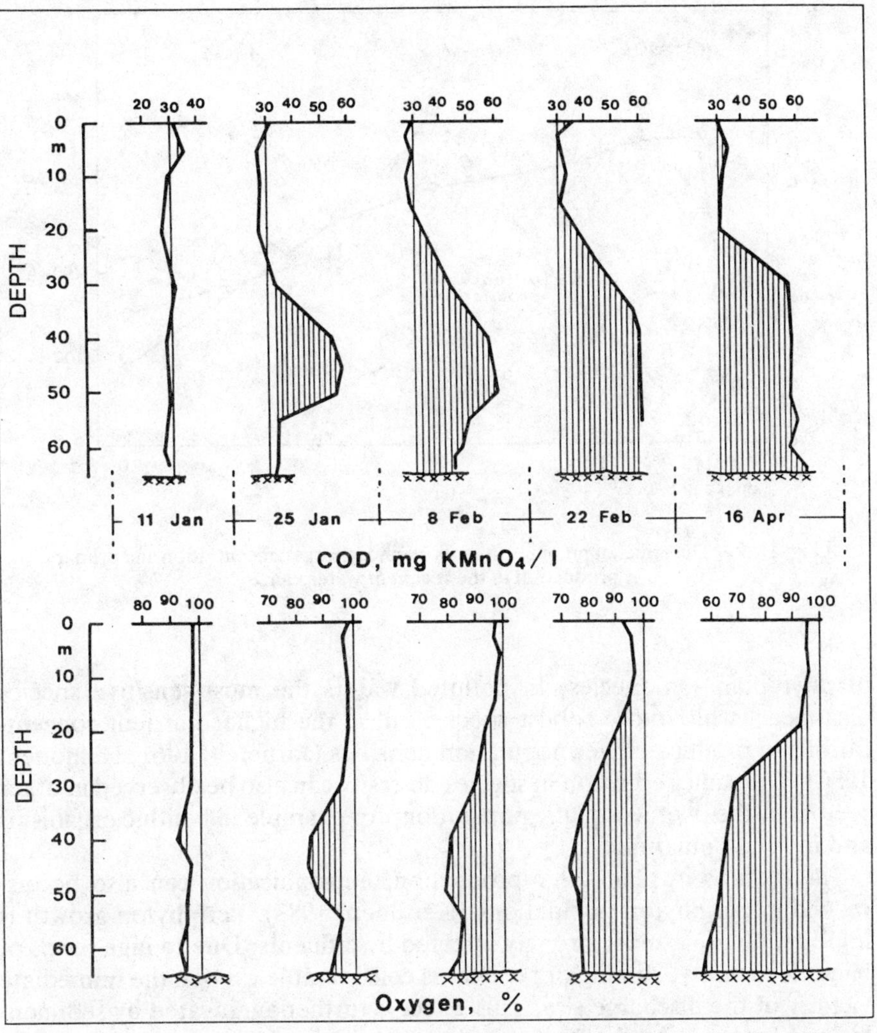

Fig. 27.2 — The effects of pulping effluents (measured as COD) in southern Lake Saimaa

diluted do the factors inhibiting production cease to operate and allow production to reach the level corresponding to the nutrient concentrations. Even then, however, because the effluents cause dark colour and small Secchi depth in the polluted waters, the layer in which primary production can take place does not extend to the normal depth. In one case, for example, the peak of primary production was observed as far away as 8 km from the polluting factory. The dilution of effluents was 1:10 and the residence time approximately one week. The values of phytoplankton biomass analysed in the area were clearly indicative of extensive eutrophication.

Primary production is dependent on the success of a very small number

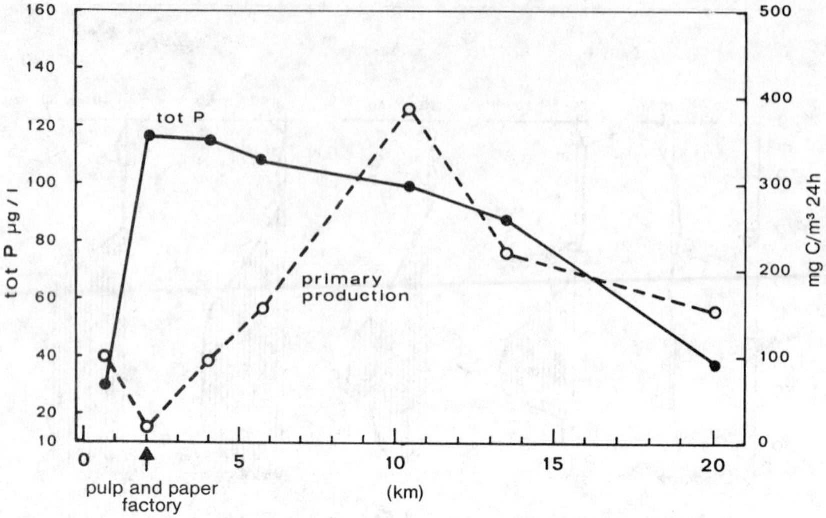

Fig. 27.3 — The effect of pulping effluents on phosphorus concentration and primary production in the recipient watercourse

of phytoplankton species. In polluted waters the most sensitive species disappear, while more robust species utilise the higher nutrient concentrations to produce higher population densities (Järnefelt, 1961; Heinonen, 1980). The same reduction in species diversity can also be observed in other sectors of the overall aquatic population, for example in benthic organisms and in fish populations.

As well as in planktonic production, eutrophication can also be discerned in periphyton populations (Kettunen, 1983). Periphyton growth is inhibited in areas most strongly affected by effluents. Due to high levels of bacterial activity, sliming of beaches is considerable even in the immediate vicinity of the discharge site. This effect is further aggravated by sedimentation of solids, usually fibres escaping from the pulping process.

27.3.3 Toxicants and toxicity

In the discussion of nutrients from forest industries above it was already noted that effluents often have an inhibitory effect on planktonic production. The most obvious toxic effects are naturally in the waters near to the factory, but they often extend to apparently completely unpolluted areas, where physical and chemical analyses reveal no other signs of the effluents. In general, it can be said that the number of toxic substances in pulping wastes is very high. Holmbom (1980) referred to over 50 identified toxic or potentially-toxic compounds analysed in effluents from Finnish kraft pulp mills. The occurrence and concentrations of the different compounds varied considerably between different mills, but the toxicity of the effluent from a single factory may also vary widely with time. Little is known of the synergistic and antagonistic reactions between different toxic compounds.

The toxicity of wastewaters has been tested in laboratory conditions using bacteria, algae, *Daphnia*, and fish as test organisms. With these routine methods it has been possible to identify the most toxic of the various process stages. The most potent toxicants produced by the pulping industry are resin acids and their derivatives, as well as certain degeneration- and chlorination-products of lignin. According to one Finnish investigation (Holmbom, 1981) the wastewaters of the debarking plant of kraft pulp mills are the most toxic. The major toxicity loading, however, is due to water from the bleaching process because the effluent volumes involved are very high.

After waste-treatment procedures the diluted effluents in the recipient watercourse are rarely acutely toxic, except in the immediate vicinity of the discharge site. Cases of widespread acute toxicity, often marked by extensive fish deaths, are usually the result of sudden and exceptional discharges. Even then the immediate cause of death is usually oxygen depletion or a major change in pH, rather than the toxic nature of the discharge. Sublethal effects, however, are rather common in the recipients of pulping mills. They may be discernible in the reproduction, growth, and behaviour of fish populations, and both histological and physiological changes in the fish may occur. A third mode of effect of toxic compounds is their accumulation first in the tissues of the primary receiver, and subsequently in the members of the ascending food chain. Examples of such accumulating substances are many metals and, particularly in the case of pulping effluents, resin acids and their degradation products and chlorinated lignin and phenol compounds. Accumulation in fish usually starts from only very low concentrations in the water. The concentration coefficient may even exceed 100. The most significant sources of long-term toxicity are the bleaching units of pulping mills.

27.3.4 Odour and taste

The first sensory observation relating to pulping mills is usually their unpleasant odour. Particularly in the vicinity of kraft mills this odour, originating mainly from a large number of sulphur compounds including mercaptan, may indeed be extremely unpleasant. As well as causing this atmospheric pollution, the compounds responsible for the odours are also discharged with effluents into the recipient watercourse. Many of the organic chlorine compounds from the bleaching process cause extreme off-odours and off-tastes in fish. The total removal of all compounds causing unpleasant odours would surely not be economically feasible, because the wastes in question can be detected even by the rather insensitive human senses at extremely low concentrations. Furthermore, many of the odoriferous compounds accumulate in the food chain and thus cause particularly strong off-flavour in fish.

27.3.5 Watercourse effects of timber acquisition

The discussion of watercourse effects resulting from the wood processing industry is usually confined only to the effects of wastewaters. This limited interpretation does not take into account the fact that incorrect methods of timber production and cutting may have widespread effects on watercourse quality.

Fertilization of forests is nowadays very widely practised. The rate of production of the forest is in this way considerably improved, but at the same time the level of leaching from the fertilized area is greatly increased. It has been estimated that approximately 1% of the total amount of phosphorus applied is washed out into the watercourse during the first year after fertilization. Later, of course, the extent of leaching is lower, but the effect of fertilization may last at least five to ten years (Kenttämies, 1981).

Increased leaching of nutrients is also caused by ditching of forest land. Ditching undoubtedly improves forest growth by altering the water economy of the soil. At the same time, however, the leaching of materials increases markedly.

Timber cutting has an effect on watercourses too. Even a 20% thinning of the tree cover increases the summer runoff by more than 20%, while clear cutting causes an increase of 85% (Heikurainen & Päivänen, 1970). As the movement of materials usually takes place in proportion with that of water, forest cutting causes a considerable increase in the leaching of nutrients to watercourses during the summer season. Acquisition of timber for the wood-processing industry thus affects watercourses in forested areas in many ways, with an overall tendency towards slow eutrophication.

27.4 FUTURE PROSPECTS

The effects of the wood-processing industry on watercourses are still considerable in Finland, despite major programmes of water protection. Chemical wood-processing is by far the greatest single polluting influence. The utilisation potential of watercourses in the vicinity of large, and even medium-sized, pulp and paper mills is considerably reduced.

Process changes within the mill have to date resulted in considerable decreases in loading. Process development with an eye to watercourse protection, or more generally environmental protection, will certainly continue in the future. However, it is also evident that increasing attention will have to be paid to the treatment of effluents. Traditional methods aimed at removing solids and oxygen-consuming organic materials will soon no longer be sufficient. Priority must also be given to the removal of nutrients and, in particular, of accumulating organic chlorine compounds and other compounds with varying toxic effects in the recipent. In the long term it will become necessary to take into account the watercourse effects of timber production, too.

REFERENCES

Heikurainen, L. & Päivänen, J. (1970) The effect of thinning, clear cutting and fertilization on the hydrology of peatland drained for forestry. *Acta Forestalia Fennica,* **104,** 1–23.

Heinonen, P. (1980) *Quantity and composition of phytoplankton in Finnish inland waters*. Publications of the Water Research Institute 37, pp. 1–91.

Heinonen, P. & Falck, P. (1971) The effect of waste water on fish. *Aqua Fennica,* 105–110.

Holmbom, B. (1980) A procedure for analysis of toxic compounds in pulp and paper mill waste waters. *Paperi ja puu* **62,** 9:523–531.

Holmbom, B. (1981) Acute toxicity to fish of kraft pulp mill waste waters. *Paperi ja puu* **63,** 11: 673–684.

Junna, J. (1984) Metsäteollisuuden jätevedet haitattomiksi puhdistamalla. (English Summary: Forest industry waste waters harmless by purifying). *Rakennustekniikka,* **6,** 495–500.

Järnefelt, H. (1961) Die Einwirkung der Sulfitablaugen auf das Planktonbild. *Verh. Internat. Verein. Limnol.* **14,** 1057–1062.

Kenttämies, K. (1981) *The effects on water quality of forest drainage and fertilization in peatlands.* Publications of the Water Research Institute 43: 24–31.

Kettunen, I. (1983) A study of the periphyton of Lake Saimaa, polluted by waste waters of the pulp industry. In: Wetzel, R. G. (ed.), *Periphyton of Freshwater Ecosystems.* 331–335. Dr. W. Junk Publishers, The Hague.

National Board of Waters, (1983) Teollisuuden vesitilasto 1979–1980 (English Summary: Industrial water statistics in 1979–1980). Vesihallitus. *Tiedotus,* **224,** 1–71.

28

The effects of afforestation on salmon and trout rivers and suggestions for their control

D. H. Mills

28.1 INTRODUCTION

The upland areas of many salmon-producing countries have been extensively afforested and it is in these areas in Great Britain that many salmon rivers begin, arising from an extensive network of small, fast-flowing tributaries which serve as the spawning and nursery areas. These upland streams are invaluable to the river system, and the degradation of their abundant, clean, cool, well-aerated and silt-free waters would lead to the loss of much of the area upon which the Atlantic salmon and trout depend for the survival of their young. It is not always realised that even the smallest of these streams, some little more than one metre wide, can be important; often salmon and sea trout will penetrate these channels to spawn. It is therefore important to appreciate how delicate the natural balance is when considering various forest management practices which could affect the aquatic environment.

28.2 THE EFFECTS AND THEIR CONTROL

28.2.1 Ploughing and drainage

Sedimentation, resulting in silting of fish spawning and nursery areas and in-filling of holding pools, can occur from erosion of plough furrows and drainage channels on steep hillsides. While drainage channels may at first prevent flooding, due to a smaller surface runoff with the land being better drained, they will, during periods of heavy or continuous rain, and when the surrounding land is saturated, produce very sudden floods of a short duration resulting in serious erosion.

Sediment yield is increased by drainage ditches, which discharge into natural watercourses from ploughed areas. This is not a temporary phenomenon, since the banks of overplanted watercourses become devoid of vegetation, thus increasing bank erosion. Erosion and sedimentation are also increased by damage to stream banks by machinery during forestry operations, and the inevitable build-up of brash or 'lop and top' and other debris in streams after thinning and clear felling.

Alignment of ploughing along the contour was once favoured and is still occasionally practised. It is now felt that this is inappropriate in wetland situations, since each furrow tends to 'pond' surface runoff and gives rise to lines of stagnant conditions. Alignments are now made to run roughly up and down the slope to improve runoff and to provide pathways for downhill seepage when the furrows become filled with plant litter. The risk of severe erosion has been shown to be small, provided cut-off drains are used and individual furrows are not used to carry water drainage from large areas lying above. The current policy of varying the lines of ploughing for landscape reasons or interrupting them on fragile areas further diminishes the risk from erosion. The alignment of the deep drains intended to collect water from the superficial furrows and from the subsoil should be nearly parallel with the contour, consistent with maintaining a suitable drain gradient (Taylor, 1970). A drain with a gentle gradient of no more than 2° is ideal and generally accords with normal forest practice. Except at the start of the drain, gradients steeper than 3° should be avoided to prevent severe erosion. The alignment of the drain should be designed to achieve the maximum interception effect with the minimum drain length. In order to catch silt where drains are leading to streams or small watercourses, they should either be tapered in depth and stopped 15 to 20 m short in order that the water discharging from these drains has to filter through the ground vegetation or, in areas of higher rainfall, a sump should be taken out just before the drain opens into the watercourse. On the very steep slopes where it is not possible to put in the cut-off drain, it is suggested that a large V-type sump drain should be cut along the bottom of the slope to trap debris which is carried down the plough furrows (Graesser, 1979). At no time should plough furrows intercept natural drains and small watercourses as this could lead to ponding and diversion of the watercourse down the furrows, resulting in accelerated erosion.

From time to time it may be necessary to clean out certain drains within the forest if waterlogging is evident. This operation should be very carefully timed so that little silt enters the watercourses to affect the stream bed and bank habitats. For this reason, it is important not to clean out drains during the period mid-October to mid-May, if possible, as during this time the eggs, and latterly the alevins, are in the gravel and would be suffocated by the silt. The time of the early autumn floods would be the most suitable for cleaning drains, but care should be taken to remove as much of the silt and leaf litter as possible from the drain and deposit it on dry land some distance away from any watercourse.

28.2.2 Planting

Changes in the flow regime of streams, as well as being affected by drainage, are further accentuated by higher water use by a forest through transpiration and it has been shown that trees can, under certain conditions, intercept substantially more water than open moorland in areas of high rainfall (Gash *et al.*, 1978).

Planting of conifers close to the edges of streams has seriously affected the stream's productivity. This has led to a paucity of insect life and fish in the streams which flow through mature forest with a dense tree canopy, which prevents the growth of ground vegetation (Mills, 1967; Smith, 1980). Basically, the protection of stream banks requires the establishment of permanent reserve strips of vegetation. Valley topography, soil and aesthetic factors will usually help to define the area and shape of reserve strip required. Before considering the scale and the nature of the bank vegetation the importance of stream-side vegetation should be stressed, as it:

- shades water, stream bed and stream bank from the sun's heat;
- provides energy to the stream through leaf fall, terrestrial insect drop and dissolved nutrients;
- gives protective cover for fish;
- protects the stream banks from channel erosion, erosion from overland runoff, and the erosive effects of precipitation impact;
- acts as a buffer against debris from overland runoff, and
- intercepts toxic materials from spraying (Burns, 1970).

Removal of stream-side vegetation can facilitate increases in the temperature of streams which may reach a lethal point for salmon and trout (Brown, 1971; Hall & Lantz, 1968).

Terrestrial insects have been shown to make up 5 to 65% of the diet of trout and salmon in some streams. Most of these insects come from stream-side vegetation and its removal reduces the number of insects available to trout and salmon, as well as to birds and some other animals.

Cover is important in providing hiding places for fish and it has been shown that the removal of bank vegetation and other shelter can destroy trout streams (Warner & Porter, 1960). It has been found that larger brown trout were 27% more numerous and weighed 44% more in a sheltered section of stream than in a non-sheltered section (Gunderson, 1968), while trout were found to be 78% more abundant in well-covered mountain streams than in an area where cover had been removed.

Vegetation on the stream banks also helps to maintain deep pools by preventing lateral erosion, which results in a wider but shallower stream channel (Johnson, 1953).

28.2.3 Reserve areas

Adverse effects can be reversed or minimised by the introduction of reserve strips, or unafforested buffer zones, between the plantation and the banks of every natural watercourse. The widths of the reserve areas of vegetation will vary with the landform, but as a rough guide the overall width (both sides) could be ten times the width of the stream, up to a maximum of 30 m overall.

Within the reserve areas tree planting should be kept to the minimum required for landscape reasons; in addition heavy machinery should be confined, as far as possible, to using defined roads and tracks.

These areas may seem unnecessarily large but they will also help to counteract the effects of precipitation interception. Furthermore, they give easy access for anglers, walkers, and staff.

Drakeford (1981) determined the width of areas on the basis of their functions, as follows:

(i) to prevent direct fall of coniferous litter into watercourses;
(ii) to allow sunlight to penetrate to the stream and its banks;
(iii) to prevent damage to stream banks by machinery and felling operations;
(iv) to allow the colonisation of the banks by natural vegetation (deciduous scrub) and so enhance food supply;
(v) to prevent direct discharge of ditches into streams, thus reducing sediment input and any acid pulse;
(vi) to allow liming of streams, stream banks and ditch discharges, should this prove necessary.

It was considered that conditions (i) to (iv) could normally be fulfilled by a zone of 5-m width on each bank. However, for (i) and (ii), the extent of the zone would depend on the height of the crop, which is determined by the windthrow hazard. In areas of low windthrow hazard, the zone width was doubled. The fulfilment of condition (v) meant the blocking of ditch entries at the stream banks. This causes flooding back from the blockage which can damage roots and so induce windthrow. The cropless buffer zone therefore has to extend as far as the flood, which depends upon topography, according to Table 28.1.

28.2.4 Stream-side vegetation

This can consist of ordinary ground vegetation and small deciduous tree species such as birch (*Betula* spp.), willow (*Salix* spp.), rowan (*Sorbus aucuparia* L.) or alder (*Alnus* spp.). Alders may also contribute substantial amounts of dissolved nutrients such as nitrates to the stream and it has been found that leaf fall from alders contained four times as much nitrogen as other deciduous species (Goldman, 1961). Alder grows vigorously on most stream sides and, to prevent excessive shading and to allow access for recreation and stream management, it should be used in well-scattered

Table 28.1 — The extent of flooding backing up blocked ditches on sloping ground

Slope (degrees)	Ditch depth (m)	Extent of flood (m)
1	1	57.3
2	1	28.6
3	1	19.1
4	1	14.3
5	1	11.4
10	1	5.8
1	0.5	28.6
2	0.5	14.3
3	0.5	9.5
4	0.5	7.2
5	0.5	5.7
10	0.5	2.8

groups, especially along streams less than 12 m wide. If willows are used they require regular basal pruning. Their root systems are effective in preventing bank erosion. A variety of other species suitable for water margins has also been suggested (Greer, 1979), including common ash (*Fraxinus excelsior* L.) and aspen (*Populus tremula* L.).

28.2.5 Fencing

Where fences are needed they should be erected well back from the stream bank and above the flood channel so that they will not catch debris or be washed out during high flows. The fences should be far enough back to allow vegetation control and stream improvements and not to interfere with angling. It must be remembered that some grazing animals can stretch their heads through a fence and graze bare the other side to a distance of about a metre. For this reason, too, fences should be well back from the outside banks of the bends as the stream may soon undermine the fence unless bank-protection devices have been constructed.

Barbed wire may be required on stock fences but, in any event, stiles should be erected at suitable places. Where livestock must have access to the water, their access should be on gently sloping ground where the stream substrate is hard and the water shallow. Their approach paths, which should be wide enough to prevent bunching up and panic, should be fenced on either side (White & Brynildson, 1975).

Where boundary or accommodation fences, or drystone walls, cross a watercourse, a water-gate is often necessary to prevent stock passage. Water-gates will vary in design but must always be independent of the fence.

They should be examined for damage after spate to ensure removal of debris of any kind which hinders stream flow and the passage of fish and, also in periods of drought, to see that they are effective barriers to stock.

28.2.6 Road construction

Road building may initially cause erosion until the exposed soil and tracks have become stabilised, and badly-positioned culverts may hinder or prohibit fish ascent. Proper drainage should be incorporated into the road construction to minimise erosion on the road surface, on the slopes and in the ditches. In order to achieve this the following steps are usually necessary.

- Water from roadside drains should be discharged into watercourses at frequent intervals, culverts being provided where necessary.
- Drainage structures should be of a sufficient size to accommodate runoff from the drainage area.
- Where ditches drain directly into permanent watercourses, a sump to trap silt and debris may be necessary.
- Special erosion-control measures may be required on roadside banks including such measures as intercepting trenches or terracing. Bare earth embankments and cuttings should be left at the natural angle of repose to encourage the establishment of vegetation.
- Culverts should be rip-rapped as necessary, to prevent erosion at both inflow and outflow ends. This is done by reinforcing the substrate at each end and reducing flow velocities.
- The outflow ends of hanging culverts should be provided with down flumes or other suitable drains where necessary. Rock or concrete aprons should be provided as stilling areas to decrease water velocity and prevent stream-bed erosion.
- Culverts or other drainage devices should be placed so as to attain a maximum drainage efficiency.
- Whenever possible culverts should be installed so as to allow passage of fish on all watercourses frequented by them.
- The inverts of culverts should be laid to be self-cleansing and properly bedded to avoid settlement. The size of the culvert will depend on the characteristics of the catchment area and the natural watercourses. Lugmayr (1984) describes an excellent structure to incorporate in the floor of culverts, and any steep slope leading up to the culverts, to produce a series of pools to facilitate the passage of fish.

Fords for heavy vehicles and forest machinery should not be used at high flows and care must always be taken to avoid oil spillage. Fords are sometimes used as spawning grounds by salmon and trout.

Where it is necessary to extract gravel from a watercourse it is essential that this is done in amounts and at a period which will minimise damage to fish, especially migratory fish. Consultations with the river authority, the riparian owner and tenants are advisable. It is essential that the stability of the stream bed is subsequently restored.

If any construction work is planned which involves the use of concrete — such as in roadworks, bank reinforcement and bridge building — care must be taken to ensure that no cement flour or raw concrete enters the watercourse. A number of serious fish mortalities have resulted from the careless use of cement.

28.2.7 Spraying

The effects on fish and stream insect populations of improperly controlled forest spraying have been well documented (Keenleyside, 1959; Symons, 1978; Holden & Bevan 1979; Turner, 1977). It is common knowledge that spraying over watercourses should be avoided if at all possible, and that the correct spray dosages should be used to minimise the damage to fish and wildlife. When agricultural chemicals are used at recommended rates and with due care, the risk to fisheries is relatively small and limited to certain types of chemicals (Holden, 1964). In the event of accidental spillage or careless disposal of concentrated chemicals, however, the problem is more serious. Exposure of fish for only a few minutes to high concentrations can result in a rapid absorption of the toxicant via the gills and death follows quickly (Frederikson *et al.*, 1975). For this reason no spraying equipment or empty, or partially empty, tins or drums of chemical should be washed out in or near any watercourse, however small, nor should empty containers be punctured and buried. Advice should be sought from the appropriate authority on the safe disposal of unwanted chemicals.

At no time should the roots or foliage of planting stock (such as young trees ready to be planted out) which have been treated with Gammacol be soaked in a watercourse just prior to planting, even if they show signs of distress in hot weather. Gammacol is extremely toxic to fish.

28.2.8 Application of fertilizers

Leaching of fertilizers into streams in forested catchments after the forests have been fertilized has been well recorded (Harriman, 1978; Moore, 1974; Anderson *et al.*, 1976). In some cases, this leaching has had little effect on water quality, and in other instances it has been suggested that it could increase stream productivity and so be beneficial. However, leaching into streams running into closed areas of water, such as lakes and reservoirs, could produce undesirable conditions of eutrophication and for this reason fertilizer programmes should be discussed with water authorities if domestic-supply reservoirs are likely to be affected.

28.2.9 Fire

In order to provide a ready supply of water for fire-fighting purposes, it is sometimes the practice where the land configuration is suitable to impound small streams and form what are known as 'fire dams'. It can be achieved with less damage to the stream environment if water is diverted into an excavated pond to one side of the stream channel. In this way the stream is kept open along its length and the passage of fish is consequently not hindered. Furthermore, the silt load in the stream is not trapped in the 'fire dam' to lower the dam's capacity and value.

28.2.10 Thinning, felling, and extraction

There has been considerable research on the effects of these operations on fish and their stream environment, and probably the effects of logging on stream quality, and salmon and trout populations in North America have received the most attention (Gibbons & Salo, 1973; Meehan, 1974; Meehan *et al.*, 1969; Ramberg, 1974). The major effects of these activities can be changes in stream flow and water temperature, sedimentation, loss of nutrients, damage to spawning grounds, blockage of streams, prevention of the movement of migratory fish, and bacterial decomposition of any bark and wood debris smothering the stream bed.

Preventing the fall of lop and top into streams is the best way to control the last three mentioned effects. *Trees should be felled away from streams* wherever possible. If tree tops and branches do enter a stream they should be removed as soon as possible, certainly no later than when the logs are extracted. Before felling teams move away the stream should be checked to ensure that the lop and top is neither in the stream nor likely to enter the stream on high flows.

When logs are to be moved care should be taken not to break the soil surface of the forest floor unduly, resulting in stream sedimentation. This can be done by:

- avoiding the use of long skid-roads on steep slopes;
- stacking logs well away from watercourses;
- not operating heavy equipment in or near streams.

28.3 CONCLUSION

As Ramberg (1974) points out, the majority of the above-mentioned forestry operations have complex effects upon aquatic environments, and this is particularly marked for clear-felling. This operation alone produces:

- a decrease in transpiration and interception resulting in a change in the flow regime;
- increased solar radiation leading to higher water temperatures and increased primary production, and
- nitrification resulting in increased primary production and eutrophication.

Drainage, too, produces a complex combination of effects not all of which are predictable. For example, there is evidence that drainage from some afforested areas is making the neighbouring water more acid and there is considerable proof (Jensen & Snekvik, 1972; Last, 1982), that this increase in their acidity is being hastened by acid rain produced from sulphur dioxide carried with the wind from industrial areas, and this is a phenomenon which has already had serious repercussions in salmon rivers in southern Norway and eastern Canada.

If the salmon and trout frequenting afforested areas are to be protected, the types of management practice outlined above must be observed and, in addition, every step must be taken to consider additional stream improvements (Mills, 1980) which will help increase the numbers of fish.

REFERENCES

Anderson, H. W., Hoover, M. S. & Reinhart, K. G. (1976) *Forests and water. Effects of forest management on floods, sedimentation and water supply.* USDA Forest Service General Technical Report, PSW18.

Brown, G. W. (1971) Water temperature in small streams as influenced by environmental factors and logging. *Proceedings of a Symposium — Forest Land Uses and Stream Environment,* pp. 175–181. Oregon State University. Corvallis.

Burns, J. E. (1970) *The importance of streamside vegetation to trout and salmon in British Columbia.* Fisheries Technical Circular No. 1 Vancouver Island Region, Fish and Wildlife Branch, Department of Recreation and Conservation, Nanaimo, British Columbia.

Drakeford, T. (1981) Management of upland streams. (An experimental fisheries management project on the afforested headwaters of the River Fleet, Kirkcudbrightshire). *Proceedings of the 12th Annual Study Course of the Institute of Fisheries Management,* University of Durham, pp. 86–92.

Frederikson, R. L., Moore, R. G. & Norris, L. A. (1975) The impact of timber harvest, fertilisation and herbicide treatment on stream water quality in Western Oregon and Washington. In: *Forest soils and forest land management; Proceedings of the Fourth North American Soils Conference,* Eds. Bernier, L. & Winget, C. H.

Gash, J. H. C., Oliver, H. R., Shuttleworth, W. J. & Stewart, J. B. (1978) Evaporation from forests. *J. Instn Wat. Engrs and Scientists,* **32,** (2), 104–110.

Gibbons, D. K. & Salo, E. O. (1973) *An annotated bibliography of the effects of logging on fish of the western United States and Canada.* USDA Forest Service General Technical Report PNW10.

Goldman, C. K. (1961) The contribution of alder trees (*Alnus terrius folia*) to the primary productivity of Castle Lake, California. *Ecology,* **42,** 282–287.

Graesser, N. (1979) How land improvements can damage Scottish fisheries. *Salmon and Trout Magazine,* **215,** 39–43.

Greer, R. B. (1979) A tree planting trial at Loch Garry (Tayside Region) aimed at habitat improvement for fish. *Scottish Forestry,* **33,** (1), 37–44.

Gunderson, D. K. (1968) Floodpain use related to stream morphology and fish populations. *J. Wildlife Management,* **32,** 507–514.

Hall, J. D. & Lantz, R. L. (1968) *Effects of logging on the habitat of Coho salmon and Cut-throat trout in coastal streams.* Oregon Agricultural Experimental Station, Technical Paper 2570.

Harriman, R. (1978) Nutrient leaching from fertilized watersheds in Scotland. *J. Appl. Ecol.*, **15,** 933–942.

Holden, A. V. (1964) The possible effects on fish of chemicals used in agriculture. *Journal and Proceedings of the Institute of Sewage Purification,* Part 4.

Holden, A. V. & Bevan, D. (Eds.) (1978) *Control of the Pine Beauty Moth by Fenitrothion in Scotland in 1978.* Forestry Commission Report, Edinburgh.

Jensen, K. W. & Snekvik, E. (1972) Low pH levels wipe out salmon and trout populations in south western Norway. *Ambio,* **1,** 223–225.

Johnson, F. W. (1953) Forests and Trout. *J. Forestry,* **51,** (8), 551–554.

Keenleyside, M. H. A. (1959) Effects of spruce budworm control on salmon and other fishes in New Brunswick. *Canadian Fish Culturist,* **24.**

Last, F. T. (1982) Effects of atmospheric sulphur compounds on natural and man-made terrestrial and aquatic ecosystems. *Agriculture and Environment,* **7,** 299–387.

Lugmayr, F. (1984) Regulierung von Fliessgewassern Vorschlage uber fischereifreundliche Einbauten, *Osterreichs Fischerei,* **37,** (7), 179–183.

Meehan, W. R. (1974) *The Forest Ecosystem of South-East Alaska 3. Fish Habitats.* USDA Forest Service General Technical Report PNW15.

Meehan, W. R., Farr, W. A., Bishop, D. M. & Patrie, J. H. (1969) *Some effects of clearcutting on salmon habitat of two southern Alaskan streams.* USDA Forest Service Research Paper PNW82.

Mills, D. H. (1967) A study of trout and young salmon populations in forest streams with a view to management. *Forestry,* **40,** (1): Supplement 85–90.

Mills, D. H. (1980) *The Management of Forest Streams.* Forestry Commission Leaflet No. 78, HMSO, London.

Moore, D. G. (1974) Impact of forest fertilization on water quality in the Douglas fir region — a summary of monitoring studies. *Proceedings, 1974 National Convention, Society of American Foresters.*

Ramberg, L. (1974) Effects of forestry operations on aquatic ecosystems. *Ecological Bulletin,* **21,** 143–149.

Smith, Barbara, D. (1980) The effects of afforestation on the trout of a small stream in southern Scotland. *Fisheries Management,* **11,** (2), 39–58.

Symons, P. E. K. (1978) Assessing and predicting effects of the fenitrothion spray programme and aquatic fauna in *Proceedings of the Symposium on Fenitrothion: The long-term effects of its use in forest ecosystems.* National Research Council of Canada, NRCC 16073.

Taylor, G. G. M. (1970) *Ploughing practice in the Forestry Commission.* Forest Record No. 73, HMSO, London.

Turner, D. J. (1977) *The safety of the herbicides 2,4-D and 2,4,5-T.* Forestry Commission Bulletin 57. HMSO, London.

Warner, K. & Porter, F. R. (1960) Experimental improvement of a bulldozed trout stream in northern Maine. *Transactions of the American Fisheries Society,* **89,** (1), 59–63.

White, R. J. & Brynildson, O. M. (1975) *Guidelines for management of trout stream habitat in Wisconsin.* Department of Natural Resources, Madison, Wisconsin. Technical Bulletin No. 39.

29

Controlling diffuse-source pollution associated with forest practices in North America

S. L. Ponce

29.1 INTRODUCTION

This chapter is intended to serve as a management tool for both water quality planners and forest managers who deal with non-point (diffuse) source pollution resulting from forest practices. Certainly, management goals may vary widely among readers of this chapter. Foresters have goals which extend beyond the control of water-quality degradation, such as timber and forage production and the provision of recreational opportunities and wildlife habitat. Water-quality planners, on the other hand, direct their attention toward the protection of water quality. This chapter is intended to serve both classes of users.

To formulate effective strategies and plans for the control of non-point source pollution, the water-quality planner and forest manager must first be able to recognise the potential for degradation of water quality in forest practices. The key to understanding the impacts on stream quality is that forest streams are not separate and distinct from the forest lands they drain, but are an integral part of the forest ecosystem (Brown, 1978). It is assumed that the reader understands the processes which affect water quality in forest streams. Research examining the effects of forest practices on stream quality is quite extensive and is summarised very well by: ASCE, 1980; Brown, 1978; US EPA, 1977; Anderson *et al.*, 1976; US EPA, 1976; ASCE, 1975; Krygier & Hall, 1971.

The objective of this chapter is to provide an overview of strategies currently used within the United States to control the degradation of water-quality from forest practices. This discussion emphasises processes and procedures rather than laws and regulations. Although laws and regulations to protect water quality vary between nations, the goals of the water-quality planner are similar.

29.2 CONTROLLING THE IMPACTS OF FOREST PRACTICES ON WATER QUALITY

The Federal Water Pollution Control Act Amendments of 1972 (P.L. 92-500) established definite goals regarding the restoration and maintenance of the physical, chemical, and biological integrity of the waters of the United States. The Act required that water-quality management and planning include a process that (a) identifies non-point sources of water pollution, and (b) establishes methods to control those sources to the extent feasible.

As a result of this Act, land management agencies throughout the United States sought to develop methodologies for identifying and assessing alternative solutions to existing, or potential, non-point source problems associated with site-specific land-management activities. The US Forest Service, in co-operation with the US Environmental Protection Agency, developed a handbook to deal specifically with the control of non-point sources of water pollution related to forest practices. The handbook is entitled 'An Approach to Water Resources Evaluation of Non-Point Silvicultural Sources (A Procedural Handbook)' (US EPA, 1980). This handbook was compiled by resource managers and scientists in the US Forest Service, American Universities, and the private sector. Much of the discussion which follows has been taken from this handbook.

29.2.1 Procedure

The general procedure presented in the handbook is illustrated in Fig. 29.1. Initially, water-quality objectives are established. Where feasible, these are set as quantitative values. These objectives provide a base against which the predicted impacts of the proposed silvicultural activities can be compared so that the need for control measures can be identified.

The silvicultural activities which are evaluated by the procedure include timber harvesting, transportation systems, and various cultural practices, such as site preparation and timber-stand improvement. The hydrologic and water-quality responses evaluated include stream flow, inorganic sediment, nutrients (nitrogen and phosphorus), heat (water temperature), organic debris and dissolved oxygen, and introduced chemicals, such as pesticides and fertilizers.

Because the character of the site largely determines the sources of non-point pollution and the effectiveness of specific control measures, good site characterisation is essential. Soil survey reports, stream survey reports, and geologic, climatic, topographic, and vegetation maps with accompanying descriptive material should be compatible with the degree of reliability expected from the analysis. The procedure has been designed to analyse third-order or smaller basins (Strahler, 1957). If larger basins are used it is difficult to detect the hydrologic response (Ponce & Meiman, 1983) and changes in water quality may not be detected.

The nature of the non-point pollutant transport process is such that the pollutant must be traced as thoroughly as possible through the entire system; therefore, the environmental factors significantly affecting its generation and transport to receiving waters have been recognised. These factors are then related to the physical and biological processes that govern the pollutant's ultimate fate. The tracing of a non-point source pollutant is critical in determining the need for, and type of, controls to employ in the forest setting because most water-quality constituents identified as pollutants also occur naturally within the system. The methodology of analysis has been structured to differentiate natural pollution sources from those which may have resulted from human activities.

It is not possible, with the current state of the art, to predict with certainty absolute values of water-quality response following forestry practices. However, comparisons of the relative change in water-quality resulting from different management practices can be made within acceptable limits. The anlaysis procedure, therefore, is based on the comparison of the water-quality objective and the relative change in water quality between existing conditions (current situation) and treatment conditions (proposed activity). The proposed activity can be implemented, of course, only if the existing water quality meets the water-quality objectives.

Figure 29.1 depicts the iterative process that may be required if the proposed silvicultural activity does not meet the water-quality objective. During this process, the resource impacts are identified, control opportunities are evaluated, and the silvicultural activity is revised as needed. In some cases, simulation may indicate that water-quality objectives will be exceeded even after all feasible controls have been applied. If this occurs, the silvicultural activity should not be considered for the area at this time. Alternative land-use management of the catchment area may be necessary.

29.2.2 Potential resource impacts

Eleven potential resource impacts have been identified. These are considered to be the most important in terms of non-point source water pollution and silvicultural activities. Each is defined below. They are listed alphabetically and not necessarily by order of importance.

- *Aerial drift and application of chemicals* This refers to any chemical pesticides, herbicides, or fertilizers allowed to fall directly on, or wash into, a receiving water.

- *Bare soil* This is the exposure of bare soil due to the reduction of vegetative ground cover, rock, and litter.
- *Channel gradient change* This is a change in the channel slope and energy relationships caused by debris jams, mass wasting, and other introduced materials.
- *Compaction* Soil compaction is the packing together of soil particles by instantaneous forces exerted at the soil surface.
- *Debris in channel* Debris in the channel refers to those obstructions in a stream channel caused by silvicultural activities, such as debris dams and fill-slope encroachment from roads.
- *Excess water* Excess water is the increase in channel flow resulting from alteration of the hydrologic processes by silvicultural activities.
- *Onsite chemical balance changes* This refers to changes in the nutrient balance caused by silvicultural activities.
- *Slope-configuration changes* This is an alteration of the land slope due to silvicultural-related activities, such as road building, log deck construction, and yarding† of trees.
- *Stream-shading changes* These occur as a result of the removal of trees and understorey vegetation that provide protection from direct solar radiation to streams and other surface waters.
- *Vegetative change* Vegetative change includes the removal of vegetative ground cover, canopy cover, or a change in vegetative type.
- *Water concentration* Water concentration occurs when water is intercepted and allowed to converge on the soil surface rather than infiltrating or spreading naturally.

29.2.3 Silvicultural activities and potential resource impacts

Table 29.1 summarises potentially-adverse resource impacts resulting from commonly-used silvicultural activities. Activities considered include methods of cutting, yarding, roads and access systems, fuel management, site preparation, and other activities.

29.2.4 Control opportunities

Table 29.2 presents a list of control opportunities that may alleviate each particular impact. Following each control opportunity is a control class: Procedural, Preventative, and/or Mitigative. *Procedural* controls are administrative actions or sanctions that result in reduced generation or transport of pollutants. *Preventative* controls apply to the pre-implementation phase of an operation. These controls are planning-orientated and involve stopping or changing a planned activity before any disturbance causing pollution is allowed to occur. *Mitigative* controls involve vegetative, chemical, or

† Yarding in this context is the hauling of trees.

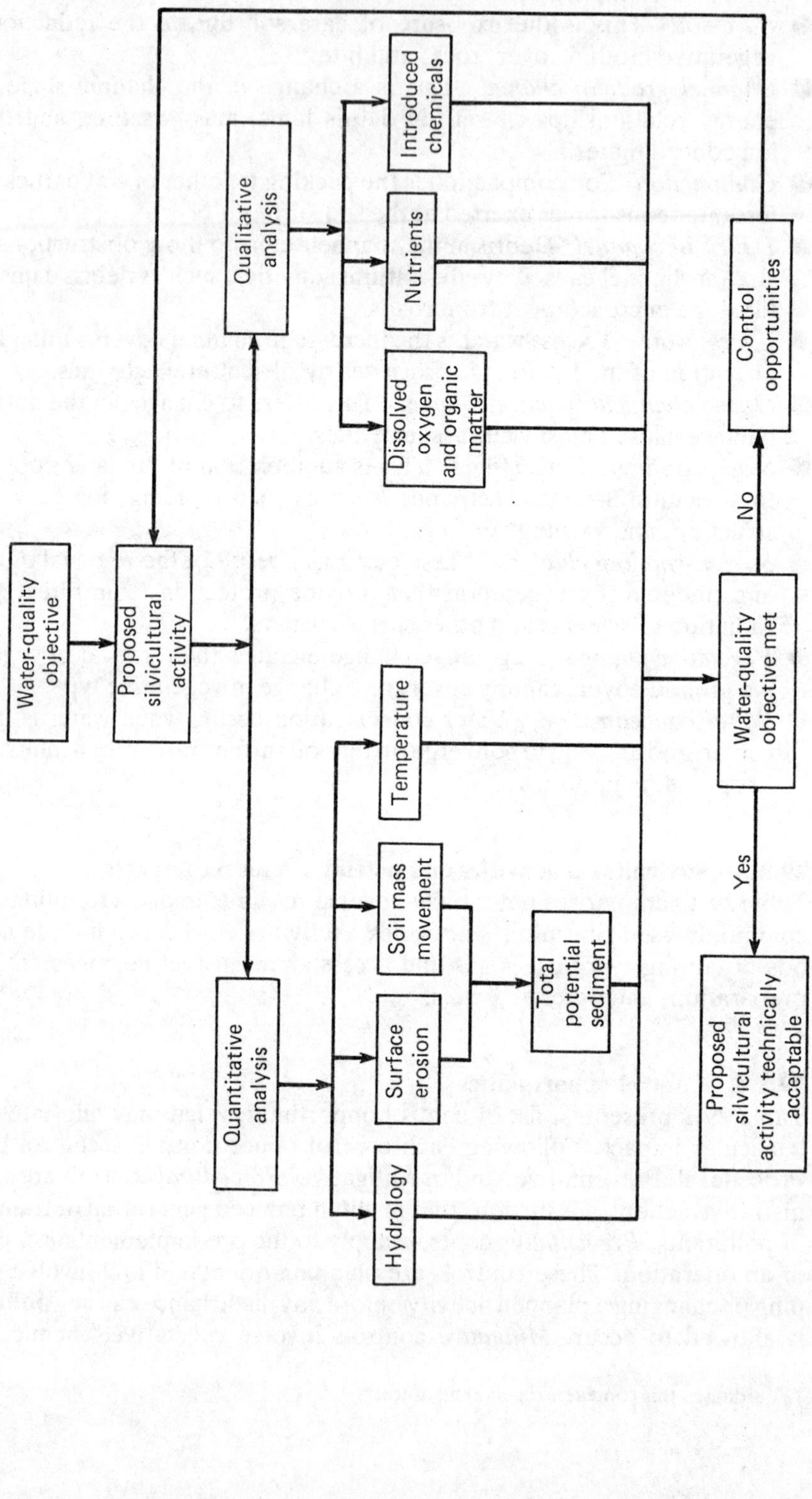

Fig. 29.1 — An overview of the procedural strategy for evaluating diffuse-source pollution derived from silvicultural activities and for assessing control opportunities (after USEPA, 1980)

Table 29.1

Silvicultural and related activities and associated potential adverse resource impacts (after USEPA, 1980)

Activity	Aerial drift and application of chemicals	Bare soil	Channel gradient changes	Compaction	Debris in channel	Excess water	On-site chemical balance changes	Slope configuration changes	Stream shading changes	Vegetative change	Water concentration
Methods of cutting											
Clear, seed tree, selection and shelterwood					●	●			●	●	
Yarding methods (see Fig. 29.2)											
Tractor skidding		●		●							●
Cable yarding (high lead)		●									●
Cable yarding (skyline)		●						●			
Cable yarding (balloon)		●									
Aerial skidding							●				
Mechanised logging (feller, buncher, etc.)		●		●							●
Construction and maintenance of roads and access systems	●	●	●	●	●			●		●	
Fuel management methods											
Burying slash		●		●				●			
Firelines and fuel breaks		●		●				●			●
Broadcast burning, hand piling and burning, machine piling and burning, prescribed underburning, and spot burning	●	●		●	●	●	●			●	●
Yarding unmerchantable material		●		●	●						
Lop and scatter					●						
Rolling chopper				●			●			●	
Chip and spread and masticate				●	●		●				
Site preparation											
Dozer stripping		●		●		●		●		●	●
Terracing		●		●		●		●			
Machine scalping		●		●							
Bedding		●									●
Ploughing		●			●			●		●	●
Disking		●			●			●		●	●
Drags		●		●						●	●
Drainage		●									●
Chemical treatment	●				●					●	
Other activities											
Mechanised planting				●							●
Release from plant competition (fire) (See broadcast burning)											
Release from plant competition (chemical)	●										
Release from plant competition (mechanical)				●							●
Thinning and clearing (hand)					●					●	
Thinning and clearing (mechanised)				●	●					●	
Fertilization	●						●			●	

Table 29.2
Potential resource impacts and control opportunities (USEPA, 1980)

Control opportunities (Type)†	All listed resource impacts	Aerial drift and appliction of chemicals	Bare soil	Channel gradient changes	Compaction	Debris in channel	Excess water	On-site chemical balance changes	Slope configuration changes	Streamside shading changes	Vegetative changes	Water concentration
1. Administrative closure of roads (MIT, PREV, PROC)			●		●							●
2. Appropriate cross-section in roads (PREV)			●						●			
3. Armouring (MIT, PREV)			●	●								●
4. Avoid roading steep slopes (PREV)			●						●			●
5. Bench cut and compact fill (MIT, PREV)						●			●			
6. Break gradient of firelines (MIT, PREV)									●			●
7. Bridges (PREV)				●		●						
8. Brush barrier filter at the toe of fill (MIT, PREV)			●			●						
9. Chemical application (PREV)		●						●				
10. Close roads after use (MIT, PROC)			●		●							●
11. Conformity to regulations (PROC)	●											
12. Control ash or dust build-up (MIT, PREV)		●						●				
13. Curbs and berms (MIT, PREV)												●
14. Cut-and-fill slope configuration (MIT)			●									●
15. Cutting block design (PREV)							●			●	●	●
16. Directional felling (PREV)			●		●	●				●	●	
17. Ditch checks (MIT)				●						●		●
18. Ditch maintenance (MIT, PROC)				●					●			●
19. Divert water on to stable areas (PREV)												●
20. Drainage above cut slopes (MIT, PREV)			●						●			●
21. Eliminate source of debris (MIT)						●						
22. Endline or fly material from waterside areas (MIT, PREV)			●		●	●						
23. Enforcement of standartds and bonding of operators (PROC)	●											
24. Fill slope design and locations (PREV, PROC)			●			●						
25. Full bench section (PREV)						●			●			
26. Haul woody material offsite (MIT)						●		●				
27. Hold water on site (MIT, PREV)			●									●
28. Identify soil and geologic characteristics and map sensitive areas (PREV, PROC)			●		●		●	●	●			●

Table 29.2 — *continued*

Potential resource impacts and control opportunities (USEPA, 1980)

Control opportunities (Type)†	All listed resource impacts	Aerial drift and appliction of chemicals	Bare soil	Channel gradient changes	Compaction	Debris in channel	Excess water	On-site chemical balance changes	Slope configuration changes	Streamside shading changes	Vegetative changes	Water concentration
29. Keep pesticides and rodenticides well away from surface runoff (PREV)		●						●				
30. Leave vegetation between strips (PREV)			●		●						●	●
31. Limit disturbed area (PROC)	●											
32. Limit equipment operation (PREV)			●		●	●			●			●
33. Locate activities producing small, woody fragments away from water (PREV)						●						
34. Locate corrals away from streams [animal skidding] (PREV)						●		●				
35. Machine or hand plant (PREV)			●		●		●	●	●		●	●
36. Maintain ground cover (PREV)						●			●		●	
37. Maintain natural watercourses (PREV)				●								●
38. Monitoring (PROC)	●											
39. Minimise convergence of firelines (PREV)												●
40. Outslope firebreak lines and terraces (PREV)							●					●
41. Oversize ditch drain (PREV)				●								●
42. Pile material in patterns								●				●
43. Prescribe and execute burns under conditions that will not result in total clean-up (PREV)			●				●					
44. Prescribe limits for the amount of area disturbed by equipment (PREV)			●		●						●	●
45. Prescribe yarding and skidding layout (PREV)			●		●				●			●
46. Prevent fire spread outside treatment areas (PREV)			●									
47. Protect bare road-surface areas with non-living material (MIT)			●			●						
48. Protect fuel storage areas (PREV)								●				
49. Reduce log length (PREV)			●									
50. Reduce logging-road density (PREV)			●		●				●			
51. Reduce road grades (PREV)												●
52. Reduce vehicle travel (PREV)					●							●

Table 29.2 — *continued*

Potential resource impacts and control opportunities (USEPA, 1980)

Control opportunities (Type)†	All listed resource impacts	Aerial drift and appliction of chemicals	Bare soil	Channel gradient changes	Compaction	Debris in channel	Excess water	On-site chemical balance changes	Slope configuration changes	Streamside shading changes	Vegetative changes	Water concentration
53. Reduction of impounded water (MIT)				●					●			●
54. Remove debris from stream (MIT)					●						●	
55. Repair and stabilise damaged areas (MIT)				●		●						●
56. Revegetate treated areas promptly as local conditions dictate (MIT)		●	●		●	●	●	●	●	●	●	●
57. Rip or scarify compacted surfaces (MIT)												●
58. Road and landing location (PREV)					●	●			●			●
59. Road ditch (MIT, PREV)												●
60. Road-drainage maintenance during storms (PROC)	●											
61. Sediment trap (MIT)												●
62. Select low impact equipment (PREV)	●											
63. Slope length (PREV)			●									●
64. Space culverts to control road ditch erosion (PREV)				●								●
65. Species selection (PREV)			●		●		●	●	●		●	●
66. Specify timing (PROC)	●											
67. Stabilizing structures or cut slopes (MIT)			●						●			
68. Timely drainage maintenance (PREV)	●											
69. Timing of chemical applications (PREV)		●									●	
70. Timing of use of off-road, heavy equipment (PREV)					●							●
71. Trash racks (PREV)												●
72. Type of site preparation treatment (PREV)			●		●		●	●	●		●	●
73. Use maximum spacing and minimum strip width in site preparation (PREV)			●				●					●
74. Waterside area (PREV)		●	●			●	●			●		●
75. Windbreaks or uncut timber to prevent wind erosion (PREV)			●									
76. Woody debris disposal sites (PREV)						●		●				

† MITigative PREVentive PROCedural

physical measures which alter the response of the water-disturbing activity after it has occurred.

The control opportunities presented in Table 29.2 are typically used in three ways:

- in the formulation of a silvicultural activity plan;
- in the prescription of mitigative controls for a previously disturbed site; and
- in the prescription of a mixture of preventative and mitigative controls for the alteration of the silvicultural activity plan so it will meet established water-quality goals.

REFERENCES

Anderson, H. W., Hoover, M. D. & Reinhart, K. G. (1976) *Forests and water: effects of forest management on floods, sedimentation, and water supply*. USDA-Forest Service, General Technical Report PSW-18/1976. Pacific Southwest Forest and Ranger Experiment Station, Berkeley, California. 115 pp.

ASCE (1975) *Watershed Management Proceedings of a symposium on watershed management,* August 11–13, Logan, Utah. Committee on Watershed Management of the Irrigation and Drainage Division of the American Society of Civil Engineers. Published by ASCE, 345 East 47th Street, New York, N.Y. 781 pp.

ASCE (1980) *Watershed Management Proceedings of a symposium on watershed management* July 21–23, Boise, Idaho. Committee on Watershed Management of the Irrigation and Drainage Division of the American Society of Civil Engineers. Published by ASCE, 345 East 47th Street, New York, N.Y. 1100 pp.

Brown, G. W. (1978) *Forestry and water quality*. Oregon State University Book Stores, Inc., Corvallis, Oregon. 74 pp.

Krygier, J. T. & Hall, J. D. Eds (1971) *Proceedings of a Symposium Forest Land Uses and Stream Environment,* October 19–21, 1970. Oregon State University, Corvallis, Oregon.

Ponce, S. L. & Meiman, J. R. (1983) Water yield augmentation through forest and range management — Issues for the future. *Water Resources Bulletin*. **19,** (3), 415–419.

Strahler, A. N. (1957) Quantitative analysis of watershed geomorphology. *Trans Am. Geophys. Union,* **38**, 913–920.

US EPA (1976) *Forest harvest, residue treatment, reforestation and protection of water quality*. Report Number EPA 910/9-76-020. Region X, 1200 Sixth Ave., Seattle, Washington. 273 p.

US EPA (1977) *Silvicultural chemicals and protection of water quality*. Report Number EPA 910/9-77-036. Region X, 1200 Sixth Ave., Seattle, Washington. 224 p.

US Forest Service (1980) *An approach to water resources evaluation of non-*

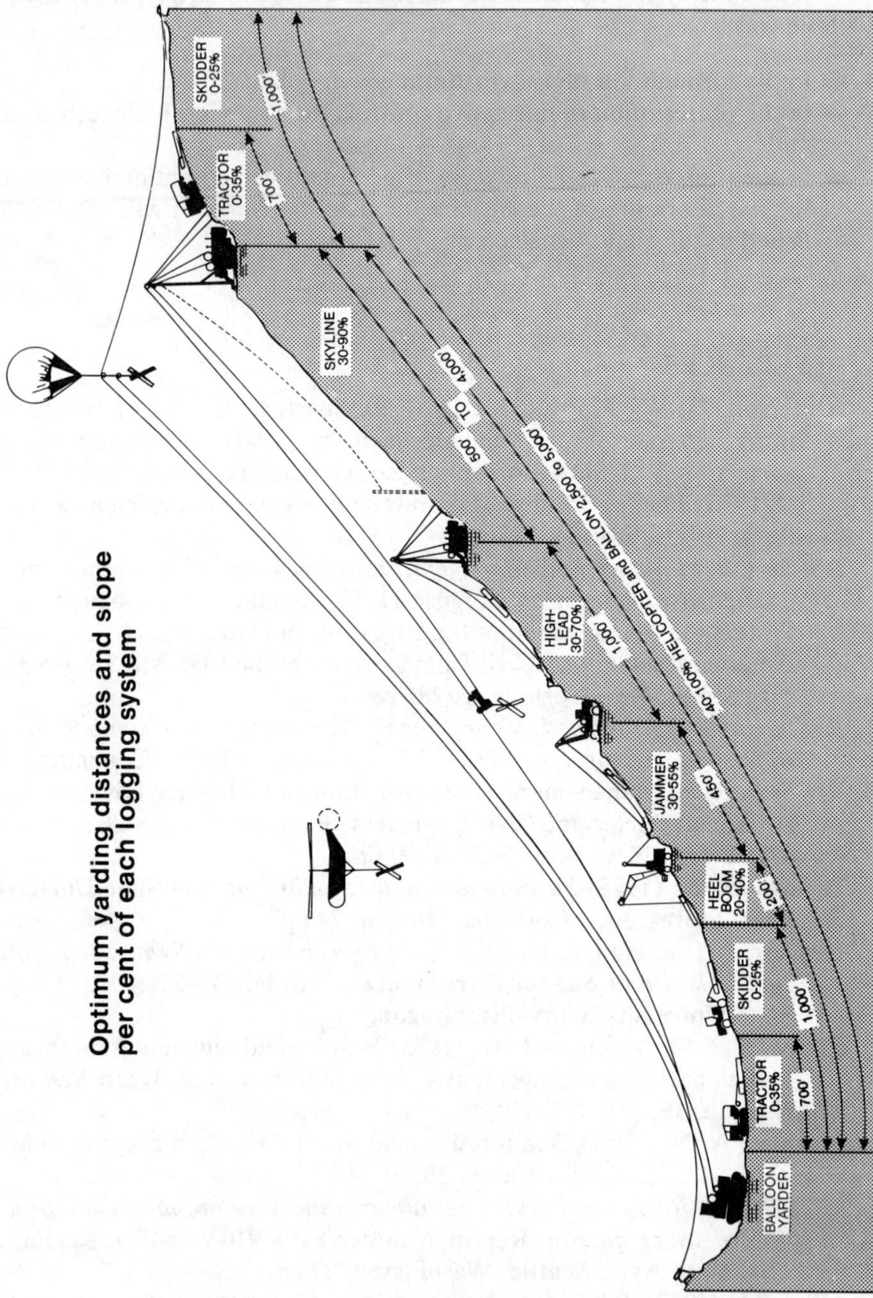

Fig. 29.2 — Optimum yarding distances (feet) and slope per cent of each logging system

point silvicultural sources (A procedural handbook). Published by USEPA, Report Number EPA-600/8-80-012. US Environmental Protection Agency, Athens, Georgia.

Discussion

Chairman: **Dr R. F. PACKHAM**
Assistant Director, Water Research Centre, Environment

The discussions on forestry fell into three broad areas — planning and managing forestry, the effects of forestry on water yield (quality and quantity) and, as a distinct issue, the effects of pulp mills.

PLANNING AND MANAGING FORESTRY

Particular concern was expressed about the afforestation of water-supply catchments but, to offset this, the value of applying ecological principles to forest management was being increasingly demonstrated.

Afforestation

—of water-supply catchments

Mr J. K. Fawell (Water Research Centre, Environment) asked for the Forestry Commission's opinion of a recent statement that the additional costs of water treatment resulting from the effects of afforestation made forestry an uneconomic use of certain catchments. **Dr Binns** agreed that the effects of forestry on existing reservoirs in terms of greater treatment cost and reduced yield made forestry appear unattractive, especially where the storage had been engineered for a precise yield. The nation must decide what it wanted from its uplands, and where forests and water-gathering grounds were to be located.

Mr L. A. Greene (Strathclyde Regional Council) suggested that afforestation was being directed towards water-supply catchments (a suggestion which **Dr Binns** refuted). Was there not enough land outside such catchments to supply the needs of forestry? The issue of water-treatment costs was taken into account in discussions with the Regional Advisory Committees on afforestation schemes. **Dr Binns** said that the Forestry Commission was now only planting a few thousand hectares annually. Most planting was by private concerns. People tried to get land to plant where they could, and each case was examined on its merits. There probably *was* enough land outside supply catchments. As far as costs of treatment were concerned Dr Binns thought that the industry would be paying for higher standards of potable water in any case. However, where a catchment was already used for supplying water it did seem that there was a disincentive to plant trees on it.

Ecological principles

Costs of conservation measures

Dr S. C. Warren (Water Research Centre, Environment) asked Dr Ponce to indicate the additional cost of timber produced by methods which were sympathetic to the needs of conservation. **Dr Ponce** could not say for sure, but he added that in his opinion most agencies and timber companies recognised that good forestry management was good river-basin management. There was a benefit in improved productivity of the land.

Mixed forests

Dr C. Newbold (Nature Conservancy Council) wondered whether the Forestry Commission was taking steps to plant mixtures of species (silver birch with Sitka spruce) to open the canopy and increase yield. The points made by **Dr Binns** in his reply were that the scope for mixing species was limited by the characteristics of the site. While on rich soils there were many possibilities, only a handful could be used in the uplands. Nevertheless there was increased interest in mixtures, partly because of the evidence that when either larch or pine was mixed with Sitka spruce the nitrogen nutrition of the latter was improved. Slow-growing species of alder (which can actually *fix* nitrogen) had attractions too. There were still management problems in the large-scale application of these ideas.

Pest control

The use of pesticides was raised by **Mr A. Dickson** (University of Strathclyde). He referred to a proposal to use fenitrothion in Scotland which was the subject of somewhat 'alarmist' press coverage. **Dr Binns** confirmed that fenitrothion was used, mostly in the North of Scotland, against Pine Beauty moth on Lodgepole pine, monitored by many agencies. Application was by controlled-size droplet techniques. Side effects seem to have been remarkably small. The Forestry Commission tried not to 'prop up' species ill-adapted to their location. The insect was a native species found on Scots pine (on which it never gave problems) but its populations had exploded in the remoter parts of Britain where Lodgepole pine had been introduced. Perphaps that species was stressed and therefore vulnerable; certainly its sensitivity to the moth raised doubts about its suitability for forestry in those areas.

Dr Newbold followed up this question with one to Professor Bruenig. In trying to achieve forests with a better ecological balance and using less herbicides was there a greater or lesser incidence of disease than in monoculture? **Professor Bruenig** had not carried out controlled field experiments on disease incidence in relation to forest structure but it was becoming increasingly clear in Auermühler forest that there were far fewer problems in the widely spaced mixed forests than in the conventional stands in the surrounding State and large co-operative forests where pesticides were used more extensively in recent years. The forest in Auermühler required hardly any control measures except some trapping to 'take the top off' the pest populations in case something serious seemed about to develop. Birds acted as useful predators and were encouraged by the lack of spraying. There could be real advantages in allowing biological self-regulating mechanisms to help in forest management.

Codes of practice
Road-building; drainage channels
Mr D. Hammerton (Clyde River Purification Board) was impressed by the care taken in preventing erosion during forestry operations in the United States. In Scotland forestry roads and drainage channels have given rise to problems, leading to damage in fisheries and fish farming. Although the Forestry Commission had developed codes of practice in collaboration with Dr Mills (see Chapter 28) there was no statutory requirement for consultation with the River Purification Boards concerning forestry roads and drains. Consultation in advance of schemes would always be helpful.

Monitoring
Dr Binns was concerned that the Forestry Commission in its slimmed-down form would be hard-pressed in attempting to perform the role of management, organisation and control that it seemed delegates required of it. He hoped that the Commission now had its own house in order but wondered how it was to ensure compliance with codes of practice in the private sector. Was this another area where the experience of the United States could be utilised? **Dr Ponce** thought that problems similar to those now facing the UK foresters were the reason behind the amendments to the Federal Water Pollution Control Act in 1972; although laws existed for the control of point-source pollution it was not considered practicable to apply point-source control methods to non-point sources. There was a measure of control of the harvesting of timber from federally-owned land, and this included most of

the western forests. The broadleaved forests in the east, on the other hand were mostly in private ownership. The Environment Protection Agency had delegated responsibility for control to the States, who in turn had delegated it to Counties. It was hard to define whether a private forester had done a good job in managing his woodlot. Public opinion and public pressure were the most effective keys to action.

Clear-felling

Mr Blackie asked Dr Ponce whether there was an upper limit placed on the area of forest which could be felled in a catchment at one time. **Dr Ponce** confirmed that there were standards. In general on third-order catchments clear-cutting was limited to a maximum of 18 ha (40 acres) on Federally managed land. He gave an example of a municipal supply catchment serving 500 000 people. Here, due to the problem of the resulting turbidity, no more than two per cent of the forest was harvested during any one year.

Best management practices

Dr Ponce continued that the principles of best management practice suggested that the best practice for a large (fourth order) catchment was the sum of all the best management practices on the smaller catchments. They had resisted any other approach.

EFFECTS OF FORESTRY

This subject area divided conveniently into water quantity and water quality, the latter hardly touching on acidification of waters in the vicinity of forests as this has been more than adequately covered in other recent symposia.

On water yield

Evapotranspiration and drainage

Mr M. Cominetti (South West Water) was interested to learn whether evapotranspiration made its effects felt on river flows, and if so was there a formula which could be applied to predict such an effect? In response **Mr Blackie** confirmed that in high rainfall areas on wet, windy uplands there was an increased loss of the water intercepted by the canopy. This reduced stream flows. It was not possible to transfer observed reductions directly from one area to another. Actual transpiration rate was somewhat less from forest than from grassland. Thus in a dry site the effect on stream flow may be very small or even reversed. Predictive methods were available for wetter areas but Calder & Newson's model (see Chapter 26) had to be finely tuned if intended for use in drier sites. **Mr Dickson** sought clarification of the decrease in runoff described in that chapter. Was it just surface runoff or the contribution from groundwater which reduced? **Mr Blackie** said that the reductions reported were in total stream flow, the losses having principally occurred from the tree canopy.

Snow

Snow formed a significant part of precipitation in the Scottish Highlands and **Mr C. D. Bayes** (Water Research Centre) asked how it affected both quantity and quality of runoff. In reply, **Mr Blackie** explained that the opportunity to study the impact of heavy falls of snow had been the second reason for the intense research effort by the Institute of Hydrology at Balquhidder. The measurement of snow inputs was a difficult problem in determining the effects on runoff quantity. Detailed studies were also in progress near Aviemore of the interception of snow in the conifer canopy and the loss rates from the intercepted snow and of the quality of snowmelt water.

On water quality

Drainage and pH

The question of drainage patterns was raised by **Dr D. H. Bache** (Strathclyde University). Should forestry drains cut across or run along contours? In the former case more rapid runoff and less chance for buffering of acidic rainfall seemed to result. **Dr Binns** emphasised that much thought had gone in to this subject. Normally two-stage systems were constructed. Short-run primary downhill drains emptied into relatively frequent secondary contour drains which collected and removed the water from the site. As the forest matured the primary system was deliberately not maintained but the secondary, permanent system was, to limit erosion and encourage percolation. More studies were needed of different methods of cultivation near streams, to encourage deep percolation and minimise the problem of acidity, natural or man-made. Ditching did seem able to encourage short-term acidification, **Dr W. R. Howells** (Welsh Water) reported, probably due to the oxidation of sulphur and nitrogen in the newly-disturbed soil. **Dr Binns** suggested that such effects would indeed be short-lived.

Mr A. I. Oxbrow (Napier College), warning against broad generalisations, cited examples of dense forest in other parts of the world where runoff was virtually non-existent, and of runoff in the English Lake District where, in 1984, the pH had been higher in water leaving forested areas than in the runoff from grassland.

Professor S. Henin was also interested in the effect of drainage on river-water quality. Waterlogged soil normally gave rise to poor quality water and when this was first drained the problem increased. Later, the soil structure improved and so did the drainage water. But how should the river quality officers handle the situation in the meantime? Secondly, in France in the past, the many small ditches draining land were maintained by hand. As the cost of manual labour rose less maintenance could be afforded and trees began to die because the soil was becoming waterlogged and surface runoff increased. Now, using machinery, the land can be drained at an acceptable price, but there will be the initial problem of water quality in the rivers. Further, there was an example where mechanised maintenance was making marked improvements to the soil inland, but the drainage water was damaging oyster beds where the rivers entered the sea.

Sediments

Comments on erosion may be found above, and congratulations to the United States on their approach to limiting such problems. **Dr Ponce** added however that the USA was not without erosion problems but their careful site evaluation prior to harvesting and their predictive models did much to alleviate the situation.

Dr Bache asked Dr Ponce which of the two mechanisms for generating sediment — surface erosion and mass wastage — predominated. He wondered how predictions were made of the rate of mass wastage. In fact, **Dr Ponce** said, there were three origins of sediment, the third being channel sources, which constituted a large component in runoff. Mass wastage could dominate; it could even be one hundred times greater during forestry operations than in control catchments. The sediments accumulated and forced the stream to spread, undercutting slopes and causing even more mass wasting. Surface erosion was predicted with a modified universal soil loss equation. Prediction of mass wasting was more of an art than a science. Experience and local knowledge was such however that with the help of aerial photography they were able to predict with reasonable accuracy.

Phosphorus and nitrogen

Mr Bayes wondered how the increasing planting of Sitka spruce would influence nitrogen fertilizer usage in future, from its present low level. **Dr Binns** confirmed that only 1300 ha of Forestry Commission trees would receive nitrogen in 1986, and that nitrogen will be needed if Sitka spruce planting continued on poor soils such as peats. Nationally the nitrogen used would be trivial but locally it could be significant. (See also the comments on the benefits from mixed forests, above.) **Dr Heinonen** requested details of the loss of phosphorus, washed out of forested areas following the application of phosphorus fertilizers. Estimates had been made, **Dr Binns** said. There were immediate losses (for example when the fertilizer fell directly into drainage channels) but the total loss did not appear to be great. It was a problem that could not be solved anyway because aerial application was the only possible means for an established crop. Dr Heinonen had shown the loss of phosphorus following clear-felling and **Dr Newbold** wanted to pursue this. What was the loading of the phosphorus, if to a stream? How long did it take for concentrations to fall to those of the control area? **Dr Heinonen** had calculated the loadings, which were elevated ten-fold above normal. Periphyton production increased after clear-felling too.

Bacteriology

Dr Ponce reported that one benefit resulting from allowing the public access to forests was an increase in bacterial quality as the wild animals moved away from the streams.

EFFECTS OF PULP MILLS

Most questions related to the effects of pulp mills on the environment but there was brief discussion of human health implications.

On the environment

Dissolved oxygen

Referring to Fig. 27.2, **Mr Hammerton** asked whether concentrations of dissolved oxygen (DO) in summer fell to lower levels than those shown in April. Also, there was an interesting increase in DO towards the bottom of the lake in the winter. He had observed such an anomaly where a cold river water had brought oxygenated water into a lake. **Dr Heinonen** said that in Lake Saimaa conditions were very poor in the winter. Ice covered the lake in early December and the wastewater would have taken the intervening two months to reach the sampling point, 40 km from the pulp mill. By that time the wastes would be travelling 10 to 20 metres above the lake bed but later they would follow the bottom contour. Conversely, in summer, the wastes travelled in the epilimnion, where they were diluted and gave no problems.

Aeration devices

Mr Cominetti recalled the successful use of surface aeration devices on the Thames estuary near paper-mill effluent discharges before the installation of treatment plant. **Dr Heinonen** was not aware of such surface devices in Finland for pulp mills but did know of the use of air-bubbles discharged at the hypolimnion to promote circulation.

BOD/COD

In answer to an earlier point by **Mr Hammerton, Dr Heinonen** said that BOD determinations, although relevant, were not made on Finnish lake waters. COD was easier to measure, and was to be considered as an indicator.

Resin acids

Mr W. Halcrow (Forth River Purification Board) had found that in one particular thermomechanical pulp mill resin acids were the most important component of the discharge. He asked what alternatives were available where activated-sludge treatment could not be used. **Dr Heinonen** had no alternatives to offer. In normal activated sludge a 30% reduction of phenolic compounds could be expected but with care in aeration and temperature control this could be increased to 80%.

On human health

Following on from this discussion **Mr Akyar** (Turkey) reported that in Turkey, Kraft pulp and paper mills seemed to be implicated in a problem of human health. Fishermen who ate fish from rivers receiving the effluents of these mills fell ill and died within about 25 days. The gills of the fish were clogged with lignin. A disease similar to methaemoglobinaemia was said to be the cause of death.

Control

Mr Hammerton noted that only about one third of the pulp and paper mills were equipped to treat their wastes. How was control achieved? Surely with such deep lakes in the receiving water system, prone to de-oxygenation, standards would need to be stringent. **Dr Heinonen** hoped that new treatment plant would eliminate the kind of problem shown in Lake Saimaa.

Recycling

On a recent visit to Finland, **Professor Bruenig** had gained the impression that there was significant recycling of potentially toxic materials and use of waste energy in pulping plants. Was this unusual? **Dr Heinonen** thought that Professor Bruenig may have been given somewhat optimistic information. However, although phosphorus loadings were still high, there were good examples of recycling fibrous material.

CONCLUSION

The last word of the discussion on forestry belongs to **Professor Bruenig** who reported discussions with an economist colleague. There was no basic conflict between maximising profit and protecting the environment. Environmental protection could be seen in the long term to be a stimulus to developing new technologies and to the most careful handling of natural resources.

General Discussion

Chairman: **Dr S. C. WARREN**
Director
Water Research Centre, Environment

CHAIRMAN'S INTRODUCTION

In opening the final discussion period of the Conference, Dr Warren explained the choice of so wide-ranging a subject as the Effects of Land Use on Fresh Waters. Only by bringing together all types of land use could we consider their competitive interactions and their effects on the whole environment.

The basic problem was of one user of land or water having an impact on another, or others. The farmer's use of nitrate affected water-supply undertakings, just as power generation could contaminate the atmosphere and damage water quality and sometimes forests. Where such activities resulted in a need for expenditure (for example in reducing nitrate in drinking water in order to comply with standards) he asked whether there should be a mechanism for transferring money from the 'polluter' (perhaps the farmer or the fertilizer manufacturer) to the water supplier. Present UK government policy was not to apply a tax on a sector specifically to cover damage caused by that sector: road tax was not applied exclusively to the construction or maintenance of roads. Nevertheless nitrate was going to add to the costs of the water undertakings. Was there a need for a framework to decide what the costs were and how these should be allocated or recovered?

A second problem arose where damage to the environment did not result in a direct financial cost to another party. Despite this, the polluter could be required to spend a great deal to reduce or eliminate the pollution. We had no means of placing a cash value on environmental quality, and this continued to be a handicap when trying to justify expense on environmental restoration.

Thirdly, Dr Warren suggested that it might be useful to discuss the nature of our means for enforcing regulations on environmental protection.

In the ensuing discussion these and other topics were taken up. They have been re-grouped below under three headings: Planning, Control, and Costs, and an attempt has been made to represent all the views expressed.

PLANNING

The need

Planning can be seen as a control measure and is treated as such in the appropriate section later in this discussion, but here it is taken in the sense of *preparation.* In a situation where proposed uses of land may conflict with one another, as indicated in the Chairman's introduction, and where neither costs nor benefits may be immediately obvious, there is clearly a need for planning — the formulation of a strategy leading to the attainment of goals — a land-use policy. Proposed changes to the European Community's Common Agricultural Policy add to this need, and the suggested current pressures for planning are headed 'What do we need to do now?'. Sadly, in the past, we have reacted to a problem only when it has reached crisis proportions. But this leads to the first question.

Is planning possible?
What are the inherent problems?

We may be faced with problems that were not only *unforseen* but also could not reasonably have been predicted. Often this *unpredictability* seems to have been due to changes in human society. The Conference learnt of such incidents. For example, in France the closure of a Citroen factory resulted in the flooding of basements in the adjacent residential area because the factory no longer abstracted water from its borehole. Similarly, in the Netherlands, an unprecedented and unpredictable increase in animal husbandry had happened in recent years, with all the consequent problems of pollution. In other cases it is the physical or chemical *complexity* of a situation which impairs our ability to predict, as in Anglian Water's realisation that groundwater catchments were so diverse in character that to assess risk required a policy of site-by-site examination. The best planning could of course be overtaken by events, and in Anglian Water such an event was cited when tetrachloroethylene pollution of groundwater occurred (cf Chapter 4), resulting in the loss of a daily supply of 6.8×10^6 litres of water.

A second problem with the planning process is that it takes *time*, as, indeed, do the solutions to many identified environmental difficulties. We may know how to minimise the damage to streams being caused by an existing forest but the sheer logistics of a rapid remedy are beyond our resources. Forestry exemplifies the complexity of the planning process.

Since only 2% of the land area of Scotland which could be afforested consists of water-supply catchments it would seem quite straightforward to apply pressure to keep forestry off all such areas. This may be the case, but there are other conflicting demands on the uplands: nature reserves, national scenic areas and areas of outstanding natural beauty, all of which have constraints placed on them in terms of forestry. None of these has yet been declared inviolable from forestry development.

A final problem facing the planner was that of *scale*. Several examples were given of this. Planning permission for a 12.5 t fish farm on Loch Lomond was opposed because of its potential to pollute the Loch. Numerous small units could have a greater (or lesser) effect on the environment than one large unit equal to their total capacity. The results were not predictable. In a study of land transformation and the effects of insiduous land use in France, construction of a new winter-sports complex was examined. This complex required that three of four hectares of land were covered in concrete for housing and roads. However, the modification that this caused to the water regime altered all the nearby torrential streams. They are now inundating the plain and 400 ha of pine forest have been damaged. Impact must be assessed on the largest relevant scale.

Whose responsibility is it to have the necessary foresight?

Bearing in mind the complex interactions between environmental quality problems from such diverse factors as forestry, lead in petrol, fish farming, and other non-point-source pollution, on which group did the responsibility for prediction lie? Was it the politician, the entrepreneur, the technician, or the civil servant? This question was answered in part by a plea that *everyone* needed to be vigilant and to try to foresee problems for the environment in time to take preventive, not remedial, action.

What planning mechanisms are available?

Environmental Impact Assessment (EIA) was available and was incorporated in a draft EC Directive, but it was generally intended to cover large-scale developments and not the kind of situation which could arise from, for example, the unrestricted growth of a single fish farm.

Risk assessment, like EIA, was well worth pursuing. On the other hand, since the assessment must include the broad issues and since it was hard to place a financial value on such things as conservation, recreation, and aesthetics, there would always be a question mark at the end of the assessment, at least at the present state of the art. This being the case it was not surprising that scientific approaches to securing action were often set aside in favour of *lobbying*. Perhaps the water industry should see this as a last resort.

Integrated land-use planning was a reality in parts of West Germany and it would be useful to exchange ideas with those concerned. The planning was carried out at the 'framework' and 'local' levels in consultation with the public and the ministries, the forest owners association and agricultural associations. Complex arrangements had been evolved to make this possible. A study of this and the UNESCO MAB project would be worthwhile. The latter was concerned with the long-term land-use planning and management of the Alps and involved collaboration between Italy, Switzerland, Germany and Austria. Modern methods of *systems analysis* and *sensitivity analysis* were being applied to the consequences of withdrawing agriculture from the mountains to the valleys, to the detriment of the latter.

The final mechanism to be considered in the planning process, and which also applies to Control, is *consultation and co-operation* (as employed in the case of West Germany, already cited). Although forestry and agriculture were largely exempt from the normal planning procedures (in which 'planning' could certainly be equated with 'attempted control'), they did operate parallel systems of voluntary consultation. For private forestry the consultation was linked with the application under the forestry grants schemes. The applicant approached the Forestry Commission, who then consulted with the appropriate range of interests — including agricultural Departments, the Nature Conservancy Council, the Red Deer Commission and the Planning Authority. It was the remit of the last-named to consult with other interested parties. In the case of difficulties reference was made (rarely at present) to the Regional Advisory Committees of each of the Forestry Commission's Conservancies. One comment on this consultation procedure was that it suffered from the disadvantage of considering applications in isolation. No-one was looking at the overall effects on the countryside as a whole. Another comment, in relation to agricultural problems, was that the system based on patient advice was under strain and it might be necessary to impose sanctions.

What do we need to know?

The proposed changes to the European Community's Common Agricultural Policy must be studied. They implied major changes to European land use including mechanisms for bringing land out of agricultural production. Should it be the uplands that were to be most affected or would it be more convenient to reduce production more dramatically in smaller areas of the richer low-lying land? In either case would the land released from farming be afforested, or would it be allowed to return to the wild by natural succession?

In West Germany 800 000 ha have already been taken out of agricultural production and a predicted further 2 to 3 million hectares would follow over the next two decades.

One consequence of abandoning farming on hill slopes was described for France. Centuries ago terraces were cut and cultivated, and were carefully drained by ditches and tunnels since the region was one receiving high rainfall. Now the land was no longer farmed and the drainage system had fallen in to decay. This gave rise to serious erosion of the hillsides with a mixture of soil and large stones being washed onto the lowland fields.

We should be concerned about organic micro-pollutants and traces of heavy metals, things that did not necessarily cause problems now but might in the future if vigilance was relaxed.

CONTROL

Introduction

Naturally at this stage in the Conference part of the discussion was a reiteration of ground covered earlier. Mention was made of the need to control the acidification of waters, pollution caused by fish farming, and the problems of excessive abstraction of water for irrigation. A need was expressed for a policy on setting bacteriological and virological standards for fresh waters used for recreation.

In the case of nitrates, a statement that the agricultural industry still needed to be convinced that lower limits to concentrations of nitrate in water were necessary, evoked the response that the limits were thoroughly supportable. Moreover, the fact that a limit might have been exceeded without fatalities was no indication that the limit was set too low! Safety factors were essential. In the UK the danger of methaemoglobinaemia was extremely low, but in Hungary and elsewhere it occurred and caused deaths.

The agricultural industry must now demonstrate a willingness to control the obvious (silage liquor, slurry) and insidious (pesticides, nitrate) forms of pollution. The fact that a suggestion had been made that the water industry should carry out research into treating farm wastes could be taken as an indictment on which the Agricultural Development and Advisory Service should seriously reflect.

Mechanisms

Methods for protecting waters from the polluting effects of various forms of land use included technical, managerial, and legislative systems.

Technical and managerial

Protection zones and policies around sensitive parts of aquifers were possible in some cases, for example the restriction of ploughing up of grassland, but the land area required was generally too large for the water undertaking to own its catchments, especially in a political climate in which the water authorities were under pressure to sell such assets.

Where an authority did not own the land it might be possible to arrange a management agreement with the land-owner, as the Nature Conservancy Council did in relation to the Wildlife and Countryside Act. Such an agreement would have to offset the loss of production from that area.

Rather than controlling the ingress of nitrate to water sources, recommendations had been made for the excess nitrate to be removed at the water-treatment works — but the costs this involved would have to be borne by someone.

Legislative

- *Local planning controls* The exemption of most forestry and agricultural practices from normal planning controls was seen as a severe disadvantage by the water authorities. Nevertheless there were some helpful mechanisms by which prior consultation or a restriction might be achieved. One of these was to seek special planning powers, not to prevent the planting of land but the provision of access roads.

● *National* The introduction of the Control of Pollution Act, Part II, with its requirement that pollution registers should be open to public scrutiny might add a new dimension of public awareness and pressure in the near future. Pressure groups and the rising interest in ecology as a political concept might also increasingly influence the situation.

● *International* EC Directives and other standards would not just operate from the point of view of human protection but for safeguarding the whole environment. The Common Agricultural Policy might bring about some control of pollution by restricting the area of land receiving fertilizer but it was more doubtful whether price controls would play much part. The traditional response by farmers to attempted control by cutting prices was to produce even more to maintain income.

General

Brief mention was made of the need for long-term strategies and for some kind of Environmental Protection Agency, perhaps helping to assess the conflicting interests of forestry, agriculture and water supply.

A continuing theme was the benefit to be derived from communication and co-operation.

To aid those whose responsbility it was to control pollution, WHO had published guidelines in two volumes. The first listed the recommended concentrations or ranges of each determinand, and the second gave the health criteria on which the recommended limits were based.

COSTS

The basis of the discussion was the validity of the 'polluter pays principle' when applied to agriculture and forestry.

Forestry

As an example, damage to the upland supply catchments of Scotland was considered. These catchments are naturally so uncontaminated that water from them needs very little, or no, treatment before being put into supply. Programmes of installing disinfection were under way but this should be all that is needed. If any other use of the catchments resulted in problems of eutrophication, turbidity, or high suspended solids, fairly expensive treatment of the water would be needed. Under current circumstances the costs would have to be charged to the consumers supplied with the treated water. This was unreasonable: private industry caused the problems, took money out of the area and put nothing back to alleviate the problems.

Decisions to modify catchment use were being taken without the water authorities' knowledge, and compromises would put them in the position of incurring very long-term costs.

Agriculture

In the Netherlands it had been proposed that a small increase would be made in the cost of animal feedstuffs (not a tax but a voluntary agreement). The money so raised would be used to help in the abatement of the problems caused by pollution from animal wastes. To accommodate those situations where, for example, a pig farmer operated in an area without problems the basic national levy (which would be very low) would apply. Where problems occurred, an additional sum would be paid.

General

Society set rules on the permissible degree of damage to the environment. If these rules were broken by a user of the land the resulting cost must be borne by the polluter. There was a need to find imaginative ways of putting the costs where they belonged.

Summing Up

H. FISH, OBE
Chairman, Natural Environment Research Council

Mr Chairman, Ladies and Gentlemen,

It is now my duty and pleasure to sum-up the proceedings of this 1985 WRc/WHO Conference on the Effects and Land Use on Fresh Waters.

It is always a good thing when telling a story, to start at the beginning and that is where I will start. This is an unusual course for me to follow, because normally I sum up Conferences before they start. In principle I normally avoid attendance at many of the sessions and whatever you may think about my summing up, I hope that you can accept that for the first time in my life at this Conference I have attended very nearly all the proceedings.

I think it is fair and appropriate to observe that the planning and operation of this Conference has been first-class. However, once again Murphy's Law somehow intervened in rendering some of us on Tuesday night, transport-less, chilled by the weather of Scotland at the ramparts of Stirling Castle, but soom warmed by the spirit of Scotland at the bar of another Stirling Castle. Those of the delegates who caught the early buses back on Tuesday night will not know much of this. Those of us who were involved can only applaud the courageous (and crafty) action of WRc in turning Murphy's Law to their advantage.

For our visitors from overseas I should explain that Murphy's Law is one of the better known exports of Ireland to England, Scotland and Wales. Murphy's Law states that if, in any human activity, anything can go wrong, it will go wrong.

We started the Conference with having the honour and pleasure of listening to the Under Secretary of the Scottish Office, Mr Michael Ancram, in opening our Conference. Mr Ancram's contribution was, according to my notes, first-class and to the point. He emphasised:

(i) the great importance of freshwater cleanliness and our duty to husband and conserve the water environment;
(ii) the international dimension of water quality standards and guidelines;
(iii) the need for reasonable flexibility in the application of standards for water quality;
(iv) the British success in improving river-water quality, for which the whole British people, and all the water authorities who serve them, can take much credit.

He also emphasised:

- that there were many problems to be solved and that the solutions of these problems had to be feasible and cost effective;
- further, that the scientific advice put forward on these problems needed to be relevant and related to the public sensitivity on these problems;
- finally, that he hoped that the outcome of this Conference would be to focus both thinking and action in improving the aquatic environment.

I am sure you will all agree with the key points as stated by the Minister, and that you feel, as I do, that our Conference has successfully focused on the thinking and action hoped for by the Minister, and subsequently endorsed by Ian Waddington of WHO in his introductory remarks.

Now we have listened to 29 very good papers at this Conference. You cannot possibly expect me to do no more than give sharp substance to the messages of those papers as tested and developed by our discussions. I could spend some two hours in summarising in detail all the many views and comments expressed at this Conference.

I must say that I felt, in considering each paper and its discussion, that there was a lot of talk about the design of old wheels and their renovation. Nevertheless underlying this regurgitation of long-standing problems, and renovations of old solutions to these problems, a number of very important messages were signalled, some unequivocal, some disputed, some flawed by special pleading, and some confused with preoccupations with the harshness of current local, regional and global economic stresses. The words balance, cost-effectiveness and cost-benefit have been frequently used, sometimes as an excuse to defend entrenched positions. Nevertheless, the truth was reasonably discernible.

The appearance of these understandable human aberrations must not be allowed to obscure the good science, the excellent surveys and accounts and the excellent contributions of the papers presented by the authors. But let me state, in turn, my analysis of each session of the proceedings.

In the first session, chaired by Frank Sanderson, we started with two papers relating to current practice of dealing with urban drainage from new towns. One paper related to Milton Keynes New Town in England, the second to the New Town of Lelystad in a polder of reclaimed land in the IJsselmeer (Zuider Zee).

The point which struck me most about the two papers was just how little interest the British Water Authorities generally take in the quality of their primary raw material — rainfall. It is true that some authorities take an interest in rainfall, particularly in the acid precipitation context, but surely it is time that the water authorities should put greater effort in to understanding rainfall contamination and its effects on surface and groundwaters.

In considering the different approaches to the drainage of the two new towns, on balance I believe the Lelystad approach to be the better one — at least the pollution sinks are within the urban area producing the pollution and hence are better open to surveillance and control. Should the next British new town in a lowland area be modelled on the Dutch system at Lelystad?

On the problems of Urbanisation we had three papers. One related to motorway runoff in which the French-drain approach to control appeared the best. The second related to waste disposal to land, which clearly spelled out that such disposal, after proper research, development and control, offered the best practicable environmental option. The third paper, describing a very difficult groundwater pollution problem in Milan and district was startling in its extent and encouraging in the sense that proposals for control of the problem had been formulated, if not yet put in hand completely.

The one paper on remedies, at the River Tame purification lakes, gave a very good example of positive action — the consequences of which are not fully understood. This system gave its problems — but minor ones compared to the overall benefit given to the improved cleanliness of the River Tame and the River Trent.

Session 2, chaired by Ian Waddington, related to Mineral Exploitation and highlighted the major improvements in water quality achieved in Scotland in relation to acid mine-wastes and in Cornwall regarding marked improvements achieved in metals and china clay pollution of Cornish rivers. On the generality of managing the liquid effluents from non-ferrous mining and mineral extraction, important points emerged.

First that in the UK, the concept of applying the approach of selecting the best practicable environment option (BPEO) of waste management lacked a mechanism for selecting such option. The suggestion by Dr Barbour that such a mechanism required the creation of a national body of top experts to determine the BPEO in different cases in the UK, was an important and attractive proposition.

Second, the point was made that legislation to control pollution in all its forms was useless without the manpower resources to apply and enforce the legislation. This is beginning to apply in England and Wales.

In respect of acid drainage problems arising from abandoned coal-mines, the most important point developed was that in giving the initial permission to create a mine, the responsible authority should, at the time of giving that permission, lay down conditions to regulate abandonment of the mine in such a way that pollution did not occur following the abandonment. Surely this is a sound policy to be applied to all new proposals for mining and mineral extraction schemes.

Sessions 3A and 3B (Part 3 of this book) related to Agriculture, where the water quality problems arising from intensive agricultural production were readily made apparent. With 13 papers presented on this subject we certainly had plenty of information to consider.

In the first group of papers, Session 3A chaired by Professor Ron Edwards, the message which came across loud and clear was that the major pollution problems arising in the freshwater context were the organic wastes from intensive animal husbandry, nitrates and pesticides. This conclusion was endorsed in the proceedings of Session 3B, chaired by Stanley Agnew. It was again in the discussion of the papers that the further major points arose.

However, some confusion developed in the discussion from a failure to make a clear distinction between point and non-point sources of agricultural pollution of water. Nevertheless, it did emerge that the majority of delegates felt:

(i) that farming *development* really should now come within the normal planning control system of the UK, through which the water authorities and the Nature Conservancy Council could make their reasonable requirements known and have them taken into account;

(ii) that while it is possible for water authorities to seek to achieve some regulation of general agricultural practices causing non-point source pollution, for the specific protection of particular sources of water supply, it was not feasible to seek the application of such control over wide regions of ground- or surface-water catchment areas;

(iii) that in respect of that part of the EEC Drinking Water Directive which relates to pesticides, the current limits were, in the context of low-toxicity herbicides, wholly illogical and unnecessary. In this context delegates felt, as far as I could judge, that the Conference should call for the particular limits to be reviewed urgently, and in consultation with the World Health Organization whose approach to the matter was much more sensible and realistic.

Thursday's session on the effects of forestry, chaired by Ron Packham, considered six papers which raised many relevant technical questions. The most important point arising was the question of whether the existing procedures for deciding how our land resources should be used, in the best overall national interest, were adequate. This question is very relevant, particularly when Professor Bruenig reminded us that, compared with Germany, we were relative newcomers to big forestry business. We do not seem to know what should be done about this, although we believe that it would be a 'good thing' to do something.

Coupling the technical aspects of forest development and management with those of agriculture brings out the point that there is much scope for improving the technical approach, with the benefit of new or revised technological developments.

I must observe that cost benefit considerations have been frequently mentioned here. I have nothing to say except to endorse that the need for application of these considerations outweighs any interest in the inaccuracies and inadequacies.

The one rather shameful thing that has emerged from the Conference is the way in which intensive dairy farming is now substantially flouting the basic intent of the Pollution Prevention Acts. Gross pollution is increasingly arising and the water authorities seem unable to combat this. While the primary responsibility for this very bad state of affairs lies with the farming industry, the reducing capability of the water authorities to take preventive and, where necessary, enforcement action because of manpower reductions is a contributing factor, while major changes in ADAS is another. Perhaps the Conference should pass on this conclusion to the DoE and MAFF with a call for a serious overall review to reverse the present, most unsatisfactory trend.

I shall leave this Conference with a much greater appreciation of the overall problems relating to land use and water resources and possible solutions of these problems. In particular I am greatly impressed with the contribution that strategic and applied research has made to our increasing and detailed understanding of the environmental processes, physical, chemical and biological, which enable us to perceive and tackle the problems. Coupled with this we are heavily dependent on applied research in to the application of new mining, agricultural and forestry technology to achieve greater environmental conservation.

I can perceive the serious need for greater national investment in this research, even though this research will have to be more selective than it has been in the past. After all is said and done, envronmental conservation relates as much to the economic development of all land and water resources as it does to conserving and improving, where necessary, the physical and genetic resources of the natural environment and the quality of human living. Equally important is the diligence and skill with which we manage our environment. The proper training and application of the manpower resources required is a critical factor in this context. Without adequate research and research training, without adequate education, training and management development, we shall steadily lose out in vital environmental conservation even as our vital national industrial base is now rapidly deteriorating. If this Conference achieves anything in highlighting these fundamental necessities then it will have been truly of national value.

Finally Chairman, may I say how much I have enjoyed the event. We have been made to work very hard and we have greatly enjoyed the overtime we worked at the social events. I hope that all delegates feel likewise. Further I want to congratulate the Water Research Centre and the World Health Organization for arranging and operating an excellent Conference. To all participants, Chairmen of Sessions, authors and discussers, I think you have done a first-class job, and the expenditure of your employers in sending you here has been thoroughly and wholly justified. We need more cost-effective efforts like this.

Appendix

Poster Paper (i)

The research programme of the Scottish Development Department: projects concerned with the effects of land use on fresh waters

T. D. Macdonald

INTRODUCTION

The Scottish Development Department (SDD) is one of the five departments of the Secretary of State for Scotland which make up the Scottish Office. Within Scotland it is the central government department responsible for administering government policy in regard to town and country planning, housing, roads and environmental services generally. These broad areas of responsibility include aspects of conservation, public water supplies and sewerage, water pollution control, flood prevention, and coast protection. Within SDD Civil Engineering and Water Services (CEWS) is the unit with responsibility for water supply and sewerage, coast protection, flood prevention, and water pollution control. In particular CEWS participates with DOE and other government departments in the formulation of the overall UK research programmes in appropriate areas, particularly water. Additionally SDD has a small research programme to deal with specific Scottish projects and part of this consists of projects promoted by CEWS and concerned with aspects of water and the environment.

THE RESEARCH PROGRAMME

Like all research sponsored by Government, the CEWS part of SDD's programme is made up of individual contracts let for work to advance the policy aims of Government. In the general sense this is applied research and not necessarily aimed at advancing overall scientific knowledge but rather throwing light on particular problems of concern. In principle, contracts are let on a competitive basis but considerations of expertise and location may point towards a particular contractor. The results of the research projects have been published for the last ten years or so in the series of Applied Research and Development Programme (ARD) reports; a list of those currently available is attached as an Annex. About half of the projects listed are concerned with aspects of wastewater treatment; two with potable water supply; three with hydrogeology and the remaining three with the monitoring and classificatin of surface fresh waters. In recent years the thrust of the programme has altered significantly. The projects currently in hand and in the process of being reported are shown in Table 1. Many of these are collaborative projects; and in such cases the

Table 1
Current and recently completed projects

Project	Other funding	Contractor	Status	Brief description of work
Low flow studies	None	Institute of Hydrology	Ongoing	Development of methods of low-flow estimation in ungauged catchments
Effects of afforestation	DOE, DEn, FC, WRc, NSHEB, BWB, Strathclyde & Central RC Forth RPB	Institute of Hydrology (main contractor) Forth RPB	Ongoing	Major process and catchment studies on effects of forest practices on quantity and quality of runoff
Loch Dee Project	DAFS, FC	Solway RPB	Ongoing	Study of effects of acid deposition and forestry on water quality and trout fishing
The use of sewage sludge in forests	None	WRc with subcontract to University of Aberdeen	Ongoing	Investigation of use of sewage sludge as forest fertilizer
Investigation of pigmented salmon	DAFS	North East RPB	Ongoing	Investigation into causes of occurrence of pigmented salmon in the River Don
Phosphorus budget for Loch Leven	NCC, Tayside RC	Institute of Terrestrial Ecology	Ongoing	Development of phosphorus budget for eutrophic lake
Natural river communities	DOE, NERC	Freshwater Biological Association	Ongoing	Major national study of invertebrates and environmental factors
Hydrogeological studies	—	British Geological Survey	Ongoing	Survey of the quality of a range of Scottish groundwaters
Effect of heavy metals on micro-organisms	—	Glasgow College of Technology	Ongoing	Investigation of microbiological processes in the presence of heavy metals
The fate of pH in distribution	—	University of Strathclyde	Complete	Study of decay of pH, raised by addition of lime, in distribution system

other sponsoring bodies are indicated. While ARD reports continue to be published on completion of certain contracts, in many cases the results have been published by the contractors in the appropriate scientific literature. It will be seen that at least six of the projects have some relevance to the theme of the Conference and in the remainder of this note a brief description will be given of each.

LOW FLOW STUDIES

This work represents the final phase of a project which began in 1980. The objective has been to develop methods of estimating a Base Flow Index, and hence low flows, and long-term average discharge in ungauged catchments. Among the factors which have been taken into account have been solid and drift geology, rainfall and the incidence of lakes and reservoirs. The availability of a proven method of assessing low flows for dilution calculations will be of assistance in pollution-control considerations.

EFFECTS OF AFFORESTATION

A number of customers have come together to fund this research which has two main elements. Studies are being carried out in two catchments near Balquhidder which have been fully instrumented to record the differences in water quality occasioned by one catchment being predominantly under heather and peat, and the other under trees. In the longer term the afforested catchment will be felled and the control planted with trees. It is hoped to continue to monitor the effects of these practices on the quality and quantity of runoff. The main contractor is also carrying out a series of individual process studies concerned with aspects of the response of the various types of vegetation. This work began in 1981 and already a number of papers have been published in the scientific press. A particular interest of CEWS has been in a small contract let with the Forth River Purification Board to provide analytical quality control for the chemical determinations being carried out by a number of bodies who have elected to carry out free-standing research at Balquhidder. This aspect is considered to be of considerable importance and where the expertise of WRc and the Forth RPB is proving valuable.

LOCH DEE PROJECT

This is again a co-operative project co-ordinated by Solway RPB. Work began in 1979 involving Solway RPB, Forestry Commission, and The Department of Agriculture and Fisheries for Scotland (DAFS) to assess the effects of acid precipitation and coniferous afforestation on surface waters; it was also hoped to investigate means of improving Loch Dee as a trout fishery. From these initial objectives a rather larger study has developed involving a number of different research organisations and already a number of scientific papers have been prepared.

THE USE OF SEWAGE SLUDGE IN FORESTS

Earlier studies concerned the application of liquid sewage sludge to growing trees which suggested that while there might be an initial benefit this might not continue in the short term. Current work is aimed at assessing whether the sludge applied is likely to have a harmful effect on root growth. This work is important in determining whether or not sewage sludge will be an appropriate forest fertilizer at a time when there is increasing pressure on other outlets for this material.

INVESTIGATION OF PIGMENTED SALMON

Recently there has been an increase in the number of coloured or pigmented salmon found in the River Don. These have been found in the lower reaches of the River which receives major discharges from the paper industry and increasing housing development. The project is aimed at discovering if this condition results from a particular chemical insult and, if it does, detecting the chemical responsible which is thought likely to be present in very low concentrations. This project is also supported by DAFS for their fishery interest.

DEVELOPMENT OF PHOSPHORUS BUDGET FOR LOCH LEVEN

This Loch is one of the most productive in the UK and has been the subject of many scientific studies, particularly of the algal blooms and general deterioration of water quality which have been a feature in recent years. The Loch is an important nature reserve and it is considered that a programme to limit the input of nutrients, particulary phosphorus, may be justified. An essential pre-requisite is to determine the sources and fate of phosphorus in the system and a collaborative study has been established with the Nature Conservancy Council and Tayside Regional Council. The contractor, The Institute of Terrestrial Ecology, will use the classical methods to determine whether or not investment in phosphorus removal at sewage works draining to the Loch would be likely to be worthwhile in terms of benefit to the water quality.

STUDY OF NATURAL RIVER COMMUNITIES

This is a long-running project funded by Government Departments and NERC to develop a system of classification and prediction of benthic invertebrate communities related to environmental variables at unpolluted sites. This work has been based on sampling by river purification boards who have returned the animals collected to FBA where they have been subjected to a rigorous taxonomic analysis. A number of new species have been found. The classification of unpolluted sites and the prediction of communities likely to be found on the basis of environmental variables is now thought to have advanced sufficiently to permit attention to move to the area of polluted waters. The CEWS involvment has been to ensure a sufficient Scottish component in the overall UK programme.

CONCLUSION

This brief note gives an indication of the direction of current research supported by SDD through CEWS contractors; further information is available from the address given at the end of the Annex.

ANNEX — APPLIED RESEARCH AND DEVELOPMENT PROGRAMME REPORTS

The ARD reports are in a series covering the fields of water supply, liquid waste disposal, and river and coastal pollution published under the Applied Research and Development Programme of the Scottish Development Department. Further information may be obtained direct from the Scottish Development Department, Civil Engineering and Water Services.

Reports in the series:

ARD1	Dewatering of sewage sludges by centrifuge	June 1974
ARD2	Groundwater resources in Scotland with particular reference to Fife and Kinross	November 1975
ARD3	Development of a water quality index	December 1976
ARD4	Dewatering of waterworks sludge	September 1977

ARD5	Pilot Plant Study of the Oxygen Activated-Sludge Process	August 1977
ARD6	Settling Velocities of Particulate Matter in Sewage	October 1980
ARD7	Treatability Studies of Two Scottish Sewages	April 1981
ARD8	Effect of Sampling Frequency on Rivers Classification	April 1981
ARD9	Application Studies of Water Quality Indices	June 1981
ARD10	Groundwater Potential of the Old Red Sandstone Formations of East Berwickshire	October 1981
ARD11	Groundwater in Scotland — a hidden asset	April 1982
ARD12	Dewatering of Sewage Sludge by Vacuum Band Filter	February 1983
ARD13	Time Domain Modelling in Sewage Treatment Control	April 1983
ARD14	Unmetered Domestic Water Consumption in Scotland	October 1984
ARD15	A Comparison of Treated Efficiencies of Covered and Uncovered Biological Filters	February 1984

Also available:

Performance of an Off Sewer Storm Sewage Tank — January 1980
(Paper presented to the Scottish Section of the Institute of Water Pollution Control)

Copies of ARD reports and of the Storm Sewage Tank paper are available free of charge from the following addres.

G. P. Winter
Room 312
Pentland House
47 Robb's Loan
Edinburgh EH14 1TY

Poster Paper (ii)

The effects of urbanisation on receiving water quality: heavy metal toxicity

J. B. Ellis, D. M. Revitt, R. B. E. Shutes and R. S. Hamilton

INTRODUCTION

Relatively few studies have investigated the effects of urban storm-water on receiving stream quality. It is known that road runoff can contribute considerable quantities of pollutants, with percentage inputs for polyaromatic hydrocarbons and heavy metals being estimated at 10 to 20% and 50 to 70%, respectively of the total loadings in receiving streams (Colwill *et al.*, 1984). Additionally, roof runoff and gully pot liquors can act as important sources of pollutants within the storm-water system. Data extracted from a number of UK studies shows that maximum and minimum pollutant concentrations in separately sewered storm runoff can vary from 21 to 582 mg/l for total suspended solids, 0.03 to 3.1 mg/l for Pb, and 29 to 200 ng/l for benzo(a)pyrene (Ellis, 1985). These ranges of event mean concentrations, although showing a high variability, are comparable to the values for combined sewer overflows (237 to 635 mg/l for total solids, 0.15 to 2.9 mg/l for Pb, and 12 to 215 ng/l for benzo(a)pyrene), giving an indication of the potentially deleterious effects of urban storm-water on receiving waters.

FIRST FLUSH EFFECTS

The impact of storm-water discharges on receiving streams can be further exaggerated by the initial peak flushing of contaminants which has been observed in a number of previous storm-water quality studies (Tucker, 1975; Ellis, 1976; Mance & Harman, 1978). This concept of first flush, which can be partially, if not primarily, attributed to the removal of dry-weather deposits from the below-ground system, is not demonstrated during all storm events. When it does occur, there is some tendency for it to approximate to the time of concentration for the system (Ellis, 1977). Ellis (1976) has observed a secondary peak of suspended solids following the initial flush of pollutants which can be related to delayed material being washed into the system from path and road surfaces, or derived from the late contributing areas. For their Stevenage New Town catchment, Mance and Harman (1978) have shown that the maximum pollutant concentrations occur within 20 minutes of storm flow commencing, whilst in the case of total heavy metals 77% of the concentrations are discharged within the first 40 minutes. In another study, Revitt *et al.* (1982) found that generally 20 to 30% of total metals were discharged within the first 20 minutes of the storm event, although for one storm a pronounced 'first flush' was observed for Cd, with 43% of the metal being discharged during this time period.

From the analysis of 22 storm-discharge events sampled at the outfall of a 245–ha urban catchment in Oxhey, Herts, Harrop (1984) has shown that the majority of pollutants exhibit a marked first-flush pattern. The notable exception was found to be suspended solids which exhibited a more even distribution of maximum concentrations on both the rising and recessional limbs of the discharge hydrograph. The available evidence, therefore, suggests that 'first flush' is a variable phenomenon and as such will impose some intractable problems for receiving-water quality as well as for conventional control procedures based on purely hydraulic criteria.

CONTRIBUTING SOURCE AREAS

Of the many types of urban surface contributing directly to storm-water runoff, the most important in terms of metal and sediment contributions is the road surface. Sediments from a variety of road surfaces have been shown to exhibit typical log normal size distributions with the degree of sorting being dependent not only on the nature of the surface, but also on the traffic density and condition (Ellis & Revitt, 1982). These factors also control the size distribution pattern for the metals which are adsorbed onto or associated with particulate material on street surfaces. Roads carrying high traffic densities exhibit the greatest metal concentrations towards the lower end of the particle size range, with Cd and Pb predominating in the 100 to 500 μm range. The removal of particulate associated metals to the gully pots occurs preferentially for the finer sediments (<250 μm) and consequently metal concentrations in sediments received at the gully pot are generally higher than those on the street surface (Harrop *et al.*, 1983). The solids which avoid entrapment in the gully pot are subjected to deposition, degradation, and resuspension processes in the below-ground system, producing fine storm-water particles with a bimodal size distribution in the 1 to 50 μm size range (Ellis *et al.*, 1982). These fine particles can be considerably enriched with heavy metals and concentrations as high as 8500 μg/g for Pb, 8000 μg/g for Zn, 1660 μg/g for Cu, and 670 μg/g for Cd, have been observed (Revitt *et al.*, 1982). In the case of Cd, this elevated level represents an increase of two orders of magnitude over the concentrations normally found in raw street sediment. However, the assessment of the impact of this considerable metal pollution load on receiving-water quality in terms of environmental mobility, bioavailability, and toxicity, requires an understanding of the physico-chemical forms in which the metals occur.

Sequential chemical extraction experiments on street surface sediments have shown that metal affinity for the 'exchangeable' phase, which includes those metals specifically surface bound and which are therefore subject to sorption–desorption processes, is independent of particle size and in the order Cd>Zn>Cu>Pb (Hamilton *et al.*, 1984). The iron and manganese oxide surface sediment fraction acts as an efficient scavenger of Pb, Zn and Cd, with Cu showing a marked affinity for the organic phase and this increases as particle size decreases. These studies suggest that the overall availability of surface sediment associated metals to the dissolved phase of runoff is Cd>Zn, Pb>Cu.

Sediment–water metal interactions

The relationship between surface sediment metal affinities and the final outfall water quality is strongly dependent on below-ground conditions. Important factors controlling the sediment–water metal partitioning include changes in the flow: suspended solids ratio, dissolved organic carbon content, ionic strength, pH, oxic/anoxic conditions, as well as biochemically mediated reactions mobilising metals.

Metal distributions in storm waters discharging to receiving waters from different catchments show that Pb is predominantly particulate associated, Zn and Cd are present mainly in the dissolved form, and Cu is evenly divided between the two phases (Morrison *et al.*, 1984). The dissolved phase can be further differentiated by choice of analytical technique into electrochemically available, chelex removable, and strongly bound fractions. The first of these represents those metals which occur as free ions and weakly complexed metals and can be considered to be bioavailable in the freshwater environment (Florence *et al.*, 1983). The importance of this potentially-toxic fraction is indicated by the fact that it can account for 43% of the total Zn, and 52% of the total Cd (Morrison *et al.*, 1984a). The solubility and ease of release of Cd from the solid phase are attributable to its ready formation of chloro complexes and provide evidence of a direct and pervasive impact on receiving waters.

Compared with road sediments, storm-water suspended solids contain metals which are dominantly in the exchangeable fraction. This is particularly true in the case of Pb, for which 85% of the metal can be in the solid phase with 74% of this present in the exchangeable fraction. There is thus a large proportion of Pb potentially available for release in a bioavailable form, especially if the receiving water possesses lower pH and higher chloride levels. The environmental mobility of Cu appears to be lower than that of other metals, although soluble Cu is found in the bioavailable fraction when readily-identifiable source material is present.

Metal distributions throughout individual storms show increases in the proportions of Cd, Cu, Pb, and Zn in the soluble phase as the storm progresses. The increase in the water-sediment ratio is obviously a controlling factor, but additionally the levels of dissolved chloride (for Cd and Pb) and dissolved organic carbon (for Cu and Zn) are important. These factors were noted to produce peak concentrations of bioavailable dissolved metals during a snowmelt event of 12.5 μg/l, 146 μg/l, 330 μg/l, and 558 μg/l, for Cd, Cu, Pb, and Zn respectively (Morrison *et al.*, 1984b).

As a consequence of its strong particulate association, the flow-weighted loadings of Pb closely follow the suspended-solids pattern throughout the storm hydrograph. Other metals show more variability, exhibiting either pronounced 'first flushes' (Cd), or secondary peaks (Cu and Zn), due to retarded inputs of enriched suspended solids. Chemographs for the dissolved metals show a delayed discharge behind peak flow, indicating a small lag time for the mobilisation of Cu, Pb, and Zn. Cadmium, in agreement with its predicted solubility, exhibits an initial loading peak which parallels the flow pattern.

The impact of metals on receiving waters is illustrated by the fact that for one recorded runoff event the bioavailable, soluble Cu, Cd, and Zn exceeded the US EPA threshold values for total metal concentrations for 77%, 42%, and 61%, of the storm flow duration, respectively (US EPA, 1983). In addition, the EPA level for significant mortality was exceeded by Cu (for 59% of the storm) and Cd (for 12% of the storm). It is significant that the toxicity of Cd decreases towards the end of the storm as the chloride loadings are depleted. These results clearly indicate the ability of stormwater-derived metals, despite their intermittent patterns of discharge, to create toxic stresses on receiving waters in urban areas.

ECOLOGICAL EFFECTS

Benthic macroinvertebrate tissue samples from species present in urbanised streams show progressively raised metal levels in individuals downstream of the urban area compared with those sampled in rural headwaters (Shutes, 1985). The mollusc *Limnea peregra* shows increases from 69 to 142 μg/g for Cu, 91 to 384 μg/g for Pb, and 275 to 619 μg/g for Zn, whilst hirudinea such as *Herpobdella octoculata*, show increases from 151 to 405 μg/g, 0 to 67 μg/g, and 627 to 1246 μg/g for Cu, Pb, and Zn respectively. The highest levels were recorded near to a motorway outfall where reduction in pH on mixing of the influent runoff with the receiving waters disrupts equilibria, causing desorption of bound metals and producing elevated metal levels downstream of the discharge point. Crustacea show particularly elevated metal tissue levels. *Gammarus pulex* is permeable and actively takes in ions to maintain its internal concentrations, facilitating trace-metal uptake. *Asellus aquaticus*, with concentrations of 43 μg/g, possess the highest levels of Cd relative to other species (1.8 to 13.8 μg/g), possibly due to the induction of a cadmium-binding protein which is thought, in some species, to modify the toxicity (Yamamura *et al.*, 1983). Whilst oxygen and BOD regimes of urbanised streams are compatible with a diverse macroinvertebrate community, biotic index scores show a progressive deterioration in water quality along urbanised streams (Shutes, 1984). There are marked reductions in species richness and abundance of individuals, with an absence of ephemeroptera and trichoptera at all stations, and *Gammarus* at downstream stations; diptera and hirudinea occur only in low numbers within heavily urbanised reaches. *Limnea* appears to be pollution tolerant and is relatively abundant throughout chalk-based urban streams. It has been suggested (Ellis, 1985a) that the metal complexing capacity of bacterial extra-cellular polymers might well provide a starting point for this toxic bioaccumulative chain, and urban rivers certainly posses a high bacterial incidence.

The variation in discharge intensity characteristic of urbanised streams results in substrate instability and is an important factor causing the loss of susceptible species such as hirudinea. Furthermore, artificial bank constraints result in the loss of marginal areas and restrict the habitats available for macroinvertebrate colonisation. The synergistic effects of these factors and the cumulative toxicity derived from metal and hydrocarbon storm-water runoff discharges, has undoubtedly caused notable changes in the ecological structure of urbanised streams.

REFERENCES

Colwill, D. M., Peters, C. J. & Perry, R. (1984) *Water Quality of Motorway Runoff.* Report 823, Transport & Road Research Lab., Crowthorne, Berks.

Ellis, J. B. (1976) Sediments and Water Quality of Urban Stormwater. *Water Services,* **80**, 730–734.

Ellis, J. B. (1977) The Characterisation of particulate solids and quality of water discharged from an urban catchment. In: *Effects of Urbanisation on the Hydrological Regime and Water Quality,* 283–291. Publ. No. 123, IAHS/Unesco, Amsterdam.

Ellis, J. B. (1985) Urban Runoff Quality and Control. In: *Advances in Water Engineering*. Tebbutt, R. H. Y. (ed.), Elsevier Ltd., London, 234–240.

Ellis, J. B. (1985a) Water and Sediment microbiology of urban rivers and their public health implications. *Public Hlth. Eng.*, **13**, 96–107.

Ellis, J. B., Hamilton, R. S. & Roberts, A. H. (1982) Sedimentary characteristics of suspensions in London stormwater. *Sediment. Geol.*, **33**, 147–154.

Ellis, J. B. & Revitt, D. M. (1982) Incidence of heavy metals in street surface sediments. *Water, Air & Soil Pollut.*, **17**, 87–100.

Florence, T. M., Lumsden, B. G. & Fardy, J. T. (1983) Evaluation of some physico-chemical techniques for the determination of the fraction of dissolved copper toxic to the Marine Diatom, *Witzschia clostenium, Analytica Chimica Acta,* **151**, 281–295.

Hamilton, R. S., Revitt, D. M. & Warren, R. S. (1984) Levels and physico-chemical associations of Cd, Cu, Pb and Zn in road sediments. *Science Total Envir.*, **33**, 59–74.

Harrop, D. O., Ellis, J. B. & Revitt, D. M. (1983) Temporal loadings of sediment and heavy metals to roadside gully pots. In: *Heavy Metals in the Environment*, Muller, G. (Ed.), CEP Ltd., Edinburgh, pp. 876–879.

Harrop, D. O. (1984) *Stormwater Runoff and Water Quality in Urban Catchments.* Ph.D Thesis, Urban Pollution Research Centre, Middlesex Polytechnic, London.

Mance, G. & Harman, M. M. I. (1978) The Quality of Urban Stormwater Runoff. In: *Urban Storm Drainage*, 603–618, Helliwell, P. (Ed.), Pentech Press, London.

Morrison, G. M. P., Revitt, D. M., Ellis, J. B., Svensson, G . & Balmer., P. (1984) Variations of dissolved and suspended solid heavy metals through an urban hydrograph. *Envir. Tech. Lett.*, **7**, 313–318.

Morrison, G. M. P., Revitt, D. M., Ellis, J. B., Svensson, G . & Balmer., P. (1984a) The Physico-Chemical Speciation of Zinc, Cadmium, Lead and Copper in Urban Stormwater. In: *Planning and Control of Urban Storm Drainage*, Balmer, P., Malmquist, P. A. & Sjoberg, A. (Eds). Chalmers Univ. of Technology, Gothenburg, pp. 989–1000.

Morrison, G. M. P., Revitt, D. M., Ellis, J. B., Balmer, P. & Svensson, G. (1984b) The transport mechanisms and phase interactions of bioavailable heavy metals in snowmelt runoff. In: *Interactions Between Sediments and Freshwater*. Sly, P. (Ed.), CEP Ltd., Edinburgh, pp. 226–229.

Revitt, D. M., Ellis, J. B. & Oldfield, F. (1982) Variation in Heavy Metals of stormwater suspended solids in a separate sewer system. In: *Urban Storm Drainage*, Yen, B. C. (Ed.), Water Resources Publ. Ltd., Colorado, pp. 49–58.

Shutes, R. B. E. (1984) The influence of surface runoff on the macroinvertebrate fauna of an urban stream. *Science Total Envir.*, **33**, 271–282.

Shutes, R. B. E. (1985) *A Comparison of the Benthic Macroinvertebrate Fauna of two North London Streams*. Research Report No. 12, Urban Pollution Research Centre, Middlesex Polytechnic, London.

Tucker, C. G. J. (1975) M.Sc. Thesis. Dept. of Civil and Structural Engineering, Trent Polytechnic. Nottingham, UK.

US Environmental Protection Agency (1985) *Toxicity Guidelines. Final Report to the Nationwide Urban Runoff Programme.* Vol. 1., Water Planning Division, Washington D,C.

Yamamura, M., Suzuki, K. T. & Kubota, K. (1983) Tolerance to cadmium and cadmium binding proteins inducted in the midge lava, *Chironomus yoshimatsui. Comp. Diochem. Physiol.* **75**, 21–24.

Poster Paper (iii)

Ferruginous discharges in eastern Scotland

T. Williamson

ORIGIN

Iron-bearing or ferruginous waters are caused by transport of soluble iron by ground or surface waters arising from the degradation of iron pyrites (FeS_2) in the rock strata. Iron pyrites is widespread in nature, but is associated particularly with coal, the mixing of which can bring about the exposure of the mineral to air moisture and set the following mechanism in action:

$$FeS_2 + H_2O + \frac{7}{2} 0_2 = e^{2+} + 2H_2^+ SO_4^{2-} .$$

Further reactions can occur, under bacteriological influence, as well as chemical, so that an acid solution of Fe^{2+}, Fe^{3+} and $SO^{2-}{}_4$ is produced. If sufficient alkaline material, for example limestone, is present in the surrounding strata, any acidity is corrected and the iron does not reach the main groundwater flow.

Water reaching the surface, after flowing underground, can be clear in appearance, but reaction with an alkaline stream and exposure to atmospheric oxygen will produce the typical 'iron ochre' precipitate:

$$Fe^{3+} + 3H_2O = Fe(OH)_3 + 3H^+ .$$

The orange-yellow solid produced is a mixture of hydrated ferric hydroxides, the composition of which is dependent on the conditions of pH, oxygen, and so on, under which it is formed.

The production of ferruginous water appears to require the following conditions.

- Pyrite-containing rocks will be exposed to a supply of oxygen and water to produce oxidation.
- Water will be present in sufficient quantity to elute the oxidation products.
- Alkaline material in the surrounding rocks will be insufficient to neutralise any acidity produced.

SOURCES

Mine waters

In coal mining, water entering the workings from overlying strata must be removed to ensure safe and efficient working conditions, normally by pumping. In working collieries water entering is quickly removed and exposed faces are kept dry, with the result that few working collieries have a ferruginous problem. Typical chemical analyses are given in Table 1.

Table 1

Location	pH	Alk	E.C.	Total Fe
Monkton Hall	7.8	315	2300	2.9
Polkemmet	7.7	360	1060	2.8
Bilston Glen	7.8	335	1390	1.7

When a colliery closes, however, the abandoned workings will contain a large surface area in the voids left below and as the water rises to reach its own level, the potential for oxidation and solution of pyrites in the system will be greater. In the early days of coal mining before the advent of pumping, driveways known as 'day-levels' were constructed to remove water quickly from the workings and it is often from these adits that water will be produced when the natural water level is reached. Water from these locations has a pH value between 6 and 7, with iron concentrations ranging from 10 to 20 mg/l, but exceptions do arise, for example Elginhaugh and Cuthill (see Table 2).

Table 2

Location	Flow (l/s)	Total iron (mg/l)	pH value	
Elginhaugh	30	88	6.1	Outcrop
Old Fordell	54	11	6.5	Outcrop
Fordell Castle	204	21	6.3	Outcrop
Minto No. 2	8	18	7.0	Outcrop
Cuthill No. 1	5	85	6.4	Outcrop
Baads	0.1	318	2.8	Bing Drainage
Foulshields	0.7	2.7	3.3	Bing Drainage

The coalfield of Central Fife was abandoned in about 1966, and the most severe effect of the water rebound occurred ten years later when a day-level at Fordell Castle, in southern Fife, began to discharge highly-ferruginous water to the

Keithing Burn. The initial discharge was at a rate of 380 l/s, with an iron concentration of 30 mg/l, and while the volume has decreased to about 150 to 170 l/s, the iron concentration has remained similar. The effect on the Keithing Burn has been considerable, with the stream being coated with orange precipitate and normal fauna being eradicated. Additionally, the town's main employer, a paper mill, had to seek an alternative source of process water, as the iron renders the water unsuitable for paper manufacturing, and is difficult to remove before use. The River Ore to the north received the break-out at several locations, which has affected not only the water quality, but also has caused problems by inundation of the foundations of buildings and minor flooding.

The initial high discharge followed by a levelling off has been noted in another day-level at Elginhaugh where the water broke-out into the River North Esk in 1966. The original iron concentration was approximately 200 mg/l, which has decreased now to 80 mg/l.

Opencast workings

The production of opencast coal in the area has been prevalent as this system is used to work out pockets of coal near the surface. As the topsoil is removed, large areas of soil are exposed to atmospheric weathering and any pyrites present will break down.

The majority of such mines operate in seams of low pyritic concentrations, but at Westfield in Fife considerable problems arose.

The site, one of the largest in Europe, has a large inflow of water which leached out ferruginous constituents to the detriment of the neighbouring Lochty Burn. The National Coal Board constructed a lime treatment plant which removed iron from the water as well as correcting acidity to some effect, although the drainage from the surrounding spoil heaps still reduced the quality of the burn. However, coaling operations have now ceased and discussions are underway on the eventual use of the site.

Spoil heaps

Drainage from spoil heaps from collieries having large quantities of pyritic material can give rise to the most acidic ferruginous discharges experienced. The chemical actions are similar to those underground, but with continuous exposure of the waste to the atmosphere and the presence of ample rainfall, the problem can be severe. The acidic nature of the waste means that vegetation which normally would alleviate the effects of rainfall cannot thrive, and continuous chemical degradation occurs.

Two locations are of particular interest.

1. Baads Colliery, West Calder

The colliery in its sixty-year life produced a spoil heap on the surface and also waste at a neighbouring site at Foulshields, where the coal was washed. Drainage from the spoil produced toxic conditions in the receiving watercourse, the West Calder Burn, for a distance of 11 km until the main river of the catchment, the Almond, was reached. In 1980, major rehabilitation of the bing was undertaken by Lothian Regional Council and the Scottish Development Department, which involved regrading, liming and planting of grass and trees. It is to be hoped that the revegetation will establish itself and stabilise the bing, thus helping to neutralise runoff.

2. Fauldheads Colliery, Harthill

This location, near the source of the River Almond, was worked for opencast coal and then was the subject of restoration, during which a spoil heap from older workings was incorporated in to the scheme. Unfortunately, this spoil proved to be acidic in character and the runoff from the site has affected the River Almond markedly, damaging the fish life as well as introducing the characteristic orange

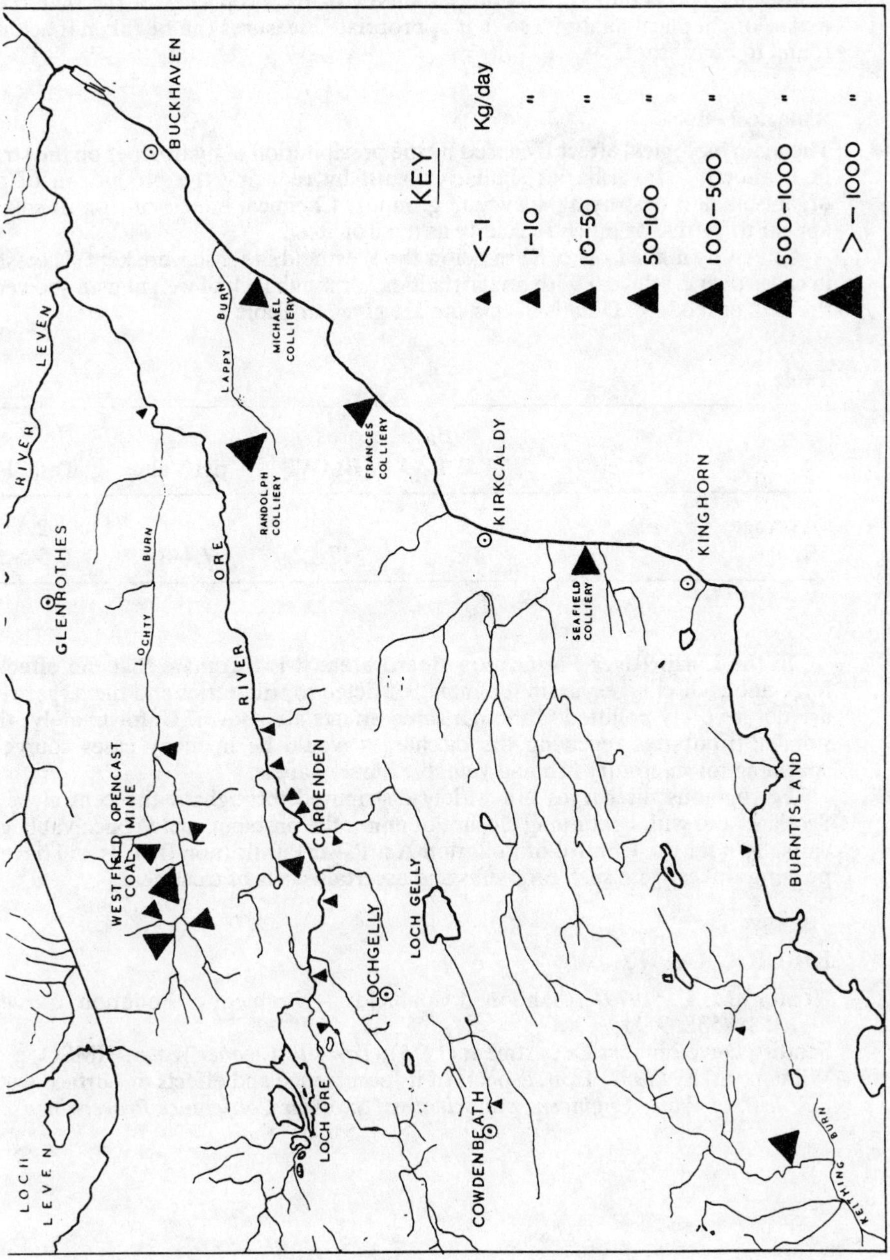

Fig. 1 — Ferruginous discharges in Fife

precipitate. Again it is hoped that restoration and rehabilitation of the site will bring an improvement in the river quality. This case is one which stresses the need in landscaping procedures, for a proper survey of the properties of the material by means of chemical analysis, so that appropriate measures can be taken if acidity is found to be present.

Biological effects

The main biological effect is caused by the precipitation of hydroxides on the stream bed, affecting the fisheries similarly to silt by reducing the production of food organisms and destroying spawning grounds. Chemical effects of iron in solution appear to be due as much to acidity as the iron itself.

In a study in the Lochty Burn below the Westfield site, fish were kept successfully in cages over a full year with no mortalities, although lack of weight gain showed up the lack of feeding. Details of this site are given in Table 3.

Table 3

	Biotic scores			
	T.B.I	BMWP	pH Value	Total Fe
Average	5	23		2.5
Range	4–5	17–27	7.4–8.4	0.5–5.5

In the Forth River Purification Board area, it is fortunate that the effects of ferruginous discharges are in the main restricted to tributaries and the larger rivers are not severely polluted, although some effects are shown. Unfortunately, these smaller tributaries receiving the discharges would be in many cases sources of spawning for migratory fish and valuable nursery areas.

Ferruginous discharges are widely distributed throughout the central belt of Scotland and with continuing closure of mines the problem could conceivably grow worse. Under the Control of Pollution Act River Purification Boards will have the power to investigate such problems and ascertain costs of treatment.

BIBLIOGRAPHY

Henton, M. P. (1979) Abandoned Coalfields — Problem of Pollution, *Surveyor,* **153,** 4538, 9–11.

Scottish Development Department (1975), *Towards Cleaner Water,* HMSO, p. 49.

Williamson, T. (1982) Iron, aspects of its occurrence and effects on surface waters. *Inst. of Water Engineers and Scientists, Summer Conference Proceedings.*

Poster Paper (iv)

Restoration of opencast coal sites: the use of sewage sludge and its effect on the quality of field drainage

C. D. Bayes and E. C. Parkinson

BACKGROUND

Each year around 2000 ha of land are restored following opencast coal mining, which are to be returned to agricultural use. The replaced top- and sub-soils generally have a poor structure, as a result of storage and handling, and an aftercare programme is implemented to rehabilitate the land. This includes the installation of field drains, with secondary treatments such as sub-soiling, and applications of inorganic fertilizer and farm-yard manures.

At a field trial on the former Westerton opencast site, Cowdenbeath, Fife, sewage sludge was incorporated into the soil at the drainage and recultivation phase of the programme. The objectives of the trial are to assess the effects on soil, herbage and field drainage, plus the operational implications and costs. The site consists of four 1-ha treatment plots receiving different sludge application rates. Each plot is drained by a separate field drainage system, consisting of five laterals at 10 m centres, with its own outfall to the watercourse, see Fig. 1. This has permitted direct monitoring of the quality of the field drainage associated with the different treatments.

SITE DESCRIPTION AND THE AFTERCARE PROGRAMME

The trial site is at an elevation of 116 to 124 m AOD with a 3% slope and it is situated in an area with less than 1000 mm rainfall/yr. The 1-ha plots consist of 200 m×50 m rectangles (see Fig. 1).

In 1977 the soils were stripped and stored and coal extraction followed in 1978 to 1979. The void was backfilled and the soils replaced with the aftercare programme commencing in 1982. Sewage sludge was applied in 1984 when the drainage was installed and the site recultivated. This stage in the programme allowed the sludge to be incorporated down the soil profile.

A summary of the aftercare programme follows.

1982 — ground limestone, basic slag and 20:10:10 fertilizer applied;

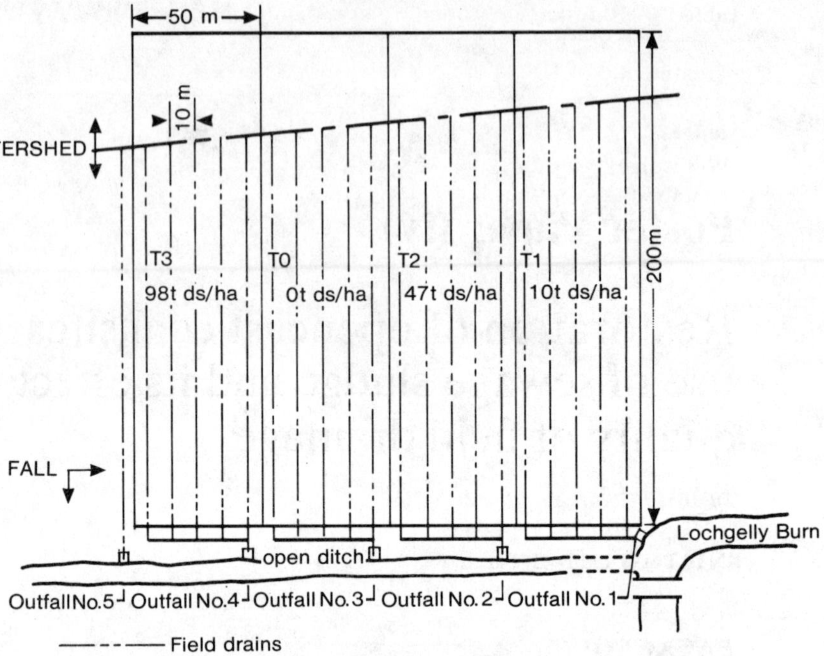

Fig. 1 — Experimental site

	—	seeded with two-year grass mixture;
1983	—	20:10:10 fertilizer applied;
1984 June	—	drainage installed at 750-mm depth;
	—	sludge applied and site recultivated;
July	—	ground limestone, basic slag and 20:10:10 fertilizer applied;
Aug	—	reseeded with permanent grass mixture.

SLUDGE AND FERTILIZER APPLICATIONS

The sludge treatment rates were 0, 10, 47, and 98 t ds/ha, and details of the additions of nutrients and potentially-toxic elements are given in Table 1.

Primary sludge cake from Dunfermline Sewage Treatment Works was used for the trial, after having been dewatered in centrifuges and stored. Sludge was applied in June 1984 with a conventional rear discharging manure spreader, capacity of $8.5\,m^3$. In addition, 20:10:10 fertilizer was applied at a rate of 0.5 t/ha to each treatment plot in July 1984. Details of the application are also given in Table 1.

EFFECTS ON THE QUALITY OF FIELD DRAINAGE

The site had a high soil moisture deficit throughout the summer/early autumn period and there were no discharges from the field drainage outfalls. In late October, when the site returned to field capacity, discharges commenced and they were monitored throughout the autumn and winter until they ceased again in early Spring 1985.

The most significant aspect of the field drainage quality was the high levels of total oxidised nitrogen (TON) when the discharges commenced in the autumn. Figure 2 shows the concentrations of TON in the T2 treatment plot drainage, 47 t ds/ha, which is typical of all the plots. In the first three-week period following the commencing of

Table 1 — Rates of application. (Sludge: June 1984; Fertilizer: July 1984)

Treatment plot		T0	T1	T2	T3
Sludge					
Wet weight	t/ha	0	30	138	228
Dry solids	t/ha	0	10	47	98
N	kg N/ha	0	181	851	1770
P	kg P_2O_5/ha	0	115	538	1120
K	kg K_2O/ha	0	14	67	145
Cu	kg/ha	0	20	93	190
Zn	kg/ha	0	24	110	230
Ni	kg/ha	0	1.0	4.7	9.8
Cd	kg/ha	0	0.17	0.80	1.7
Cr	kg/ha	0	3.2	15	31
Pb	kg/ha	0	3.5	16	34
20:10:10 Compound fertilizer					
N (as NH_4)	kg N/ha	57.5	57.5	57.5	57.5
N (as NO_3)	kg N/ha	42.5	42.5	42.5	42.5
P	kg P_2O_5/ha	50	50	50	50
K (as KCl)	kg K_2O/ha	50	50	50	50

the flow, the TON levels were highly elevated, up to 130 mg N/l. They then fell sharply to less than 5 mg N/l, and progressively decreased to the pre-application level.

A similar pattern was found for chloride, which was applied as potassium chloride with the compound fertilizer.

However, if the levels of TON are compared across the plots (Fig. 3) it is apparent that sewage sludge is not the prime cause of the elevated levels. The three-week mean TON of the control plot drainage, 0 t ds/ha, was 56 mg N/l.

All plots had received compound fertilizers and the sludge-treated plots showed an increase in TON concentration of 9, 29, and 39% respectively for the 10, 47, and 98 t ds/ha applications. At the highest sludge application rate, 98 t ds/ha, (1800 kg organic N/ha), the field drainage TON concentration increased by only 22 to 78 mg N/l.

The BOD, ammoniacal nitrogen, and orthophosphate concentrations in the field drainage discharges are not significantly different across the 0, 10, and 47 t ds/ha treatments, but each parameter is elevated at the 98 t ds/ha rate, see Table 2.

No evidence was found of leaching of toxic metals into the field drainage discharges, with the exception of a slight increase in zinc concentrations at the 47 and 98 t ds/ha rates. Thus, although the sewage sludge applications provided a source of organic nitrogen that resulted in elevated TON levels in the field drainage, the source was secondary to the 0.5 t/ha compound fertilizer application.

Levels of TON are unlikely to be as high from fertilizer applications to established grassland and this will be monitored in autumn 1985.

ACKNOWLEDGEMENTS

The authors gratefully acknowledge the collaboration in the trial of the Opencast Executive — National Coal Board, Fife Regional Council, the Department of Agriculture and Fisheries for Scotland, and the East of Scotland College of Agriculture.

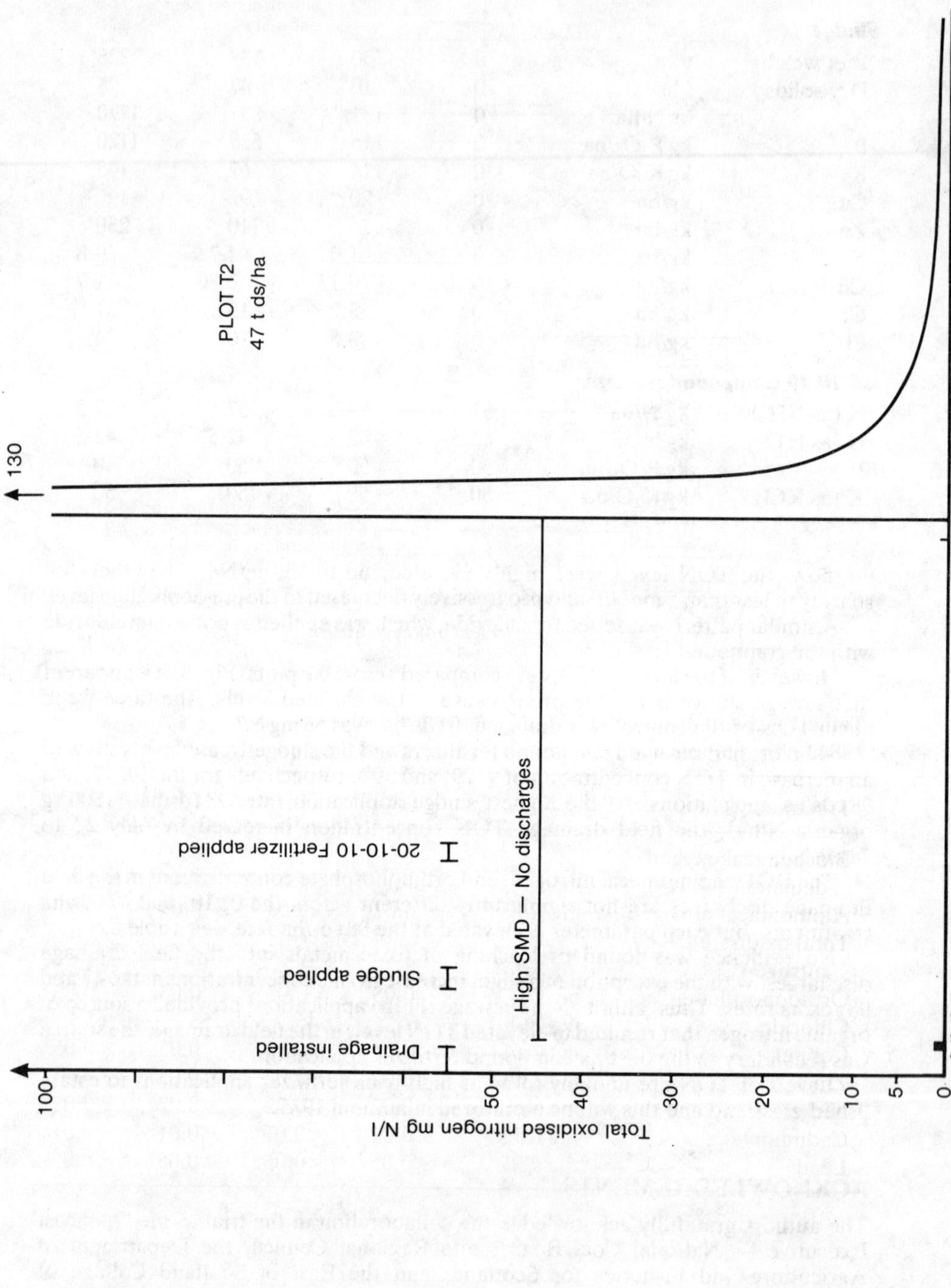

Fig. 2. — Total oxidised nitrogen in the field drainage from Plot T2

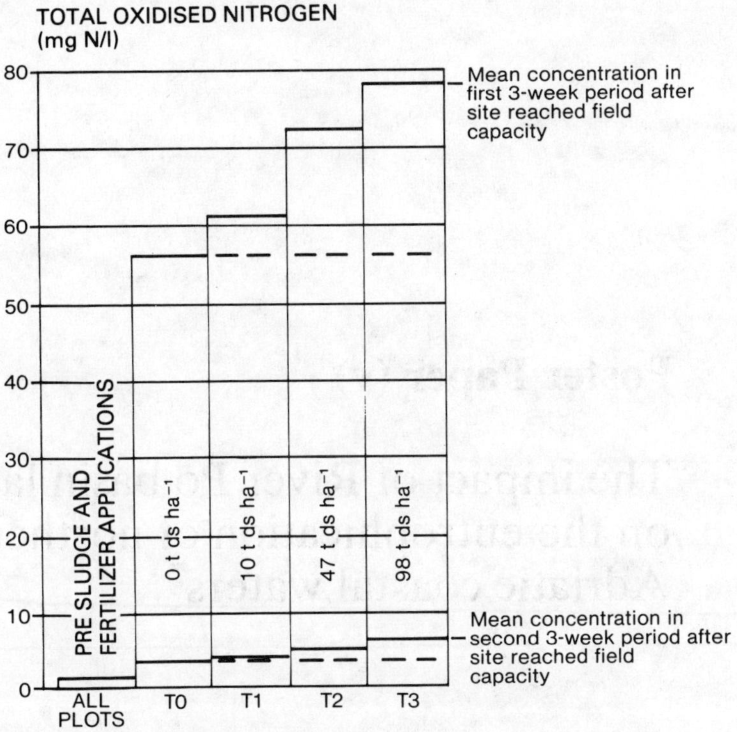

Fig. 3. — Field drainage quality

Table 2 — Field drainage quality (Mean concentrations in six-week period Oct/Nov 1984, after site reached field capacity)

Treatment plot		TO	T1	T2	T3
Biochemical oxygen demand (+ATU)	mg/l	1.1	1.0	1.4	2.6
Ammoniacal nitrogen	mg N/l	0.02	0.04	0.04	0.27
Total oxidised nitrogen	mg N/l	38	41	49	54
Orthophosphate	mg P/l	0.03	0.03	0.03	0.05
Chloride	mg Cl/l	37	46	53	65
Copper	mg Cu/l	<0.05	<0.05	<0.05	<0.05
Zinc	mg Zn/l	0.02	0.02	0.03	0.03
Nickel	mg Ni/l	<0.02	<0.02	<0.02	<0.02
Cadmium	mg Cd/l	<0.01	<0.01	<0.01	<0.01
Lead	mg Pb/l	<0.05	<0.05	<0.05	<0.05

Poster Paper (v)

The impact of River Po basin land use on the eutrophication of northern Adriatic coastal waters

G. Chiaudani, G. Premazzi, and M. Vighi

INTRODUCTION

In recent years the impact of eutrophication and the associated water quality deterioration has emerged as the principal water problem in Italy. An extensive survey on the Italian inland waters has provided evidence showing that over 80% of the lakes and impoundments are affected by detrimental changes in natural trophic conditions. As regards marine coastal waters, eutrophication phenomena seem to be restricted to particularly sensitive areas such as river mouths, closed and crowded bays, and harbours (Chiaudani *et al.*, 1978; Chiaudani *et al.*, 1982; Chiaudani & Vighi, 1982).

However, the massive mortality of benthic animals that occurred in September 1975 along the 130-km coast of the Emilia-Romagna Region (Northern Adriatic Sea), as a consequence of anoxic conditions following an unusually heavy phytoplankton bloom, revealed the increasing impact of excessive fertilization and an impending danger to fisheries and touristic activities. The fishery yield corresponds to more than 20% of the total Italian catch and tourism records more than 40 million annual presences.

As a consequence, eutrophication in this area can be considered the most relevant water-quality problem not only in Italy, but in the whole Mediterranean Sea.

The Adriatic Sea has a surface of 132 000 km^2 (1/20 of the Mediterranean Sea). It receives relevant amounts of freshwater discharges, about 5300 m^3/s, corresponding to about one third of the total Mediterranean river contribution. Through the main rivers (Po, Adige), a pollution load from 25 million inhabitants (that is about one half of the whole Italian population) is conveyed to the Northern Adriatic Sea. In addition the industrial load which represents about 75% of the Italian total, must be considered.

CAUSES OF THE EUTROPHICATION PROCESSES

Different approaches were utilised to identify the limiting nutrient factor for algal growth at ecosystem level, examining the N/P ratio; at population level, verifying the effect of enrichment with different nutrients, both on algal monocultures or natural mixed population bioassays. The results of algal bioassays, carried out on a significant number of water samples, demonstrate that phosphorus can be considered in the majority of cases to be the primary limiting factor in this area (Chiaudani *et al.*, 1980).

In order to plan technical and administrative interventions, the national Authorities promoted integrated researches, with particular care to evaluate the phosphorus load to the surface waters since the control of eutrophication can be considered, at present, as synonymous to the control of phosphorus loadings. Man-made sources of phosphorus may be divided into two catgories: point, and diffuse sources. A point source may be industrial waste, sewage, waste of animal husbandry or storm sewer outlet. Any soluble substance containing phosphorus and deposited on the drainage basin, such as fertilizers, animal droppings, plant remains, waste products or eroded soil from land use, can contribute to the non-point sources.

In the River Po basin, the total P load and the relative weight of the different sources have been evaluated by means of a theoretical approach on the basis of statistical data.

The domestic point loadings were obtained by evaluating both the metabolic contribution and the consumption of polyphosphates, which are added as builders in synthetic detergents. Agricultural loadings were evaluated by considering the consumption of synthetic fertilizers. Moreover, animal production was considered and the amount of animal manure, deposited on cultivated land, as a natural fertilizer, or directly discharged in sewages, was calculated. It was calculated that about 10% of animal manure is directly discharged into surface waters, resulting in a point source, while 60% is deposited on cultivated land, and the remainder is treated or discharged in septic tanks.

For all types of cultivated land, a mean loss coefficient of 3% of the total amount of phosphorus applied to the soil, was adopted.

Industrial loadings, in relation to eutrophication, are not as relevant as for other kinds of water pollution and are restricted to some particular activities (such as the food industry). A precise evaluation of industrial phosphorus loading is not easy but, according to an estimate by the OECD (1982), it could be considered as high as about 10% of domestic point loading.

The contribution of non-cultivated lands was evaluated by assuming a mean phosphorus-loss coefficient obtained in function of the different geological characteristics of the soil and of the different vegetation. Atmospheric load (wet and dry precipitation) is quite difficult to quantify, but in general it could be considered negligible in respect of the principal point sources. In this evaluation atmospheric load has not been considered.

According to all these considerations, the following coefficients were adopted to evaluate the theoretical phosphorus load for the River Po drainage area.

Domestic loads:	0.64 kg P per capita, per year (metabolic and detergent phosphorus)
Cultivated lands:	3% of total phsophorus applied (synthetic and natural fertilizers)

Non-cultivated lands: 0.1 kg P/ha per year
Industrial loads: 10% of domestic loads
Animal farms (directly discharged):

- Horses: 0.590 kg P per capita per year
- Cattle: 0.455 kg P per capita per year
- Pigs: 0.280 kg P per capita per year
- Sheep: 0.075 kg P per capita per year
- Poultry: 0.003 kg P per capita per year

The reliability of this theoretical approach in calculating phosphorus loadings was verified by applying it to the drainage basins of the River Po and the River Adige. The comparison between the theoretical evaluation and the experimental data was very satisfactory (Chiaudani *et al.*, 1978).

Table 1 summarises phosphorus loadings in the Po Valley deriving from different sources, and the percentage in respect to the total load. Point sources represent about 78% of the total phosphorus load, indicating a significant increase of urbanisation and industrialisation in the Po drainage area.

Table 1 — Phosphorus loadings from different sources in the River Po catchment area (Data for the period 1980–82)

Source	t P/year	%
Domestic: detergents	4758	30.5
metabolic	3916	25.1
Cultivated lands	3136	20.1
Animal farms	2668	17.1
Industry	858	5.5
Non-cultivated soils	264	1.7
Total P loading:		100.0
— experimental	15600	
— theoretical	17582	

BENEFITS AND COSTS OF PHOSPHORUS-CONTROL STRATEGIES

For effective control of eutrophication, the identification of the main sources of phosphorus load which may reach a waterbody is fundamental. Under normal circumstances, the first priority would be to control point sources because they are the easiest to control. These measures include essentially wastewater treatment with chemical precipitation of phosphorus, and the reduction or replacement of polyphosphates in detergents. In situations where point-source reduction cannot improve water quality, diffuse source-load control should be given high priority. This could mean the modification of cultivation and land use practices.

However, the development and implementation of phosphorus-load control measures is not purely a scientific and technical problem. Economic considerations play a dominant role in this matter; the measure of success of eutrophication control is the most economical achievement of the trophic response in a waterbody. Therefore, it is essential to estimate realistically the cost of the control measures for different P sources so that control priorities may be established on a sound economic basis. A simple model for the prediction of the trophic level in Northern Adriatic coastal waters was set up (Chiaudani *et al.*, 1982a). Figure 1.1 shows the predicted trophic conditions deriving from different phosphorus reduction measures in the River Po drainage basin in comparison to the present trophic status.

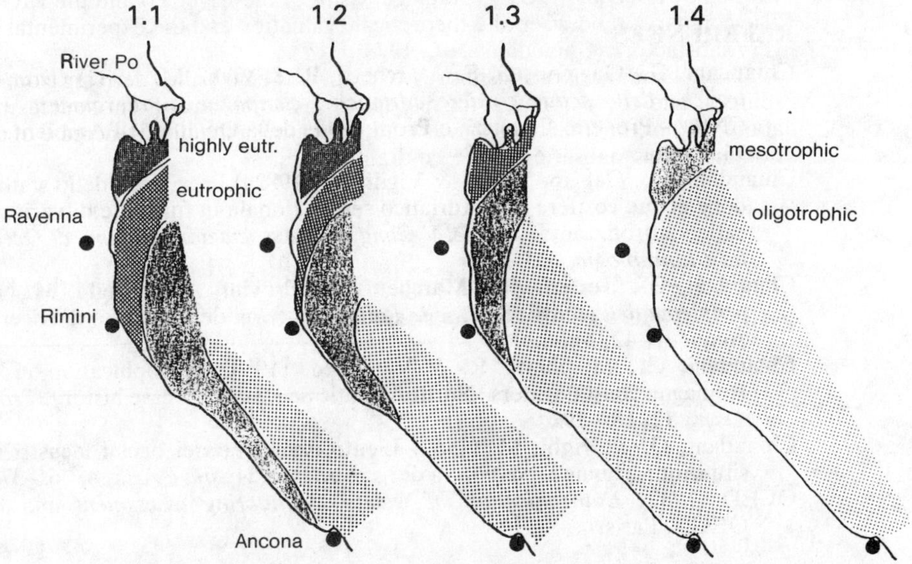

Fig. 1 — Comparison between the present trophic status of Northern Adriatic coastal waters (1.1) and the predicted trophic levels deriving from different phosphorus reduction strategies: 40% reduction of diffuse sources (1.2); 95% reduction of urban point sources (1.3); 95% reduction of all point sources and 40% reduction of diffuse sources (1.4)

The authors would like to point out the following:

- 40% reduction in P loads from diffuse-source control measures, i.e. the maximum level realistically achievable, results in a slight improvement of the water quality (Fig. 1.2);
- 95% reduction of urban point sources results in an acceptable trophic condition of a large part of Italian Northern Adriatic coastal waters. Therefore, this intervention deserves the first priority in the Po Valley restoration programme to reduce eutrophication in marine waters (Fig. 1.3);
- 40% reduction of rural diffuse sources and 95% reduction of all point sources results in a trophic status of the waters, nearer to its natural conditions (Fig. 1.4).

On the basis of the estimated costs for different phosphorus-reduction alternatives (OECD, 1982), the policy against diffuse sources (Fig. 1.2) would involve considerable expenditure: the unit of cost of P reduction to $174/kg, the total costs would be more or less $225 million per year. Total costs for the P reduction from urban point sources could be of the order of $60 million (policy 1.3), while for a complete intervention on all P loads, the estimated costs would be as high as $320 million per year.

The financial support to take 'down process' measures to control point sources of phosphorus for 1985 to 1986 is $500 million for completing a network of treatment plants.

As regards agricultural load, the Authorities are now conducting a public awareness programme aimed at improving farming techniques.

REFERENCES

Chiaudani, G., Gaggino, G. F., Marchetti, R. & Vighi, M. (1982) *Caratteristiche trofiche delle acque costiere adriatiche: campagne di rilevamento 1978–80.* CNR — Progetto Finalizzato Promozione della Qualitá dell'Ambiente. AQ/2/14, Roma.

Chiaudani, G., Gaggino, G. F. & Vighi, M. (1982a) Previsione dello stato trofico delle acque costiere dell'Adriatico settentrionale in funzione di variazioni del carico eutrofizzante. *Atti del V Cong. dell'Associazione Italiana di Oceanografia e Limnologia,* 323–339.

Chiaudani, G., Gerletti, M., Marchetti, R., Provini, A. & Vighi, M. (1978) *Il problema dell'eutrofizzazione in Italia.* Quaderni dello Istituto di Ricerca sulle Acque, N.42, Milano.

Chiaudani, G., Marchetti, R. & Vighi, M. (1980) Eutrophication in Emilia-Romagna coastal waters (North Adriatic Sea, Italy): a case history. *Prog. Wat. Tech.,* **12,** 185–192.

Chiaudani, G. & Vighi, M. (1982) L'eutrofizzazione dei bacini lacustri italiani: situazione attuale e possibilità di risanamento. *Acqua e Aria,* **4,** 361–378.

OCED (1982) *Eutrophication of waters. Monitoring assessment and control.* OECD, Paris.

Poster Paper (vi)

Phosphorus reduction at major sewage treatment works to improve water quality in Lough Neagh, Northern Ireland

A. V. Gray

INTRODUCTION

Deterioration in the water quality of Lough Neagh was first taken seriously when in 1967 a massive algal bloom of *Anabaena flos-aquae* occurred. The effects caused problems in three major areas:

(i) at water-treatment plants, which served domestic and industrial users, through severe blockages in filters;
(ii) in the bottom water of the lough drastically-reduced oxygen levels resulted in substantial fish kills;
(iii) along considerable lengths of the shore-line and associated waters where thick masses of algae had been accumulated by wind action, there was reduced recreational use.

Concerted Government action was initiated to manage the vast natural resource of Lough Neagh for water abstraction, nature conservancy, protection of the eel fishery, and the various amenity uses (Sutton, 1976). Scientific investigations highlighted phosphorus as being the limiting nutrient for algal growth in the Lough (Smith, 1976). Furthermore a desk study revealed that a major portion of the phosphorus entering the lough annually was originating at the sewage-treatment works in the catchment area (Wood & Gibson, 1973).

Major field surveys were then undertaken (Wilson, 1977) and it was established that almost 55% of the phosphorus available to stimulate algal growth originated from sewage effluents, with some 33% coming from just ten major discharges (Foy *et al.*, 1982).

A consideration of the available methods which could be adopted for controlling eutrophication led to the conclusion that chemical treatment at the main sewage-treatment works within the catchment area would probably be the most effective and least expensive option.

CHEMICAL-TREATMENT METHODS

Advice obtained from the former Water Pollution Research Laboratory indicated that chemical treatment of biologically-treated effluents would be the preferred method. Furthermore since a bulk supply of lime was locally available it was the most appropriate chemical compared with either iron or aluminium salts which would have had to be imported.

The results of laboratory and pilot-plant scale experiments showed lime slurry addition to secondary sewage effluents using pH control was an appropriate means of obtaining the requisite phosphorus removal (Gray, 1976). Attempts to find solutions to problems with an unacceptably high pH in the final effluents, which thus required considerable acid neutralisation prior to discharge, coupled with large additional volumes of chemical sludge for disposal, led to the discovery of the Split Effluent treatment process. This enabled the same phosphorus removals to be achieved with 35% to 50% less lime and a slightly thinner chemical sludge was produced which had somewhat enhanced thickening characteristics (Gray, 1981).

A comparison of various coagulants in terms of their cost effectiveness to remove phosphorus from various secondary sewage effluents indicated that lime dosing using the Split Effluent method was preferred for percolating-filter plants, whereas ferrous sulphate dosed *in situ* at the mixed-liquor stage should be adopted at activated sludge plants (Sutton, 1976). Pilot-scale and full-scale plant trials with lime were then undertaken. However, the major escalation in the price of lime over a few years due to the cessation of the local supply, coupled with increased transportation costs as a result of the oil crisis, led to a re-appraisal of the preferred coagulant.

At the same time another coagulant, ferric aluminium sulphate (FAS liquor), which was manufactured from locally-available ferruginous bauxite admixed with sulphuric acid, became available for sewage treatment. An in-depth evaluation of this coagulant, culminating in a two-and-a-half-month field trial at two different sewage treatment works, resulted in this coagulant being chosen as the preferred chemical to effect phosphorus removal at the ten major STW in the Lough Neagh catchment area (Gray, 1985). The chemical dosing plants were then designed and constructed. They were simple in concept, consisting of a lined glass-fibre holding tank leading to a break pressure tank and thence to two metering pumps, each set at an appropriate dosing rate and each controlled by a time clock.

During the subsequent commissioning trials of the chemical-treatment facilities at each STW, the dosing was set in such a way as to be approximately phosphorus-load related; the FAS liquor was, however, added to the crude sewage rather than to the secondary biologically-treated effluent. The basic reasoning for this was due mainly to concern having been expressed about the very large quantities of additional chemical sludge which would be produced from secondary stage dosing, coupled with final effluents which could have pHs in the pH 5 to 5.5 range. By dosing at the primary stage it was thought that the larger compressive forces within the sludge would result in the volumetic increase of sludge being minimised, and also that the pH of the primary settled sewage would be increased from the pH 5 to 5.5 range to about pH 7 in the secondary biological treatment stage, so rendering it more acceptable for immediate discharge (Fig. 1).

The commissioning trials at each STW lasted at least five weeks and usually somewhat longer. By the end of the trials each of the chemical treatment facilities had been shown to be effective in maintaining a phosphorus concentration in the final effluent of less than or equal to 1 mg P/1 for at least ten to 14 days. The maintenance and operational control of the plants was successively handed over to DOE site staff and each of the STWs in question (Fig. 2).

Some unexpected problems appeared at a number of the STWs during the commissioning trials. At Banbridge STW a local creamery discharged a trade effluent and increased the normal daily maximum of about 17 mg P/1 to 57 mg P/1. One particular shock loading of this type which lasted for around 24 hours upset the

STW purification for over five days. At Armagh the contents of septic tanks were received in a very uneven distribution throughout the week and on occasions led to substantially-reduced protozoan populations within the activated sludge, as well as increased total phosphorus concentrations in the final effluent. At Seagoe, sewage was received at the STW from a nearby pumping ststion, in a pulsating flow of high frequency. Fortunately the dosing method employed was found to be effective in reducing the phosphorus from these sewages.

Overall it was estimated that prior to chemical treatment a total of about 118 tonnes of phosphorus was being discharged from the ten major STWs in question. Subsequent to the commissioning trials it was calculated that this had been reduced to 22 tonnes per annum, a reduction of 96 tonnes per annum.

Long-term monitoring of the effectieness of these chemical-treatment facilities has been undertaken (Gray, 1984). At some plants considerable difficulties, mainly of an operational nature, have been experienced. The following are typical.

- Power failures
- Mechanial breakdown of equipment
- FAS liquor supplied being outside specification and producing blockages in supply lines (typically pre-Easter 1984)
- Lack of supply of FAS liquor (mainly October/November 1981)
- Insufficient operational control (poor maintenance of recommended dose rates)
- Interference with normal biological processes (long-term effects)
- Normal STW civil engineering structures unable to cope with the consequences of FAS dosing, for example, extra sludge produced not being accommodated within the limited sludge storage capacity

Clearly when one or more of these problems has upset chemical treatment for a prolonged period the compliance with the required 1 mg P/1 effluent standard has not been achieved. Furthermore at a few percolating-filter plants the FAS dosing has been changed from pre-primary to post-secondary treatment but pre-secondary settlement. Major blockages within the filters occurred due to a combination of reasons: these included the particular character of the crude sewage being received at these works. To prevent too acidic secondary effluents (with pHs less than 6) it has been necessary to reduce the volume of FAS liquor added and subsequently to dose polyelectrolyte at the pre-secondary settlement stage to prevent excessive solids carryover in the final effluents at these few STWs.

BIOLOGICAL-TREATMENT METHODS

In contrast to the chemical-treatment approach it would be clearly advantageous if a biological process could be constrained to effect phosphorus removal to the required low-concentration level. In theory then the cost of the chemical coagulant otherwise required should be saved along with the costs of handling the extra chemical sludge generated by the dosing process.

Interest in such a process was first studied on a full-scale plant in San Antonio, Texas, and reported by Vacher *et al.*, in 1967. Further studies in South Africa mainly by Barnard in 1972 to 1975 indicated that control of the biological processes could be achieved to effect much greater than normal biological phosphorus removals and thereby produce final effluents with only small residual phosphorus concentrations. This process which was termed 'luxury uptake' of phosphorus was thought to be somewhat dependent on relatively high ambient temperatures, such as those associated with warm temperate climates (for example, Texas, USA, and South Africa). Not surprisingly, therefore, the possibility of employing biological treatment to provide phosphorus removal at the STWs in Northern Ireland was never seriously considered. Despite this some evidence has emerged during normal field work to suggest that this phenomenon may have potential in the Northern Ireland context (Gray, 1984a).

During the summer of 1979, when a routine survey at one particular STW was being carried out to measure the phosphorus loading across the plant, it was noted that the results were substantially different from any obtained previously. For a short period conditions within the STW were such that very low phosphorus concentrations were being discharged in the final effluent despite 'normal' levels occurring within the primary settled sewage (0.5 to 0.8 mg P/1 and approximately 3 mg P/1 respectively). Clearly, enhanced biological phosphorus reduction was occurring since the sewage at that plant was generally weak. Later on in the survey very large phosphorus concentrations (11 to 12 mg P/1) were recorded in the final effluent, indicating a major phosphorus release from the activated sludge due to changes in the purification process. This was viewed as a transient feature (Fig. 3).

Subsequently in the autumn and winter of 1979 when field trials using the FAS liquor were being carried out it became apparent that enhanced biological phosphorus removal was occurring. Detailed analysis of the results indicated that bioflocculation could not explain the degree of phosphorus removal which was occurring across the secondary biological treatment stage. As the appraisal of the chemical treatment was the primary objective of the field trials, no further interest was taken in what again was viewed as a transient phenomenon.

When about two years later (in the autumn of 1982) the background monitoring was started at the beginning of the commissioning trials for the FAS dosing plant at the same STW, it was noted that substantially greater than normal phosphorus removal was occurring across the secondary biological treatment stage. It appeared to be occurring fairly consistently and since normal climatic conditions were pertaining monitoring was continued in an attempt to highlight the causative features. This was not possible as several operational changes were occurring randomly on site and within the local sewerage system (Fig. 4).

At the same time as this monitoring was being carried out a laboratory-scale reactor was set up and operated on a batch mode basis, using a 24-hour cycle. The typical daily sequence of the batch-treatment process was:

- addition of untreated sewage to the activated sludge;
- gentle mixing for 3 hours without aeration (anoxic period);
- aeration for 6 to 9 hours with mixing;
- quiescent settlement for 7.5 hours;
- withdrawal of the supernatant liquor from the settled activated sludge.

The results achieved from this laboratory work indicated that consistent phosphorus reduction could be obtained along with high quality effluents having low phosphorus concentrations (that is less than 1 mg P/1). The process was shown to operate consistently with crude sewage sampled from three different STWs over a wide geographical area. For the removals achieved the P/COD ratio was 0.03 suggesting a P/BOD ratio of 0.06, that is considerably larger than the 0.01 assumed in normal purification.

FUTURE APPLICATION

Four of the ten major STWs in the Lough Neagh catchment area are activated-sludge plants and are thus potentially capable of having a 'retrofit' and being operated in such a manner as to achieve phosphorus reduction by biological means. Chemical costs based on the recommended dosing rates for the STWs immediately following the commissioning trials of the coagulant addition facilities were about £200 000 per annum. Such savings are equivalent to a capital investment of £4m at 5% interest rate, or £10.6m if amortised at 5% over a 20-year period. Clearly there appears to be an economic incentive to investigate the possibility of adopting a biological phosphorus-removal process at the four major activated-sludge plants in the Lough Neagh catchment area.

EFFECT ON LOUGH NEAGH

Work carried out by staff of the Northern Ireland Department of Agriculture at the Freshwater Biological Investigations Unit (FBIU) has indicated that about 280 tonnes of phosphorus which is available for stimulating algal growth enters the Lough each year (Foy *et al.*, 1982). Reducing this figure by about 100 tonnes from the ten major point sources would be expected to produce a considerable improvement. It has been predicted, using a correlation produced by the FBIU, that a 30% to 40% improvement in water quality could be reasonably expected. Results of chlorophyll *a* measurement by the FBIU in recent years have indicated an improving situation (Foy, R. A. & Smith R. V., private communication).

The results of continued monitoring should indicate the usefulness of the approach that has been adopted in Northern Ireland to preserving Lough Neagh as such a valuable natural asset.

ACKNOWLEDGEMENTS

The author wishes to thank the Director of the Industrial Science Division of the Department of Economic Development and the Directorate of the Water Service of the Department of the Environment (Northern Ireland) for permission to publish this chapter. Thanks are also due to the many colleagues who were involved in various aspects of the laboratory and field work referred to in this chapter. The views expressed are those of the author and do not necessarily represent government opinion, except where stated.

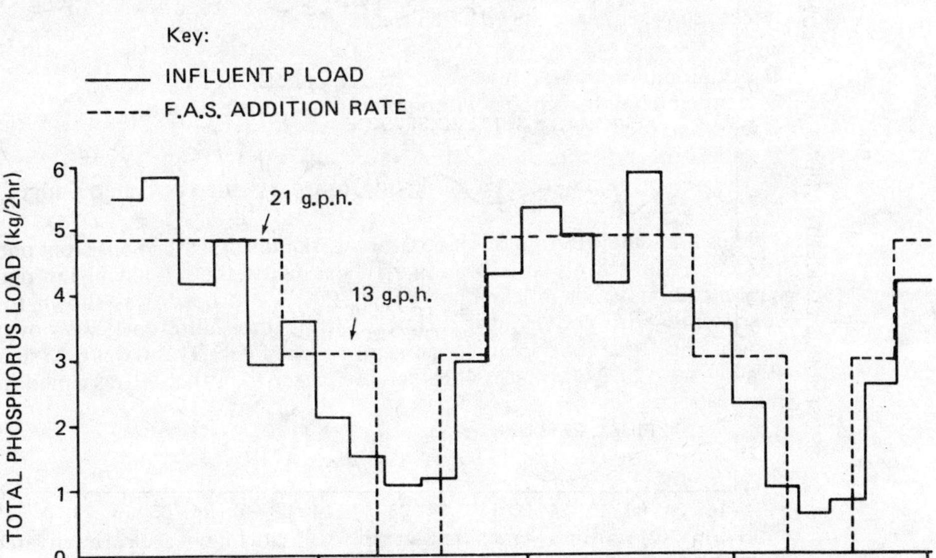

Fig. 1 — Banbridge S.T.W.

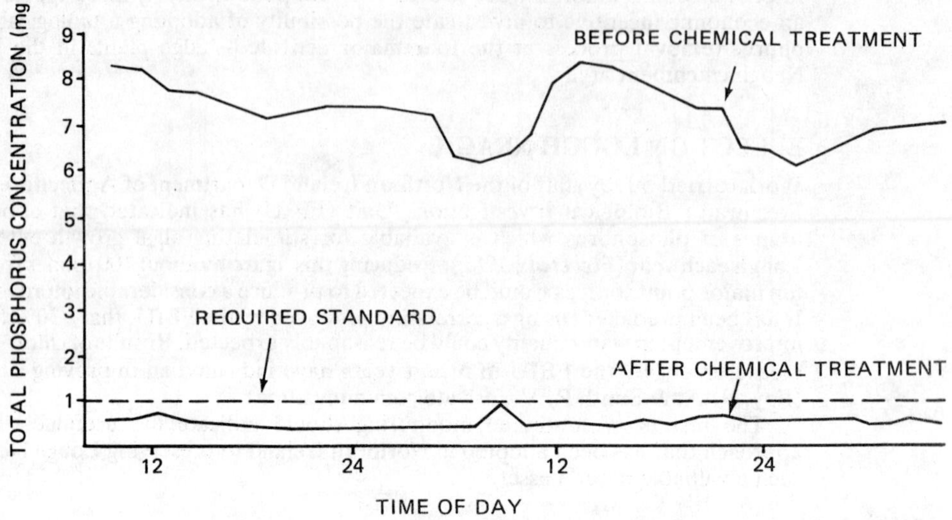

Fig. 2 — Banbridge S.T.W. final effluent quality

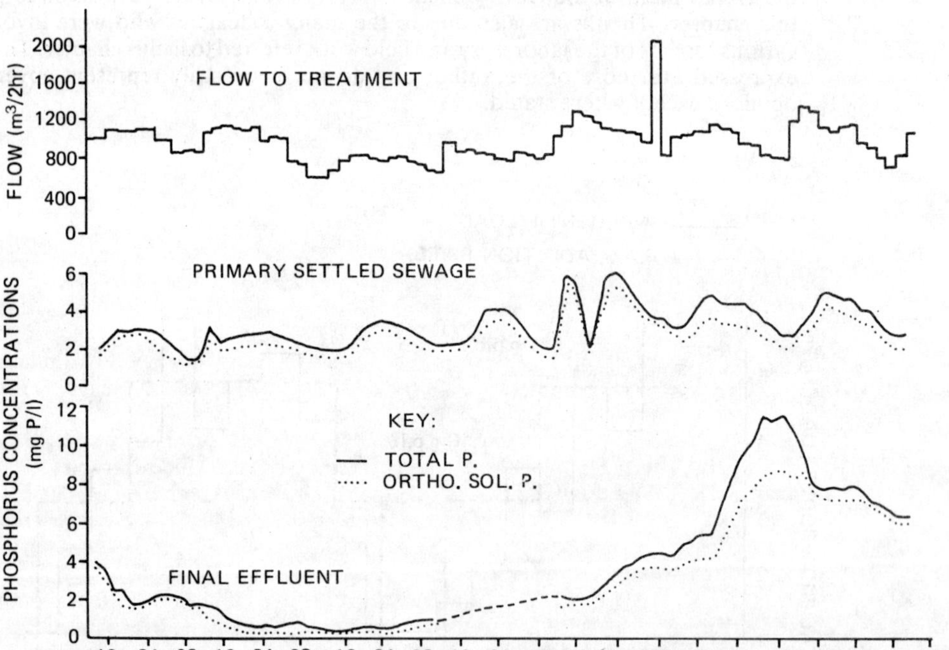

Fig. 3 — Ballynacor S.T.W survey June/July 1979

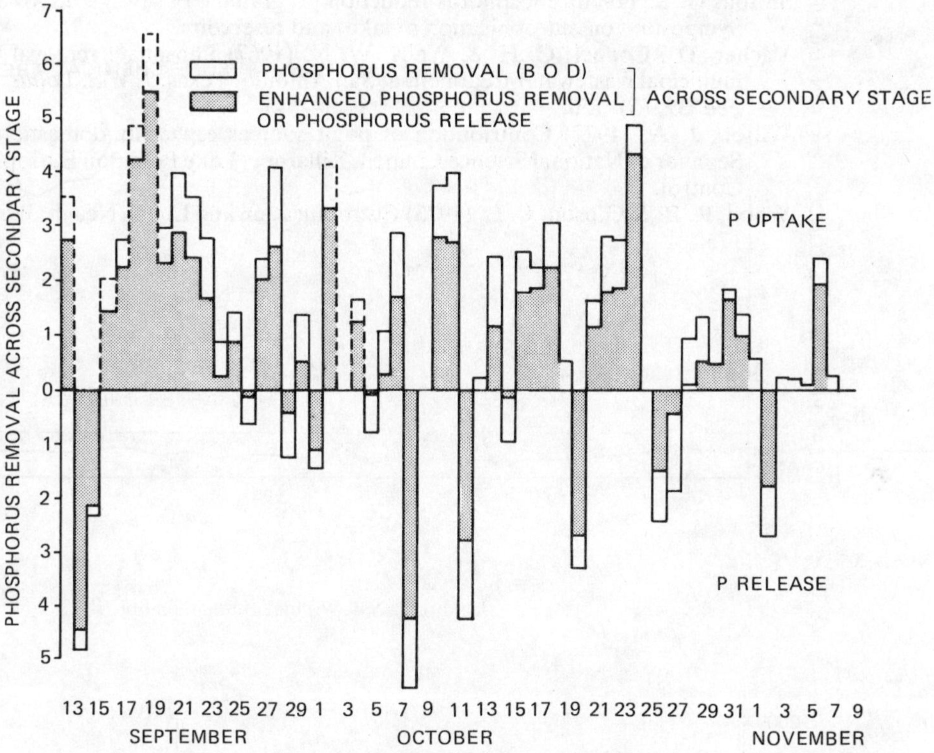

Fig. 4 — Ballynacor S.T.W. 1982 results

REFERENCES

Barnard, J. L. (1973) Biological denitrification, *J. Wat. Pollut. Control Fed.* **72,** 6.

Ibid (1974) Cut P and N without chemicals. *Wat. Wastes Eng (GB).* **11,** 7.

Ibid (1975) Biological nutrient removal without the addition of chemicals. Wat. Res. **9,** 485.

Foy, R. A., Smith, R. V., Stevens R. J. & Stewart, D. A. (1982) Identification of factors affecting nitrogen and phosphorus loadings to Lough Neagh. *J. Environ. Management,* **15,** 109.

Gray, A. V. (1976) Chemical treatment for phosphorus reduction at sewage disposal works. IPHE Symposium on eutrophication of lakes and reservoirs.

Gray, A. V. (1981) Phosphorus reduction studies. *Wat. Pollut. Control,* **80,** (3), 333.

Gray, A. V. (1984) Monitoring and Water Quality — Part II — Sewage Treatment Works. Ulster Anglers Federation Conference, Greenmount, Antrim, on Lough Neagh and its Rivers.

Gray, A. V. (1984a) Northern Ireland experience of enhanced biological phosphorus removal. IAWPRC Post Conference Seminar, Paris, on Enhanced Biological Phosphorus Removal from Wastewater.

Gray, A. V. (1985) Removal of phosphate at sewage treatment works and the implications on phosphate loadings into Lough Neagh. *J. Instn. Wat. Engrs and Scientists.* **39,** (2), 137.

Smith, R. V. (1976) Lough Neagh investigations. IPHE Symposium on eutrophication of lakes and reservoirs.

Sutton, G. K. (1976) Phosphorus reduction programme of sewage disposal. IPHE Symposium on eutrophication of lakes and reservoirs.

Vacher, D., Connel, C. H. & Wells, W. N. (1967) Phosphate removal through municipal wastewater treatment at San Antonio, Texas. *J. Wat. Pollut. Control Fed.* **39**, (5), 750.

Wilson, J. A. (1977) Contribution of point sources especially domestic sewage. Seminar of National Science Council, Killarney. Lake Pollution Eutrophication Control.

Wood, R. B. & Gibson, C. E. (1973) Eutrophication and Lough Neagh. *Wat Res,* **7,** 173.

Poster Paper (vii)

Leaching resulting from land application of sewage sludge

M. Melkas and M. Melanen

The present Finnish guidelines recommend spreading sewage sludge on non-soil and mixing it into the soil immediately. Application on frozen soil may cause harmful leaching of constituents to recipients.

TEST ARRANGEMENTS

Movement of nutrients, heavy metals, organic matter, and micro-organisms, resulting from application of dewatered municipal sewage sludge on flat clay soil was studied in a four-year experiment by the National Board of Waters (Finland) and the Agricultural Research Centre (Melanen *et al.*, 1985).

The year 1979 was a calibration year and the treatments (16 to 20t/ha of dry matter) were given in 1980. Sludge was spread both on snow (treatment b) and on non-frozen soil (treatment c) in spring. Limed sludge (pH 8.4) (treatment d) was also spread on non-frozen soil. The reference plots were not fertilized (treatment a). No fertilization was applied in 1981; in 1982, 350 kg/ha of NPK fertilizer (16-7-13) was applied to all plots. The test plots were 609 to 620 m^2 in area and each treatment was repeated four times. The crop cultivated was barley. The amounts of the three major nutrients spread with the sludges were as follows (in kg/ha):

	N_{tot}	P_{tot}	K
sludge on snow (treatment b)	1300–1400	500–600	130–150
sludge on non-frozen soil (treatment c)	750–950	400–500	80–100
limed sludge on non-frozen soil (treatment d)	450–550	450–500	25–30

HYDROLOGICAL CONDITIONS

Two of the years — 1981 and 1982 — were hydrologically exceptional: the 1981 mean runoff has an average return period of about 30 years; the 1982 spring runoff is exceeded only once in 20 to 30 years on an average.

Subdrainage runoff casued 75 to 90% of total runoff. Most surface and subdrainage runoff came during the snow-melting period.

LEACHING AND CROP YIELD

A variety of elements and characteristics were analysed both on subdrainage and surface waters. Nutrients and heavy metals (N_{tot}, N_{NH_4}, N_{NO_2}, N_{NO_3}, P_{tot}, P_{PO_4}, K, Ca, Mg, S_{SO_4}, Na, Cl, Fe, Mn, Zn, Cu, Cr, Ni, Cd, Hg, Pb), conductivity, pH, organic matter (COD_{Mn}), suspended solids (SS), and indicator micro-organisms (coliform bacteria and enterococci).

Two seasonal peaks were generally observed in concentrations: one in the snow-melting period and another in the autumn.

Approximately 70 to 90% of nitrogen was in nitrate form. The nitrogen content of subdrainage water was higher than that of surface water. The effect of the treatments could usually be seen in both median and flow-weighted mean concentrations. Raised contents were still observed in autumn 1982. The effect of the limed sludge (treatment d) was smaller than that of the other sludges. Raised contents of ammonium nitrogen were observed in both subdrainage and surface waters resulting from the application on snow (treatment b): the monthly flow-weighted means were 1.0 to 1.7 mg/l N_{NH_4} (the contents in the other plots were below 0.1 mg/l N_{NH_4}).

Approximately 30 to 50% of phosphorus was in phosphate form. The phosphorus content was generally higher in the surface water than in the subdrainage water. Raised concentrations were observed after the treatments, especially in surface waters.

Only a slight increase was observed in the flow-weighted means of potassium, and it was mostly caused by runoff. The small increase in COD_{Mn} was obviously also mostly explained by runoff.

The median concentrations of heavy metals remained at the detection limit of analysis. Applications thus did not seem to cause any harmful leaching of heavy metals.

A statistical evaluation of bacteria in runoff was not possible because of too small a number of samples. Faecal coliforms seemed to disappear in a short time after the applications.

The effect of the treatments was also visible in the annual leaching rates (Table 1). On average, 90% of total nitrogen, almost 80% of total phosphorus, and over 80% of potassium came through subdrainage. The exceptionally rainy years had a considerable effect on leaching rates.

The loss of nitrogen given in the sludge treatments varied from 3 to 23% in 1980 to 1982; only 0.1 to 0.2% of phosphorus was lost through leaching.

Minor differences were observed in the average three-year barley yields of the various treatments.

CONCLUSIONS

A four-year experiment on application of dewatered sewage sludge on clay soil yielded the following main findings:

- Climatic and weather conditions (total precipitation, snow accumulation, soil frost penetration, and so on) have a considerable effect of leaching. (Two of the years studied were hydrologically exceptional.)
- Most of the total leaching, generally over 80 to 90%, is through subdrainage. The snow-melting period has a dominant role with regard to annual leaching.
- No harmful leaching of heavy metals can be expected when normal sludge is applied in favourable conditions according to the present guidelines.
- The most profound effect of the treatments is on nitrogen leaching; the duration of the treatment effect is over two years. Limed sludge causes less leaching than the other sludges.
- In the long run, sludge applications on snow and on non-frozen soil yield equal

Table 1 — Annual subdrainage leaching (four replicates per treatment) (Melanen *et al.*, 1985)

Hydrological year†	Treatment	Leaching (kg/km²/yr) N_{tot}	N_{NO_3}	P_{tot}	P_{PO_4}	K
1978–1979	a	1600	1300	12	3.5	1100
	b	1800	1600	13	4.0	1400
	c	1800	1500	13	4.0	1700
	d	2000	1800	13	4.3	1300
1979–1980	a	1200	930	11	4.6	1200
	b	1900	1400	15	5.2	1300
	c	1900	1700	12	5.4	1700
	d	1300	1100	11	5.0	1100
1980–1981	a	5500	5700	55	22	4000
	b	16000	16000	83	41	4800
	c	24000	23000	79	35	6400
	d	7900	6600	83	40	4000
1981–1982	a	1700	1400	54	15	3900
	b	4000	3800	74	33	5300
	c	4900	4600	82	29	6400
	d	2700	2500	81	29	4000

† From 1 November to 31 October

nitrogen leaching; however, raised contents of ammonium nitrogen in runoff waters can be observed over a period after spreading on snow.

- Minor differences can be found in crop yields (barley) of the different treatments.

The smaller leaching caused by limed sludge is partly explained by the lower nitrogen content of this sludge. Slow snow-melting after the application of sludge on frozen soil may have reduced the adverse effects of this treatment.

REFERENCE

Melanen, M. Jaakkola, A., Melkas, M., Ahtiainen, M. & Matinvesi, J. (1985) *Leaching resulting from land application of sewage sludge and slurry*. Helsinki, National Board of Waters, Finland. Publications of the Water Research Institute 61. In preparation.

Poster Paper (viii)

The environmental impact of cage culture on Scottish fresh waters

M. J. Phillips, M. C. M. Beveridge and J. A. Stewart

INTRODUCTION

Cage culture is a method of farming fish in an enclosed mesh or netting structure held within a larger body of water. It is a technique which has become increasingly popular over the past 15 years and has now been used for culturing over 70 species of fish in 35 countries (Coche, 1983; Beveridge, 1984).

The popularity of cage culture is largely due to its low capital cost in comparison with many traditional land-based systems and the relative ease with which suitable water can be used for farming by the simple addition of one or more cages. These advantages and an abundance of high-quality fresh waters have led to cage farming becoming particularly popular in Scotland (Table 1). Rainbow trout (*Salmo gairdneri*) is the main species produced in Scottish freshwater cages, largely for direct human consumption, with a small percentage for the restocking of angling waters. A number of Atlantic salmon (*Salmo salar*) smolts are also produced for transfer to and ongrowing in the sea, but the overall weight of production is small in comparison with rainbow trout.

Table 1 — Production of rainbow trout in cages in Scottish freshwater lochs

Year	Production (tonnes)	Percentage of total rainbow trout production
1979	178	15
1980	470	29
1981	721	34
1982	728	36
1983	840	42
1984	932	45

A cage farm, unlike most other forms of aquaculture, is an integral part of a natural aquatic ecosystem and there are inevitably some interactions between the cages and external environment (Fig. 1).

The recent expansion and considerable future potential has, therefore, led to some concern, expressed by statutory bodies, conservationists, anglers and other water users, as well as the fish farmers themselves, about the environmental effects of cage farming. There have been objections to cage developments on aesthetic grounds (Beveridge & Muir, 1982), but most concern has focused on the impact of cage wastes on the receiving water body. These environmental issues are particularly important in Scotland where the nutrient-poor nature of many lochs may make them particularly vulnerable to eutrophication, but have recently assumed wider importance in Europe and America, from all those concerned with the management of fresh waters. The aim of the present paper is to discuss some of the wastes produced from cage culture in Scottish waters, with particular reference to rainbow-trout production.

WASTE PRODUCTION FROM RAINBOW TROUT

Total waste loadings

The total waste loading of the main elements wasted, carbon, nitrogen, and phosphorus, can be predicted from a knowledge of element input in fish food and element output in fish flesh, for example;

$$E_w = E_{food} - E_{fish}$$

where E_w = Total waste loading of element (kg)
E_{food} = input of element with food (kg)
E_{fish} = output of element with fish harvested (kg).

A survey of rainbow-trout cage farms in Scotland shows that 1.5 to 2.0 tonnes of artificial fish diet are fed to produce 1.0 tonne of fish. These figures can be used to predict the range of waste loadings encountered (Table 2).

Solid and soluble waste loadings

The total waste loadings give an indication of the total loading of an element, but the environmental impact also depends on the proportion of solid and soluble wastes produced, which require field measurements of sedimentation rate and nutrient changes in farmed lochs. This solid input is in practice easier to quantify and reveals that the major percentage of carbon and nitrogen is soluble waste, whereas approximately 50% of phosphorus is discharged as solids (Table 3).

Phosphorus and wastes

The phosphorus component of the waste is the most important in Scottish lochs because it is phosphorus which controls the primary productivity, or trophic status, of most fresh waters.

Phosphorus is an essential mineral which fish require in their diet for normal growth and metabolism. Rainbow trout require 7 to 8 mg/g dietary phosphorus (Beveridge, 1984), but most commonly used diets contain, on average, 17 mg/g (Tacon & De Silva, 1983; Phillips, in prep.); consequently there is a considerable excess of phosphorus which is discharged as waste.

A theoretical analysis of the passage of phosphorus through a rainbow-trout cage producing one tonne of fish is illustrated below (Fig. 2). This analysis agrees well with actual sedimentation rates, and again suggests that approximately 50% of phosphorus is discharged as solids. The high losses also suggest that reduction in phosphorus levels in feed could considerably reduce the environmental impact.

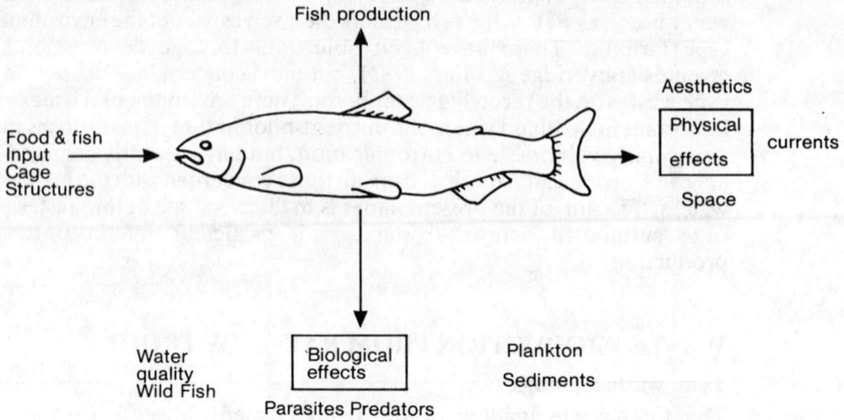

Fig. 1 — The impacts of intensive cage culture on the environment

Table 2 — Total waste loadings from rainbow trout cage culture (kg waste per tonne fish production), indicating the range of loadings with two extreme feed input values (1.5 to 2.0 tonnes)

Element	Input with feed (kg)	Output with feed (kg)	Waste loading (kg)
C	589.8–786.4	154.4	435.4–632.0
N	105.3–144.0	27.2	78.1–116.8
P	25.5– 34.0	4.8	20.7– 29.2

Table 3 — Solid waste loadings in relation to total waste loadings: Data from an 85-day study of sedimentation rate below one rainbow trout cage (Phillips, in prep.)

	Total loading (kg/tonne fish produced)	% solid waste
Carbon	400.1	24.1
Nitrogen	74.3	14.1
Phosphorus	20.1	51.2
Total solids	292.9	100.0

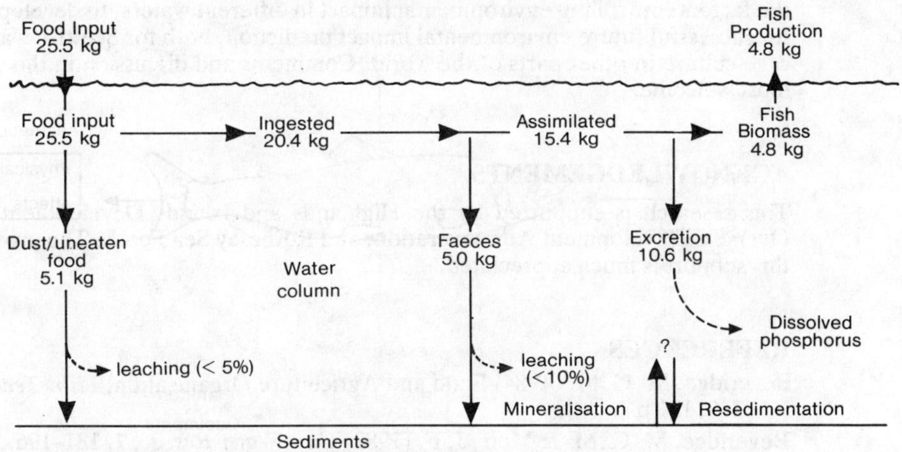

Fig. 2 — Theorectical phosphorous losses to the environment assuming 1–5 tonnes of food are used to produce 1.0 tonne of fish. Modified from Beveridge (1984) and Philips & Beveridge (in prep.)

MODELLING OF ENVIRONMENTAL IMPACT

One of the main aims of this research is to produce predictive models for the assessment of the environmental impact of freshwater cage culture.

Present research has concentrated on the use of the original Dillon and Rigler (1975) model to predict changes in water phosphorus concentration in response to an additional phosphorus loading from cage culture (see Beveridge, 1984). The approach is as follows.

1. Measure retention time, loch volume, and baseline water phosphorus concentration.
2. Decide on acceptable final phosphorus concentration in water, taking account of phosphorus/chlorophyll relations and the requirements of the water body for other uses (such as potable water, conservation).
3. Input phosphorus production per tonne of fish farmed, with an immediate loss of 50% phosphorus to the sediments.
4. Output acceptable tonnage to be farmed in water in relation to trophic status.

This type of approach has been successfully used to provide conservative predictions of environmental change, but there are problems.

1. The relationship between phosphorus concentration and 'acceptable' change is poorly understood in many waters, particularly in relation to suitable water quality for the farming of salmonids.
2. The general model has not been extensively tested in the range of lochs encountered (for example, waters high in humic acids respond less well to phosphorus fertilization than clear water bodies).
3. The role of sediment recycling of phosphorus in different waters is poorly understood.
4. There are large variations in waste output between farms, making it necessary to use pessimistic figures in determining environmental impact.

The aim of the present research is to provide improved data on waste output and the factors controlling environmental impact in different waters, to develop models for successful future environmental impact prediction, both for Scottish waters and cage culture in other parts of the world. Comments and discussion in this area are most welcome.

ACKNOWLEDGEMENTS

This research is supported by the Highlands and Islands Development Board, Overseas Development Administration, and Rothesay Sea Foods. The provision of this support is much appreciated.

REFERENCES

Beveridge, M. C. M. (1984) Food and Agriculture Organisation, *Fish. Tech. Pap.*, **255**, 131 p.

Beveridge, M. C. M. & Muir, J. F. (1982) *Can. Water Res. J.*, **7**, 181–196.

Coche, A. G. (1983), Food and Agriculture Organisation. *Fish Circ.*, (714) Rev. 1: 61p.

Dillon, P. J. & Rigler, F. H. (1975) *J. Fish. Res. Bd Can.*, **32** (9), 1519–1531.

Tacon, A. G. J. & De Silva, S. S. (1983) *Aquaculture*, **31**, 11–20.

Poster Paper (ix)

The effects of reducing the urban load of phosphorus on Lough Neagh algal populations

R. V. Smith

INTRODUCTION

The enrichment of fresh waters with nutrients leading to high levels of algae is a major pollution problem in Northern Ireland. In contrast with England and Wales, which in general do not possess large lowland lakes which drain significant urban areas, Northern Ireland has the largest body of water in the British Isles, namely Lough Neagh, situated centrally in the province draining a catchment of 4453.4 km^2 with an estimated human population of 358 500. Lough Neagh gave particular cause for concern in the summer of 1967 when a large crop of the blue-green alga *Anabaena flos-aquae* formed a surface-water bloom. The algae were blown downwards into Antrim Bay and their accumulation interfered with three of the Lough's important uses, namely water abstraction, fisheries, and recreation. Antrim Bay is the site of the water intake for Belfast and is also an important recreational area for Antrim. The algae caused the death of eels held in cages, perhaps by algal toxins, but more probably by the decaying mass of algae removing the oxygen from the water and so suffocating the fish.

This spectacular growth of algae may be interpreted as a response to abnormally-high loadings of nutrient to the lake. A lake and its catchment area function as part of a single system and the enrichment of a lake with nutrients arises because of an intensification of human activity in the lake catchment area. Human activity can induce loadings of nutrients that are comparable or exceed in magnitude the natural nutrient loading rates from catchment systems to lakes.

SOLUBLE REACTIVE PHOSPHORUS: THE CRITICAL NUTRIENT

The amount of N and soluble reactive phosphorus (SRP) entering Lough Neagh from the major rivers, direct drainage, and sewage treatment works, have been measured for the years 1971 to 1981. To determine if there was a relationship between these loadings and algal levels in Lough Neagh a correlation matrix was calculated between mean chlorophyll *a* concentrations for different periods of the year and nutrient inputs. In general the results showed that mean chlorophyll *a* in the Lough (a measure of algal concentrations) was highly correlated with SRP input but not with the input of nitrate. The most highly-significant correlation ($r^2 = 0.84$, $P < 0.001$***) was found between the October to September loading of SRP (the hydraulic year) and the mean chlorophyll *a* in Lough Neagh for the period June to September (Fig. 1). Nitrate loading for the same period showed no relationship with

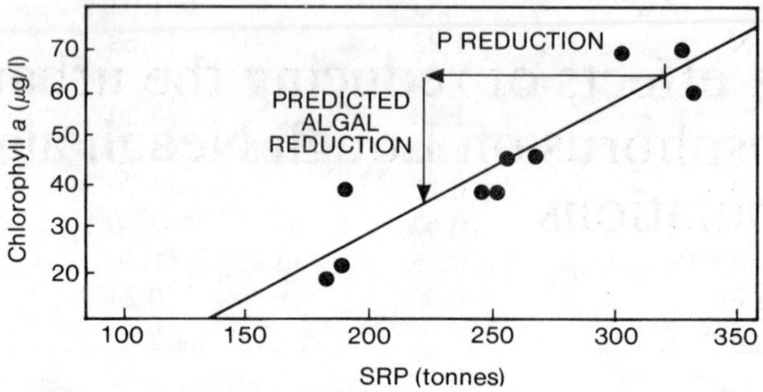

Fig. 1 — The relationship between chlorophyll *a* in the Lough and the SRP input together with the predicted effect of a 100 tonnes reduction.

chlorophyll ($r^2 = 0.203$). These results strongly imply a causal relationship between algal levels in the Lough and the SRP input.

WHERE DOES THE SRP ORIGINATE?

The status of SRP as the critical nutrient in Lough Neagh appears to be supported by the available data. However, the next important consideration is what proportion of SRP and other nutrient inputs to Lough Neagh originate from point sources such as sewage-treatment works. Smith (1976) made an intensive budget study of one river catchment, the Main, which drains to Lough Neagh. The observed outputs of SRP from the Main were in good agreement with the predicted outputs based on land drainage measurements and intensive surveys of sewage treatment works. The budget indicated that 62% of the SRP output was derived from sewage disposal works effluent (see Table 1). It was observed that less than 10% of the loading of nitrate in the river was derived from sewage effluent. For nutrients other than N and P sewage effluent made an insignificant contribution to the river output. Therefore SRP is the only critical nutrient which is amenable to a reduction in river loading by nutrient removal from a sewage treatment works.

An alternative statistical analysis of the data on land use and population variables which had been collected for the years 1971 to 1979 was undertaken to see if SRP loadings from different rivers could be correlated with urban population density (Foy *et al.*, 1982). The SRP regression equation is given below:

$$\text{SRP loss rate} = 0.451 + 0.573\ \text{URBAN POPULATION} + 0.000043\ \text{CREAMERY OUTPUT} - 0.127\ (\text{FLOW})^{-1}.$$

URBAN POPULATION density was the most highly-correlated variable reflecting the high P content of domestic sewage. The regression coefficient of 0.573 was

Table 1
Relative contribution of effluent from sewage-disposal works to the observed nutrient output from the R. Main

Nutrient	Sewage effluent loading (kg/ha yr)	River loading (kg/ha yr)	Contribution of sewage effluent (%)
Soluble reactive P	0.40–0.53	0.65	62.00–82.00
Total P	0.53	1.08	49.00
Nitrate N	1.22	13.68	9.00
Total N	1.66	21.02	8.00
Cl	1.50	135.20	1.00
Na	1.00	95.70	1.00
K	0.45	11.70	4.00
Ca	0.67	132.10	0.50
Mg	0.25	66.50	0.40
Soluble Fe	0.00	0.34	0.00
Soluble Mn	0.00	0.13	0.00

equivalant to a daily per capita of 1.6 g P/person day, a figure in line with direct measurements of domestic sewage in the catchment area (Alexander & Stevens, 1976). FLOW^{-1} was also strongly correlated with SRP loss rates suggesting increased leaching from the soil under high flow conditions. However, differences in land use or stocking density were not found to be significant.

Using the 1979 URBAN POPULATION and CREAMERY OUTPUT variables for the whole catchment and the 1971 to 1979 mean flow value, an SRP budget has been constructed (Fig. 2) which shows that 54% of the SRP originates from sewage-

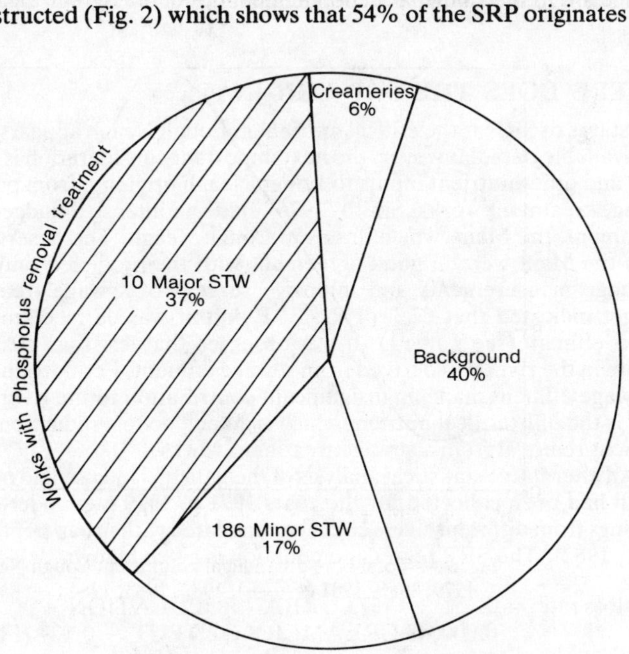

Fig. 2 — Inputs of SRP to Lough Neagh.

treatment works, 6% from creamery discharges, and the remainder as a background contribution the main component of which is land drainage. The only practical way of reducing the SRP input is to reduce the loading from sewage-treatment works but removal at all works is difficult because there are 196 in the catchment. However, the ten major sewage-treatment works serving large populations account for 37% of the input of SRP to Lough Neagh.

THE PHOSPHORUS REDUCTION PROGRAMME

The most effective strategy for reducing the SRP load to Lough Neagh is the introduction of P-reduction treatment at the sewage-treatment works. In 1982 drawing on these findings the Department of the Environment (Northern Ireland) completed a programme to install P reduction at all of the ten major works which together contribute about 37% of the load to the Lough. Most of the SRP in the influent sewage is precipitated out at the primary settlement stage by dosing the incoming sewage with DIFCO (an aluminoferric liquor available locally and developed by the Industrial Science Division of the Department of Commerce) in amounts such that effluent finally discharged to the rivers contains no more than 1.0 mg/l total P.

THE EFFECTS OF THE PHOSPHORUS REDUCTION PROGRAMME

About 60 tonnes of phosphorus was removed in 1983 and 1984 compared with the target of 100 tonnes. Nevertheless, the reduction of P in the sewage-treatment works significantly reduced the amount of P in Lough Neagh. Mean concentrations fell consistently: 63 μg/l in 1981, 58 μg/l in 1982, 49 μg/l in 1983, and 40 μg/l in 1984. Most significant has been the reduction in the crops of blue-green algae (Fig. 3) which has

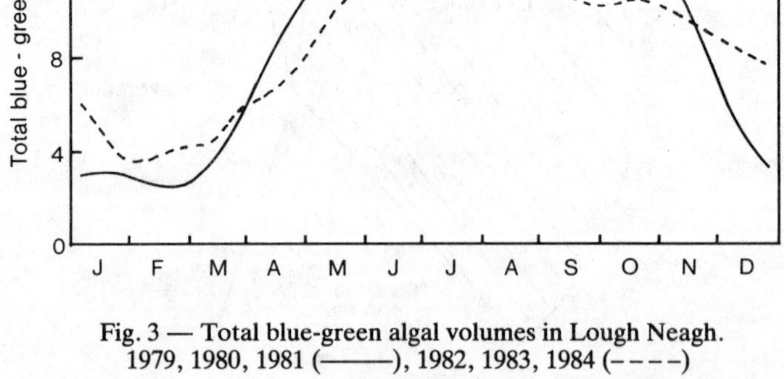

Fig. 3 — Total blue-green algal volumes in Lough Neagh.
1979, 1980, 1981 (———), 1982, 1983, 1984 (- - - -)

led to important benefits at Dunore Point Water-Treatment Works which supplies water to Belfast. It is understood that Dunore Point experienced severe water-extraction problems in the years 1979 to 1981 because of the large populations of blue-green algae. Indeed if algal levels had been as high as the period 1979 to 1981, the Works would have had great difficulty in maintaining the optimum output during the drought period of 1984.

REFERENCES

Alexander, G. C. & Stevens, R. J. (1976) Per capita phosphorus loading from domestic sewage. *Water Res.* **10**, 757–764.

Foy, R. H., Smith, R. V., Stevens, R. J. & Stewart, D. A. (1982) Identification of factors affecting nitrogen and phosphorus loadings to Lough Neagh. *J. Environ. Managements.* **15**, 109–129.

Smith, R. V. (1976) *Nutrient budget of the River Main, Co. Antrim.* Technical Bulletin of the Ministry of Agriculture, Fisheries and Food, London, No. 32, pp. 315–339.

Poster Paper (x)

The effect of overgrazing the Mambilla Plateau, Nigeria, on the quality of the River Donga

Lilian M. Evison

INTRODUCTION

The Mambilla Plateau is a high level area (1000 to 2000 m) of almost 4000 km^2 in south west Nigeria, bordering on the Republic of Cameroun. It is proposed that a dam could be constructed on the River Donga on the plateau to create a high-head hydro-electric power generation scheme. The information described in this chapter was collected with a view to determining the secondary impacts of the proposed reservoir on the area.

The area is characterised by a vast series of convex crested ridges, falling through steep sideslopes to flat or gently concave valley bottoms. To the north of the area are basalts with drainage generally towards the south west forming minor tributaries of the River Donga. The remainder of the plateau is mainly on granite and gneissic parent rocks of the Basement Complex. The area is drained by a valley system of the east-west and north-south flowing upper reaches of the Donga River and its main tributaries in the Balewa, Mbaso, Luggungo, Ndarup, and Warwar.

The climate divides into a wet and a dry season, with rains between March and November (220 to 240 days) with the heaviest rainfall in July, August, and September. The long-term mean annual rainfall is 1800 mm, with high intensities (up to 100 mm/h) sustained for short periods of ≅10 mins. The air temperature is relatively cool because of the altitude, and ranges from 17 °C to 21 °C at Gembu, the main town near the River Donga. In the absence of cloud, the range of sunshine hours over the year is small, being 11.7 hours in December and 12.9 hours in June at latitude 7° north. Cloud cover during the rainy season and harmattan haze during the dry months reduce the hours of sunshine to 9 hours in December and 3 hours in July at Gembu.

The soils of the plateau are remarkably uniform humic ferrisols resulting from soil formation under high rainfall and high altitude grassland. Typically they are dark reddish-brown sandy-clay loams at the surface (0 to 20 cm), and clay loam upper subsoils (20 to 60 cm) and lower subsoils (to 150 to 180 cm). The soils are moderately acid (pH 5.5 to 6.0) and are chemically poor with cation exchange capacity between 15 and 25 meq./100 g soil.

The area was originally thought to be covered by Montane Forest, but today forest is found only as relicts in steep valleys and in an emasculated form in valley bottoms where most of the trees have been cut out for fuel, and only palms and other species not used as fuel remain. At the present time the Montane vegetation of the high Plateau is essentially a fire induced pro-climax grassland. In protected areas it is essentially a tall tufted grass community dominated by *Hyparrhenia diplandra* with *Andropogon auriculatus* and *Setaria anceps* as sub-dominants. Below this higher grass cover, several short grasses, recumbent species and herbs occur between the tussocks. The entire Mambilla Plateau is today so overstocked with cattle that most of the grassland has been pushed back to early successional stages, leaving bare ground which is very susceptible to sheet erosion.

POPULATION AND LIVESTOCK

At the present time the road system on the plateau is rudimentary with an unsurfaced road, almost impassable in the wet season, rising up the escarpment from Serti on the plains to Mayo Selbe. The road continues to Nguroje, the main cattle market on the plateau, and on to the town of Gembu. A network of tracks, sometimes passable by Land Rover, covers the remainder of the plateau, particularly to Mayo Ndaga in the west and to Dorofi in the south on the border with the Republic of Cameroun. Altogether a total of 25 300 households live on the plateau, with an estimated population of 128 000 people.

A total of 22 000 households are engaged in agricultural activities — the settled agriculturalists who cultivate mainly in the fertile valley bottoms, and the Fulani pastoralists, who range over the plateau with their herds of cattle. The nomadic Fulani clans have been increasing the size of their herds over the 50-year period for which records are available (Table 1). In addition to these numbers, there is an

Table 1
Cattle population on Mambilla Plateau

1930	18 000
1949	120 000
1954	161 000
1963	235 000
1974	305 000
1976	375 000
1981	450 000

annual influx of 6000 to 15 000 cattle from the Camerouns for sale and export. These cattle contribute a little to the overgrazing problem, and to a considerable extent to the localised erosion along the cattle-movement routes. It is estimated that the sustained carrying capacity of the plateau is 200 000 cattle, and this figure was exceeded in the 1950s, accelerating soil erosion problems since then.

There are three main markets on the plateau at Dorofi, Mayo Ndaga, and Nguroje, and cattle sales at each are from 200 to 300 per week in the dry season, and 500 to 600 in the rainy season. Total annual sales are of the order of 80 000 with 75% of these exported on the hoof. The main cattle routes are north then north west from Dorofi to Nguroje, and west from Mayo Ndaga to Nguroje where there is a form of licensing for all stock leaving the plateau.

EROSION

Both natural erosion and man-made erosion occur on the plateau, but it is sometimes difficult to distinguish the cause. Many of the conventional slip failures are undoubtedly of natural origin, and in some parts of the River Donga catchment undercutting of stream banks is very common, possibly due to increased river flow resulting from overgrazing. There are many gulley-type scars, at least some of which have been triggered off by overgrazing or cattle tracks. Both the overgrazing and cattle tracks are unevenly distributed on the plateau, so that some valleys and hilltops are very badly affected by erosion, while neighbouring areas are in much better condition. Cultivation practices also contribute to erosion problems. At present about 20 000 ha of land are cultivated, but this area is increasing, often on slopes of 20% with little or no attempt at conservation practices, so these soils are very vulnerable to erosion.

QUALITY OF THE RIVER DONGA

The limited amount of data available suggests that the River Donga used to be a 'black river' with a suspended-solids concentration of only 240 mg/l at Nyam Hurole in 1955. By 1977 the river was red with silt at all times of the year, with a mean suspended-solids concentration of 5000 mg/l.

In August and September 1982 water samples were collected from as many parts of the River Donga system as were accessible during the rainy season. Conductivity and pH were measured at the time of collection with EIL portable instruments, and a range of chemical analyses were made on a Hach DREL/4 portable laboratory, usually within 24 hours of collection. Most of the analyses conform to the APHA Standard Methods (1975). In addition some samples were also transported back to England for comprehensive analysis in a conventional laboratory within seven days. Some of the results collected are summarised in Table 2.

The results indicate that the River Donga is a very soft, medium bicarbonate water, highly unsaturated with respect to calcium carbonate. The conductivity of the water is extremely low (average 19 μS/cm, total dissolved solids $\cong$23 mg/l) with very low concentrations of cations. The water is slightly acidic, with iron and manganese being the only metals present in high concentrations. In fact, the chemical composition of the river water was very similar to rain water, which was also analysed.

The most significant feature of the River Donga analyses was the high suspended-solids concentrations, particularly at the Tepkwar site near Gembu. It is thought that the high suspended-solids values found at Tepkwar (1450 mg/l) probably mainly originate from the Warwar tributary. Unfortunately, it was not possible to sample the River Warwar because of the road conditions at the time. This valley is close to the cattle-market town of Dorofi on the border of the Camerouns, so is subject to heavy cattle-grazing pressure and well-defined cattle tracks. It is notable that very low suspended-solids values were found in the Rivers Kusuku and Sum Sum which drain from the basalts to the north of the Mambilla Plateau. Nyivu is a village on the plains below the escarpment, and it is notable that much of the suspended-solids load of the river had been deposited by here, giving an average concentration of only 360 mg/l.

Table 2
Summary of Rivers Donga, Sum Sum, and Kusuku, analyses
(August to September 1982)

(no. of samples in brackets)	R. Donga & tribs. u/s Tepkwar (10)	R. Donga at Tepkwar 30/8 (1)	R. Donga at Tepkwar Laboratory analysis (5)	R. Donga at Nyivu (30)	R. Kusuku (9)	R. Sum Sum (4)
pH	6.7	6.64	6.25	6.8	6.7	7.0
Equilibrium pH value (pHs)	—	–	8.5	—	—	—
Langelier Index (pH-pHs)	—	—	−2.25	—	—	—
Conductivity (μS/cm)	21	15	19	20.5	17.5	15
Colour (P units)‡	341	—	—	577	142	35
Turbidity (FTU)	214	330	674	188	47	17
Total dissolved solids (mg/l)†	—	—	23	—	—	—
Suspended solids	385	1450	1442	360	39	7
Volatile suspended solids	—	—	74	—	—	—
Carbonate hardness	9.5	7	2.4	6	6	11
Non-carbonate hardness	1	5	Nil	8	5	1
Total hardness	10.5	12	2.4	14	11	12
Alk. to phenolphthalein	14	15	Nil	14	10.5	10.5
Total alkalinity	—	–	6.2	—	—	—
Calcium	—	—	0.5	—	—	—
Magnesium	—	—	0.49	—	—	—
Sodium	—	—	2.3	—	—	—
Potassium	—	—	1.1	—	—	—
Ammoniacal N	1.0	—	0.10	—	—	—
Nitrate and nitrite N	0.0	0.9	0.14	0.575	0.5	0.5
Free CO_2	—	—	8.2	—	—	—
O_2 absorbed (4 hr at 27 °C)	—	—	11.6	—	—	—
Chloride	—	—	1.5	—	—	—
Sulphate	6.5	7	1.0	6.75	24	21
Total phosphate	—	—	0.96	—	—	—
Ortho phosphate	0.66	0.075	<0.01	0.19	0.08	0.12
Reactive silica	—	8.5	8.8	13	10	7.5
Total aluminium	—	—	20.1	—	—	—
Aluminium (filtered)	—	—	0.01	—	—	—
Total iron	1.6	3.6	16.5	3.35	0.6	0.46
Iron (filtered)	—	—	0.096	—	—	—
Total Manganese	0.0	0.4	0.4	0.9	0.1	0
Manganese (filtered)	—	—	<0.02	—	—	—
Zinc	—	—	0.054	—	—	—
Zinc (filtered)	—	—	0.02	—	—	—
Copper	—	0.05	0.036	0.18	0.05	0.06
Copper (filtered)	—	—	<0.01	—	—	—
Lead	—	—	0.034	—	—	—
Lead (filtered)	—	—	<0.02	—	—	—

† All results in this section are mg/l.
‡ Platinum units.
All samples were analysed using the Hach DREL/4 portable water chemistry laboratory, except for the five laboratory samples.

Laboratory Quiescent Settlement tests confirmed that 95% of the solids settled out in 24 hours, but thereafter only very slow settlement occurred. Particle-size analysis indicated that about 40% of the sediment was of diameter greater than 10 μm, and 60% had diameters ranging from 1 μm to 10 μm.

Measurements of the rates of flow and sediment concentration in the River Donga suggest that the sediment load at Tepkwar is about 5 million tonnes per year at present, and the effects of increased cultivation, overgrazing and of cattle movement are likely to lead to a substantial increase in gross erosion and gully erosion, unless the following steps are taken.

- To limit cattle numbers, ideally to 50% of present numbers.
- To introduce land use planning so that each type of land would be used for the form of agriculture for which it is most suitable, instead of the present dominance of grazing.
- To introduce better management of the cattle and the grazing land.

Poster Paper (xi)

The use of herbicides on water catchments

J. K. Fawell

INTRODUCTION

Herbicides are a useful and powerful tool both in the general management of land and its commercial exploitation. The responsibility of those exploiting land is to do so efficiently, while the water industry has a responsibility for protecting water quality. These responsibilities are not incompatible provided that herbicides, in common with other biologically-active chemicals, are used with care and respect.

In general the three main areas of potential problems with herbicides are as follows.

Spills

Incidents caused by relatively large quantities of material, particularly concentrates, entering water, usually by carelessness.

Contamination

Acute problems caused by careless use of herbicides such as overspraying water-courses, drains or standing water, direct runoff from sprayed or treated areas, and careless disposal of empty containers or washings from equipment.

Leaching

The slow wash-out of herbicides from treated areas over a long period of time.

Problems in the first two categories can be prevented, or at least greatly reduced, by sensible precautions in the use of herbicides. The third category is more difficult and controversial. Any predictions about the fate of herbicides in soil require detailed knowledge of the behaviour and degradation of each compound and formulation in a wide range of soil types and conditions. Although research is necessary on the behaviour of herbicides in the field before this can be quantified, the problems from leaching appear to be very much less than those caused by spills and careless use.

Knowledge of which product to use under differing circumstances, good product information, proper education and training of operators and supervisors and good supervision are essential for minimising spills and careless use.

Communication between the user, water undertaking, and agricultural advisory department is an essential step in the process of safe use of herbicides, especially knowledge of each other's problems, responsibilities, and view point.

Some key points in limiting problems are as follows.

- Use responsible staff who are trained in the use of herbicides.
- Make sure all staff involved in using herbicides know what they are using and the rules relating to that product.
- Read the instructions *before* using herbicides and each time they are used.
- Mix dilutions away from sensitive areas and take precautions to ensure that any spillage is contained.
- Make up just sufficient for the job.
- Use the correct equipment for appliction. Make sure it is calibrated and carefully maintained.
- Prevent drift and runoff. Do not spray when heavy rain is expected or in high wind.
- Store concentrates and any excess dilutions properly.
- Dispose of empty containers and waste chemicals with great care.
- Where any possibility of contaminating water exists, it is preferable to use formulations cleared for use in and around watercourses and lakes.
- Certain herbicides may cause taint. When they are to be used around reservoirs it may be necessary to treat a small part at a time in order to ensure maximum dilution of any runoff.
- Take advice.
- Establish good liaison between land owner and water supply organisations.

Poster Paper (xii)

Soil injection of organic manures

J. E. Hall

INTRODUCTION

Over 90 million tonnes of organic manures are spread on farmland annually (77 m tonnes from housed livestock, and 15 m tonnes as sewage sludge) with a theoretical nutrient value of well over £200m. Careful management is required not only to realise the full benefits of these nutrients, but also to avoid pollution either as odours and aerosols generated during surface spreading, or more seriously through the surface runoff of slurry into watercourses. Spreading of slurry, in fact, accounts for half of the odour-related complaints from the public and the incidents of serious watercourse pollution from surface runoff have risen in recent years.

Soil injection can overcome many of the potential environmental problems associated with applying organic manures to farmland, as well as providing a number of additional agronomic benefits for the farmer (see Table 1). For the water

Table 1
The benefits and constraints of soil injection

Benefits	*Constraints*
Runoff control	Soil type
Odour control	Soil moisture
Visual aesthetics	Sward damage
Pathogens buried	Weather dependent
No crop taint	Uneven growth
Soil loosening	Stony soil
Improved nutrient management	Higher power requirement
Operational flexibility	Operational complexity

authorities the major advantage is operational flexibility since it may release land otherwise barred to surface spreading of sewage sludges through proximity to housing; injection may also result in reduced sludge-treatment costs as in some instances the transport and injection of liquid sludge is cheaper than dewatering, transporting, and spreading of cake sludge, and furthermore, the injection of unstabilised sludge may be a cheaper option to stabilising prior to disposal. This latter point is particularly relevant with regard to the proposed European Commission Directive on the use of sewage sludge in agriculture (Commission of the European Communities, 1982) which would prohibit the spreading of unstabilised sludge unless immediately incorporated or injected; about one third of the sludge spread on land in the UK is unstabilised.

As the potential benefits of soil injection are becoming increasingly recognised by the water authorities, the Water Research Centre instigated an extensive research programme undertaken by Silsoe College (partly funded by the Science and Engineering Research Council) to determine the optimal design and operational criteria for soil injection. The research concentrated on the exacting demands of grassland where the quality of surface finish is paramount unlike on arable soil where finish is not a concern. Concurrent with this research WRc is collating operational experiences throughout the water industry and will publish a code of practice for the technique.

INJECTOR TINE

Current soil-injection equipment owes much to the development of narrow tines for injecting anhydrous ammonia and to improvements in soil-loosening equipment. Ammonia injection causes little disturbance and is adequate for incorporating volumes of only 1.0 to 2.5 m3/ha, but the amounts of slurry which are normaly applied are up to 100 times greater, and much greater soil disturbance is required to ensure complete incorporation.

The addition of wings to subsoiling equipment was found to increase significantly the area of soil disturbance with only small increases in the draught force required (Negi *et al.*, 1978). The amount of soil disturbance and hence the surface roughness may be controlled by altering the width of the implement tip and the lift height of the wings; wide tips and high lift wings give greater disturbance than narrow tips and low lift wings. The voids thus created in the soil are sufficiently large and well distributed to absorb large volumes of slurry which are introduced under the soil surface either by a pipe fixed to the rear of the tine, or by using a hollow tine down which the sludge may be pumped. Figure 1 indicates the principles involved for simple, winged, and

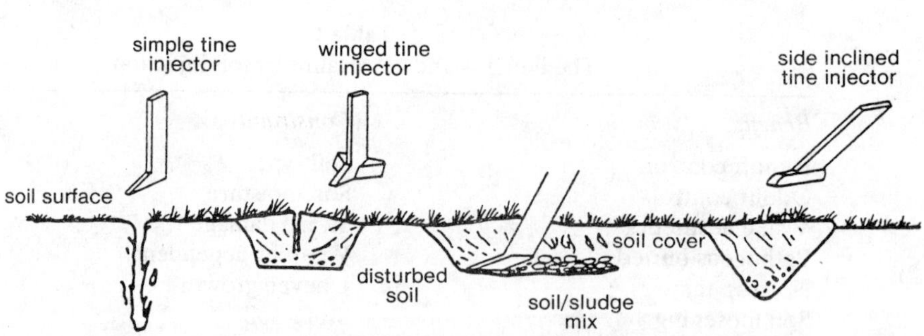

Fig. 1 — Principles of soil injection

side-inclined tines. Although the side-inclined tine does not have wing attachments, and hence it will work at a comparable depth to simple tines, its shape offers a lower draught force requirement and less surface disturbance.

Although successful injection of slurry may be achieved with simple, non-winged tines, such tines cause relatively little soil disturbance and the working depth has to be greater in order to achieve the same injection rates as winged tines operating at 150 to 200 mm. The slurry tends to remain in the slot formed by simple tines which can result in poor control of surface runoff on sloping ground and subsequent crop growth is generally more uneven. There is also a greater risk of simple tines working beyond their critical depth where the effect of the tine changes from soil loosening to soil compaction.

Winged tines require a higher draught force than simple tines for a given working depth, but this is more than compensated for by being able to incorporate slurry effectively at shallower depths and therefore at lower draughts than simple tines. Furthermore higher rates of application can be achieved with winged tines for a given depth compared with simple tines. The maximum practicable application rate is around 140 m^3/ha injected at 150 to 200 mm. Whilst higher rates can be achieved the amount of soil cover over the slurry-soil mix is reduced with greater risks of slurry coming to the surface.

Injection into arable land is far less critical than into grassland as surface finish is not important and the tines may be positioned closer together so that they interact, resulting in greater disturbance and more even distribution of slurry. Generally the maximum disturbance created by simple tines is obtained at a spacing 1.5 times their working depth; for winged tines the spacing will be greater depending on wing width. On grassland it is important that the zones of soil disturbance do not interact as this could result in excessive root damage, particularly in dry soils. For tines with 300-mm wings operating at 150 mm depth, optimum spacing in grassland is 650 mm under most conditions reducing to 500 mm when injecting heavy soils in the spring. Obviously if injecting into a row crop like maize, tine spacing must suit that of the rows and care is required to avoid root disturbance.

Grassland injectors have very narrow tips and low-lift wings and operate with castor-type disc coulters to minimise soil heave and sward damage. Following press wheels produce a level finish and close the slot to seal the slurry in without exerting sufficient pressure to force the slurry out. Where no press wheel is provided, the field should be rolled as soon as the land is trafficable and before stock are returned. The angle of rake of the injector tine also has some influence on the quality of surface finish and tends to be a compromise between the lower draught requirements of a forward-inclined tine, but greater pursing of the turf around the injector shank, and the backward-inclined tine where draught is higher but sward finish is superior.

BENEFITS OF INJECTION

Surface runoff

Pollution of watercourses through surface runoff generally results from one or more of the following factors: excessive rates of application, heavy rainfall following spreading, and sloping ground. Whilst a grass crop or cultivation can help limit the extent of runoff from sloping ground (Vetter & Steffans, 1981), soil injection can significantly reduce the risks of this occurring provided there is sufficient soil cover to contain the slurry. Ross *et al.*, (1979) in monitoring runoff quality for a range of pollution parameters from grass and cultivated contour plots which had received cow slurry to the surface or injected, found that injection 'essentially eliminated any pollutant yield in the run-off'.

Slurries with low dry-solids contents will be much more susceptible to run-off than thick slurries, particularly as the slope increases, and the concern is that the injection slots may act as drains. Injection along the contour can prevent this, but operationally this is not always possible and on gradients of over 10% it is normal practice to inject down the slope. Winged and inclinded tines, because they cause a much greater distribution of soil voids, are far better than simple tines in preventing run-out from the injection slots on sloping ground. However, Godwin and Warner

(1984) have shown that application rates should be decreased for sludges with less than 5% dry solids as the slope increases. Above 5% dry solids, slope had no effect on sludge movement when injected with winged tines (Fig. 2).

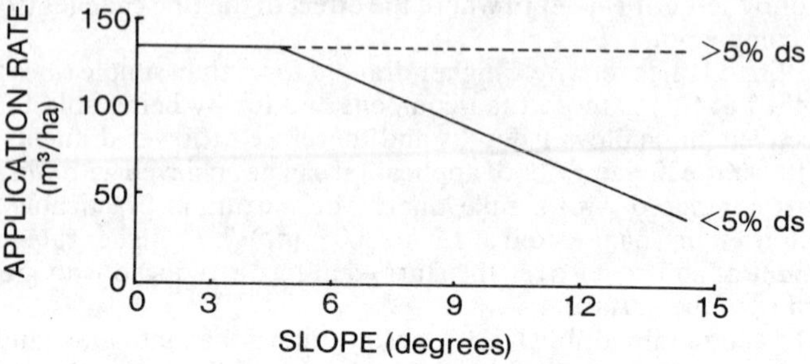

Fig. 2 — Effect of slope upon the maximum application rate

Whilst the risks to groundwater quality are probably no more or no less than for surface spreading provided injection is in the rooting zone, injection into dry, cracked soils, or into drained land with shallow gravel backfill, is inadvisable due to the possible risks of sludge running directly into land drains and watercourses.

Odour

Complete control of odours in the field requires care to ensure that no slurry is released at the surface, either by injecting at too high a rate for the conditions, or through releasing slurry before the injection tines have fully penetrated the soil, and at the end of the run lifting the equipment before shutting off the slurry flow. Competent and well-trained operators are essential.

The major benefit from odour control is that it releases land otherwise barred to slurry spreading through its proximity to housing. This is equally important to both the farmer and water authority. For the farmer it removes pressure from other areas of the farm and allows more efficient utilisation of the nutrients in the slurry, and for the water authority it can result in substantial savings in transport costs as farm land within urban areas is generally nearer to the sewage treatment works.

Sludges and slurries which have been stabilised, particularly by anaerobic digestion, have no offensive smell, but injection may still be desirable to prevent aerosol production during spreading, and for visual and aesthetic reasons.

Implications for livestock

Both livestock slurry and sewage sludge can contain pathogenic organisms, and guidelines (Department of the Environment/National Water Council, 1981) for the use of sewage sludge on pasture provide 'no grazing' intervals for ruminants following surface spreading to ensure there is no risk of infection; these are six months for unstabilised, and three weeks for stabilised, sludges. If complete burial of sludge can be guaranteed then the return of grazing animals to injected pasture would be controlled more by the state of the sward rather than the potential pathogen content of the sludge. Andrews *et al.* (1983) and Dickson and Tribe (1984) have confirmed that the soil cover over injected sludge is an effective barrier to infection and that numbers of organisms fall rapidly following injection.

Good farming practice would require sludge to be injected after grazing or mowing, but before grass regrowth, and then to allow at least three weeks before returning stock to land. This allows sufficient time for the liquid to dissipate, for the injection pathway to seal thereby preventing access by stock, and for the grass to respond to the nutrients. This suggests that a 'no grazing' period of three weeks for all sludges and slurries would be appropriate provided injection is satisfactory. If deemed to be unsatisfactory because there is some material on the surface, then the 'no grazing' period for surface applications should be adhered to. This is an important aspect of injection particularly in the light of the proposed EC Directive on the use of sewage sludge in agriculture which, if implemented, would not only prohibit the surface spreading of unstabilised sludges unless they were immediately cultivated in or injected, but would also require a six-week no grazing period for stabilised sludges.

Injection benefits livestock in two other ways. Firstly it is well known that cattle do not relish grazing on land surface treated with their own slurry, and experiments have been conducted to show that grass injected with slurry was more acceptable (Pain & Broom, 1978). Secondly, sewage sludges and some livestock slurries contain heavy metals depending on the extent of industrial discharge to the sewer or the type of feed additive (such as Cu for pigs). The concern is that if sludge is applied to permanent pasture regularly over a period of time, the levels of metals at the surface may give rise to an elevated intake of metal by grazing animals by direct ingestion. Injection will ensure any metals present are well distributed within the soil profile and thus diluted.

Nutrient management

Livestock slurries and digested sewage sludge contain appreciable quantities of ammonia, often up to 80% of the total nitrogen. This is not only a valuable nutrient readily available for crop uptake, it is also highly volatile. Farmers in the UK are, by and large, unaware of the very high losses of ammonia that can occur when surface spreading particularly on sunny and windy days (it can be as high as 50% within a week of application); however, in parts of Europe, particularly Holland, soil injection is encouraged to prevent such losses to the atmosphere. A number of researchers have shown yield advantages following injection where equal quantities of slurry have been injected or surface applied (reviewed by Hall, 1985) although on grassland this benefit may be reduced by yield losses caused by local tine damage to the sward (Godwin & Warner, 1984).

Cultivation effects

Injection tines perform a soil-loosening function as this is the necessary action to create the voids to absorb slurry. The extent of loosening depends on soil conditions and the design of the injector. In grassland, injection can provide useful soil loosening, particularly in permanant pasture where compaction from the hooves of grazing animals and from vehicles can cause problems through poor drainage and restricted grass-root growth. The constraint on the amount of loosening achievable is that the injected surface must be left clean and level. In arable soils, this constraint does not apply as subsequent cultivations will remove any unevenness caused by the injection tines. The tine spacing can therefore be altered to achieve maximum soil disturbance and so could perform primary cultivations particularly where stubble is injected in minimal cultivation systems.

CONSTRAINTS ON INJECTION

Soil moisture condition offers the most significant constraint on injection. On arable soils this is generally not a problem as land availability for injection is usually restricted to stubble when ground conditions are good and quality of finish is unimportant. Grassland is potentially available throughout the year apart from conservation and grazing periods, but for the full benefits of injection to be realised, adverse conditions, when significant sward and soil damage can occur, need to be recognised and avoided.

On very wet soils, there are the easily-recognisable problems of excessive sinkage and wheelslip. Sward damage may occur earlier when compared with surface spreading of slurries under such conditions through the extra draught requirements of the injectors, but this can be alleviated to some extent by fitting flotation tyres. A further constraint on heavy soils when wet, is that anaerobic conditions could prevail in the slurry channel delaying root growth, and hence crop response and will probably result in significant denitrification losses. Operationally this means avoiding injecting into the soils above field capacity moisture content or when heavy rain is forecast.

The more difficult problem is in hard, dry conditions when trafficability is not a problem but injection can result in damage to the soil-root contact and cause excessive soil heave. This is most likely to occur when the soil enters a drying phase in early summer and the soil moisture deficit exceeds 40 mm (1.5 inches), although much depends on prevailing weather and sward conditions and soil type. Much higher soil moisture deficits can be tolerated when injecting in the autumn as surface conditions are generally moist and there is the greater probability of rainfall

A further constraint is found in stony soils. This will result in increased wear of the injector tines, and whilst injection into stony arable soils is successful, when injecting into grassland stones may be brought to the surface and could present a hazard to mowing equipment if not rolled in. In soils with very large stones, the injectors are protected from damage by shear bolts or automatic spring release mechanisms, but this will result in disrupted turf and slurry appearing on the surface when activated.

There are two further constraints offered by soil type; firstly heavy-textured soils tend to have a much narrower range of moisture content over which the soil can be injected, and secondly such soils tend to have up to double the draught power requirement of light-textured soils. Power requirement is a constrtaint in so far as it affects the choice of injector to suit existing tractors on the farm, or it requires the purchase of a higher-powered tractor. The range of power requirements quoted by injection equipment manufacturers is 80 to 120 hp for a tractor-tanker-injector combination (Water Research Centre, 1984), and four-wheel drive is desirable. Although many farms now have tractors in this range, tractors with lower horsepower can be used by mounting the injection equipment directly onto the tractor with slurry being supplied by a slurry tanker drawn alongside, or indeed by pipe from the field headland.

CONCLUSIONS

Soil injection of farm slurries and sewage sludges eliminates the odour and visual problems associated with surface spreading; it can also control surface runoff and help prevent watercourse pollution. Thus injection can meet with environmental quality objectives and potentially provide savings in sludge treatment and disposal costs. The additional attraction for the farmer is that it can avoid crop taint and pathogen transfer on pastures, and can provide soil loosening with better nutrient management.

Against these benefits must be put the extra weather dependency of injection, the constraint of certain soil types, and that in grassland there may be an uneven response and small short-term reductions in yield compared with surface spreading. There is also the issue of increased operational complexity and the capital cost of equipment. Although it is clear that injection is a more environmentally-acceptable method of spreading liquid manures, it should not be viewed as a panacea but primarily as a means of increasing the flexibility of slurry and sludge spreading operations.

REFERENCES

Andrews, D. A., Mawer, S. L. & Matthews, P. J. (1983) Survival of salmonellae in sewage sludge injected into soil. *Effl. and Wat. Treat J.* February 1983, pp. 72–74.

Commission of the European Communities (1982) Proposal for a Council Directive on the Use of Sewage Sludge in Agriculture.

Department of the Environment/National Water Council (1981) Report of the sub-committee on the disposal of sewage sludges. DoE/NWC Standing Technical Committee Report 20.

Dickinson, P. H. & Tribe, H. T. (1984) A note on the fate of salmonellae. Presumptive coliforms and faecal streptococci in raw sewage sludge buried in soils. Work conducted under contract to WRc, unpublished.

Godwin, R. J. & Warner, N. (1984) *Soil injection of sewage sludge.* Contract report to WRc.

Hall, J. E. *Machinery spreading: soil injection as a barrier to odour dispersion.* Presented to FAO/EEC joint workshop of 'Odour prevention and control of organic sludges and livestock farming', NIAE, Silsoe, April 1985.

Negi, S. C., McKeyes, E. Godwin, R. J. & Ogilvie, J. R. (1978) Design and performance of a liquid manure injector. *Trans Am. Soc. Agric. Engrs.* **21**, 963–966.

Pain, B. F. & Broom, D. M. (1978) The effects of injected and surface-spread slurry on the intake and graxing behaviour of dairy cows. *Anim. Prod.* **26**, 75–83.

Ross, I. J., Sizemore, S., Bowden, J. P. & Haan. C. T. (1979) Quality or run-off from land receiving surface application and injection of liquid dairy manure. *Trans. Am. Soc. Agric. Engrs.* **22**, 1058–1062.

Vetter, H. & Steffens, G. (1982) Surface run-off. In: *Nitrogen losses and surface run-off from lanspreading of manures.* Brogan, J. C. (ed.). Nijhoff/Junk, The Hague, pp. 70–75.

Water Research Centre (1984) *Application of sewage sludge to land. A directory of equipment.* Available from WRc Environment, Henley Road, Medmenham, Marlow, Bucks.

Poster Paper (xiii)

Effect of cultivations on removal of rainfall and nitrate from a mole drained site

G. L. Harris and P. Colbourne

INTRODUCTION

The experiment is a joint venture between the Ministry of Agriculture, Fisheries and Food (MAFF) Field Drainage Experimental Unit and the Agricultural and Food Research Council (AFRC) Letcombe Laboratory, on a heavy clay of the Denchworth soil series, located at Brimstone Farm, near Faringdon, Oxfordshire.

The requirement for this experiment arose from the increased use of minimal cultivations for preparing land for winter cereals. Earlier work by Letcombe Laboratory had shown that the use of minimal cultivations can result in changes in soil structure enabling more rapid water movement to depth.

Work by the Field Drainage Experimental Unit has shown that in heavy-textured soils, water moves laterally through the plough layer if no attempt has been made to modify the subsoil (Trafford & Rycroft, 1973). Even where mole drainage has been introduced the main routing for water from the soil surface has been shown by Goss *et al.* (1983) to be through large fissures linking the disturbed topsoil with the mole channel. There is also evidence that in untilled soils earthworm activity and cracking (Ellis *et al.*, 1979) in dry weather can lead to the formation of continuous vertical channels which persist from year to year. Enhanced infiltration of water can result although other work has suggested that after closure of the main cracks, surface runoff after winter rainfall events may be more rapid and of larger quantities on minimal tilled land. The effect of minimal tillage on the routing, quantity and rate of water movement to the underdrainage system and ditch network therefore needed further examination.

The agricultural influence on water quality was reviewed by Cooke (1976) who showed the need for further studies on seasonal variation on the amount of pollutants such as nitrate-N in surface waters. Influences of cultivation and management practices including field drainage were particularly important.

The experiment commenced in 1978 and is planned to run ten years so that the full effects of any changes that occur as a result of the different cultivation treatments can be fully evaluated.

EXPERIMENTAL DESIGN

Site details are reported by Cannell *et al.* (1984). The full experiment has ten plots with intensive drainage and ten plots without drainage. In the first two years all plots were tine cultivated to 25-cm depth and uniformity established. In subsequent years cultivation treatments of conventional ploughing or minimal tillage (direct drilling or shallow tine) were imposed on the drainage treatments (Fig. 1). Winter wheat was grown in all years except one when a break crop of winter oats was introduced.

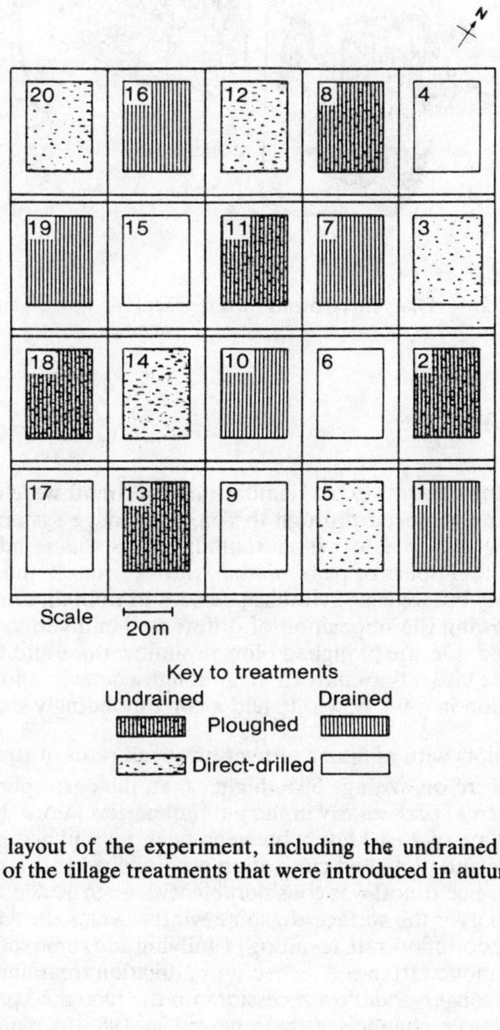

Fig. 1 — Plot layout of the experiment, including the undrained plots and the location of the tillage treatments that were introduced in autumn 1980

MEASUREMENTS

The detailed measurements are described by Harris *et al.* (1984). Hydrologic measurements include the drainage water occurring as surface runoff, plough layer flow or equivalent depth, and flow from the mole and pipe drainage systems. Flow measurements are made using a V-notch weir system and associated autographic recorder (Fig. 2). Samples of drainwater for analysis of nitrate-N are taken from a U-bend water trap which prevented mixing with water already in the weir system.

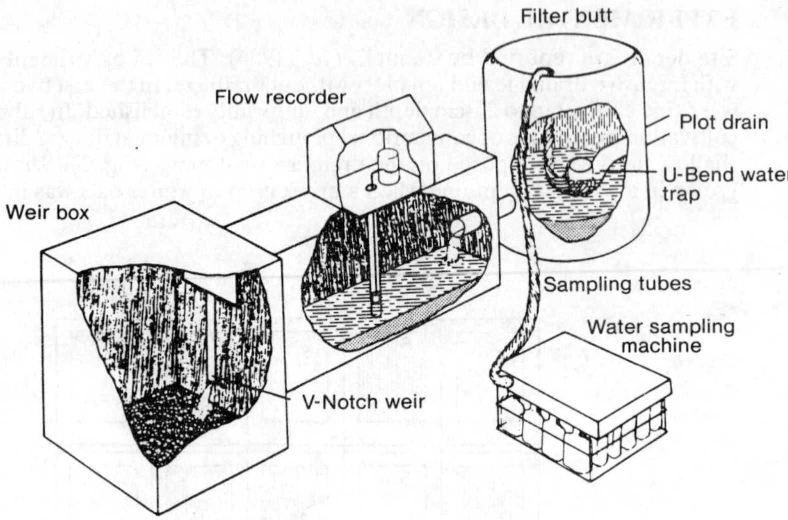

Fig. 2 — Diagram showing arrangement of flow recorder and water sampling machine

RESULTS

Runoff

During the uniformity trial similar rates of runoff were recorded from all plots with 90% being removed through the mole drainage system (Harris *et al.*, 1984). The drainflow response to winter rainfall events was rapid, peak drainflow occurring within 2 to 3 hours of peak rainfall. Surface runoff and plough layer flow occurred frequently but, except in intense periods of rainfall, amounts were very low.

Following the imposition of differential cultivation a marked change in runoff was noted. On the ploughed plots drainflow rates and timings were broadly in line with those under tine cultivation, although a deterioration in the mole channels led to a reduction in peak flow rate and a correspondingly small increase in surface layer flow.

On plots with minimal cultivations peak rates of runoff from the mole drainage system were on average 30% higher than the corresponding ploughed plots and in some events, particularly in the autumn period, more than 50% higher. In addition the lag time of 2 to 3 hours between peak rainfall and peak drainflow observed for both tine and ploughed cultivation was reduced to 1.5 to 2 hours for minimal tilled land. Surface runoff was considerable with up to 30% of winter rainfall being rapidly removed over the surface. In some winter events the surface runoff peak exceeded the peak drainflow rate resulting in gullying and some soil erosion. Total quantities of water removed from the respective cultivation treatments were broadly similar due to much longer drainflow recessions on the ploughed plots.

The mole channels were renewed in 1982 to maintain an effective drainage system. Subsequently both quantities and rates of drainflow from the ploughed plots were compatible with new moles and tine cultivation. The renewed cracking caused by the remoling led to a reduction in surface layer flow on all ploughed plots. On the direct drilled plots, little change was observed in the surface layer flow, but the higher peak rates compared to the ploughed plots observed before remoling were accentuated. This led to very high peak drainflow rates well in excess of those that normally discharge into the river courses. Total quantities of plot flow remained similar for both cultivation treatments.

Soil structure was examined, and in line with other Letcombe work was found to be better developed under minimal tillage with more continuous and extensive vertical cracking both over the moles and at mid-mole spacing. No difference in other parameters such as deterioration of the mole channels themselves or earthworm burrows was found, indicating that the difference in drainflow was due to this structural development. This work is supported by Leeds-Harrison *et al.* (1982) who found substantially peakier mole drainflow from plots with extensive cracking associated with mole drainage compared to moles with no cracking.

The direct consequence of this rapid removal of excess water through the drainage system is that higher peak flow rates in watercourses and river systems must be expected where considerable land is subject to minimal tillage. Soil erosion may also be a problem.

Nitrate

After tine cultivation in 1978 and 1979 nitrate concentrations in the mole drains were 50 to 95 mg N/l in the autumn, declining to 1 to 5 mg N/l by March. Nitrate losses were 40 to 60 kg N/ha; over 90% via the mole drains. These concentrations are consistent with the range reported by Williams (1976) for drainage water from fields at Rothamsted Experimental Station and its associated outstations. The concentrations in the surface water runoff were always low but increased slightly towards the end of the drainage season in association with spring fertlizer applications and stimulation of mineralisation of soil organic nitrogen as soil temperatures increased (Harris *et al.*, 1984).

Ploughing and minimal cultivations were compared from 1980 to 1984 and losses tended to be greater after ploughing. Nitrate losses ranged from 3 to 75 kg N/ha, reflecting variable rainfall distribution. In 1981–1982, 27 kg N/ha were leached from minimal tilled soil, and 43 kg N/ha from ploughed soil with peak monthly losses of 7 and 14 kg N/ha respectively during October, December, and January, from the mole drains. Interflow losses had peaks in October and January which amounted to 3 kg N/ha in total. There was a small amount of loss in surface runoff from the minimal tilled soil during October to January and a little in April, adding up to 1.2 kg N/ha. On average 20% of the nitrate loss was by surface layer flow in minimal tilled soil but only 8% in the ploughed soil.

The concentration of nitrate in water leaving the soil was well above the WHO and EEC recommended limits for nitrate in drinking water (Royal Commission, 1979) for much of the drainage season. Concentrations were somewhat lower in water draining from minimal tilled soil with a maximum of about 35 mg N/l in the autumn compared with 55 mg N/l in water draining from ploughed soil. This reflected faster conduction of water through the minimal tilled soil and greater mineralisation in the ploughed soil.

In all cases these high concentrations of nitrate are transported rapidly via the mole drainage system into the watercourses.

SUMMARY

An experiment was established to examine the effect of different cultivation treatments on the removal of excess rainfall and nitrate from a mole drained clay soil. Under tine cultivation and ploughing surface runoff was low, but with direct drilling surface flow removed up to 30% of the winter rainfall. Under minimal tillage peak drainflow rates gave an average increase of 30% in the winter. This would result in a considerable increase in peak flow in watercourses draining such land. Under all cultivation treatments concentrations of nitrate-N in the drainage system were highest in the autumn and lowest by the end of March. These ranged from 30 to 50 mg N/l in autumn for the ploughed plots, whereas under direct drilling, for the same period they were 15 to 30 mg N/l. Total nitrate-N removed in the drainage water from direct drilled plots was lower.

ACKNOWLEDGEMENTS

The authors thank all MAFF and AFRC colleagues who helped with the collection and analysis of data. The work was funded by the Minstry of Agriculture, Fisheries and Food.

REFERENCES

Cannel, R. Q., Goss, M. J., Harris, G. L., Jarvis, M. G., Douglas, J. T., Howse, K. R. and Le Grice, S. (1984) A study of mole drainage with simplified cultivation for autumn sown crops on a clay soil. 1. Background, experiment and site details, drainage systems, measurements of drainflow and summary of results, 1978–80. *J. Agric. Science.* Cambridge, **102**, 359–559.

Cooke, G. W. (1976) A review of the effects of agriculture on the chemical composition and quality of surface and underground waters. In: *Agriculture and Water Quality*, Ministry of Agriculture, Fisheries and Food, Technical Bulletin No. 32, pp 5–57. HMSO, London.

Ellis, F. B., Elliot, J. G., Pollard, F., Cannell, R. Q. and Barnes, B. T. (1979) Comparison of direct drilling, reduced cultivation and ploughing on the growth of cereals. 3. Winter wheat and spring barley on a calcareous clay soil. *J. Agric. Science.* Cambridge, **93**, 391–401.

Goss, M. J., Harris, G. L. and Howse, K. R. (1983) Function of mole drains in a clay soil. *Agricultural Water Management.* **6**, 27–30.

Harris, G. L., Goss, M. J., Dowdell, R. J., Howse, K. R. and Morgan, P. (1984) A study of mole drainage with simplified cultivation for autumn sown crops on a clay soil. 2. Soil water regimes, water balance and nutrient loss in drainwater, 1978–80. *J. Agric. Science.* Cambridge, **102**, 561–581.

Leeds-Harrison, P., Spoor, G. and Godwin, R. J. (1982) Waterflow to mole drains. *J. Agricultural Enging Res.* **27**, 81–91.

Royal Commission (1979) *Agriculture and Pollution.* Royal Commission on Environmental Pollution, 7th report, HMSO, London.

Trafford, B. D. and Rycroft, D. W. (1973) Observations on the soil water regimes in a drained clay soil. *J. Soil Sci.*, **24**, 380–386.

Williams (1976) The chemical composition of rain, land drainage and borehole water from Rothamsted, Broom's Barn, Saxmundham, and Woburn Experimental Stations. In: *Agriculture and Water Quality*. Ministry of Agriculture, Fisheries and Food, Technical Bulletin No. 32, pp 174–200. HMSO, London.

Poster Paper (xiv)

Commercial forest felling and drainage-water chemistry

J. K. Adamson

POSTER ABSTRACT

At Kershope Forest in Cumbria, water is collected for analysis from six ditch systems each of which drains a separate plot of approximately 2 ha. All plots are on a Peaty Gley soil which was planted with Sitka spruce. Three plots have now been felled. Sampling began two years before felling and continues at weekly intervals on standing and felled plots.

Increases in the concentration of dissolved NO_3-N, NH_4-N, PO_4-P and K result from felling. Higher concentrations are compounded by higher discharge rates. Changes in the decomposition of forest organic matter, reduced trapping of aerosols, and cessation of root uptake are probable factors.

To determine the sources of nutrient fluxes and likely effects on the next forest rotation, water is sampled at all stages in its passage through the standing and felled forest systems (including four soil horizons).

This work forms part of a programme of field and laboratory work to determine the effects of felling on soil processes. A parallel site is at Beddgelert Forest in north Wales. Staff involved are from the Institute of Terrestrial Ecology's Merlewood and Bangor Research Stations and the Forestry Commission Research Stations.

Poster Paper (xv)

Land use and reservoir gathering grounds

A. M. C. Edwards

INTRODUCTION

In 1984/85 Yorkshire Water supplied on average 1363 Ml/d of potable water to a population of approximately 4.5 million people. Direct-supply reservoirs provided 50 % of the water, and the remainder came in nearly equal proportions from groundwater sources and abstractions from lowland rivers. Great importance is attached to maintaining the quality of the water provided for public supply. A number of cases have occurred, however, in recent years where the quality of raw water from direct-supply reservoirs has been adversely affected by land-use changes.

The Yorkshire Grid has been developed to optimise the use of water from the various sources in terms of demand, reliable yield, and cost. The importance of the direct-supply reservoirs in Yorkshire is further emphasised by the fact that the unit costs of providing drinking water are much lower compared with water drawn from the lowland rivers; this is because less sophisticated treatment is required and because the water generally gravitates into supply. Figure 1 shows the main sources of supply and the reservoir gathering grounds.

PROTECTION OF RESERVOIR GATHERING GROUNDS

Yorkshire Water has 91 direct-supply reservoirs located in the Pennines, and another three in the North York Moors. The oldest one in service was constructed in 1818 and the most recent, Winscar, became operational in 1982. In addition there are 20 reservoirs used exclusively for compensation water or river regulation. Grimwith

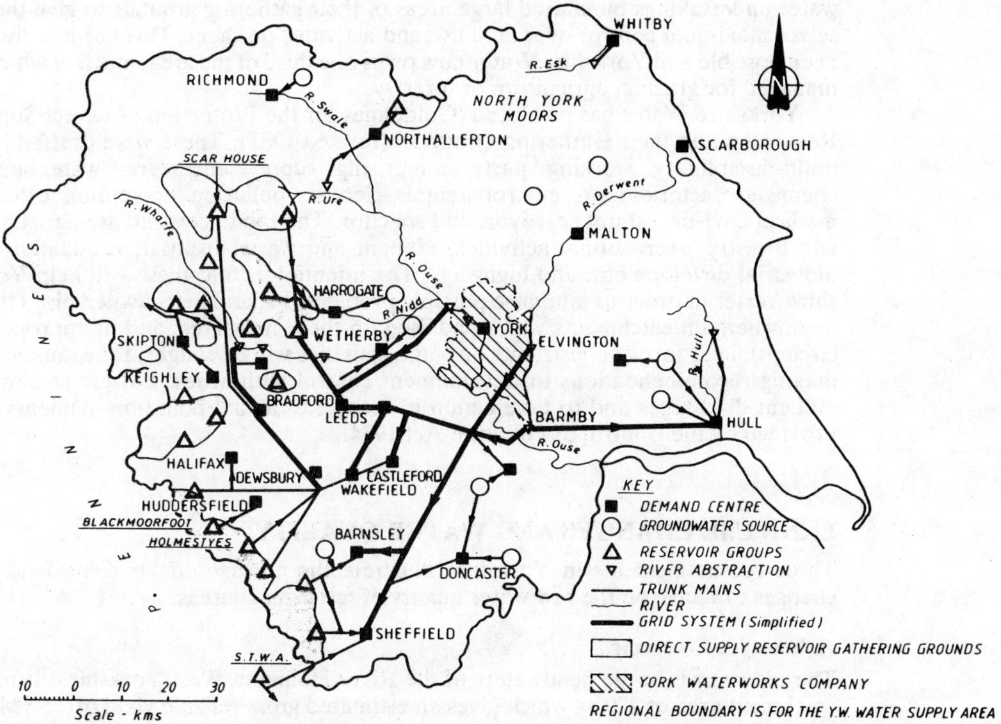

Fig. 1 — Yorkshire Water — principal water supply sources and reservoir gathering grounds

Reservoir was used for the regulation of the River Wharfe for the first time following enlargement in the drought summer of 1984.

The gathering grounds of the direct-supply reservoirs amount to 53 690ha, of which 17 200 ha is owned by Yorkshire Water. The land use of the gathering grounds is:

Moorland	68%
Pasture and farming	28%
Forestry	4%

The generally accepted lines of defence for the quality of water put into supply from impounding reservoirs are:

(i) Surveillance and protection of the gathering grounds and the reservoirs so as to minimise harmful contamination of the raw water by reducing or eliminating potential sources of pollution
(ii) Long-term storage of raw water in reservoirs
(iii) Monitoring the quality of the raw water prior to treatment
(iv) Provision of treatment adequate in relation to the raw-water quality and the risk and type of contamination to that raw water
(v) The provision of buffer storage between the treatment works and the consumer.

Not all of these lines of defence are available for all sources. Traditionally, many water undertakings purchased large areas of their gathering grounds to give themselves maximum control over land use and activities on them. This has not always been possible and Yorkshire Water now owns one third of the area, much of which is managed for grazing, agriculture or forestry.

Yorkshire Water has produced 'Guidelines for the Protection of Direct Supply Reservoirs and their Gathering Grounds' (revised 1985). These were drafted by a multi-disciplinary working party comprising supply engineers, water-supply chemists, bacteriologist, environmental scientist, pollution prevention officers, medical advisor, estates surveyor, and solicitor. The topics covered are agriculture and forestry, recreational activities, effluent and waste disposal, residential and industrial development, and highways. The intention is that these will help Yorkshire Water in order to minimise possible problems on land in its ownership. Other landowners on catchments will be advised on these measures, and in appropriate circumstances formal constraints on some activities will be sought, for example, by making recommendations to development control authorities. Powers to control effluent discharges and to take action in respect to actual pollution incidents are provided by the Control of Pollution Act 1974.

LAND USE CHANGES AND WATER QUALITY

Three recent incidents in Yorkshire illustrate the adverse effects some land-use changes can have on the raw water quality of reservoir sources.

Holmestyes Reservoir

The reservoir is in the headwaters of the River Holme in West Yorkshire. It has a catchment area of 220 ha which gives an estimated gross reliable yield of 2.95 Ml/d, after allowing for the compensation water requirement of 0.89 Ml/d. Water is supplied to approximately 10 000 people in the Holmfirth area.

Some afforestation took place in the 1950s. In September 1980 58 ha were deep-furrow ploughed, generally up and down the slope. Pollution of the reservoir with fine colloidal material occurred following 80 mm of rain in two days in October 1980. The treatment plant, consisting of three pressure filters, failed to remove the turbidity to an acceptable level and the reservoir had to be taken out of supply. Temporary water supplies for Holmfirth were arranged from other sources.

Attempts were made to improve the quality of the runoff into the reservoir but these met with only limited success. A package-type flotation plant was thus constructed to provide an initial stage of treatment prior to filtration. The reservoir was brought back in to service after 11 months. The capital cost of the new plant was £143 000 and the additional running costs during the period the reservoir was out of use were £34 600. The turbidity problems have as yet not abated to a level where the use of the flotation plant can be discontinued; its additional running costs are in the order of £20 000 per annum.

More information on the erosive processes is given in Burt *et al.* (1983), and on the water-supply aspects by Austin & Brown (1982). A similar type of incident in Wales has been described by Stretton (1984).

Blackmoorfoot

Blackmoorfoot reservoir is one of a number which supply water to the Huddersfield area. It has a catchment area of 818 ha of which about one third is in Yorkshire Water's ownership. Between 1976 and 1983 the agricultural quality of part of this land was improved by the intensive application of fertilizers and lime along with field-drainage works. This had a gradual deteremental effect on the quality of water in the reservoir, which changed from being oligotrophic to relatively eutrophic with attendant algal problems. This has led to taste and odour complaints from consumers. Bacterial levels in the reservoir also increased. The average pH rose from 5.3 in 1975/76, to 7.0 in 1983/84. Iron and aluminium levels have thus dropped and, although they occasionally caused colour problems in the past, the benefit of 'free' coagulant has been lost. More chemicals have thus to be added. The throughput of the pressure filters has been reduced by the need for increased backwashing resulting from the algae. The unit treatment costs have risen more than fourfold as shown in Table 1.

Table 1

	1975/76	1978/79	1983/84
Average pH	5.3	6.4	7.0
Water treated (Ml/a)	6594	5841	4818
Chemicals used:			
Aluminium sulphate (tonnes)	40	50	90
Magnafloc (tonnes)	—	—	7
Total chemical cost £	5663	7079	18 413
Unit chemical cost £/Ml	0.86	1.21	3.82

Costs at 1983/84 price base

Scar House

The reservoir is at the head of the River Nidd and supplies an average of 66 Ml/d to the Bradford area. Moorland gripping (that is open field drains) was undertaken in 1980 on land not owned by Yorkshire Water. This is thought to be the cause of the increase in the colour of the raw water. The dosage of alum, lime, and polyelectrolyte has had to be increased and more sludge is produced. It is estmated that the annual costs have increased by £26 500 for chemicals, and by £7500 for waterworks sludge disposal at 1983/84 prices.

REFERENCES

Austin, R. & Brown, D. (1982) Solids contamination resulting from drainage works in an upland catchment, and its removal by flotation, *J. Inst. Wat. Eng. Sci*, **36,** (4), 281–288.

Burt, T. P., Donohoe, M. A. & Vann A. R. (1983) The effects of forestry drainage operations on upland sediment yields: the results of a storm-based study. *Earth Surface Processes and Landforms*, **8**, 339–346.

Stretton, C. (1984) Water supply and forestry — a conflict of interests: Cray reservoir, a case study. *J. Inst. Wat. Eng. Sci*, **38,** (4), 323–330.

Poster Paper (xvi)

Increased streamflow after clearfelling a spruce plantation

D. G. Pyatt and A. R. Anderson

INTRODUCTION

When upland forests are clearfelled there is an increase in streamflow but the size of the increase depends on the nature of the forest, the soil, and the climate (Anderson *et al.*, 1976). The Forestry Commission felled a Sitka spruce (*Picea sitchensis* (Bong) Carr) plantation 20-m tall on three 2-ha plots of a drainage experiment at Kershope Forest, Cumbria, and compared the water balance with a control plot before and after felling. Changes in the chemical composition of the water from its receipt as precipitation to its departure as drainflow were monitored by the Institute of Terrestrial Ecology.

METHOD

The precipitation (P) was measured outside the forest in five gauges, including one at ground level which caught 2.5% more on average than gauges 30 cm above ground level. The results from the ground-level gauge are quoted below. Throughfall (T) was collected in 90 cylindrical gauges of 11 cm diameter, and streamflow (S) was collected from 30 trees. Interception loss was calculated as the difference P−(T+S). The plots are contiguous, clearly defined by perimeter drains, and the soil is a peaty gley with a watertight clay base; other losses of water are therefore thought to be mainly by transpiration and evaporation from the soil. There is no ground vegetation under the spruce stand.

RESULTS

In 1981 and 1982 (before felling) interception loss accounted for 38% of the precipitation, drainflow for 49%, and transpiration and other losses for 13% (Table 1).

Table 1
Water balance (mean of three plots) (mm)

Year	Precipitation P	Throughfall T	Stemflow S	Drainflow (runoff) R	T+S−R	P−R
1981	1403	740	122	692	170	711
1982	1568	841	137	769	209	799

Felling of the three plots of Table 1 took place during 1983 leaving a ground surface virtually devoid of live vegetation through 1984, but covered in a thick layer of dead branches and tops. In 1984 drainflow represented 66% of precipitation, the other 34% being ascribed mainly to evaporation from the harvesting debris and soil surface (Table 2).

Table 2
Water balance of plots before and after clearfelling

		Mean of three plots felled during 1983		Control plot (not felled)	
Year	Precipitation P	Drainflow R	P−R	Drainflow Rc	P−Rc
1982	1568	769	799	710	858
1983	1278	622	656	462	816
1984	1259	837	422	433	826

The water balance of the control plot during 1983 and 1984 was affected by the rainfall being lower than in 1982. Total evaporation and other losses (P−Rc) were slightly reduced but represented an increased proportion of precipitation. Drainflow was markedly reduced in absolute terms and as a proportion of precipitation. If this effect is allowed for in the 1984 results for the felled plots, the 'expected' drainflow if felling had not taken place would have been 477 mm. Thus felling may have increased drainflow by 360 mm.

CONCLUSIONS

The increased drainflow resulting from clearfelling forest at Kershope is towards the upper end of the range of values in world literature. This was expected because of the high interception loss from the plantation, because the vegetation change affected the whole plot, and because the plot is watertight.

Measurements are continuing in 1985 and it is possible that the increase in drainflow will be greater than in 1984 if the dead branches decay and if it is a wetter year.

REFERENCE

Anderson, H. W., Hoover, M. D. & Reinhart K. G. (1976) *Forests and water: effects of forest management on floods, sedimentation and water supply*. Forest Service Technical Report PSW-18. Berkeley, Calif. USA.

Poster Paper (xvii)

Seasonal variation in water losses from Law's forest lysimeter at Stocks Reservoir

P. D. Walsh and **Susan Walker**

INTRODUCTION

Analysis of data from the forest lysimeter in the Stocks Reservoir catchment demonstrates a seasonal variation in losses. Losses can be subdivided into interception by the tree canopy, and evapotranspiration. Interception is the evaporation from the wetted surface of the canopy immediately after rainfall. Evapotranspiration is the evaporation from the ground, ground plant cover, and water transpired through the leaves.

IMPACT OF AFFORESTATION ON WATER YIELD

The impact of forests on water yield from catchments has been recognised since the mid-fifties, when Law (1956) showed that for 1955 to 1956, 290 mm/yr less rainfall (20% of annual runoff) was being converted to runoff from the Stocks lysimeter of Sitka spruce. Subsequent work by the Institute of Hydrology has established the mechanisms that lead to the reduction in water yield in forested catchments (Newson, 1979). Further recent analyses of 11 years of data from the lysimeter used by Law showed losses of 381 mm/yr due to evapotranspiration, and 569 mm/yr due to interception by the canopy. This compares with a total loss of 403 mm/yr, on average, from the reservoir catchment prior to afforestation (Calder *et al.*, 1982). Though Stocks lysimeter results initially invited critcism due to the small-scale nature of the experiment, re-examination of the data now reveals favourable comparison with results from the more recent catchment-sized experiments on the Plynlimon catchments in mid-Wales.

STOCKS LYSIMETER

Stocks reservoir in the Pennines has a catchment area of 37.5 km^2 which varies in altitude from 180 to 540 m AOD, with an average annual rainfall of 1650 mm. In 1953 the Forestry Commission, who lease 22% of the catchment, began planting Sitka spruce, *Picae sitchensis.* A lysimeter was established in a small area of older trees, planted in 1929. Underlying the lysimeter is natural clay which prevents downward drainage. Concrete cut-off walls were built to prevent lateral drainage. The lysimeter covers an area of 0.05 ha and is at an elevation of 183 m AOD.

Gross rainfall reaching the lysimeter was estimated as the mean of three ground-level gauges, sited away from the plantation. Five stem-flow gauges were located on the lysimeter. Throughfall, the precipitation reaching the ground, was estimated as the mean of 10 to 20 ground-level gauges. Flow meters were used to measure runoff out of the lysimeter. Interception was calculated as the difference between gross rainfall and throughfall. Evapotranspiration was calculated as the difference between throughfall and lysimeter runoff. The total loss from the lysimeter is the sum of interception and evapotranspiration.

SEASONAL PATTERN OF LOSSES

Figure 1, which shows the average variation of gross rainfall and losses through the year, reveals a high level of evapotranspiration in the summer and early autumn. Interception is at its highest in the autumn and winter when it is the most significant contributor to losses. Since interception losses from the trees are attributable to an increased evaporation rate from a wetted tree canopy (Calder, 1979), this is found to occur during the wetter part of the year.

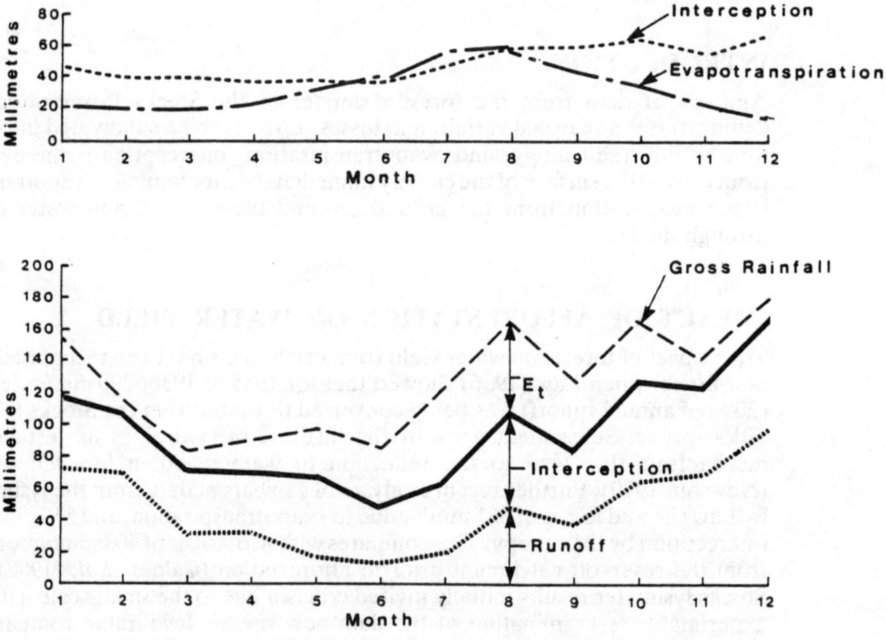

Fig. 1 — Average seasonal variation of interception and evapotranspiration

The significant correlation (r = 0.88) between gross rainfall and interception is shown in Fig. 2. This is based on 11 years of monthly data. A plot (Fig. 3) of evapotranspiration against gross rainfall (r = 0.39) is not significant at the 95% level.

Analysis by time of year reveals a seasonal pattern. Interception and gross rainfall are significantly correlated in all months with correlation coefficients from 0.68 to 0.96. Figure 4 shows regression lines based on data for pairs of consecutive months which all have correlation coefficients which are significant at the 95% level. The relationship between evapotranspiration and rainfall is highly significant in July ($r = 0.94$) but not for the months October to December (Fig. 5). There is no worthwhile improvement if data for consecutive months are combined.

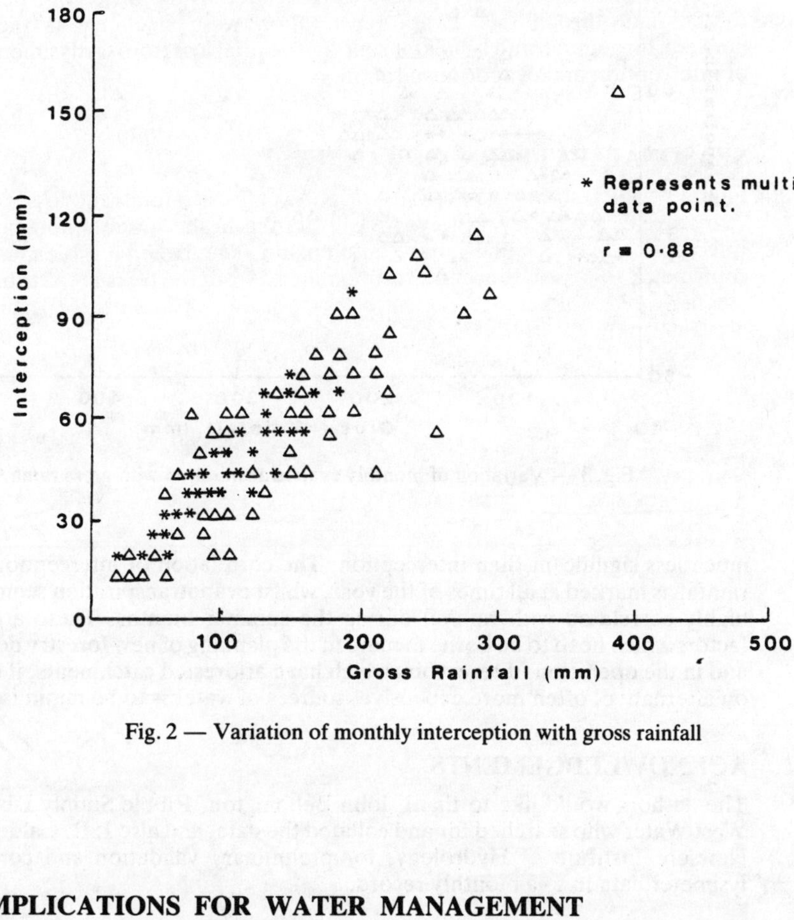

Fig. 2 — Variation of monthly interception with gross rainfall

IMPLICATIONS FOR WATER MANAGEMENT

The seasonal pattern of water losses has significance for the management of reservoirs, particularly large ones with long critical periods. At the end of a dry period, in addition to rainfall being taken up by any soil-moisture deficit, some rainfall on a forested area would be lost by interception by the canopy. Other implications of the afforestation of catchments on water supplies were discussed by Walsh (1980).

CONCLUSION

Analysis of lysimeter data from the Stocks Reservoir catchment has confirmed a marked seasonal pattern in losses. Interception is at its highest during the winter months. Evapotranspiration reaches a peak in summer, but for much of the year is

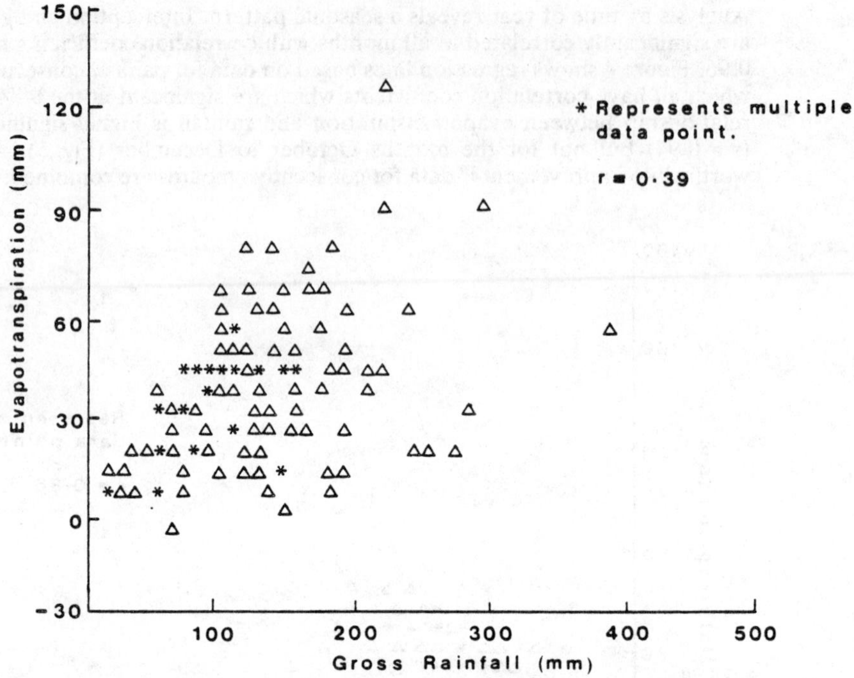

Fig. 3 — Variation of monthly evapotranspiration with gross rainfall

much less significant than interception. The correlation of interception with gross rainfall is marked at all times of the year, whilst evapotranspiration seems only to be highly correlated with rainfall during the summer months. These are important factors which need to be borne in mind in the planning of new forestry developments and in the operation of reservoirs which have afforested catchments, if expenditure on alternative, often more expensive, sources of water is to be minimised.

ACKNOWLEDGEMENTS

The authors would like to thank John Bebbington, Ribble Supply District, North West Water who searched for and collated the data, and also I. R. Calder and M. D. Newson, Institute of Hydrology, for preliminary validation and compilation of lysimeter data in to a monthly record.

REFERENCES

Calder, I. R. (1979) Do trees use more water than grass? *Water Services*. January, pp. 11–14.

Calder, I. R., Newson, M. D. & Walsh, P. D. (1982) The application of catchment, lysimeter and hydrometeorological studies of coniferous afforestation in Britain to land use planning and water management. *Proceedings of International Symposium on Hydrological Research Basins and their use in Water Resources Planning*. Berne.

Law, F. (1956) The effect of afforestation upon the yield of water catchment areas. *Jour. Brit. Watworks Assoc.*, **38**, 489–494.

Newson, M. D. (1979) The results of ten years experimental study on Plynlimon, mid-Wales, and their importance to the water industry. *J. Inst. Wat. Engrs and Scientists*, **33**, 321–333.

Walsh, P. D. (1980) The impact of catchment afforestation on water supply interests. *Aqua*, **4**, 0082–0085.

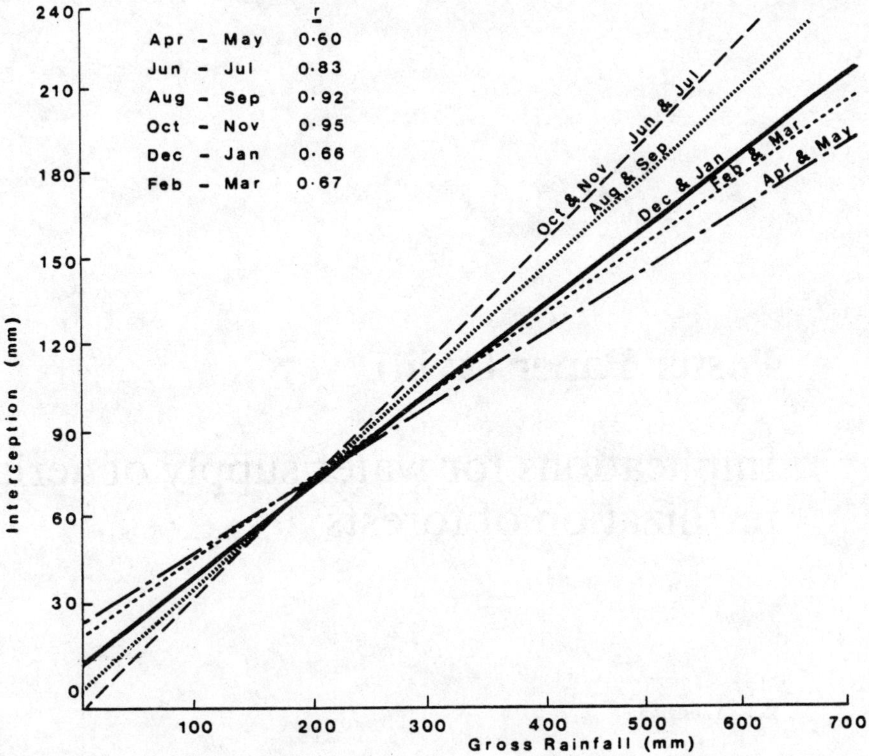

Fig. 4 — Regression lines for interception versus gross rainfall

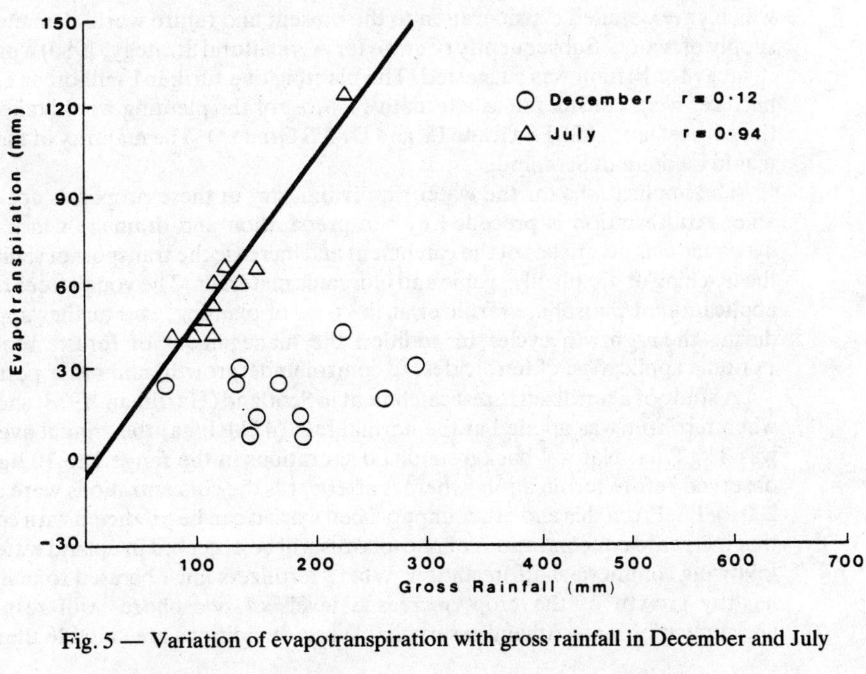

Fig. 5 — Variation of evapotranspiration with gross rainfall in December and July

Poster Paper (xviii)

Implications for water supply of aerial fertilization of forests

R. E. Youngman

In 1977 the Forestry Commission published a review (Forestry Commission, 1977) which gave detailed consideration to the present and future world demand for, and supply of, wood. Subsequently (Centre for Agricultural Strategy, 1980) a production strategy for Britain was suggested. The planting of a further 1 million or 1.8 million hectares were considered as alternatives. Most of the planting would take place on the poorest land, MAFF Grade IV and DAFS Grade D. The majority of the planting would be done in Scotland.

The implications for the water supply industry of these proposals are considerable. Afforestation is preceded by site preparation and drainage which alter the hydraulic characteristics of the catchment and increase the transport of sediment and the leaching of dissolved organic and inorganic materials. The young trees receive an application of phosphate fertilizer at the time of planting, and further applications during their growth cycle. In addition the management of forests requires the periodic application of herbicides, to control undergrowth, and other pesticides.

A study of a fertilized forest catchment in Scotland (Harriman, 1978) showed that when fertilizer was applied at the normal rate (47 kg P/ha) the annual average loss was 2 kg P/ha. Natural background concentrations in the range 1 to 10 μg P/l were observed before fertilization, whereas afterwards the concentrations were as high as 200 μg P/l. From this and other unpublished work it can be predicted with confidence that increased concentrations of phosphorus will be observed in upland watercourses following commercial afforestation. Where fertilizers must be used to maintain the healthy growth of the crop, increased levels of phosphorus will result in the enrichment of many upland reservoirs. Where conditions are suitable there will be increased growth of algae as it has been shown that concentrations as low as 10 μg P/l are sufficient to stimulate troublesome algal growth.

A study at the WRc (Critchley *et al.*, 1980) showed that more then 260 reservoirs in upland areas are potentially at risk of suffering algal problems if the proposal to afforest an extra 1.8 million hectates in implemented. Estimates of the number of works which would experience treatment difficulties due to algae and the treatment changes which would be required allow an assessment of the capital costs which might be incurred. These lie between £37m and £104m (1982, Q4 prices) depending upon the existing treatment facilities and the severity of the algal problems experienced or anticipated.

A joint Institute of Hydrology/Water Research Centre colloquium held at Wallingford in March 1980 suggested, *inter alia*, that further research was needed to allow a better prediction of nutrient transport from upland forests subject to aerial fertilizer application, and to predict the algal response of waters to increases in nutrient concentrations.

Early in 1981 two adjacent areas of ten-year-old Sitka spruce plantations in Glenorchy Forest, situated near Dalmally in Argyll, were chosen for study. An unfertilized subcatchment acted as a control for another which was to be fertilized in accordance with normal forest practice in 1982.

Flat V stream-gauging weirs were constructed on the burns leaving the chosen subcatchments and water-level recorders were installed as were automatic water samplers. Samples can be taken at intervals from 15 minutes to 8 hours. This allows the detailed study of nutrient flux during storm and other events. All samples are preserved to prevent loss of phosphorus between sampling and analysis.

Detailed meteorological data have been made available by the installation of a Didcot Automatic Weather Station. The nutrient content of the rainfall is measured every two weeks.

Fertilizer was applied to the 'treated' area in July 1982. Chemical analytical data are available for the periods before, during, and after application and are continuing.

The project is now entering its second phase. Algal bioassays will be carried out in three lochs selected for their low nutrient content and range of colour from 20 to 200 Hazen (low to high). Columns of water in each loch will be isolated within plastic bags supported on floating rafts. These will be enriched with phosphorus (and nitrogen) to levels suggested by the results from the Glenorchy Forest studies. Unenriched columns and the lochs themselves will act as controls. Changes in the algal content of the water which may result from the enrichment will be quantified at weekly intervals in month-long experiments. Three such experiments (spring, summer, and autumn) are planned.

The information obtained from the Glenorchy Forest study and the algal bioassay tests will be used to produce a report for the water industry. This will be done in consultation with those at the Centre who have expert knowledge of eutrophication and the various water-treatment-process engineering options and their costs. The report will assess treatment requirements which may arise if the Forestry Commission were to afforest an extra 1.8 milion hectares of upland Britain.

REFERENCES

Centre for Agricultural Strategy (1980) *Strategy for the UK forest industry*. CAS Report 6. Reading. Centre for Agricultural Strategy.

Critchley, R. F., McNaughton, J. & Youngman, R. E. (1980) *Afforestation of upland catchments: economic implications for the water industry*. WRc Internal Report 19–M.

Forestry Commission (1977) *The wood production outlook in Britain. A review*. Edinburgh. The Forestry Commission.

Harriman, R. (1978) Nutrient leaching from fertilized forest watersheds in Scotland. *J. Appl. Ecol.*, **15**, 933–942.

Index

Page numbers in **bold** type refer to major entries; page numbers in *italics* refer to illustrations

B

M